HUMAN FACTORS IN ORGANIZATIONAL DESIGN AND MANAGEMENT - II

Second Symposium on
Human Factors in Organizational Design and Management
held in Vancouver, B.C., Canada, 19 – 21 August, 1986

Co-sponsored by:
The Human Factors Society
Japan Ergonomics Research Society
Human Factors Association of Canada
International Ergonomics Association
HFS Organizational Design and Management Technical Group

Symposium Chair:
Hal W. Hendrick

Technical Program Chair:
Ogden Brown, Jr.

NORTH-HOLLAND
AMSTERDAM · NEW YORK · OXFORD · TOKYO

HUMAN FACTORS IN ORGANIZATIONAL DESIGN AND MANAGEMENT - II

Proceedings of the Second Symposium held in Vancouver, B.C., Canada, 19–21 August, 1986

Edited by

O. Brown, Jr. and H. W. Hendrick
University of Southern California
Los Angeles
California
U.S.A.

1986

NORTH-HOLLAND
AMSTERDAM · NEW YORK · OXFORD · TOKYO

ISBN: 0 444 70076 5

Published by:
ELSEVIER SCIENCE PUBLISHERS B.V.
P.O. Box 1991
1000 BZ Amsterdam
The Netherlands

Sole distributors for the U.S.A. and Canada:
ELSEVIER SCIENCE PUBLISHING COMPANY, INC.
52 Vanderbilt Avenue
New York, N.Y. 10017
U.S.A.

Library of Congress Cataloging-in-Publication Data

Symposium on Human Factors in Organizational Design and Management (2nd : 1986 : Vancouver, B.C.)
Human factors in organizational design and management-II.

"Second Symposium on Human Factors in Organizational Design and Management held in Vancouver, B.C., Canada, 19-21 August, 1986, co-sponsored by the Human Factors Society ... [et al.]"--CIP prelim. p.
Includes bibliographies and index.
1. Organizational effectiveness--Congresses. 2. Quality of work life--Congresses. 3. Human engineering--Congresses. I. Brown, O. (Ogden) II. Hendrick, Hal W. III. Human Factors Society. IV. Title.
HD58.9.S95 1986 658.3'142 86-16519
ISBN 0-444-70076-5

PRINTED IN THE NETHERLANDS

PREFACE

This book contains a series of papers which were presented during the Second International Symposium on Human Factors in Organizational Design and Management, held in Vancouver, Canada, on 19–21 August, 1986. The Symposium was sponsored jointly by the International Ergonomics Society, the Human Factors Society, the Society's Technical Group on Organizational Design and Management, the Human Factors Association of Canada, and the Japan Ergonomics Research Society. The Symposium provided a unique opportunity for researchers, practitioners and students of human factors/ergonomics, organizational behavior and related fields to come together from around the world and share their expertise in this rapidly expanding interdisciplinary area.

New technology, a changing demographic composition, changing values and attitudes of workforces, and a renewed emphasis on both productivity and the quality of work life have created a need for a true macroergonomic systems approach to the design of organizational and managerial systems. These factors also have heightened the need to consider organizational design elements and managerial processes in the application of human factors/ergonomics to the design of specific subsystems, jobs and workstations. Although a number of new methods have been developed and utilized to study and apply human factors on a macroergonomic level, only recently has there been communication of these methods of research and applications across continents. The purpose of this Symposium was to foster this technical exchange process.

Included herein are a broad selection of papers on theory, methodology, research findings, reviews and case studies from leading professionals throughout the world. These papers provide the reader with a good insight into this aspect of organizational design, and a 'feel' for human factors at the macroergonomic level.

Ogden Brown, Jr. and Hal W. Hendrick
Editors

SECOND INTERNATIONAL SYMPOSIUM ON HUMAN FACTORS IN ORGANIZATIONAL DESIGN AND MANAGEMENT

SPONSORS
Human Factors Society
Japan Ergonomics Research Society
Human Factors Association of Canada
International Ergonomics Association
HFS Organizational Design and Management Technical Group

ORGANIZING COMMITTEE
Hal W. Hendrick, Chair
Ogden Brown, Jr.
Andrew S. Imada
Kageyu Noro
Sadao Sugiyama
Ulrika Wallersteiner
Michelle M. Robertson

TECHNICAL PROGRAM COMMITTEE
Ogden Brown, Jr., Chair
Marjorie Bauman
Sara J. Czaja
Susan Dray
F. Lawrence Ficks
Dorothy Finley
Glen R. Gallaway
Peter A. Hancock
Martin G. Helander
Hal W. Hendrick
Andrew S. Imada
Najmedin Meshkati
Kageyu Noro
David B.D. Smith
James W. Suzansky
Hideo Tanaka
Yoshimi Yokomizo

HUMAN FACTORS SOCIETY

The Human Factors Society, established in 1957, is an interdisciplinary organization of professional people involved in the human factors field. The society promotes the discovery, exchange, and application of knowledge concerning the relationship of people to their machines and their environments. It furthers the assignment of appropriate roles to humans and machines in systems. It advocates the consideration of operators, maintainers, and users in the design of equipment and facilities. The Society supports the development of working and living environments that are comfortable and safe. It encourages the appropriate education and training of those who conceive, design, develop, manufacture, test, manage, and participate in systems.

Meetings and publications are sponsored by the Society. An annual meeting is held in the fall. It includes a technical program, consisting of research reports, panel discussions, and workshops.

Publications include the bimonthly journal *Human Factors* and the monthly *Human Factors Society Bulletin*.

The Human Factors Society has chapters throughout the United States which sponsor local meetings and publications.

The society's technical groups each focus on a particular aspect of the human factors field. The technical groups also hold meetings, publish newsletters, and sponsor technical sessions at the annual meeting.

For information write to:-

Human Factors Society
P.O. Box 1369
Santa Monica
CA 90406
U.S.A.
Phone: (213) 394-1811

INTERNATIONAL ERGONOMICS ASSOCIATION

The International Ergonomics Association is an association of ergonomics and human factors societies around the world. Developing out of the Ergonomics Research Society, which was founded in 1949, the international organization has continued to flourish and now has a membership from some 15 federated ergonomics societies in both hemispheres, a central secretariat, and an executive council of international representation.

The International Ergonomics Association brings together organizations and persons concerned with ergonomics or human factors; i.e., the relations between man and his occupation, equipment, and environment in the widest sense, including work, play, leisure, home, and travel situations. To further these aims, it establishes international contacts among those acting in this field, promotes knowledge and practice of ergonomics on an international basis especially in areas where no Federated Society exists, cooperates with international associations to encourage the practical application of ergonomics in industry and other areas, and promotes scientific research by qualified persons in this field.

Apart from the international congress, held every three years, the IEA also sponsors joint conferences of an international nature which are organized by one or more of its member federated societies.

TECHNICAL GROUP ON ORGANIZATIONAL DESIGN AND MANAGEMENT

HUMAN FACTORS SOCIETY

The Technical Group on Organizational Design and Management (ODAM) is concerned with improving organizational effectiveness and the quality of work life through an integration and joint optimization of psychosocial, cultural, and technological factors with human–machine performance interface factors in the design of jobs, workstations, organizations, and related management systems through a sociotechnical systems approach.

JAPAN ERGONOMICS RESEARCH SOCIETY

The Japan Ergonomics Research Society is an affiliated society of the International Ergonomics Association.

HUMAN FACTORS ASSOCIATION OF CANADA

The Human Factors Association of Canada is an affiliated society of the International Ergonomics Association.

CONTENTS

SYSTEM SIMULATION, MODELING AND ANALYSIS

ERGONOMIC FACTORS IN HUMAN–COMPUTER INTERACTION

ORGANIZATION AND MANAGEMENT OF HUMAN FACTORS

ERGONOMIC CONSIDERATIONS IN TECHNOLOGY TRANSFER AND ORGANIZATIONAL DEVELOPMENT: CROSS CULTURAL IMPLICATIONS

TRANSLATING RESEARCH ON NEW TECHNOLOGY INTO PUBLIC POLICY

PARTICIPATORY ERGONOMICS, QUALITY OF WORK LIFE AND ORGANIZATIONAL EFFECTIVENESS

MACROERGONOMICS

---

*These papers arrived too late to be included under their appropriate session headings. Their correct position is given by the superscript numbering at the end of their titles; 1) System Simulation, Modeling and Analysis; 2) Ergonomic Factors in Human–Computer Interaction; 3) Training Systems.

AGING AND STRESS

Human Factors in Organizational Design and Management - II
O. Brown, Jr. and H.W. Hendrick (Editors)

HUMAN AGING: AREAS OF CHALLENGE FOR HUMAN FACTORS IN THE MICRO AND MACRO DESIGN OF ORGANIZATIONAL SYSTEMS

David B. D. Smith

Human Factors Department
and
Ethel Percy Andrus Gerontology Center
University of Southern California

The labor force in many industrial nations is aging. The worker 40-60 years of age will be a dominate force in the labor market of the future. Factors in workload, technology and obsolescence are reviewed for the aging employee. It is argued that these issues should be of concern in the human factors design of the organization and work enviroment.

INTRODUCTION

Aging has been an important social and economic force in industrially developed countries for a decade or more. If fertility declines as projected in developing countries these nations will also experience the aging of their populace. For these reasons it is important to assess the implications of this greying of the population for human factors design in systems. This paper reviews some general age issues and research findings relevant to human factors, as regards work and age.

Here the term "older worker" refers to persons 40-60 years of age; those people in what is commonly called middle age. Persons still employed in the traditional retirement years over 60-65 are designated as "elderly workers". The choice of 60 years is not totally arbitrary. There is a general sense in which age declines in ability and other work factors are more pronounced after the age of 60. Both sectors of the overall population in many countries are growing in real numbers and as a per cent of the population.

The emergence of age issues across the adult years calls for a life span view of aging. In this view development and change appear as life long processes, not confined to early or later years. Baltes, Reese and Lipsitt (1980) proposed three sets of factors influencing the life course. Age-associated events (ontogenetic) are the biological and environmental determinants closely related to chronological age. Examples are physical and sensory changes, formal education and the family. These influences tend to impact all members of a society in a similar way.

History-associated events are specific environmental factors related to history that change the course of adult development. Technological change, war, legislation and shifts in demography and occupational structure are instances. The direct impact of such events are likely to be specific to an age segment of the population. Significant life events are those specific historical features, such as, an accident, unemployment and medical and psychological trauma.

These set of events impact the adult life span and cause the great inter-individual variability observed in middle age and later life. The variety of events and their interaction, as suggested by the model, indicates aging as a multidimensional process that requires multidisciplinary study and application. The task for human factors is to identify the nature and course of these changes as they suggest implications for research and for design of the person-machine and organizational system.

DEMOGRAPHICS OF THE WORK FORCE

What we do for work or what we have done largely determines our place in society, in our homes, and our self-esteem. The longitudinal study of Palmore and Jeffers (1971) shows the importance of work's interaction with age. They found job satisfaction the best indicator of longevity, even better than a physician's judgement of health status.

What about the elderly, those persons over 65? In the U.S. thirty years ago half of all elderly worked. In the 1980's this has dropped to less than 20 per cent (Taeuber, 1984), the total number remaining about constant.

Data prepared for the 1981 White House Conference on Aging concluded that "the outlook is for a dynamic labor shortage caused by sharply reduced labor force growth, with the shortage becoming progressively worse over the next quarter of a century". The report then concluded that an increased participation by elderly workers would be an excellent solution to this short fall. In essence, the future does not strongly argue for increases in elderly (65+) seeking employment. Those remaining past the point of retirement will represent the most capable. Rather than competency, the issue for the elderly may well be how to retain or entice these people to work.

A different situation obtains for the older worker (40-60 years). In the U.S. the median age of the labor force is about 40. By the year 2000 approximately 50% of employed persons will be in the age range from 35-54 (Robinson, 1985). Two factors have lead to this concentration of middle aged workers. The first reflects cyclical changes in birth rates producing a "bulge" of middle age persons relative to younger and older ages. The second is a decrease in the participation rate of workers over 55 years old; this is a combination of early retirement and growth in unemployment of ages 55-64. For the U.S. this means that the middle-aged (older) worker will be a significant if not the dominant labor force development in the years to come. These trends for older and elderly worker are not wholly unique to the U.S., but are present in varying degrees in other industrialized nations as well.

Recent publications on work and age of interest to human factors include: Older Employees (Rosen and Jerdee, 1984); The Aging Employee (Yolles, Krensky, Kieffer and Carone, 1984); Age, Health and Employment (Birren, Robinson and Livingston, 1986), Management and the Older Workforce (Humple and Lyons, 1983). Major works that address more directly human factors concerns are, Aging and Technological Advances (Robinson, Livingston, and Birren, 1984) and a special issue of the journal Human Factors published in 1981.

PERFORMANCE, WORKLOAD AND THE OLDER WORKER

Research on human performance under different conditions is the empirical foundations of human factors. Welford (1980) presented a capacity model for describing performance as a function of age. This model consists of three elements. Capacities are the physical and mental resources available to the individual and are limited by such factors as arousal level, individual differences and age. Demands are the external and internal workloads imposed on capacities. Finally, strategies represent the means by which capacities are allocated by the person to satisfy demands. According to the model a decline in capacity occurs with age. At some point demands can exceed the capacity resulting in age decrement. The older person may use different methods to compensate, e.g, working longer, shifting jobs, sacrificing speed for accuracy or using special experience.

A quarter century of research in gerontology shows that the older adult progressively with age does reveal a decline in capacity as measured by laboratory tests of memory, learning, perception, psychomotor skills and other functions mediated by the central nervous system. This research in gerontology in its inception grew directly out of concerns for applied issues of work. One approach has been to try to develop a functional age index or profile of capacity.

Functional Age Studies of performance usually find the variance greater within an age group than between age groups; so some older people will perform better than many younger ones. Thus, chronological age characterizes average changes effectively but less so when applied to the individual. This led McFarland (1943) to suggest the concept of functional age as a means of assessing a worker's capacity to perform a job. In this approach a person could be functionally younger or older than their chronological age (McFarland, 1973).

The concept of functional age has received wide popular usage and attempts to devise a single index or profile of functional age for work assessment has been attempted (Dirken, 1972; Heron and Chown, 1967). A useful critique of the concept of functional age is given by Salthouse (1986).

In a recent example, Braune and Wickens (1986) tested groups 20-60 years of age on an "information processing performance battery using aviation-relevant task structures". The stated purpose was to develop a functional age profile. Presumably, such an index could be used as part of selection and assessment criteria for pilots. Whether such a profile has practical implications remains to be seen (Salthouse, 1986), however, the results illustrate some issues for human factors research and application. First, age trends are not uniform. This study found abilities that declined across all age groups, abilities that declined beyond 40, and abilities that showed no decline. Second, it shows the potential for age as a variable in human factors/performance research. Finally, quite apart from issues of selection or assessment the results of such research may have implications for the design of jobs and environments which can best take account of peoples capabilities as they age.

Workload Demands with Age In the "post-industrial" society the need for human power in jobs is disappearing. Within a given occupation, work demanding physical strength or rapid pace are not usually held by older workers. Middle age workers leave such labor before decreases in strength or aerobic capacity decline to levels that would be challenged by the job. In addition, management practice and union contracts are apt to call for transfering older employees to lighter work. Indeed modern jobs may provide inadequate exercise; thus Shephard (1980) has argued for the need for job related programs to restore fitness lost through sedentary work and leisure.

In jobs of modern systems, task demands most often occur in the information processing and psychomotor skills area, so called "mental work load". This has received considerable empirical study in human factors, though largely with respect to specialized environments such as aviation. In general these studies have not concerned themselves with age. A good deal of laboratory research from gerontology indicates that the difficulty of a task plays an important role in determining whether performance change is observed with age and, in general, task loading influences older ages proportionately more than younger ages (Cerella, Poon and Williams, 1980). This disproportionate effect of load occurs for most tasks that decline with age, but is especially pronounced where the measure is response speed.

However, the application of these laboratory findings to most work situations is dubious. The reasons for this are several. First, these studies have mostly involved a comparison of extreme ages not the age range (40-60) relevant to today's older worker. Second, the few studies that have been done with specific work groups indicate that age differences in performance and variability in performance is considerably less in older persons than that observed in general laboratory studies. Szafran (1962) measured psychophysiological and behavioral performance of pilots under various load. Those over 40 showed, under standard load conditions, little performance loss compared to younger pilots . Even under increased task loading only minimal functional loss occurred with age. Apparently people who are healthy, trained, motivated and performing tasks that are an habitual part of their activity maintain such psychomotor and information processing skill throughout their work career.

There is evidence for quantitative and qualitative differences in career demand and work environments as people age. For example, health related problems become important to the 45-64 year old person. While younger ages are more likely to mention problems at work and in the family (French and Caplan, 1973). Other issues that surface in the middle years are career progression, career ceilings, and obsolescence. Job insecurity and preservation of benefits are a major threat to some middle aged employees. Thus, mobility rates of young people are higher than for older individuals.

Cooper (1984) has marshalled evidence that "for middle-aged individuals there is a tendency to work long hours and to take on to much work". There is support for the thesis that such career and self imposed work demands in middle years are linked to low self-esteem, job dissatisfaction, emotional stress and cardio-vascular disease. These issues of middle age stress need to be part of decisions on how to structure work systems and organizations. For the older employee issues surrounding performance, stress and quality of work life should not be addressed without mention of the likely impacts of technological change and innovation, so it is to this topic that we now turn.

TECHNOLOGY ISSUES FOR THE OLDER WORKER

Hoos (1984) has pointed out that when we look for the impact of technology in the workplace "we find ourselves in the position of the six blind men who went to see the elephant" all came away with a different conclusion. Indeed few things are certain about how technology will affect jobs and work.

Consequences of Technology One consequence of new technology is to change the nature of jobs and tasks. Examples are industrial automation and robotics. Though positive outcomes have been argued, in practice these technologies usually result in de-skilling, task routinization, a reduction in need for individual judgement, paced demands and the lose of social contact and support (Osterberg and Enqvist, 1984).

In other work enviroments, such as nuclear and conventional power plants, aviation and military weapon systems, many jobs are now of the command control variety. Monitoring and information processing tasks and long periods of low activity, occasionally punctuated by high levels of demand, characterize these jobs. Computer-based office information systems also require the learning of new and different skills and adaptation to special office and social environments.

The quality of work life is likely to be an additional consequence of technological innovation. Rousseau (1978) reported a relationship between certain technologies (e.g. mass production) and job satisfaction, absenteeism, turnover and stress. Amick, Weisman and Smith (1984) have extended this association to health outcomes.

Technology and Age There is reason to believe that the demands of technological innovation will be disproportionately settled on older workers, especially males over the age of 45 (Hendricks, 1984). Technological change will likely influence both the option(s) to work and quality of work for older employees. Both positive and negative outcomes are possible.

On the negative side technology may limit the older person's job opportunities. One estimate is that technological advances will cause the loss of 10-14 per cent of existing jobs (Shephard, 1984). Displaced by new technology, or any other reason, the older person stays unemployed longer than younger workers, and is more likely to remain permanently unemployed (Taeuber, 1984). Also he/she is less apt to be selected for retraining and is less likely to seek retraining.

A second negative outcome could be on the traditional advantage that the older person has had in experience and wisdom. These have always been the one sure strategy the older employee had to maintain parity and compensate for any decline in other areas. Organizations have valued experience, especially in areas of problem solving and decision making. The development of expert systems and decision support technologies portends the day when the heuristics of experience and wisdom can be available at any age. A recent survey conducted in business showed a decline in the value placed upon experience.

A positive impact may be through assistive support or by substitution for those job elements for which an older person is disadvantaged. A significant number of older people face some kind of physical disability that limits the kind of work that can be performed. In the U.S., twenty-five per cent of individuals 55 to 64 cannot work regularly or at all because of disability (Social Security Administration, 1982). The handicapped and disability area has not received significant human factors attention over the years, as a review of the meeting papers at the Human Factors Convention and International Ergonomics Convention will show. Even for those persons without a chronic disability, new technologies may be of use where there are declines in physical strength, information processing skills or tolerance to stress.

Also on the positive side, technology may offer the potential of extending the working life. In the one case, this may be accomplished by reducing the hours needed at work, by increasing part-time job opportunities and by offering work at home. When polled employees state (51%) they wish to continue to work after retirement age, at least part time. Nollen (1982) reported a study from 68 organizations with permanent part-time worker programs showing that these employees perform better than full-time counterparts. They were also better on measures indicative of job satisfaction, such as, turnover and absenteeism, probably the result of less fatigue.

Elsewhere it has been argued that the most pervasive option to meet the dynamic age changes in the work force will be permanent part-time employment (Smith, 1984). In fact, the percentage of persons over 45 in the U.S working part-time rose sharply between 1969 and 1979. Study is needed on how to make use of emerging technology and to structure part-time and off-site work appropriate to the needs of the aging worker.

<u>Obsolescence and Training</u> In recent decades, the U.S. has undergone a revolution of technology and growth in specialized bodies of knowledge. The issue of obsolescence for significant segments of the work force has arisen from this revolution. The percentage of persons in highly technical and professional occupations in the U.S. has more than doubled since 1950. A number of recent reviews have dealt with the importance of training for obsolescence and technological change (Sterns, 1986; Wexley, 1984). These suggest three major issues.

Sterns (1986) has argued that "the major issue in the training of older workers is assuring equitable access to training opportunities". Obsolescence is no respecter of age. However, the older worker is peculiarly limited in solving this problem by negative age stereotypes held by his employer and even by himself. A growing older work force may change these values; research has long shown them to be inadequate.

A second major issue is the organizational environment for training. Dubin (1976) has emphasized as a critical factor the need to motivate employees to seek retraining. He identifies as the environmental dimensions the potential for job progression, past training success, and an organization's design, incentives and supervisor's attitude.

The structure of the training program is the third issue. The Belbins (1972) illustrated the value of adapting the method of training in industrial skills to the age of the trainee. For the older worker they emphasized self-paced tasks, the avoidance of interference and distraction, familiarity, meaningfulness of context, experiential learning, and help in organization and memory. Recent research in problem solving and learning has shown that older subjects have better task strategies in their repertoire than they often exhibit, and given appropriate training show marked improvement in performance (Denney, 1974).

Training, the design of training, and the role of training in systems development have long been of interest to human factors specialists. This expertise needs to be applied in matching training programs, job design, equipment interface and organizational structure to the older work force.

SUMMARY

The older employee will be a growing factor in the labor force and needs to be a concern in human factors design of the organization and work environment. With the exception of anthropometry, human factors research and application has tended to ignore age as a variable of concern. A thesis of this review is that this should not be true for the future.

REFERENCES

[1] Amick, B.C. III, Weisman, C.S. and Smith, M.J., The implications of technological change for worker health: The role of the job redesign model in public health, in: Hendrick, H.W., Brown, O. (eds.), Human Factors in Organizational Design and Management (Elsevier, North Holland, 1984).

[2] Baltes, P.B., Reese, H.W., and Lipsitt, L.P., Life-span developmental psychology, Annual Review of Psychology 31 (1980) 65-110.

[3] Birren, J.E., Robinson, P.K., Livingston, J.E. (eds.), Age, Health and Employment (Prentice-Hall, Englewood Cliffs, N.J., 1986).

[4] Belbin, E., and Belbin, R.I., Problems in Adult Retraining (Heineman Pub., London, 1972).

[5] Braune, R. and Wickins, C., The functional age profile: An objective decision criterion for the assessment of pilot performance capacities and capabilities, Human Factors 27 (1985) 681-693.

[6] Cerella, J., Poon, L.W. and Williams, D. M. Age and the complexity hypothesis, in: Poon L.W. (ed.), Aging in the 1980's (American Psychological Association, Washington, D.C., 1980).

[7] Cooper, C.L., Sources of occupational stress among older workers, in: Robinson, P.K., Livingston, J. and Birren, J.E., Aging and Technological Advances (Plenum Press, New York, 1984).

[8] Denney, N.W., Classification abilities in the elderly. Journal of Gerontology 29 (1974) 304-314.

[9] Dirken, J.M., Functinal Age of Industrial Workers (Wolters-Noordhoff, Gronigen, Netherlands, 1972).

[10] French, J.R.P. and Caplan, R.D., Organizational stress and individual strain, in: Marrow, A.J. (ed.), The Failure of Success (AMACOM, New York, 1973).

[11] Hendricks, J., Impact of technological change on middle-

aged and older workers: Parallels drawn from a structural perspective, in: Robinson, P.K., Livingston, J.and Birren, J.E. (eds.), Aging and Technological Advances (Plenum Press, New York, 1984).

[12] Heron, A. and Chown, S., Age and Function (Churchill, London, 1967)

[13] Hoos, I. Aging in a technological society. Disiderata and dilemmas in decision-making, in: Robinson, P.K., Livingston, J. and Birren, J.E. (eds.), Aging and Technological Advances (Plenum Press, New York, 1984).

[14] Humple, C.S., and Lyons, M., Management and the Older Workforce: Policies and Programs (American Management Association, New York, 1983).

[15] Kahnemen, D., Attention and Effort (Prentice-Hall, Englewood Cliffs, 1973).

[16] McFarland, R.A., The older worker in industry, Harvard Business Review, Summer (1943) 505-520.

[17] McFarland, R.A., The need for functional age measurements in industrial gerontology, Industrial Gerontology 19 (1973) 1-19.

[18] National Science Foundation, Science at the Bicentennial, (N.S.F., Washington, D.C., 1976).

[19] Ostberg, O. and Enqvist, J., Robotics in the workplace: Robot factors, human factors and humane factors, in: Hendrick, H. W. and Brown, O. (eds.), Human Factors in Organizational Design and Management, (Elsevier, North Holland, 1984).

[20] Palmore, E. B. and Jeffers, F. (eds.), Prediction of Life Span (Heath-Lexington, Boston, 1971).

[21] Robinson, P. K., Age, health and job performance, in: Birren, J.E., Robinson, P.K. and Livingston, J.E. (eds.), Age Health & Employment (Prentice-Hall, Englewood Cliffs, N.J., 1986).

[22] Robinson, P.K., Livingston, J. and Birren, J.E., Aging and Technological Advances, NATO Conference Series III (Plenum Press, New York, 1984).

[23] Rosen, B. and Jerdee, T.H., Older Employees: New Roles for Valued Resources (Dow Jones-Irwin, Homewood, Ill., 1984).

[24] Rousseau, D., Characteristics of departments, positions and individuals: contexts for attitude and behavior, Administrative Science Quarterly 26 (1978) 521-540.

[25] Salthouse, T.A., Functional age, in: Birren, J. E., Robin-

son, P.K. and Livingston, J.E., Age, Health & Employment (Prentice-Hall, Englewood Cliffs, 1986).

[26] Shephard, R. J., Work physiology and activity patterns, in: Milan F.A. (ed.), The Human Biology of Circumpolar Populations (Cambridge University Press, London, 1980).

[27] Shephard, R.J., Technological change and the aging of working capacity, in: Robinson, P.K., Livingston, J. and Birren, J.E. (eds.), Aging and Technological Advances (Plenum Press, New York, 1984)

[28] Smith, D.B.D., Some observations on the greying of the labor force: implications for human factors. Paper presented at the First Symposium on Human Factors in Organization Design and Management (Honolulu, 1984).

[29] Social Security Administration, 1978 survey of disability and work (U.S. DHHS, Washington, D.C., 1982).

[30] Sterns, H.L., Training and retraining adult and older adult workers, in: Birren, J.E., Robinson, P.K. and Livingston, J.E. (eds.), (Prentice-Hall, Englewood Cliffs, (1986).

[31] Szafran, J. Psychophysiological studies of aging in pilots, in: Talland, G. A. (ed.), Human Aging and Behavior (Academic Press, New York, 1968)

[32] Taeuber, C., Older workers: force of the future, in: Robinson, P.K., Livingston, J. and Birren, J.E. (eds.), Aging and Technological Advances (Plenum Press, New York, 1984).

[33] Yolles, S.F., Krensky, L.W., Kieffer, S.N., and Carone, P.A. (eds.), The Aging Employee (Human Science Press, New York, 1984).

[34] Welford, A.T., Where do we go from here? in: Poon, L.W., Aging in the 1980's (American Psychological Association, Washington D.C., 1980).

[35] Wexley, K.N., Personnel training, Annual Review of Psychology 35 (1984) 519-551.

Human Factors in Organizational Design and Management - II
O. Brown, Jr. and H.W. Hendrick (Editors)

Measurement of the Older Worker's Vocational Ability and Its Application

Yoshimi Yokomizo and Akinori Komatsubara
Department of Industrial Engineering and Management
Waseda University, Tokyo 160 Japan

Because of the recent Micro Electronics Revolution, not a little older workers have been replaced to new technology workshops. To obtain successful fruits in such job replacements, we developed and conducted tests and questionnaire surveys to measure and evaluate the older workers vocational ability and the influences of replacement on their daily lives.

1. Introduction

In Japan, new technologies brought by the recent so-called Micro Electronics Revolution have remarkably increased. By this revolution, the blue-collar workers who engaged in operation of old-fashioned machines or equipments have been replaced to workshops with automated or computerized machines, and they are required to be familiar with the new technology as programming, robot teaching and so on.

On the other hand, in Japan the population of the aged has been increasing and the birth rate has been decreasing. Therefore as companies must employ older workers, the older worker's adaptability to the new technology has become a serious problem and companies must measure and evaluate the each older worker's vocational ability and the vocational profile.

Though companies have usually decided the replacement for the workshops with new technology by merit rating, sometimes the replacement have failed. Therefore we studied the effectiveness of merit rating in relation with workers' vocational ability measured by objective test methods. Moreover we studied the influences of the replacement on daily lives of workers.

2. Merit Rating and Vocational Ability

Method

We developed the test battery for the measurement of older workers vocational ability. This test battery evaluate five important vocational evaluation points ; health, physical ability, psychological or intellectual ability, personal history or experience, and personality. This test battery consists of 25 tests shown in table 1.

We conducted this test battery at a certain steel company and measured the vocational ability on about 50 blue-collar workers of both the old and the young. The results were analyzed to know the individual vocational profile of each worker. The data of merit rating of them were also gathered. The individual vocational profile and the data of merit rating were compared.

Result

We drew the profile curves of each tested worker and imvestigated the relation of the test results and the merit rating. Fig. 1 shows the example of profile curve. The following points were turned out as the result.

(1) If we imagine the personal profile based on the results of the tests, the profiles are similar to the evaluation by the merit rating.

(2) Motor ability of hand and arm much differs between the workers of 40+ years and 50+ years.

(3) There are large individual differences on intellectual ability of the older workers. This difference has relativity with the educational background, the vocational history and the personality.

(4) Those who show high score on logical thinking ability obtain the better personal evaluation on merit rating.

3. Influences of Replacement on Daily Lives

In order to know the influences of replacement on daily lives, we conducted questionnaire survey on 110 blue-collar workers at another steel company, who experienced replacements.

The questionnaire involves survey items of objective task evaluation, subjective task evaluation, and condition of daily lives. The testees were required to answer these items at the time of before-replacement and after-replacement. We made interviews with some of the testees and their supervisors or managers. The following tendency were obtained as the results.

(1) If the replacement is from the direct workshops to clerical or new technology workshops, most of the workers will complain because their skill will not be useful at the new workshop, and sometimes the replacement will fail. In such cases not only workers meet trouble at workshops but their daily lives will tend to become troublesome. However some workers can adapt themselves to the replacement, and if we examine their personal profile, they usually show high willingness or high job motivation.

(2) If the replacement is from the direct workshops to direct workshops, few failures of replacement happened.

4. Discussion

We can say that merit rating can show the worker's overall vocational profiles but it can not show the vocational ability required for the new technology workshops. Therefore if the replacement is conducted based only on their merit rating, sometimes the replacement will fail. Moreover in such cases workers's daily lives are also influenced and their lives will tend to be troublesome.

Reference

Yoshimi Yokomizo, Measurement of the Older Worker's Capability, Ergonomics, 28, 1985

Yoshimi Yokomizo et al., Measurement and Evaluation of Vocational Ability of Older Workers, IEA 85, 1985

Table 1 Contents of the Test Battery

Item 1	Health (questionnaire survey)
	Chronic disease
	Recent health condition
	Recent fatigue or physical disorder
Item 2	Physical Ability (measurement test)
	Subjective evaluation with evaluation scales
	Far and near sight
	Audition
	Balance ability
	Lung capacity
Item 3	Psychological Ability (measurement test)
	Attention ability
	Cybanetical Control Number Test
	Simple judgement
	Logical thinking
	Short and long term memory test
	Motor skill
	Spatial perception
	Clerical perception
	Numerical perception
	Verbal perception
Item 4	History and Experience
	Educational history
	Vocational history
	Social life history
Item 5	Sociability (questionnaire and interview)
	Sociability
	Vocational aptitude
	Vocational attitude
	Personality

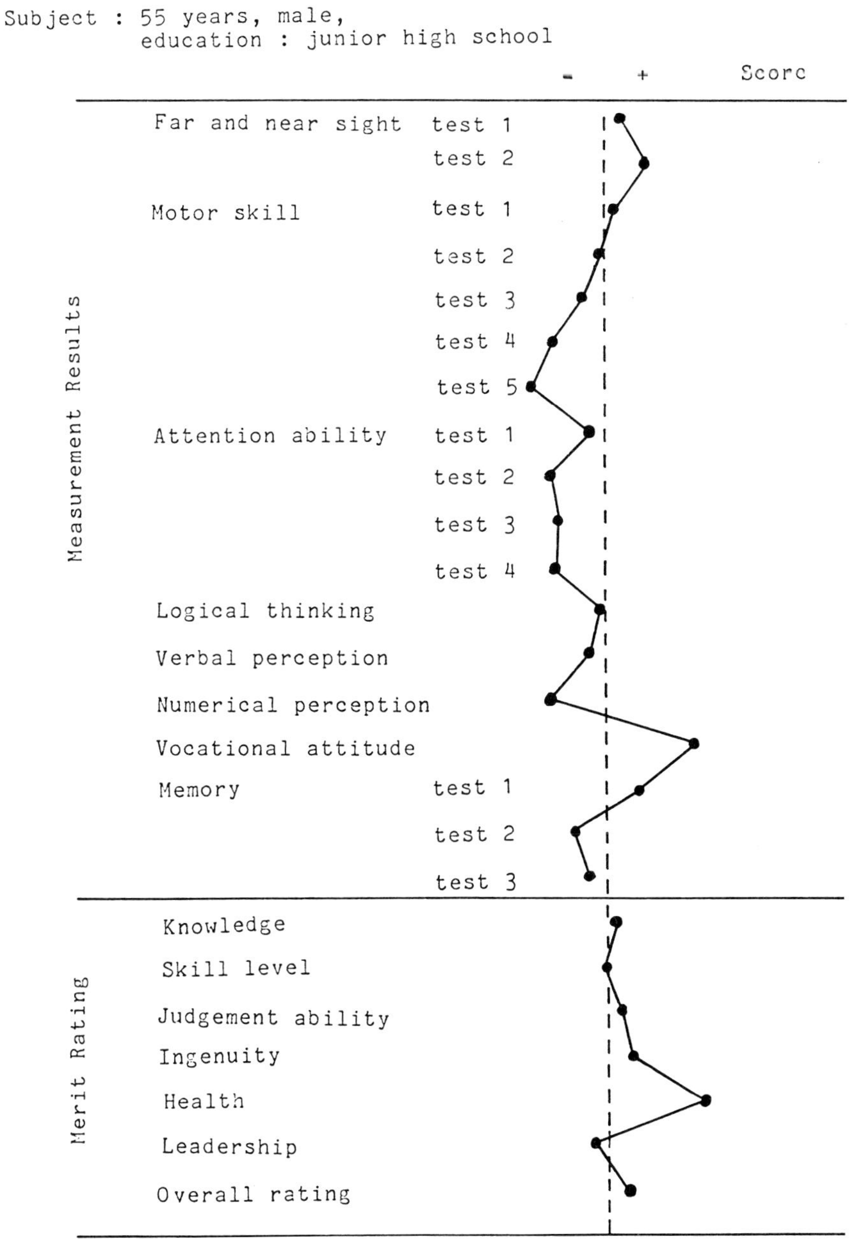

Fig. 1 Example of Vocational Profile obtained by the Test Battery and Merit Rating

Human Factors in Organizational Design and Management - II
O. Brown, Jr. and H.W. Hendrick (Editors)

LEISURE PROGRAMS FOR OLDER ADULTS - ORGANIZATIONAL HUMAN FACTORS CONSIDERATIONS

David R. Stirling,[1] Thomas J. Smith, School of Kinesiology
Simon Fraser University
Burnaby, B.C.
and
Stephen W. Meehan[1]
Active Health for Seniors
White Rock, B.C.

The planning and implementation of leisure education and recreation programs for older adults presents a challenge for health and recreation specialists, particularly for seniors living in apartment complexes in large urban centres. In this study of five hundred and eighty-two (582) seniors, biweekly exercise program and transportation services were the most successful in attracting older adults to participate. A number of the strategies used to provide recreation programs for this group of older adults is discussed.

INTRODUCTION

The planning and implementation of leisure education and recreation programs for older adults has become a challenge to many recreation and health care professionals. In 1981, 11% of the Canadian population was reported as elderly, i.e. persons over the age of 65 Statistics Canada, (1984). This represents some 2,347,000 individuals. By the turn of the century this will increase to over 18% Epp, (1985). Currently 70% of our elderly in Canada live in private households, as opposed to living in nursing homes and institutions, and of these elderly persons approximately half live in private apartments. Table 1 shows a comparison of living accommodation of retired and elderly persons living in five major centers across Canada. With this demographic shift in the population, i.e. larger portion of the population being elderly and living on their own, it becomes apparent that there will be an ever growing market for services and products for the elderly.

[1] Dr. Stirling is the Director of Active Health for Seniors White Rock Baptist Church, White Rock British Columbia. This project was funded by Employment and Immigration Canada Grant 5802 DX5.

Seniors living in apartment complexes are not easily contacted by physical activity and health care professionals. For security and privacy reasons, few organizations are permitted entry into these dwelling places. For the younger adult resident, such arrangements present less of a problem as their lifestyle and work patterns typically are focused outside of their residence. For the older adult, however, residential security and privacy constraints can lead to serious isolation and loneliness. This report describes some human factors and organizational management approaches which have proved successful in activating seniors despite these obstacles.

TABLE 1.

Comparison of retired and elderly persons living in six cities in Canada. Shown are the number of 'free living' individuals living in apartment block complexes and those in private homes.

City	Total Populations	Apartments		Private homes	
		Number of persons	% of total population	Number of persons	% of total population
Toronto	2,998,700	107,395	3.6%	78,860	2.3
Montreal	2,828,300	176,420	6.2%	45,520	1.6
Halifax	277,700	8,350	3.0%	7,735	2.8
Winnipeg	584,800	26,540	4.5%	23,620	4.0%
Victoria	233,500	21,715	9.3%	12,950	5.5%
White Rock	13,550	3,085	23%	1,165	8.6%

Statistics Canada, 1981 census

METHODS

The objective of the study was to identify and characterize the most effective human factors strategies for contacting and activating older adults living in apartment complexes in the city of White Rock, B.C. A 16-week leisure activation program was developed for 582 older adults living in six apartment blocks. Initially, 192 individuals registered for the program, while 136 remained in the program at its conclusion.

Once the apartment manager and the resident's committee had been contacted different aspects of the activation program were presented and developed. The program developed featured five distinct segments: (1) transportation services, which ranged from transporting seniors to various sites of historical and aesthetic interest, to trips for shopping or a doctor's appointment; (2) a telephone tree, which involved a number of individuals who would phone other seniors once a day every day over the sixteen week study period; (3) exercise classes, which participants attended biweekly in their apartment block complex over the sixteen weeks; (4) audio tapes, which enabled participants to receive and listen to audio tapes on various subjects, ranging from music to poetry and/or prose; and (5) courses and workshops to encourage special interests.

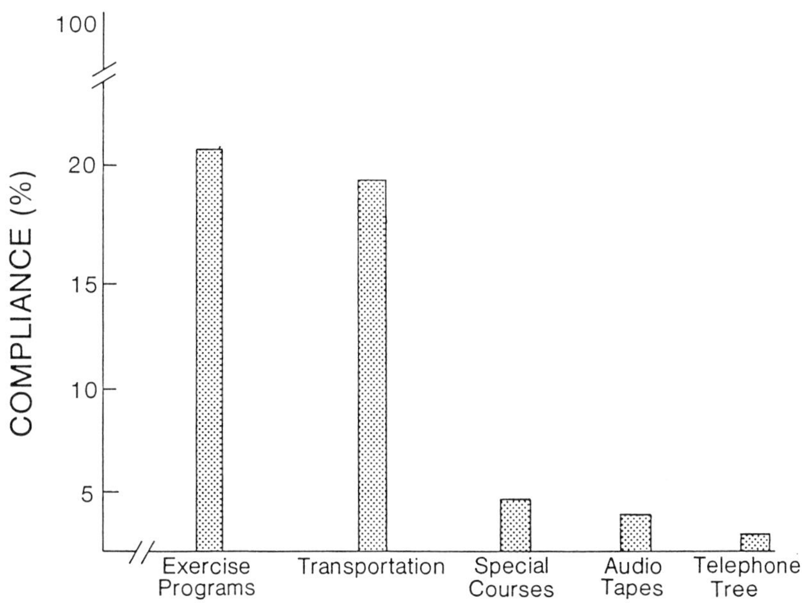

Figure 1. Compliance of Older Adults participating in Recreational Programs (compliance=total number of persons participating/total number of persons contacted x 100)

RESULTS AND DISCUSSION

The data reported in Figure 2 indicate that the exercise and the transportation program segments were the most effective strategies for attracting and activating the older adults involved in the study. The tapes and special courses or seminars were marginally effective, with the telephone network being least effective, in attracting older adults. With respect to activating and maintaining contact with seniors throughout the program, exercise was the most effective program strategy employed. Little difference was observed between the transportation, tapes and special courses. The older adults showed the least interest in participation in a telephone network program. All program segments, excluding exercise, experienced about a 33% attrition rate.

A survey of participants, completing the program, was conducted in an attempt to identify human factors aspects which contributed most strongly to the overall success of the activation program. Those factors cited most prominently in the survey are summarized below.

1. Time of Operation and Duration of Sessions

Sessions which ran between 10.00 a.m. and 3.00 p.m., and which lasted no more than 70 minutes, were the most effective and appealing. It was observed that interest and attention waned towards the end of some special sessions which lasted longer than 70 minutes. The 40 minute exercise classes appeared to be optimal for this age group.

2. Central location

Older adults appreciate being able to participate in programs that are conducted geographically close to their living residence, and which require little or no transportation. Programs offered within the local apartment blocks provided an important participation factor. Consequently, we feel that programs to or within the living residence (in this instance, the apartment blocks) will enhance participation.

3. Minimal Financial Cost

Most older adults are on fixed incomes. Equitable programs with fees close to service cost will attract willing participation. Most older adults appreciate the privilege of paying their way

5. Positive Physical and Social Environment

Negative physical or social environments deter older adults from getting involved in activity programs. Important factors include room decor and climate, and social atmosphere. For example, the meeting hall should be within an ambient temperature of 23-26°C, and colorfully decorated. Opportunity for social interaction is important. Programs that provided a period of easy conversation and light refreshment were positively received.

6. Recognition of the Participants

Every individual appreciates being recognized and called by name Older adults are no exception. If the leader can recognize the state of well being of the participants, and monitor their daily changes, the older adults appear to respond readily. Followup of each program segment also was important in order to maintain interest. Several approaches to followup were used in the various program segments. For example, in the special courses offered, persons attending any of the previous courses received a personal invitation to a new course being offered. At the conclusion of each program session the teacher/instructor tried to talk to each person in attendance seeking their impressions of the program and expressing interest in their general well being. In the exercise program if someone was ill or not feeling well the instructor will seek these people out either at the end of a class or prior to the next class again expressing interest in their general health and their feelings towards the programs offered. In segments where this occurred regularly (i.e. exercise), the retention rate was high. In segments where this did not occur, (i.e. the telephone network), the retention rate was low.

7. Minimal Equipment and Prerequisite Requirements

Programs that have a number of prerequisite and equipment requirements tend to have fewer participants.

8. Contact with Younger People

Most older adults live in close contact with other older adults. As a consequence, they look forward to meeting and interacting with younger adults as part of their program activities.

9. Privacy

Access to most apartment blocks in not achieved easily. However, once contact has been established, the respect for the privacy and ownership of the facilities was highly regarded. For example, visits by outside persons were all pre-planned, use of special equipment pre-arranged, parking was pre-arranged, no equipment was left at the apartments and the facilities used were carefully maintained.

10. Program Advertisement and Administration

Older adults appreciate participating in programs with minimal administrative details. The privilege of running programs in the apartment block facilitated the advertisement process. Personal invitations to each phase of the leisure activities program were easily accomplished, using the apartment mailbox system.

Although the factors specified above for the most part seem self-evidence, their identification represents one of the few reported cases in which a leisure-recreation program for older adults has been systematically evaluated in terms of its human factors attributes, based on responses from the participants themselves. The results clearly indicate that program success rests on far more than intrinsic content. Our primary conclusion from the above information is that program success in more critically determined by how carefully and comprehensively it is designed and tailored to meet the human factor needs of older adults and the older adult community. Our results are not based on a case-control design. The only evidence supporting our conclusions are the survey reports of the participants. If our conclusions are correct, the message for physical activity and health care professionals is that human factors considerations play a prominent role in designing programs for older adults.

Our findings call into question some elements of one of the principle theories of human-environment interaction and aging, the ecological theory of adaptive behavior and aging, Lawton and Nahemow, (1973), Lawton and Simon, (1968). This theory defines aging as as process of continual dynamic adaptation, in which both the individual and the environmental changes as well as the changes in his/her ability to reach a level of comfort generally decreases because of these adaptive changes. As a result, the aging individual tends to refocus control of the interaction process more on external environment factors, and begins to interact more passively with the environment. This "environment docility" hypothesis assumes that the environment becomes increasingly important as a person ages.

Our findings support the premise that environmental factors, i.e. human factors attributes, are critically important to older adults, insofar as leisure and recreational programming is concerned. We suggest, that the process of human-environment interaction and its changes with aging are more accurately and realistically interpreted in terms of behavioral cybernetic theory. This theory (Smith and Smith, 1986, 1985; Smith, 1972) assumes that human behavior, like all other modes and levels of function and organization, is a cybernetic process, in which the individual actively and continuously controls motorsensory behavior through a hierarchy of closed-loop feedback and feedforward control mechanisms.

In behavioral cybernetic terms, the regulatory imperatives which define behavior in the older adult are no less prominent than those manifested in younger individuals, in the sense that life itself is a product of stimulus control and environmental control. We believe that what changes with aging is control emphasis and control priorities. As aging inevitably degrades sensory acuity, motor performance capabilities, and motor sensory and physiological integration, the older individual quite naturally endeavors to exert enhanced control over external environmental factors as a means of compensating for these intrinsic deficits. We believe, this shift in regulatory emphasis is an active cybernetic, not a passive adaptive process. It serves to sustain the regulatory competence of the older adult in maintaining active behavioral control over the external environment.

From this perspective, the positive program attributes identified earlier can all be interpreted as active environmental control regulatory preferences. The expressed preferences for daytime sessions, minimal travel requirements, and positive environment and social interactive factors may be readily explained as feedback control products of the reduced mobility, ingrained circadian rhythm, and established social tracking characteristics of older adults. Behavior and participation in the program in accord with these preferences reflects an active control process.

One of the key observations in support of this behavioral cybernetic interpretation is the somewhat unexpected finding that the biweekly exercise program proved to be the most effective strategy for activating older adults (refer to Figure 1). Such a preference is not consistent with 'environment docility,' passive adaptation, and/or a preoccupation with comfort among the elderly. A basic tenet of cybernetic theory, is that behavioral performance, energy metabolism, and behavioral-physiologic integration are all feedback regulated by motor activity. Physical activity represents a ubiquitous strategy by means which individuals achieve and sustain optimal behavioral-physiologic integration. Our results suggest that both older and younger adults alike will take advantage of appropriate exercise opportunities when they are present.

We believe that the greater appeal of the exercise classes, relative to the other program segments, can be explained because their design embodied several human factors features. In particular, the classes offered ample opportunity for social interaction with other participants (class sizes ranged from 8 to 14 persons) and with a younger instructor. Such opportunities were less for the transportation, audio tape and special course program segments.

A recent review of the literature revealed no prior information describing factors which motivate older adults to join and participate in leisure and recreation programs. We believe that the major contribution of this report is the specification of a number of human factors considerations that recreation and health care professionals should deal with in the organizational design of such programs. Further research into effective leisure program strategies for older adults, and their human factors aspects, is needed.

REFERENCES

Epp, J. (1985). Care for an Aging Population. Keynote Address, Canadian Association on Gerontology, 14th Annual Scientific and Education Meeting, Hamilton, Ontario

Lawton, M.P., and Nahemow, L. (1973). Ecology and the aging process. In C. Eisdorfer and M.P. Lawton (Eds.). The Psychology of Aging. New York: VanNostrand Reinhold.

Lawton, M.P., and Simon, B.B. (1968). The ecology of social relationships in housing for the elderly. Gerontologist 8: 105-115.

Leaf, A. (1973). Unusual Longevity: The Common Denominators. Hospital Practice 8(10):75.

Shephard, R.J. (1978). Physical Activity and Aging. London: Croom Helm Publishers.

Smith, K.U. (1972). Cybernetic psychology. In R.N. Singer (Ed.). The Psychomotor Domain. Movement Behaviors. Philadelphia: Lea and Febiger, pp. 285-348.

Smith, T.J., and Smith, K.U. (1986). Feedback-control mechanisms of human behavior. In G. Salvendy (Ed.). Handbook of Human Factors/Ergonomics. Volume 2. New York: P. Wiley. In press.

Smith, T.J. and Smith, K.U. (1985). Cybernetic factors in motor performance and development. In D. Goodman, R.W. Wilberg, and I.M. Franks (Eds.) Differing Perspectives in Motor Learning and Control. Amsterdam: North Holland, pp. 239-283.

Statistics Canada (1984). The Elderly in Canada. Catalogue 99-932

Stirling, D.R., Miller, G., Barker, P., Rowden, G., Meehan, S. and Ralston, M. (1985). Exercise: An effective strategy to activate seniors in physical activity. In B. McPherson, (Ed.). Sport and Aging. Human Kinetics. Baltimore: Publisher.

Wallersteiner, U. (1985). Human factors and the challenge of aging. Proceedings of the 18th Annual Meeting of the Human Factors Association of Canada. Hull, Quebec: H.F.A.C., pp. 81-84.

ROBOTICS

Human Factors in Organizational Design and Management - II
O. Brown, Jr. and H.W. Hendrick (Editors)

ROBOTIC VEHICLES: A REVIEW

H. McIlvaine Parsons

Essex Corporation
333 North Fairfax Street
Alexandria, Virginia 22314
U.S.A.

Mobile robots, which offer design challenges to human factors scientists and engineers, have been proliferating in a variety of settings and degrees of automation. They are found in the air, in space, undersea, in nuclear facilities, on the battlefield, in bomb disposal, in the factory, in the laboratory, and elsewhere. Human factors interests include task/function allocation, human engineering design, skill requirements, safety, team concerns, and system development. In a newly conceptualized battlefield system, robotic vehicles would function as decoy tanks.

INTRODUCTION

A robotic vehicle has no operator on board. But are humans entirely absent? For some robotic vehicles the only human involvement is programming what the vehicle will do, perhaps also starting it on its way and stopping it or dealing with mishaps (and maintenance). In that sense some mobile robots resemble stationary industrial robots. They include automatic guided vehicles in factories, missiles and drone aircraft in warfare, some "toy" robots, and a number of experimental vehicles in university laboratories. Most robotic vehicles are guided (and their manipulators controlled) by human operators at the end of a communication (and perhaps power) umbilical or radio link. If the vehicles must function in complex and unpredictable environments it may be too difficult to program their performance in advance — though the artificial intelligentsia believes that will become possible. These vehicles may have some autonomy, however, if they have an on-board computer and various sensors (e.g., visual, acoustic, infra-red). There can be a variety of symbiotic relationships between the vehicle and its remote operators, in various forms of supervisory control. Optimizing such relationships is one of the human factors challenges such vehicles present. Increasingly, these vehicles are being called telerobots. They fall between stationary teleoperators and industrial robots, both of which are characterized by having artificial arms. In a sense these telerobots have artificial legs. Though some experimental and even operational mobile robots actually have legs, mostly they have wheels or tracks (like a tank). In many cases robot vehicles have been envisioned or developed because they can be operated in hostile environments.

What may be the functions of a robotic vehicle? First of all, of course, it must find its way to some destination or explore some area, and at the same time avoid obstacles of various sorts that might stop or injure it, or that it might injure. But mobile robots do not move around just to joyride (except for some "toy" robots, perhaps). They may carry some burden, whether a pallet or a warhead. With cameras they may investigate places or objects, whether to inspect some nuclear-plant tubing or conduct military reconnaissance. With arm manipulators resembling those of an industrial robot, they may help repair a satellite or disarm a bomb. They may fire weapons like a tank or a sentry. They can also simply pretend to be what they are not — a function with which this paper will conclude.

CATEGORIES OF MOBILE ROBOTS

Till now there has been no comprehensive review of robotic vehicles, though Martin and Kuban (1985) have edited a volume which surveys the field, Giralt (1984) examined vehicle automation, and Martin and Hamel (1984) related robots to teleoperators. The following review categorizes robotic vehicles by the environments in which they operate.

Air

Only lately have remotely piloted vehicles (RPVs) begun to be called robotic. An instance is the Aquila, a reconnaissance aircraft in production for the U.S. Army. The Israelis' Mastiff and Scout enjoyed considerable success in their last conflict with the Syrians. Cruise missiles could also be termed robotic, in fact all kinds of missiles — air to ground (or ship), air to air, ground (or ship) to air, and ground to ground (e.g., anti-tank missiles, especially as these have progressed from remote guidance by wire to self-guidance). IR and other sensors have given missiles increased autonomy, making them seem more "robotic." Target drones have existed for many years. One of the earliest — if not the first — ground-directed missile was the Bomarc, produced almost thirty years ago in connection with the SAGE system of continental air defense. Probably the most autonomous development is the Navy's Phalanx Close-In Weapon System (CIWS) which the AEGIS Weapons Control System coordinates with other ship self-defense systems. Although the AEGIS supervisor activates the Phalanx under condition 1 (general quarters), there is no other human intervention in the combat mode; the Phalanx with its own radar detects a close-in missile attacking its ship and fires a 20mm gatling gun at it at 3,000 rounds per minute. (A torpedo might also be viewed as a robotic vehicle.)

Space

By a strict interpretation, only the Soviets have fielded a robotic vehicle as such in space — the wheeled Lunakhod lunar manipulator in the 1960s. The U.S. Lunar Surveyor in that time period was a stationary device placed on the moon's surface, as were the two Viking landers with surface sampler subsystems sent to Mars in 1976. NASA's manned Space Shuttle carried the Remote Manipulator System (RMS) attached to that space vehicle. This six-degrees-of-freedom arm has two automatic (pre- and post-launch programmed) modes, a manual augmented mode, a manual single joint mode, and a direct drive (hard-wired) mode; manual modes are controlled through two 3-DOF joysticks, one for translational and one for rotational control. NASA and Rockwell International have been developing a true robotic unmanned vehicle, Robin, that would assist an orbital maneuvering vehicle (OMS) in satellite repair by replacement of units; a mobile three-arm manipulator (MRMS) on the Space Station would work with Robin. Martin Marietta has been working on a satellite servicing manipulator arm and Grumman Aerospace on a "telepresence work station" with a "dexterous manipulator system." In short, NASA has a great interest in telerobots in space, as alternatives to EVA (extra vehicular activity). A major issue is whether telerobots should be controlled from a space station or from the earth, with the latter's communication delays.

Undersea

There are several types of unmanned submersibles for subsea operations. One is towed along the ocean bottom or is self-propelled, primarily for search and inspection. The type ranges from "flying eyeballs," underwater cameras, to bigger vehicles with additional sensors but all are relatively small; they look, for example, for wrecks or mineral deposits. Examples include the Ametek Straza Division's Scorpi, Hydro Products RCV 150, and NAVFAC's Snoopy. Another type is a bottom crawler with one or more manipulators and remote control and power supply, for construction, repair, or rescue/recovery, for example the British Seadog that buries cables, MBA Associates Scarab for undersea telephone cable repair, a mine neutralization vehicle (MNV), and the "nozzle plug" (NP) used by NASA to help recover solid rocket boosters. A future, more sophisticated and autonomous category will maneuver around obstacles and build or repair major structures; the Naval Ocean Systems Center (NOSC) has been developing for this purpose a Work Systems Package (WSP). Also in the future is the small, fully autonomous EAVE-EAST (Experimental Autonomous Vehicle) under development since 1979 at the University of New Hampshire's Marine Systems Engineering Laboratory. EAVE-WEST is a free-swimming testbed with surface supervisory control for survey, search and recovery of pipelines and structures, developed by NOSC and the U.S. Geological Survey. Petroleum and offshore mining companies have been using scores of robotic undersea vehicles; for example, Exxon Research and Production has a submersible petroleum production platform and maintenance teleoperator system. The Woods Hole Oceanographic Institution operates the submersible Alvin and with a French group last year discovered the wreck of the Titanic; it used the Argo, a large tethered reconnaissance vehicle with sonar and panoramic television; the Jason, a small vehicle with high resolution TV and robotic arms, swims out from the Argo.

Undersea robotic vehicles have a considerable history. A General Mills stationary Model 150 manipulator was fitted in 1961 for manned deep sea operations. The Navy began developing "cable controlled underwater recovery vehicles" (CURVs) in the 1960s. CURV III recovered a nuclear weapon off the Spanish coast in 1965 and rescued the manned submersible Pisces off Ireland in 1976. The late Jean Vertut in the 1970s developed Eric I and Eric II for the French Navy; the latter was enclosed in Pagode as an undersea garage. In England, Heriot-Watt University created two tethered submersibles, Angus 001 and Angus 002, with Angus 003 as a garage for a free swimmer. The Japanese have produced underwater vehicles for surveys and emplacing bridge footings.

Nuclear Facilities

Extensive development and use of stationary teleoperators in nuclear power plants and R&D facilities have been followed by increasing interest in mobile telerobots. But even in the 1950s the Westinghouse Hanford Co. produced Louie, a teleoperated transporter, which later was used at Three Mile Island, and in 1966 the Programmed and Research Systems Corporation's Herman appeared at Oak Ridge National Laboratory. In 1967 Jean Vertut began the Virgule project for the French Atomic Energy Commission that led in the 1970s to the four-wheeled, radio-controlled rescue vehicle, MA23, with two MA22 arms. Also in the 1970s, the British fielded Roman — a tracked Remotely Operated Mobile Manipulator, and the Federal Republic of Germany a Nuclear Emergency Brigade with the MF2 vehicle that succeeded a manipulator crane vehicle and was followed by the MF3, a variable-geometry vehicle. (The Japanese Atomic Energy Research Institute more recently produced a wheeled Mobile Work Station, and Hitachi in 1983 developed for nuclear work a vehicle with five legs and a wheel on each.) In the U.S. in 1979 Jan Hodges developed Fred, a six-wheeled, 400 lbs., remotely controlled vehicle with a water spray for decontamination, and the Department of Energy in 1983 used the new 25-lbs. Sisi at Three Mile Island. Westinghouse has produced a manipulator arm called Rosa (also for underwater use).

The Nuclear Regulatory Commission has examined surveillance and inspection tasks a mobile robot can perform (e.g., steam generator tube inspection, crack detection, radiation level monitoring) and the Electric Power Research Institute 22 maintenance tasks (e.g., pipe cutting, welding, reactor cavity cleaning, flange unbolting/rebolting). A number of new robotic vehicles have been undergoing test by NRC, EPRI, and utilities, such as Batelle's Recomp, a Remote Technology Corporation vehicle, Odetics' six-legged Odex, and Automation Technology Corporation's Surveyor. The 270 kg RTC vehicle can hoist a color TV camera 5 m above the floor and send a picture to a control station; it can be preprogrammed or operated manually by joystick. The wireless 200 lbs. Surveyor, formerly the ARD IRIS (Industrial Remote Inspection System), has a telescoping arm and two cameras that provide 3D television displays. Its developer has published several experiments which investigated narrow vs. wide-angled lenses for turning corners and going through doorways, direct vs. remote viewing in approaching or moving away from the remote operator, and two-dimensional vs. three-dimensional viewing in maneuvering the robotic vehicle through an obstacle course. In addition, Carnegie-Mellon University created a 1,000 lbs., six-wheeled remote recon vehicle (RRV), the Rover, with a two-person control crew, one operating the umbilical reel, and a second RRV, the Moose, that could delivery 1,200 hammer blows per minute in scabbling for decontamination; the latter has been developed further by Pentek Inc. The Center for Engineering Systems Advanced Research of Oak Ridge National Laboratory has been developing an experimental self-powered mobile robot system called Hermies (now in the Hermies II version) with a wheel-driven chassis, dual manipulator arms, on board distributed processors, and a directionally controlled sensor platform. GCA/PaR Systems has built 15 remotely operated vehicles.

Battlefield

A major project has been undertaken by the Defense Advanced Research Projects Agency (DARPA) as one-third of its Strategic Computing Initiative. An "autonomous land vehicle" (ALV) is envisioned for Army use that can operate cross-country on its own over pathless, varied terrain, presumably for reconnaissance. Original plans were very ambitious in expectations as to what new developments in computers, sensors, and artificial intelligence could accomplish for such a vehicle, and when. The goals may have been somewhat modified; otherwise the project might meet the same fate that befell machine translation twenty years ago, when a National Academy of Sciences review committee's report abruptly terminated years of federal funding, hyperbole, and wishful thinking about miracles just ahead. (Machine translation is presently successful, e.g., at the Pan American Health Organization, but in a symbiotic relationship with human translators, who on a word processor post-edit the computer's output.) A prototype vehicle has been tested on a well-marked roadway (without the vehicular traffic that would be the analogy of a landscape or battlefield full of unpredictable obstacles, including hostile forces). Currently the Army's Engineer Topographic Laboratory (ETL) is lending some technical support, and contractors include Martin-Marietta and General Dynamics. ETL had undertaken one of the first conceptual studies of a robotic recon vehicle with support from the Jet Propulsion Laboratory, but this vehicle was remotely controlled by a navigator/commander who planned its course with computer-generated maps and a driver who steered the vehicle with television displays of obstacles observed by cameras on the vehicle. Another conceptual study came from the Rand Corporation. In TEARS (Tank

Effectiveness Augmentation by Remote Subsystems), robotic tanks ("Demons") would detect targets and fire weapons at them. More recently, Essex Corporation has conceptualized a complementary (or independent) system, DRIFT (Deception by Robotic Imitation of Friendly Tanks). More will be said about DRIFT and its "Sirens" later.

The U.S. Army itself (with support from various contractors) has considered a variety of functions for robotic vehicles for the battlefield (and the Marine Corps has also done studies), such as anti-armor, reconnaissance, decontamination, refueling, counter-mining (removal), sentries, and explosive ordnance disposal (EOD); robotic EOD is also a Navy concern as the primary military department for that function. At one point a robot was proposed as part of a tank crew to load munitions but it was felt that "hard" automation (if any) would be better. A transportable robot (not strictly a robotic vehicle) to transfer munitions for repackaging behind the lines has been funded for development and test bed demonstration through the U.S. Army Human Engineering Laboratory at the Aberdeen Proving Ground (Maryland). HEL is also creating a robotics laboratory that has a stationary robot and will have a teleoperated vehicle with a teleoperated manipulator, in its Soldier-Robot Interface Program (SRIP). The Army's Tank and Automotive Command (TACOM) has been developing test beds for robotic vehicles through General Dynamics and FMC Corporation. A forthcoming Army project will be a teleoperated anti-armor vehicle (small, light, unarmored, expendable) that might have multiple purposes, including EOD or might embody Rand's Demons or Essex's Sirens). In addition, Grumman on its own has developed a missile-firing robotic vehicle, the Ranger, and Robot Defense Systems has produced the larger Prowler.

Bomb Disposal

In contrast to battlefield (or shipboard) EOD involving hostile or friendly duds, robotic bomb disposal is conducted primarily in the United States by civilian federal agencies (e.g., the FBI) and municipal police and fire departments, though military personnel also may be involved, especially overseas. The locale is likely to be some built structure rather than open terrain, so a mobile bomb disposal device may have to be small enough to go through doorways and agile enough to climb stairs. A mobile bomb disposal robot is teleoperated and generally carries a camera for television transmission to a remote two-person (usually) crew, a manipulator or grapple and hoist or dragger to move or remove the bomb (or an automobile), and often a shotgun (directed at the bomb rather than individuals). Perhaps the most widely distributed telerobot of this type (said to be 500 in forty countries) is the tracked English (Morfax Ltd.) Wheelbarrow Mark 7, also used by the British Army for EOD (as is the case for other bomb disposal vehicles). Also from England have come the tracked Hunter (SAS R&D Services, in the U.S. in St. Paul, MN), and the wheeled Hadrian (Monitor Robotics). Ireland has produced the wheeled Hobo L2A1 (Kentree), Canada the wheeled RMI-3 (Remote Mobile Investigation Unit, by Pedsco), and the Federal Republic of Germany the tracked MF3 Ferngelenktes Manipulator Fahrzeug (Blocher-Motor KG). In the United States, Bendix fielded the wheeled Rover in 1973, and Ion Track Instruments (ITI) has the tracked RoVeh, TeleOperator System Corp. (TOS) the tracked Telemac, and Robot Defense Systems the wheeled Prowler (already mentioned under Battlefield systems).

Factory

Robotic vehicles have been proliferating in industrial plants as Automatic Guided Vehicles (AGVs), essentially to move raw materials or manufactured components from one location to another, in warehouses or on the factory floor, replacing trucks and forklifts and even conveyors, pallet jacks, or manual operations. They consist of tow tractors pulling trailers or trains of vehicles; pallet trucks ("stop and drop"); unit loads (that also lift or lower to various load decks); fork trucks (interfacing with racks or stands); and specialty vehicles for assembly processes. Those traversing a floor follow an imbedded wire guide or a tape or painted line (optically) or have inductive sensors. Path control comes from a central computer or an on-board computer or both, with collision avoidance by the host computer, programmable controllers, hard-wired reset logic, beacons, infra-red detection (the lead vehicle has a target on its rear), ultrasonic detection, or pressure bumpers that stop the vehicle. More flexible systems have vehicles that follow a prememorized path, with accurate positioning sensors or IR, ultrasonic, laser, radio, microwave, or vision sensors that pick up fixed or moving "targets" that enable, for example, trigonometric calculations. More in the future, and closer to battlefield recon robots, are "terrain comparison" systems that perform scene analysis and compare a stored reference scene with sensed data.

Laboratory

Following Stanford University's pioneering "Shakey" have been many experimental mobile robots built to investigate autonomous movement, including obstacle avoidance, with or without tethers (for power). ORNL's Hermies has already been mentioned. Others have been Stanford's AI Lab Cart; Carnegie-Mellon University's Rover, followed by the tethered three-wheeled Neptune and untethered and larger six-wheeled Terragator (and plans for Pluto and Uranus) as well as a one-legged hopping machine; the Jet Propulsion Laboratory's Rober; and a two-wheels-plus-caster vehicle at George Washington University. In England the University of Warwick converted a jitney and forklift and Queen Mary College developed two mobile robot carts, and a laboratory in Rennes, France, has created the Vesa. There have also been a number of legged and crawling mobile robots. In the U.S., Odetics has produced the six-legged Odex and Ohio State University has been developing the six-legged Hexapod. (General Electric created a very large Quadraped in 1966 for the U.S. Army as a potential device for Vietnam.) Overseas, the Soviets are said to have developed two six-legged robots, and the Japanese are credited with a biped, a four-legged robot, a five-legged one for nuclear facilities, and an eight-legged robotic vehicle said to be in actual use for underwater construction. The Japanese have also worked on crawlers; for example, Hitachi has shown one whose shape changes in ascending or descending stairs. No doubt there are other laboratory robotic vehicles not mentioned here.

Miscellaneous

Uses for robotic vehicles have been expanding, and it is difficult to stay current with them all, especially those in development. Some examples can demonstrate the diversity of the field. Back in 1973 the U.S. Navy Civil Engineering Laboratory at Port Hueneme worked on a small fire-fighting tractor, and currently the Naval Surface Weapons Center is developing a teleoperated mobile robot to fight fires on the decks of aircraft carriers (with direct vision rather than television). The Navy has also supported work on teleoperated devices that can move along the sides, surfaces or propellers of ships for inspection, cleaning, grinding, and painting. The Marine Robot RM3 produced in France also cleans hulls underwater; its legs have vacuum and magnetic cups to hold it to vertical surfaces. In Japan, Mitsubishi has an underwater robot to clean vertical, ceiling, or bottom surfaces; it is held against a surface by the reaction force of water ejected perpendicular to that surface. The Automatic Technology Corporation uses a similar method for its new, mobile Scavenger submersible robot scrubbing and filtering system for electric and chemical plants and water processing facilities. In Japan the Kimatsu Technical Research Center has an eight-legged robot "Recus" (Remotely Controlled Underwater Surveyor) for rubble removal and excavation, and there are excavator robots elsewhere. Use of mobile robots for coal mining has been investigated in the U.S. and England. Back in the 1970s the Japanese gave a laboratory demonstration of automated garbage collection. They also have described a mobile "Guide dog robot" and a robotic wheelchair. Mobile robots are being planned for sweeping the Paris Metro; the Japanese are considering the same for their subways. Denning Mobile Robots has proposed that prisons use its 385 lbs. mobile robot as a prison guard, and other developers have suggested their robotic vehicles could function as guards or sentries. There have also, to be sure, been proposals to use "toy" mobile robots for household chores, such as serving cocktails, cleaning floors or windows, or vacuuming carpets, but these have not been shown as yet to be cost-effective, nor have there been sufficient assurances they would not harm children or even adults. Possibly the most intensive study of a robotic vehicle for home use has been a University of Cincinnati investigation of a programmed or teleoperated lawnmower.

HUMAN FACTORS INVOLVEMENT

Robotic vehicles for some settings, such as bomb disposal, seem to have lacked applications of human factors engineering entirely, much as it might be needed. However, this discipline is becoming increasingly involved in other robotic fields. One impetus has been the extension of teleoperation from stationary manipulators to remotely controlled vehicles carrying remotely controlled sensors or manipulators. Thomas B. Sheridan at the Massachusetts Institute of Technology, whose elucidation of supervisor control with stationary undersea teleoperators was supported by the Office of Naval Research, has been conducting research for NASA on space teleoperation (cutely called by some "telepresence"). Among his former students, Dana R. Yoerger at Woods Hole Oceanographic Institution has helped design the Argus/Jason vehicles and participated in the discovery of the sunken Titanic, and Thurston L. Brooks is now involved in NASA projects at the Jet Propulsion Laboratory, where Antal K. Bejczy has long been doing research in teleoperation controllers; there was also some human factors support for the ETL study mentioned earlier. From one perspective a particularly significant human factors involvement in robotic vehicles is the Soldier-Robot Interface Program, mentioned earlier, at the Army's Human Engineering Laboratory, which the Essex Corporation is supporting along with Oak Ridge National Laboratory and Odetics. This research program will include generic experimental studies of remote vehicular and manipulator control. In other human factors research, Ross L. Pepper and his associates at NOSC have been investigating stereoscopic television for robotic vehicles, Azad M. Madni and his associates at Perceptronics have investigated battlefield robotics, Dennis B. Beringer and others then at the University of Wisconsin conducted studies involving the Ohio State University Hexapod, and there has been a considerable human factors participation in the Navy's carrier firefighting project. The human factors studies involving the ARD IRIS/ATC Surveyor have already been mentioned; Eugene B. Silverman, president of ARD (Advanced Resource Development Corporation) and ATC, is a human factors psychologist.

Targets of Human Factors R&D

Human factors R&D in robotic vehicles can and should be directed at the following questions:

Task/function allocation. What should remote human operators do and what should be done by programmed software in combination with on-board sensors? What forms should supervisory control take in a robotic vehicular system, contingent on the mission and environment?

Human engineering design. What design should control devices (e.g., joysticks) have (e.g., how many degrees of freedom)? How should these be allocated among various functions such as vehicle control, manipulator control, and sensor control? Should closed-circuit television employ stereoscopic or monocular vision, and if the former, in what kind of display system, and in either case with what camera placement on the vehicle and manipulator? What sorts of map displays should be provided for navigation, and if these are computer-driven, how should the software be designed for effective operator-computer interaction?

Skill requirements. What are the operational and maintenance skills required? Can these be combined in the same individual? Should the same individual have different maintenance skills, with multi-skill training and selection for versatility? Such questions are especially pertinent to battlefield robots and will affect job (MOS) and military training structure.

Safety. What kinds of hazards will remote operators encounter, and how can they be safeguarded?

Team concerns. How many remote operators will be needed, contingent on the number of mobile vehicles to be controlled and their functions and on vehicle-control communications? What sort of coordination will be needed between navigating, steering, and operation of sensors, manipulators, and (possibly) weapons?

System development. Might human factors scientists or engineers envision new systems or subsystems, perhaps with special needs for human-machine interaction?

DRIFT

The DRIFT concept is a partial answer to the last question. DRIFT can be viewed as a new sociotechnical system. It consists of an assemblage of humans and machines working together, with potential variations in the design of the assemblage to make it effective and many interactions between the humans involved as well as between the humans and machines. Many military systems could be thus regarded as sociotechnical systems, but most are associated with an anti-system (adversary) and differ in that respect from sociotechnical systems in industry or mining. That association increases, to be sure, the interactions between the humans involved, though not in a cooperative manner.

DRIFT as proposed to the Army (Parsons, Mavor, and Shields, 1984) and as noted earlier would consist of remotely controlled (and potentially autonomous) unmanned vehicles imitating friendly tanks to deceive, primarily, hostile tank units, as a kind of "ground chaff." Essex suggested that decoy tanks could constitute a useful new weapon system for the Army, which has been indicating increasing interest in the concept. Because the "Sirens" would be unarmored, lack sensors, weapons, and electronics for target detection and fire control, and might have minimal communications, they should be cheaper than actual robotic tanks ((Rand's "Demons") and might be fielded in large quantities. In addition to the military advantage DRIFT might yield on the battlefield through feints and expenditure of enemy munitions, it might advance the progress of full robotic automation since no other functions must be combined with locomotion, which is also simplified because it would make little difference if a few vehicles were blocked by obstacles, and obviously as decoys they would not require concealment. It seems most likely that the Soviets, as masters of deceit, have already adopted this type of system. Indeed, one can envision a battlefield full of opposing decoys along with real tanks whose crews try to figure out what to attack. (Sounds like Star Wars.)

The DRIFT concept may have a particular interest for engineering psychologists because a Siren would "attack" not another machine and its operators to destroy them so much as those operators directly through their senses and sensors to fool them. A necessary human factors involvement in developing DRIFT will be designing Sirens with visual, auditory, radar, and infra-red signatures so they do deceive the adversary. It may have been this appeal that helped generated the concept, which had not emerged in the plethora of suggestions within the Army and from its contractors about ways of using robots on the battlefield. As further possible reasons why it was missing, decoys do not seem to fit the categories of either a weapons system or a reconnaissance system, nor as mechanical contrivances have they greatly interested electronics specialists in countermeasures and counter-countermeasures, who did suggest that robotic vehicles might be used to emit electronic noise (jamming).

Where was the institutional memory? Imitation tanks were employed by American forces in World War II. Inflated rubber tanks in and near Dover helped convince Hitler that the Channel crossing would come near Calais. An engineer battalion accompanied the actual crossing in Normandy and performed successful deceptions with inflated tanks and guns all the way to the Rhine. In a later war, the North Koreans deceived American troops into attacking rubber-inflated tanks, and when the Americans caught on, the North Koreans covered real tanks to make them look like the fakes — and were not attacked. Countermeasures, counter-countermeasures, counter-counter-countermeasures — ad infinitum!

REFERENCES

1. Giralt, G. (1984). Robotics and artificial intelligence. NATO ASI Series, Vol. F11, pp. 365-393. Berlin: Springer-Verlag.

2. Martin, H. L. and Hamel, W. R. (1984). Joining teleoperation with robotics for remote manipulation in hostile environments. In Proceedings of the Robots 8 Conference. Dearborn, MI: Society of Manufacturing Engineers.

3. Martin, H. L. and Kuban, D. P., Eds. (1985). Teleoperated robotics in hostile environments. Dearborn, MI: Robotics International of SME.

4. Parsons, H. M., Mavor, A. S., and Shields, N. Jr. (1984). Human factors engineering issues in Army field robotics applications. Report for the U.S. Army Human Engineering Laboratory. Alexandria, VA: Essex Corporation. (Available only to government agencies and their contractors.)

Human Factors in Organizational Design and Management - II
O. Brown, Jr. and H.W. Hendrick (Editors)

OPTIMIZATION OF HYBRID PRODUCTION SYSTEMS: THE INTEGRATION OF ROBOTS INTO HUMAN-OCCUPIED WORK ENVIRONMENTS

M. Rahimi and P.A. Hancock

Safety Science Department and
Human Factors Department
Institute of Safety and Systems Management
University of Southern California
Los Angeles, California 90089-0021
U.S.A.

ABSTRACT

A model is presented to assist managers in transitioning to and operation of hybrid automated manufacturing systems. Hybrid systems are defined as those in which both human and machine interact, each as cooperative intelligent entities. Specific elements in the model are discussed with reference to examples from human-robot cell production. Potential for prediction from the model is examined and presented as a decision aid in a managers resource allocation decisions. Areas which require further understanding in hybrid systems are considered.

I. INTRODUCTION

A definition of a "system" is an aggregation of interacting objects or components to fulfill a common purpose which transcends their individual purpose. With the recent emphasis on the development of "automated systems", this definition assumes even greater importance. Central to contemporary developments, although the United States is among the most productive countries in the world, both European and Pacific rim countries have introduced a significant industrial productivity challenge in the recent years. A major component of this productivity challenge has stemed from developments and applications of advanced technology. There is every indication that for the United States, increase in productivity has and will rely on fundamentally similar technological premises. Significant commitments have already been made to the use of computers and microprocessor-based factory automation. This rapid evolution of manual work into fully automated work systems has been faced with problems and misconceptions when applied to already operational systems. Increasingly, managers, designers, manufacturers, and users of these systems are realizing the difficulty of achieving a transitional process.

Basically, three evolutionary trends are evident in the current factory automation: (1) a widespread need and appreciation of the practicality of computerized manufacturing and robotics, (2) the increasing power and simplification of computerized systems and interfaces, and (3), the increasing realization of the impact of automation on machines, workers, and compatibility between humans and machines. The obvious progress toward factory automation has been to introduce islands of automated systems (e.g., robots) -- a great majority of these cells being independent of each other. The immediate task of managers is to weld these elements into an integrated system. In terms of hardware, making use of more powerful and low-cost computers to perform more complex jobs and provide efficient communications will make such an integrated factory system economically viable. It is this perspective sincerity factory automation that has been the center of recent attention . However, without a broader understanding of several interacting factors involved in design and implementation of automated systems, these moderately focused efforts will be only modularly effective. This paper attempts to introduce a broader "systems" approach to the consideration of factory automation (e.g., robotics). Subsequently, it focuses on some of the system constituents such as management, jobs, operators, and technology.

II. A MODEL FOR HYBRID SYSTEM DEVELOPMENT

II.1 Structure of the Model

In the present model we have represented factors of increasing importance in an hierarchic tree structure. The apex of the structure represents overall productivity and is immediately decomposed into three major elements relating to the worker-machine environment, human factors in management, and personnel resource management, see Figure 1. In continuing with the pattern of factor identification, each of the

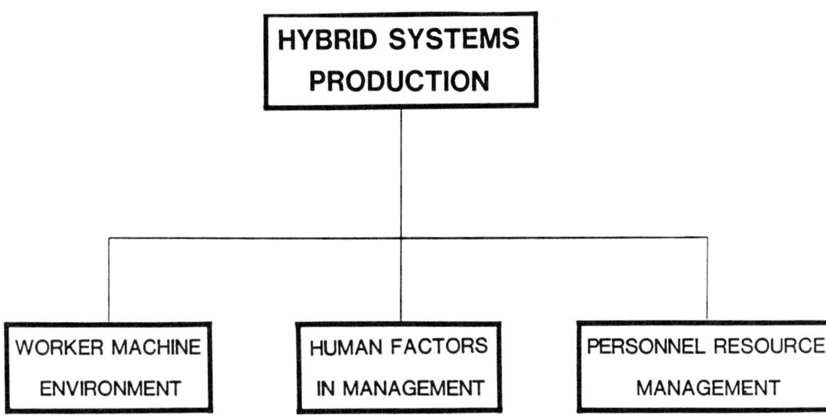

Figure 1: First decomposition for a hybrid automated production system

major elements assumes the role of apex in the following figures which are used to identify subsidiary components in each content area. Details of these illustrations can be seen in Figures 2, 3, and 4 which precede discussion of the elements they contain.

Through the adoption of this hierarchy, the manager can use the simple structure we have presented to locate individual problem areas. For example, training and personnel placement issues are considered under personnel resource management. These considerations can be then integrated into an on-going picture of the transition process as it applies to a specific organization. While the structure presents a static representation of factors, it is their dynamic variation under the impact of particular driving forces that is of particular interest and this nature of the model is developed briefly below. In reference to the systems approach, this model incorporate several factors which are essential in assessing effects of technology on humans in the work environment. Figure 1 shows the first layer of the schematic block diagram for this model.

II.2 On Prediction and Dynamic Change

As has been presented previously, the structure of the model is a hierarchic representation of factors involved in hybrid system production. As such, it presents content areas of concern which define loci of transition in the optimization of overall production. One most important issue is how to present dynamic change in order to extract predictive information for the managers use. To achieve this, as a first pass approximation, three forms of information are needed. First, the current state of the system as defined by given values attached to each content area. Second, the system specific parameters which define optimality for each specific system considered. Finally, what direct and interactive which effect may be achieved through resource expenditure as the internal time of the system proceeds.

These latter two factors, a defined goal state, and an envelope of potential time varying change are the more important, and as the model is a multidimensional representation of qualitatively differing elements, thus the desired dynamic information may be expressed as a matrix of informational elements. Initially, all identified factors should be considered subject to improvement through resource investment, whilst only obvious interactive effects (e.g., effect of feedback/remuneration on training/performance) should be considered operative. By simulation is possible to suggest the most beneficial pattern of resource investment for the manager to receive maximal return during the transition process into hybrid production. Clearly, the model may be revised and the simulation improved with repetitive iterations giving greater insights into alternate interactions and differential gain values for existing elements.

III. WORKER-MACHINE ENVIRONMENT

In Figure 2, as illustrated below, we have subdivided the major element into its component sections which are of particular importance to the manager.

III.1 Software and Hardware Reliability

Automated manufacturing systems are being increasingly dominated by their software content. Software engineering is a rapidly developing field in which well understood and universally accepted design concepts have yet to be established. The field has recognized the importance of incorporating software reliability considerations into the initial stages of the design--a systems approach. That is, software development is a part of the

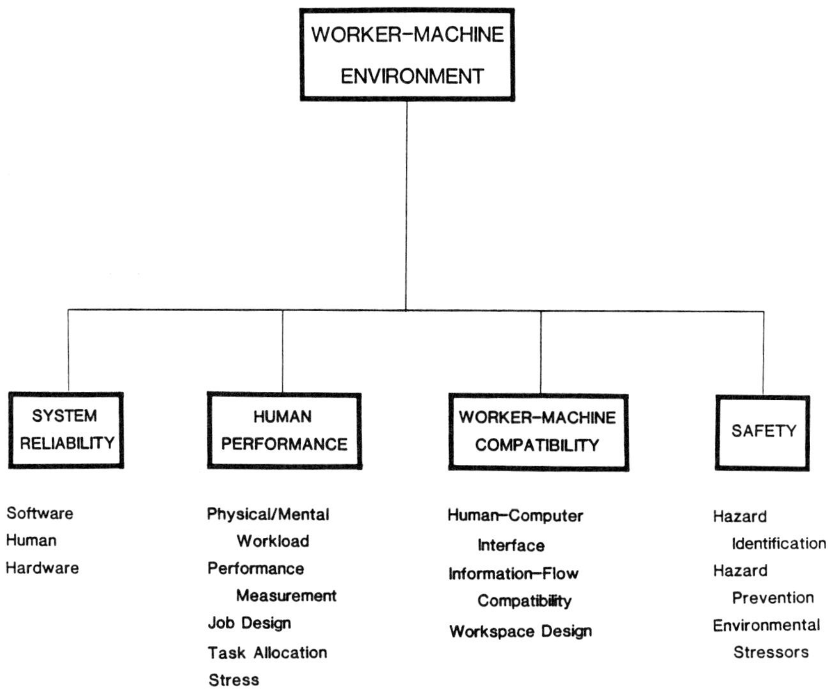

Figure 2: Components of worker-machine environment and related subdivisions

overall system development process. This recognizes that complex computing systems are comprised of interacting hardware and software elements which are subject to a wide spectrum of failure modes and errors. Software reliability, like hardware reliability, includes steps that start during the conceptualization of every system and continues through system analysis, software design, coding, testing, and system operation and modification. The emphasis is placed on the initial analysis and design where most of the software flaws and errors are introduced. It is estimated that 60% of all software errors are committed before coding even begins (Lakner and Anderson, 1985). Unless special attention is given to each step of program development and to the elimination of errors and unreliability, the progressive stages of software development can result in a very inefficient and even dangerous system. The most cost effective approach is toward detection and elimination of errors as early in the program development as possible. The cost of software error detection and elimination increases exponentially as the software is progressively integrated into the overall system during operational phases.

Developing a total combined hardware/software system reliability and failure hazard rate is a difficult task. The knowledge obtained from hardware reliability and safety methodologies and models does not always apply to software subsystems. The following are some of the major differences to be noted:

1. Software has no bathtub hazard rate curve (Kline, 1980).

2. Software will not wear out.

3. Software engineering, reliability, and safety are relatively new fields.

4. Little data are available for software reliability and safety evaluation.

5. Software is design oriented, whereas, hardware is affected by design, production, operation, and maintenance.

6. Redundancy in software procedures, codes, etc..., may not have the same effect as hardware redundancy.

7. Software retrofitting to increase reliability and safety has a higher probability of creating complexities when interacting with other parts of the system.

8. It is more difficult to maintain software modularity for higher reliability. Characteristically, software components are highly interdependent.

9. In software, Mean Time Between Failure (MTBF) relates to time between detection of errors not occurence of undesired events or failures. Failure rate distribution for software is initially high, and will continually decrease with operating time.

10. System failure is normally traced back through hardware first.

11. Software fault detection is a function of test time, type of test and choice of test data.

12. Software has a higher potential for monetary savings.

Software reliability is gaining importance in systems with humans acting as a critical subsystem. The issue of safety of humans and very expensive peripheral systems can be integrated into the discipline of system safety. System safety concepts and methodologies can be used to simultaniously identify and control hazards associated with system constituents. It identifies the mechanism for considering software as a system constituent. Development of "software system safety" was emphasized by the release of military standard MIL-STD-882B. Rhoads (1985) points out that software system hazard is a result of conflict between instructions and data encountered versus that expected for a given condition and time. With the emphasis of software in advanced technology systems, hazards are generated through conflicts, not failures. The purpose of software system safety is to identify such conflicts, and to focus on those unexpected outputs which could become hazardous. System safety focuses and analyzes only the critical elements of the three subsystems of software, hardware, and human, in addition to their interactions. Some of the techniques suggested and used in software system safety analysis are:

Fault Hazard Analysis
Preliminary Hazard Analysis
Operating Hazard Analysis
Failure Modes and Effects Analysis
Fault Tree Analysis
Safety Cross-Check Analysis
Sneak Analysis

Some of these techniques have been used to analyze military software subsystems (Kattuah, 1985; Cazden, 1985). Also, a report by the Electronic Industries Association (1984) has outlined a series of recommendations for performing software safety analysis. This report emphasizes a multifaceted analysis approach of management considerations, technical aspects, and detailed procedures.

III.2 Human Reliability

In order to project productivity in hybrid operation the manager requires information on overall system reliability. The key to this overall estimate is the accurate assessment of human reliability and its appropriate incorporation into an overall equation. There are two distinct approaches to this problem which has seen the development of some controversy. One position holds that the human may legitimately be inserted into overall calculation and that appropriate estimates may be derived accordingly. An alternate view proposes that the human must be treated as a qualitatively different component in which behavior varies not only in numerical magnitude but in fundamental content. Under this latter perspective the simple combination of two qualitatively different functions leads to more complex calculation and difficulty in producing an effective estimate. The interested reader may wish to examine the controversy in more detail and should refer to the review work of Adams (1982). An alternate view emphasizing direct human reliability estimation incorporation can be found in the work of Swain (1970).

The major problem in understanding human reliability is the lack of a coherent theory of tasks and a theory of error. At our present state of advance, we have little knowledge in the area of what constitutes an elementary unit of a task and what descriptive taxonomy might describe the plethora of human error types. While these insights are unavailable, a veridical model of human reliability cannot be fully elaborated. However, because of operational pressure, estimates of human reliability are important in the practical operation of complex systems and best estimates are presented accordingly.

III.3 Physical and Mental Workload

Despite the knowledge that musculo-skeletal injuries continue to be the greatest source of industrial lost-time, specifically lack injuries the transfer within work of an emphasis from physical to mental workload is recognized. In a hybrid cell, the physical workload will largely devolve to the more capable robot component, whilst elements such as problem-solving may fall more to the human worker. In some of our previous work (Hancock, Meshkati, & Robertson, 1985; Meshkati, Hancock, & Robertson, 1984; Wierwille, Rahimi, and Casali, 1986) we have recognized the difficulty in quantitatively evaluating mental workload. There is a necessity to posses such information in that excessive or insufficient mental load may be instrumental in either long or short-term dysfunction of the human worker, e.g., job burnout and boredom.

In a series of articles on adaptive machine systems, we (Hancock & Chignell, 1986; Chignell & Hancock, 1985) have sought an approach by which productivity may be optimized while human mental workload is maintained within acceptable limits. Many restrictions concerning speed of operation are inherent in human limitations while flexibility of goal-directed action acts to curtail robot abilities. It is the purpose of the hybrid cell to utilize the best capabilities of each element while eliminating the potential limitations intrinsic to their individual functioning. In order to achieve this harmonious and most probably optimum long term state, the on-line quantitative evaluation of mental workload is a necessary requirement.

III.4 Environmental Stressors

Environmental stressors, in the context of hybrid production, are values of the surrounding physical environment which act to disturb the output efficiency of the operational cell. Due to the scale of fragility, such environmental conditions are far more liable to impact the human than the robot. Environmental stress can encompass physical characteristics such as temperature, vibration, radiation and variations in gaseous constituency of the atmosphere. In typical contemporary production areas tolerance limits are set quite liberally, particularly with respect to our ignorance, long-term repetitive exposures. In other specialized production areas, e.g., clean-rooms, tolerances are much smaller in that they might affect product as well as process and processor.

In the work of Hancock and Chignell (1985), adaptability criteria for setting stress tolerance standards have been discussed. These principles advocated stress levels that do not violate individually scaled tolerance functions. In the referenced work, both normal and failure modes of activity are identified and ways in which thresholds of adaptability might be modelled are considered as methods through which stress monitoring might be achieved. We should not omit reference to Salvendy's insights a freely-chosen and optimal long term operation rate in which it is suggested that individuals are best able to regulate long-term work rate. In an analogous manner, through the use of the concept of comfort, we have suggested that stress levels should not violate the subjectively sought zone of comfort of the human operator.

III.5 Information-Flow at the Interface

One problem of immediacy in the development of hybrid production systems is the nature and method of information transmission across the ilnterface. One major problem is that robot/machine structure has been developed somewhat independent of interactive considerations. How is the machine to present information? Should there be visual, verbal or dual-channel transmission? Is this to be presented in parallel with the operation at hand or serially contingent upon minimum-cycle time of each constituent of the cell. We have argued that both the physical and cognitive interface need to consider the principle of compatibility in order to minimize potential operator-machine information handling conflicts.

A manner in which to envisage such cooperation is through the use of direct manipulators. These prosthetics directly replicate the movement of the operator at a remote point. In this case, dependent upon lever arms and rates of movement gain, the operator and machine represent one entity. The question of interface integration is how to physically realize this direct interaction and yet to retain the two intelligent entities as cooperative operators.

When the somewhat mechanistic problems of the physical interface are resolved, there remains the larger and less tractable questions concerning the cognitive model of the operator. In an adaptive manner, it is important that through analysis of operator behavior that the flexible machine element should mould its behavior in response to the interactions generated (Chignell & Hancock, 1986). These processes are iterative and require a form of machine learning. This continues of importance as the operator becomes progressively familiar with the capability of the system at hand.
Unless the cell is required to perform purely repetitive actions, and this would be a clear case for direct automation, then the cell should be viewed as a problem-solving team. The cognitive solution would be a direct result of cooperative action while the physical action to achieve success is delagated largely to the robot. This latter argument militates for the same degree of parallel processing on behalf of the machine element.

III.6 Workspace Design

The consideration of humans in new automated systems developments have raised new questions and concerns for system designers. The roles of Human Factors in system development have been emphasized (Meister, 1982). A broad model for integrating humans and machines and their separate levels is given by McCormick and Sanders (1982):

1. Statement of Objectives
2. Separation of Functions
3. Allocation of Functions
   a. Human
   b. Hardware
   c. Human/Hardware Interface Design
4. System Integration

With introduction of automation (e.g., robots, flexible manufacturing systems, and CAD/CAM), another layer of complexity is added to this integration scheme. Separation of functions and allocation of tasks have been the focus of a number of researches (Paul and Nof, 1979; Nof, Knight, and Salvendy, 1980; Brennan, 1984). The human/robot function allocation, for example, has been primarily based on "Fitt's like lists" investigating the capabilities of humans and robots. Robot-man charts have been developed which compare abilities in the categories of (1) Action and Manipulation, (2) Brain and Control, (3) Energy and Utilities, (4) Interface, and (5) Miscellaneous Factors. Kamali, Moodie, and Salvendy (1982) have extended the robot-man chart into a table comparing the abilities of automation, robots, and humans as part of a framework to aid the manufacturing engineers in job design of combined systems. Due to lack of specificity, this approach has not found widespread applications. For example, the memory capability of a robot processing unit may be much greater than a human operator. However, the human flexibility in utilizing memory capacity makes this simple comparison untenable.

With the capabilities of current robots within a hybrid workstations, perform tasks with special accuracy requirements, movement/space limitations, and certain safety requirements. Conversely, humans perform tasks which are unstructured, require complex reasoning and decision making, and require supervisory control. These tasks are the major parts of jobs such as maintenance, trouble shooting, back-up, inspection, and supervision.

In regards to the role of the human in flexible manufacturing systems (FMS), the areas of significant attention as indicated by Ranky, 1983; Salvendy and Nof, 1984 are:

1. Compatible software for human operators.

2. Structured methods to transfer information between CAD and CAM and the human input/output into the system.

3. Humans role in maintenance, trouble shooting and diagnostics, and repair.

4. Interaction between human operator and computerized decision support systems.

5. Optimal allocation of machines in a FMS for each supervisor.

The crucial role of advances in Artificial Intelligence and Expert Systems is helping address some of these questions.

III.7 Hazard Identification and Prevention

A number of studies have pointed out the importance of looking into factors which cause robotic accidents (e.g., Backstrom and Harms-Ringdahl, 1983; Sugimoto and Kawaguchi, 1983). It is becoming apparent that the existing industrial structural organizations and systems do not allow for the rapid implementation of totally automated factories in the immediate future. In the past few years, robots have been introduced into the factory plan with little attention to their interactive role with humans. As the capabilities and limitations of robots versus humans become more clear, an increasing number of robots will be introduced with hybrid functional relationships. Industrial and Human Factors Engineers should play an important role in the design of such hybrid workstations. Ironically, the introduction of robot automation to eliminate human involvement has led to a series of accidents. Thus, the number of accidents and death caused by robots are increasing. A description of these causes help manufacturers and users of these automated systems plan for such a hybrid workstations. The following is a list of some of possible sources of robot accidents (Rahimi, 1984):

Machine Factors

-Robot component failures (mechanical, electrical, etc.)
-Software control inadequacy, error, and reliability
-Unexpected halts and stops
-Improper design of teach control panel
-Inadequate safety devices
-High speed of robot arm
-Human-robot work environment

Operator Related Factors

-Entering danger zone of a halted robot for:
  .trouble shooting
  .repair
  .testing
  .maintenance
  .teaching
-Improper initial robot start-up procedure
-Work-piece adjustment and positioning
-Operator unaware of robot movement
-Intentional disabling of dafety devices
-Collision path for robot and human tasks
-Lack of training

The safety engineering solution to the above list of safety problems can be obtained using the following five approaches:

1. Developing a complete robot sensory detection capability for human and nonhuman proximity detection.

2. Improving hardware and software control design and reliability for robots.

3. Incorporating ergonomic considerations for design of hybrid workstations.

4. Developing a comprehensive safety training for individuals associated with robot operation and maintenance.

Several "system safety" methodologies have been suggested (Rahimi, 1986) to model and analyze petentially dangerous factors contributing to robotic accidents. In addition to Energy Barrier Analysis, other potentially useful techniques include:

Preliminary Hazard Analysis
Subsystem Analysis
Failure Mode and Effect Criticality Analysis
Critical Incident Technique
Task Analysis
System Simulation
Fault Tree analysis
Management Oversight and Risk Tree Analysis

Strengths and weaknesses of each methodology must be considered when applied to robotic workstations.

IV. HUMAN FACTORS IN MANAGEMENT

As with the previous major element we have decomposed Human Factors In Management into a number of subsidiary components and these are presented in Figure 3 below.

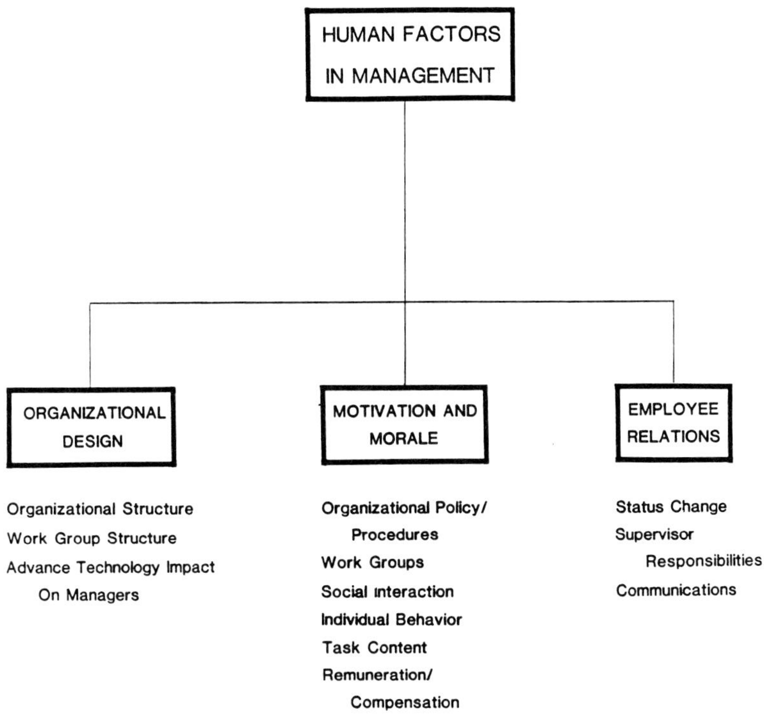

Figure 3: Components of human factors in management and related subdivisions.

IV.1 Advance Technology Impact on Managers

Technology has clearly affected the role of both blue and white collar work groups. Evidence points to the notion that technology has been and will continue to chase the nature of jobs skill requirements. Outside of certain industries--most notably metals fabrication and steel production in particular--the number of management jobs is actually growing in manufacturing. New managers are necessary to handle new work and organizational structure resulted from automation:

1. Automation increases capital requirements and adds to the complexity of accounting chores.

2. Automation increases fixed costs, adding to pressure on sales staffs to bring a steady stream of orders--raising competition between certain employee groups.

3. Flexible production systems allow flexible marketing and advertizing. Small product changes are aimed at segments of consumer demands.

4. Production flexibility requires a closer track of materials, work in progress, and finished goods.

5. Automated systems (e.g., robots) can produce new products which otherwise would not be produced. For example, products with production processes which are hazardous to human operators.

In general, increased automation is associated with increased complexity of the organizational system components. A major component of this complexity relates to the type and amount of information that managers need to handle. There is a natural tendency to produce the information using computer information management systems thus a need to train the managers in handling such information.

More importantly, flexibility in manufacturing can drastically shorten "thinking time" for managers. There are shorter cycles for product design, inventory ordering, and financial and capital recovery. This reshaping in the managers skills and capabilities must begin with certain formal training and education. Increasigly, more MBA programs are offering courses in manufacturing management. Degrees in Industrial Management or Engineering are more in demand than before. The issue is how to integrate an educational package which best responds to this emerging and changing field.

V. PERSONNEL RESOURCE MANAGEMENT

In this final section we presented the subsidiary components of Personnel Resource Management in Figure 4 below.

V.1 Education, Simulation, and Integration

What are the managers first steps in the construction and instantiation of a hybrid production system based upon a cooperative cell unit? The manager must first phase in the familiarization process in order to achieve optimal integration. Education is the first element of this process. The human operator needs same minimal information concerning the capability of the machnine and this process portion of education should be concern with both knowledge and hands-on supervised activity (training). To provide information concerning boundary conditions of machine operation, the operator should be able to explore the potentialities through a computer-based simulation. In the final portion of this phasing in, the operator is allowed to integrate with the actual machine system. This process is most important and will be subject to more detailed exposition in some following work.

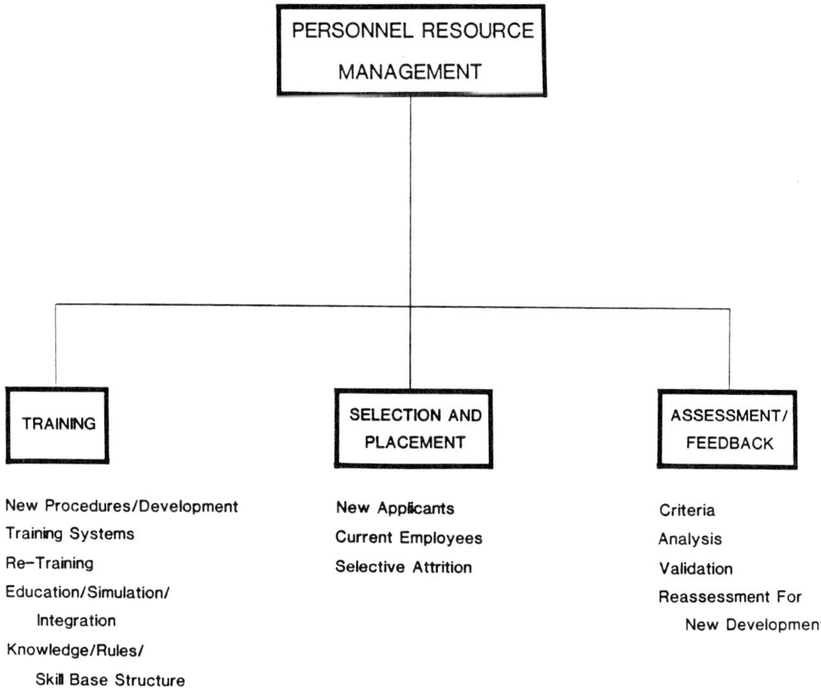

Figure 4: Components of personnel resource management and related subdivisions.

V.2 Skill, Knowledge, and Rules Based Operation

One major question for the manager concerns just how the interactive operation should occur. Is it sufficient that the operator simply follow a set of operational rules? For the hybrid cell to operate in a problem-solving mode, rules-based operation would provide insufficient flexibility for allowing the generation of novel solution. The hybrid cell would be enhanced if a combination of skill and knowledge base operation were employed. Such operation is not merely the procedure of the operator but also includes how the machine interacts with the human entity. Further discussion of skill, rules and knowledge base operation may be found in Rassmussen and Rouse (1977).

V.3 Current Employees and Attrition

Rigorous scientific data are very limited on the effects of automation on employment. There appears to be some evidence that automation has reduced the number of manufacturing jobs available for general worker population and this trend appears to be accelerating. Only 20% of American nonfarm workers are in manufacturing now, compared to 30% in 1965 and 34% in 1950.

And, the increasing automation will eliminate 100,000 to 200,000 factory floor jobs and create close to 60,000 jobs (Ross, 1986). The technological changes in the Canadian railway system indicate a similar trend (Peitchinis, 1978). The main employment effect of the rapid trend toward automation is one of massive job redistribution. Ranky (1983), based on Bureau of Labor Statistic projections, showed that the decline in number of machine tool operators is 12% and for set-up workers and tool and die makers is 24%. However, the projected increase in maintenance electricians is 28%, computer service technicians 93%, electrical engineers 27%, and industrial and production engineers 26%. An example for such a job skill shift is a welding operator. When the job is robotized, the operator will require less skill, education and training. Thus, this job will probably be downgraded. However, there will be a shift from operating to maintenance of robots. As a result, the proportion of jobs lost will be increased and the retraining for the new skill requirements will be essential. These sections consider only a restricted sample of those we have identified in figures 2, 3, and 4. Further consideration of these topics are the subject of current work in our laboratory.

VI. Summary

In the present paper, we have presented a structured "systems" model for the integration of hybrid automated production systems into full scale operation. Description of the model consists of the hierarchic identification of procedures involved in the integration process. We do not claim to have exhaustively identified these factors nor thier potential for interaction. Rather, we have presented the main items of concern and have selected among these to provide information only on specific areas. We have focused upon the content rather than the operation of the model in that the specific dynamic changess depend upon the situational importnace of the elements identified by individual managers. However, it would be possible to generate predictions from the model given the individual managers system concerns. Thus the model would act as a mangerial decision aid in terms of resource allocation to achieve optimal transition between current and hybrid production and to retain optimal productivity once transition has been achieved.

VII. REFERENCES

Adams, J.A. (1982) Issues in human reliability. Human Factors, 24, 1-10.

Brennan, L. (1984) The influence of new technology on the allocation of functions decision. Proceedings, First International Conference on Human Factors in Manufacturing Automation, London, U.K., 79-86.

Cazden, D.B. (1985) Software Sneak Analysis as a Development Support Tool. Proceedings, Seventh International System Safety Conference. 2.6-2-1 to 2.6-2-6.

Chignell, M.H., and Hancock, P.A. (1985) Knowledge-based load leveling and task allocation in human-machine systems. In: Proceedings of the Twenty-First Annual Conference on Manual Control Columbus, OH.

Chignell, M.H., and Hancock, P.A. (1986) Integration of the cogtnitive and physical aspects of the human-machine interface. Paper submitted to the Annual Meeting of the Human Factors Society, Dayton, OH.

Electronic Industries Association (1984) A method of software safety analysis. Safety Engineering Bulletin No. 6., EIA G-48.

Hancock, P.A., & Chignell, M.H. (1985) The principle of maximal adaptability in setting stress tolerance standards. In: R.E. Eberts and C.G. Eberts (Eds.). Trends in Ergonomics/Human Factors II, North Holland: Amsterdam.

Hancock, P.A., and Chignell, M.H. (1986) Input information requirements for an adaptive human-machine system. In: Proceedings of the Tenth Conference on Psychology in the Department of Defense, Colorado Springs, CO.

Hancock, P.A., Meshkati, N., and Robertson, M.M. (1985) Physiological reflections of mental workload. Aviation, Space and Environmental Medicine, 56, 1110-4.

Kamali, J., Moodie, C.L., and Salvendy, G. (1982) A framework for integrated assembly systems: Humans, automation and robots. International Journal of Production Research, 20, 431-448.

Kattuah, F.E. (1985) Applicability of system safety methods to software/firmware safety. Proceedings, Seventh International System Safety Conference. 2.6-4-1 to 2.6-4-18.

Kline, M.B. (1980) Software and hardware reliability and maintainability: What are the differences? Annual Reliability and Maintainability Symposium. 179-185.

Lakner, A.A. and Anderson, R.T. (1985) Reliability Engineering for Nuclear and Other High Technology Systems. Elsevier Applied Science Publishers, Essex, England.

Meshkati, N., Hancock, P.A., and Robertson, M.M. (1984) The measurement of human mental workload in dynamic organizational systems: An effective guide for job design. In: H. Hendrick and O. Brown (Eds.). Human Factors in

Organization Design and Management, North Holland: Amsterdam.

Nof, S.Y., Knight, J.L., and Salvendy, G. (1980) Effective utilization of industrial robots - A job and skill analysis approach. AIIE Transactions, 12, 216-225.

Paul, R.P. and Nof, S.Y. (1979) Work methods - A comparison between robot and human task performance. International Journal of Production Research, 17, 277-303.

Peitchinis, S.G. (1978) Technology and Employment in Industry and Services. Six Countries Workshop, Paris, France, November

Rahimi, M. (1984) System safety approach to robot safety. Proceedings, Human Factors Annual Meeting, 102-106.

Rahimi, M. (1986) System safety for robots: an energy barrier analysis. Journal of Occupational Accidents. (In press)

Ranky, P.G. (1983) The Design and Operation of Flexible Manufacturing Systems. IFS Publications, Bedford, England.

Rasmussen, J., and Rouse, W.B. (1977) Human detection and diagnosis of system failures. Plenum Press: New York.

Ross, S.S. (1986) Robots in the U.S.A. Careers, April, 78-81.

Salvendy, G. and Nof, S.Y. (1984) Integration of human operators in computerized manufacturing: Some considerations. Proceedings, Annual International Industrial Engineering Conference, 179-188.

Swain, A.D. (1970) Development of a human error rate data bank. Abuquerque, NM: Sandia Corporation, Sandia Corp. Report SC-R-70-4286, July.

Wierwille, W.W., Rahimi, M., and Casali, J.G. (1985) Evaluation of sixteen measures of mental workload using a simulated flight task emphasizing mediational activity, Human Factors, 27, 5, 489-502.

POSSIBILISTIC LINEAR REGRESSION MODELS

Human Factors in Organizational Design and Management - II
O. Brown, Jr. and H.W. Hendrick (Editors)

POSSIBILITY ANALYSIS OF RELOCATING CENTRAL GOVERNMENT OFFICES BY POSSIBILISTIC LINEAR REGRESSION MODEL

S. TSUTSUI*, I. HAYASHI** AND H. TANAKA**

* Department of Management and Information Science,
Faculty of Commerce, Hannan University,
5-4-33 Amamihigashi, Matsubara-shi, Osaka 580 JAPAN
**Depertment of Industrial Engineering,
College of Engineering, University of Osaka Prefecture,
4-804 Mozu-umemachi, Sakai-shi 591 JAPAN

1. INTRODUCTION

The congestion of cities is today becoming a serious problem in industrial countries. This is particularly true in Tokyo, the capital of Japan. In general, some degree of functional concentration in a single city is desirable for efficiency in carrying out certain social activities. However, it would seem that congestion in Tokyo today has almost reached the critical point.

Many proposals have been offered as possible solutions. One of them is the idea of relocating central government offices. However, this invariably involves several difficult problems. This paper examines the influence that relocating those offices would have on face-to-face communication (hereafter referred to as FTFC) activities between the offices and the citizens. In line with this, the social surveys concerning FTFC at central government offices and an analysis of them using possibilistic linear regression model[2)] were carried out.

2. STUDY OF FTFC AT CENTRAL GOVERNMENT OFFICES[3)]

To obtain a rough idea of FTFC activities observed in central government offices, the number of visitors on a certain day was counted at eleven office building. The tally was taken on December 11, 1975. Total visitors to all buildings totaled 55,798.

To analyze FTFC at central government offices, one building was selected and questionnaires were given to visitors. The survey date, March 4, 1976, was decided with care to insure the actual FTFC activities would be normal on that day. Fig.1 shows the distribution of visitors' working places.

3. DISTANCE FACTOR OF FTFC ACTIVITIES

AS seen in Fig.1, about 75 percent of all visitors come from the three central wards of Tokyo. Thus, the question is how those people would be affected by relocating central government offices. It was assumed that at present it is not inconvenient for those people, either in time or distance, to visit central government offices. In this chapter this issue will be discussed by making clear the relation between distance and the characteristics of FTFC. To express the measure of distance, "time distance" is used in addition to "physical distance."

3.1 Distance factor with frequency of visit

Average visiting frequency per annum is shown in Fig. 2. This is calculated from the answers to questions of visiting frequency. Visiting frequency is seen to decline proportionately as distance becomes greater as seen in Fig. 2. This means that visiting frequency is deeply entwined with the distance factor.

3.2 Distance factors according to type of office that visitors come from

It is understood from Fig. 3 that the proportion of visitors from the head office declines with the increase in distance (or change of zone) as far as with in the "half-day travel zone." This tendency reveals the concentration of headquarter functions in metropolitan areas. That tendency disappears beyond the "half-day travel zone." These facts allow the following interpretations concerning factors determining the location of the head office (nerve center in management). There is a communication zone centered around central government offices and extending to the "half-day travel zone" limit. Here, headquarters functions are concentrated and oriented toward the core of the city for reasons of greater convenience in communication with the central business district (CBD) in the three central wards. On the other hand, once out of that zone, head offices no longer seek convenience in communicating with CBD, but settle in the chosen location with little regard to ease of communication with CBD.

4. ANALYSIS OF DISTANCE FACTOR BY POSSIBILISTIC LINEAR REGRESSION MODEL

4.1 Possibilistic linear regression model[2)]

Possibilistic linear regression model (hereafter referred to as PLRM) based on fuzzy sets theory[1)] has been proposed by Tanaka et al[2)]. In this section, the distance factor relative to the number of visitors is analyzed by PLRM. In the conventional regression model, deviations between the observed values and the estimated values are supposed to be due to observation errors. On the contrary, in PLRM it is assumed that these deviations depend on the indefiniteness of the system structure, and are regarded as possibility of the system parameters. In our analysis, the variation of the number of visitors relative to distance is assumed to result from possibility of each visitor's human behavior for FTFC activities. Thus, the spread of the number of visitors is regarded as possibility of system parameters. Possibility can be represented by a fuzzy number. A fuzzy number $\underset{\sim}{A}$ models a vague number as "approximate α," and is given as a membership function $\mu_{\underset{\sim}{A}}(a)$ with degree of [0,1] and this degree can be interpreted as possibility. Here, we define membership function $\mu_{\underset{\sim}{A}}(a)$ of $\underset{\sim}{A}$ as a triangular fuzzy number denoted by $\underset{\sim}{A} = (\alpha, c)$ as

$$\mu_{\underset{\sim}{A}}(a) = 1 - |a - \alpha| / c \tag{1}$$

Where, α is the center and c is the width of the fuzzy number $\underset{\sim}{A}$.

Let us consider the following possibilistic linear function with parameters of fuzzy numbers $A_j = (c_j, c_j)(j = 1, 2, --, n)$.

$$\underset{\sim}{Y} = f(\mathbf{x}, \underset{\sim}{\mathbf{A}})$$

$$= \tilde{A}_1 x_1 + \tilde{A}_2 x_2 + \cdots + \tilde{A}_n x_n \quad (2)$$

In PLRM, the problem is to determine fuzzy parameters $\tilde{A}_j$ so that the observed data may be contained in the estimated fuzzy number $\tilde{Y}$ within a certain degree h ($0 \leq h < 1$). Possibility of $\tilde{Y}$ can be calculated from possibility of $\tilde{A}_j (j = 1, 2, \cdots, n)$ and membership function of $\tilde{Y}$ is represented by the "extension principle" as[1)]

$$\mu_Y(y) = 1 - |y - \mathbf{x}^t \boldsymbol{\alpha}| / \mathbf{c}^t |\mathbf{x}| \quad (3)$$

In PLRM, the following assumptions are taken:

(1) Given an obserbed value (x_i, y_i) and threshold degree h, then

$$\mu_{\tilde{Y}_i}(y_i) \geq h \qquad i = 1, 2, \cdots, N \quad (4)$$

(2) The index of fuzziness of PLRM is

$$S = \sum_i \mathbf{c}^t |\mathbf{x}_i| \qquad i = 1, 2, \cdots, N \quad (5)$$

Here, S show the sum of width of $\tilde{Y}_i$.

Under the above assumptions, our problem is to obtain fuzzy parameters $\tilde{A}_j$ that minimize S in the equation (5) subject to the constraints equation (4). This problem can be easily solved by LP approach.

4.2 Results and considerations

The following possibilistic linear function is adopted as a model:

$$\tilde{Y}_i = \tilde{A}_0 + \tilde{A}_1 x_i^{-1} + \tilde{A}_2 X_i^{-2} + \tilde{A}_3 x_i^{-3} \quad (6)$$

where, $\tilde{Y}_i$ is the estimated fuzzy number which explain the number of visitors from area i per ten thousand working population in area i, x_i is the distance from area i to the government offices in km and $\tilde{A}_0$ - $\tilde{A}_3$ are fuzzy numbers.

$\tilde{A}_0$ represents number of visitors who might not depend on distance x_i. Therefore, those visitors' FTFC activities seem very important. $\tilde{A}_1 x_i^{-1}$, $\tilde{A}_2 x_i^{-2}$ and $\tilde{A}_3 x_i^{-3}$ represent the number of visitors who might depend on distance x_i.

The results of PLRM is shown in Fig.4 and Table 1. These results can be summarized as follows:

(1) As seen in Fig.4, the estimated fuzzy numbers by PLRM can be represented with upper limit line, lower limit line and central line. Thus, possibility of the number of visitors is estimated such that observed data should be inside of the upper and lower limit regression lines. The central regression line shows the highest possibility of the number of visitors; on that line $\mu_{\tilde{Y}}(y) = 1$ holds.

(2) Since $\tilde{A}_3$ has the largest value among $\tilde{A}_0 - \tilde{A}_3$ and $\tilde{A}_3 x_i^{-3}$ account for a large portion of $\tilde{Y}_i$ in the near area from government offices, $\tilde{Y}_i$ decrease sharply in proportion to x_i^{-3} as distance become greater in the near government offices.

(3) As seen in Fig.4, there is a major breakpoint between 20 and 50 km (the boundary between "half-day travel zone" and "one-day travel zone"), beyond which changes in distance hardly affect the frequency of visits.

5. MODEL PATTERN OF INFLUENCE ON FTFC ACTIVITIES OF RELOCATING CENTRAL GOVERNMENT OFFICES

A model pattern which explains the influence on FTFC activities of relocating central government offices can be considered in the following manner. The interrelation between central government office functions and type of business office determines the potential for FTFC. The actual frequency of FTFC activities, on the other hand, is dependent on "distance-resistance characteristics." Relocation of the business office to an area proximal to central government offices, and a restructuring of the business will take place if the business office recognizes that its FTFC activities with government offices have become an impediment to normal operation due to remote location.

Both distance-resistance characteristics and considerations behind location of enterprise can clarified by studying the distance factor of the FTFC examined in Chapter 3 and 4. Consequently, the effect of the relocation of central government offices upon FTFC activities can be estimated.

The results of Section 3.1 and 4.2 can be considered aspects of the distance-resistance factor. Based on the factors that emerge, it is estimated that the change in the number of visitors due to central government office relocation would be as shown in Table 2.

From the results discussed in Section 3.2 and 4.2, it can be considered that the work of an office, whose main business include FTFC with central government offices, will be greatly inconvenienced if the central government offices are located beyond the "half-day travel zone." This gives some idea of considerations behind location of enterprise.

6. CONCLUSIONS

On the basis of the results of this study, it may be suggested that it would be most desirable to relocate to a zone where, first, "relocation will not exert a sudden or crucial impact on its external relations and environment." This means that the relocation zone should allow FTFC with 90 percent of the total visitors from business offices at the center of Tokyo. Second, it is desirable to relocate to a place where the "remoteness of the place is such that it gradually promotes or necessitates relocation of business offices." In other words, people from the city center should feel a certain degree of inconvenience in visiting the new location. To meet the above requirements, the "half-day travel zone," involving a distance of 30 to 40 km, or a time distance of one-half hour, is an optimal zone for relocation.

REFERENCES

1) L.A. Zadeh,"Fuzzy sets as a basis for a theory of possibility," Int.J. Fuzzy set and Systems, Vol. 1, pp. 3- 28(1978).
2) H. Tanaka, S. Uejima and K. Asai, "Linear regression analysis with fuzzy model," IEEE Trans., Vol. SMC-12, pp. 903-907(1982).
3) S. Tsutsui, S. Murayama and H. Misaizu, "Relocating government offices and its influence on face-to-face communication: A case study," IEEE Trans., Vol. SMC-10, No.3, pp.134-142(1980).

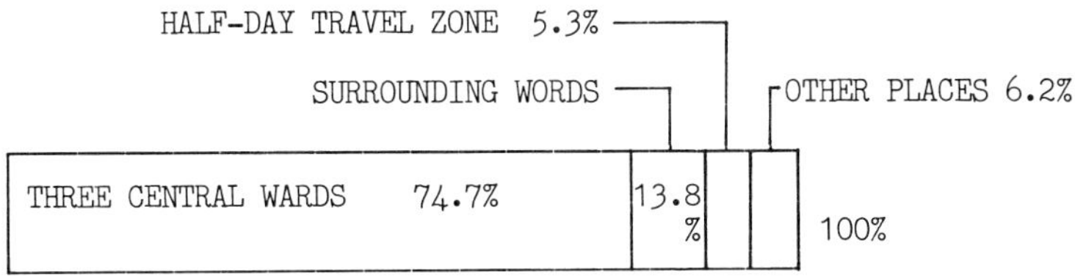

Fig.1 Location of visitors working place

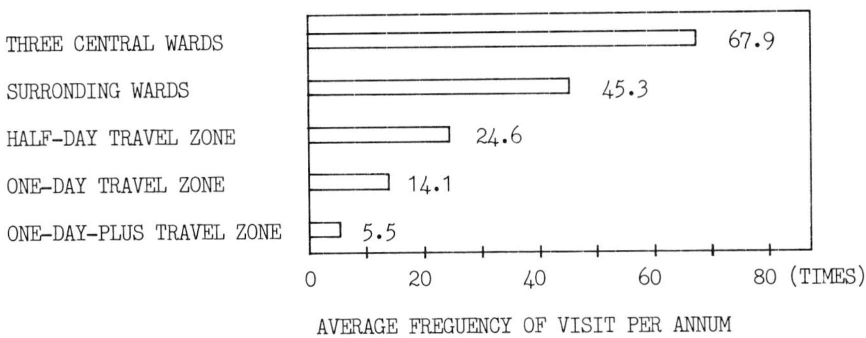

Fig.2 Distance versus visiting frequency

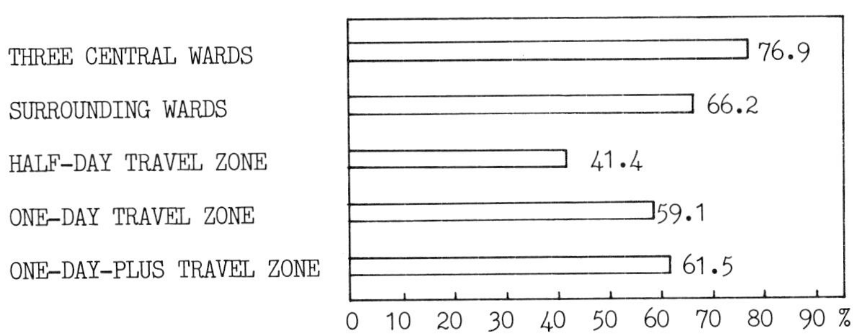

Fig.3 Proprtion of visitors who belong to the head offices of their organization

Table 1 Estimated fuzzy parameters

A_j	A_0	A_1	A_2	A_3
(α_j, c_j)	(0.043,0.028)	(4.64,4.58)	(0,0)	(480.3,350.4)

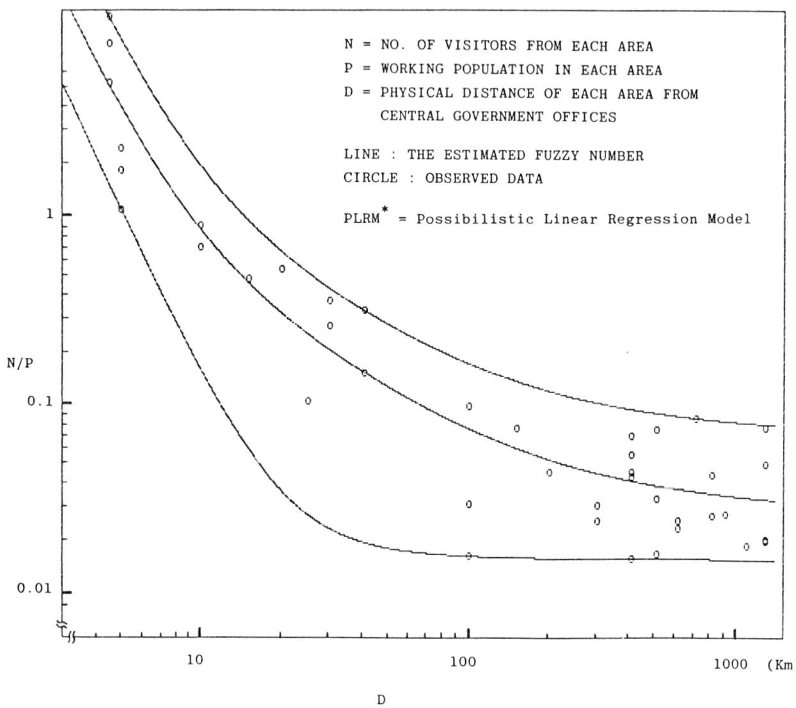

Fig.4 Distance characteristics of number of visitors by PLRM*

Table 2 Influence of relocation of central government offices on number of visitors

Located to	by Visiting frequency factor	by PLRM
Present state	100.0%	100.0%
Half-day travel zone	36.4%	x 0.20-28.1%
One-day travel zone	20.9%	xx 0.14-11.3%
One-day-plus travel zone	8.1%	xxx 0.13-6.20%

x Tachikawa, xx Tukuba, xxx Kyoto are assumed

Human Factors in Organizational Design and Management - II
O. Brown, Jr. and H.W. Hendrick (Editors)

IDENTIFICATION OF LEARNING CURVE BY POSSIBILISTIC LINEAR REGRESSION MODEL

T.SHIMOMURA*, H.TANAKA** & J.WATADA***

*Department of Industrial Engineering, Osaka Electro-Communication University, 572 Osaka, Neyagawa, Hatsu-cho 18-8 (JAPAN)
**Department of Industrial Engineering, College of Engineering, University of Osaka Prefecture, 591 Osaka, Sakai, Mozu-Umemachi 4-804 (JAPAN)
***Faculty of Business Administration, Ryukoku University, Fukakusa, Fushimi, Kyoto 612 (JAPAN)

It is natural to interpret the learning phenomena as the possibility of human ability. The Fuzzy regression model is employed to analyze the learning phenomena. This approach enables us to understand two characteristics of a learning effect. One is the trend of the learning effect. The other is the width or amount of possibility at each time. The Fuzzy regression model has been applied to analyze the learning phenomena in data entry jobs using a Japanese wordprocesser.

INTRODUCTION

The learning ability of a person or the adjustability of an organization causes the decrease of time per cycle or of the cost per cycle for a large number of repeats. This is called a learning phenomenon. The decrease of time or the cost through repetition influences on determing production standards or on estimating the potential for cost reductions of high-volume production[1]. Therefore, the identification of learning plays an essential role in industries.

The learning phenomenon is depicted in terms of the trend of learning values named a learning curve and of the deviation of learning values from the trend[2]. Human learning process itself might come from the possibilities of human ability. Therefore, this paper intends to discuss the human learning ability within the framework of the possibilistic concepts. The learning curve will be identified through a possibilistic regression model[3] whose parameters are fuzzy numbers and which expresses possibility of learning curve. The concepts of possibility, proposed by L. A, Zadeh[4], has widened applicable fields in which Fuzzy Set Theory can be applied. The possibility is closely related to a fuzzy number. In the possibilistic linear model, the deviation of learning curve relating to human nature can be obtained as possibilistic intervals. Thus, this approach might be fitting to a vague phenomenon.

POSSIBILISTIC REGRESSION MODEL

Fuzzy regression analysis[4] has been proposed on the basis of the Fuzzy Set Theory. In this fuzzy regression model, the difference between observed and estimated values is not considered as statistical error, but is assumed to result from the fuzziness of a system structure itself. Regarding the deviation between the data as the reflection of fuzziness on system parameters, i.e. in the possibilistic linear regression model, spread of data is supposed to be due to the possibility of the parameters of the linear function.

Let us determine the following possibilistic linear function.

$$Y = A_1 x_1 + \cdots + A_k x_k \qquad ; j = 1, \cdots, k \tag{1}$$

The problem here is to determine the fuzzy parameter vector $\mathbb{A}$ so that the output data $y_i; 1, \cdots, N$, may be conteined in the estimated fuzzy set $Y_i = \mathbb{A}\mathbf{x}_i$ at more than a certain degree h.

The following are supposed to obtain a possibilistic linear regression model:

(i) A possibilistic linear model is written in Eqn.1.

(ii) Fuzzy parameter A_j is a triangular fuzzy number denoted as(α_j, c_j). where α_j and c_j stand for the center and width of fuzzy set A_j, respectively and the membership function of fuzzy number A_j is written as

$$f_{A_j}(a_j) = 1 - \frac{|a_j - \alpha_j|}{c_j}$$

Fig.1 illustrates a triangular fuzzy number A_j.

(iii) Given input-output relations $(\mathbf{x}_i, y_i)$ and a threshold h where $i = 1, \cdots, N$, it must hold that

$$f_{Y_i}^{*}(y_i) \geq h \qquad ; i = 1, \cdots, N \tag{2}$$

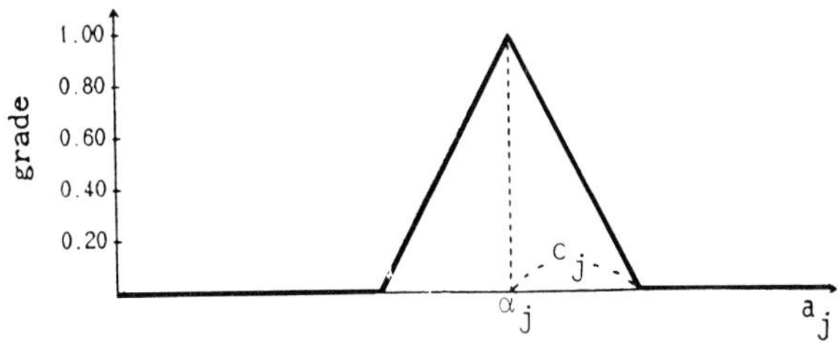

Figure 1
Fuzzy number $A_j=(\alpha_j, c_j)$

This means that the estimated fuzzy set containes the observed values y_i at least at the degree of more than h.

(iv) The criterion for evaluating the possibilistic linear model is

$$S=\sum_{i=1}^{N}\sum_{j=1}^{k} c_j |X_{ij}| \tag{3}$$

where $\sum_{j=1}^{k} c_j |X_{ij}|$ is width of the Y_i and therefore S is the sum of width of $\{Y_i\}$.

The possibilistic linear regression model is obtained by fuzzy parameters $A=(\alpha, c)$ that minimizes S in (iv) subject to the constraints (iii). Given a threshold h, this results in the following linear programming problem.

$$\min_{\alpha, c} S=\min_{\alpha, c}\sum_{i=1}^{N}\sum_{j=1}^{k} c_j |x_{ij}| \tag{4}$$

subject to

$$(1-h)\sum_{j=1}^{k} c_j |x_{ij}| + x_i\alpha \geqq y_i \qquad ;i=1,\cdots,N \tag{5}$$

$$(1-h)\sum_{j=1}^{k} c_j |x_{ij}| - x_i\alpha \geqq -y_i \qquad ;i=1,\cdots,N \tag{6}$$

$$c_1, c_2, \cdots, c_k \geqq 0 \tag{7}$$

Possibilistic parameters $A_j = (\alpha_j, c_j)$, j=1,...,K, can be easily obtained by solving the above linear programming problem.

LEARNING CURVE IN DATA ENTRY JOB

Data entry is an important job concerning information processing. The success of information processing is dependent upon the accuracy of the data entry. Here we analyze the learning phenomenon in data entry operations for word processing.

The word processor employed was Fujitu OASIS-100 and the subjects were three students who had no experience on any type of word processor including a type writer. One cycle performance of the job consisted of the input of 200 letters of Japanese alphabet through an English-Japanese alphabet conversion. That is, every Japanese phonetic alphabet letter can be expressed by using two or one English letters. When English letters are input in the word processor, every two/one English alphabet letters are automatically converted into the corresponding Japanese alphabet letter throuh the English -Japanese alphabet conversion function.

The time of each operation cycle was measured in 30 complete cycles. The subjicts had ten minutes rest interval between cycles. In the experiment mistyped characters were not counted as input characters. After the experiment the corrected time t is calculated by the following equation.

$$t=\frac{T}{a}\cdot 200$$

where T=meaning times per cycle

a=the number of correct letters per cycle

RESULTS

As illustrated in Fig.2, the cumulative mean of the data shows the decreasing hyperbolic relation of learning curves. Therefore, we assumed a possibilistic regression function as

$$Y=A_0+A_1x^{-n} \tag{8}$$

Because of the sufficient number of the observed data, the possibilistic regression model was analyzed at h=0. For subjects A, B & C, the obtained possibilistic learning curves are as follows:

$$Y_A^* = (209.6,\ \ 16.7) + (267.4,\ \ 56.8)\ x^{-0.41} \tag{9}$$

$$Y_B^* = (162.6,\ \ 17.7) + (411.4,\ \ 42.2)\ x^{-0.45} \tag{10}$$

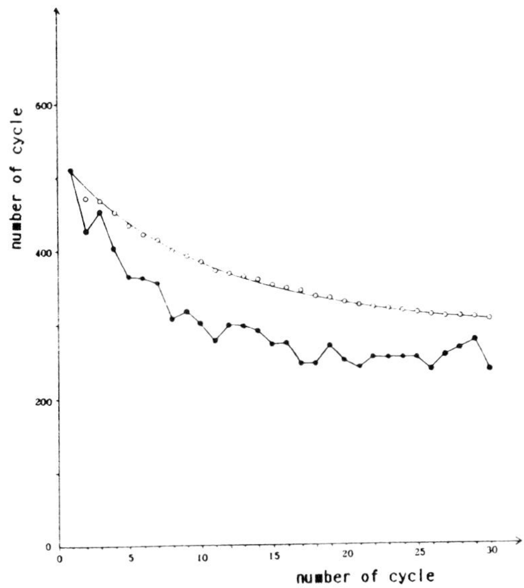

Figure 2
Learning Curve of Cumulative Mean

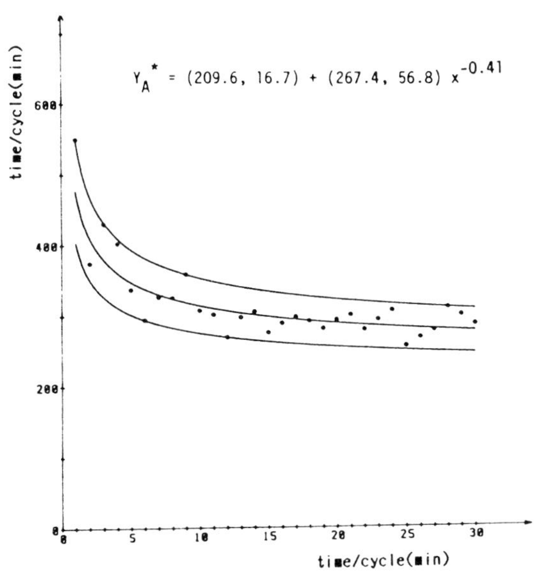

Figure 3
Possibilistic Learning Curve of Subject A

$$Y_C^* = (139.5,\ 30.1) + (312.6,\ 0.0)\ x^{-0.30} \tag{11}$$

Fig.3 illustrates a possibilistic learning curve for subject A. As shown in Fig.3 the estimation of each cycle time is calculated in a fuzzy number which expresses a possibilistic interval. The learning effect could be recognized in the data entry job through the decreasing possibilistic function obtained from the data each of three objects. Especially, subjects B have a large learning effect. On the other hand, the result of subject C in the data entry job has fluctuated widely over the trials but it dose not show a sufficient learning effect. Furthermore, subject A and B have a learning effect even on the possibility interval. That is, as they became experienced in the job, they could perform the task in a stabler speed than the subject C.

CONCLUDING REMARKS

In this paper, the learning phenomenon has been discussed using a possibilistic model.

Emphasis should be placed on the learning phenomenon which is brought out through the possibility of human ability. Therefore, the possibilistic regression model can be employed to analyze the learning phenomenon within the framework of the possibility concept.

This possibilistic approach enables us to grasp the two characteristics of learning effects. One is related to the trend of learning effect and the other is related to the possibility interval at each time.

REFERENCES

[1] W.M.Hancock and F.H.Bayha, The Learning Curve, in Handbook of Industrial Engineering ed. by G.Salvendy, John Wiley, New York, 1982, pp 4.3.1-4.3.13.

[2] S.Nakayama and R.Sakai, The Evaluation on the Achievement of Workers,J. of Japan Industrial Management Association, Vol. 32, No. 5, (1981), pp 372-377(in Japanese).

[3] L.A.Zadeh, Fuzzy Sets as a Basis for a Theory of Possibility, Int.J. Fuzzy Sets and System, vol. 1, (1978), pp 3-28.

[4] H.Tanaka, S.Uejima and K.Asai, Linear Regression Analysis with Fuzzy Model, IEEE Trans. on SMC, 12 (1984), pp 903-907.

Human Factors in Organizational Design and Management - II
O. Brown, Jr. and H.W. Hendrick (Editors)

MEASUREMENT OF OFFICE WORK BY POSSIBILISTIC LINEAR REGRESSION MODELS

H.TANAKA*, I.HAYASHI* & J.WATADA**

* Department of Industrial Engineering, College of Engineering, University of Osaka Prefecture, 591 Osaka, Sakai, Mozu-Umemachi 4-804 (JAPAN)

** Faculty of Business Administration, Ryukoku University, Fukakusa, Fushimi, Kyoto 612 (JAPAN)

Fuzziness must be considered in systems where human estimation is influential. We have defined a possibilistic linear regression model which can deal with fuzziness as the concept of possibility. It is the purpose of this paper to apply a possibilistic linear regression model to analyze the measurement problem of office work in a bank. This analysis can estimate the number of office workers in fuzzy numbers.

INTRODUCTION

Possibility concepts and its measure defined by L.A. Zadeh can deal with vagueness included in phenomenon of indefiniteness, which could not be explained in terms of probability. Fuzzy regression model has been proposed by H.Tanaka et al[1)] and discussed in detail in references[2),3)]. Possibilistic regression model is an interpretation of fuzzy regression models in terms of possibilistic measure. This means that a possibilistic model is governed by a possibility measure in the same way as a probabilistic model is governed by a probability measure.

The objective in this paper is to discuss a possibilistic regression model with constraints on its parameters and to apply it to measurement of office work. This model is named a constraint possibilistic regression model. In real problems we have knowledge on parameters of models with which experts can provide us. In these cases it is natural and proper to analyze the model using the knowledge obtained from experts. The structure of possibilistic regression model makes it easy to solve the constraint problem on parameters. The constraint possibilistic regression model is applied to analyze the measurement problem of office work. In office work measurement , standard time is important not only to determine the number of workers, but also to improve the work efficiency. In this application we apply our method to the case of a bank office to determine the standard time of several unit jobs and estimate the number of office workers.

POSSIBILISTIC LINEAR SYSTEMS

First, let us define a possibilistic linear system

$$Y=A_1x_1 +\dots + A_nx_n \quad (1)$$

where A_i is a fuzzy parameter called a fuzzy number.

Definition 1.

A symmetric fuzzy number A_j denoted as $(\alpha_j,c_j)_L$ is defined by

$$A_j(a_j) = L(|a_j-\alpha_j|/c_j) \quad (2)$$

where a reference function L(x) satisfies (i) L(x)=L(-x), (ii) L(0)=1 and (iii) L(x) is strictly decreasing on $[0,\infty)$.

As examples of Definition 1, $L(x)=\max(0, 1-|x|^p)$, $L(x)=e^{-|x|^p}$ and $L(x)=1/(1+|x|^p)$ are shown by Dubois and Prade, where $p>0$.

A fuzzy function whose parameters are fuzzy number is denoted by $Y=f(\mathbf{x},A_1,\dots, A_n)$. The membership function of Y is defined by the extension principle as

$$\mu_Y(y) = \sup_{\{\mathbf{a},\mathbf{x}|y=f(\mathbf{x},\mathbf{a})\}} \mu_{A_1}(a_1)\wedge \dots \wedge \mu_{A_n}(a_n) \quad (3)$$

where $\mu_Y(y)=0$, for $\{\mathbf{a} ,\mathbf{x}|y=f(\mathbf{x},\mathbf{a})\}=\phi$. This is called the extension principle.

Using the extension principle, we have

$$Y=A_1x_1 +\dots+ A_nx_n \Leftrightarrow \mu_Y(y) = L((y-\mathbf{x}^t\boldsymbol{\alpha})/\mathbf{c}^t|\mathbf{x}|) \quad (4)$$

Since fuzzy sets can regarded as a possibility distribution, we call (1) a possibilistic linear system. The inclusion relation of fuzzy numbers with the degree h is defined as

$$A \supsetsim_h B \Leftrightarrow [A]_h \supset [B]_h \quad (5)$$

where $[A]_h=\{x|\mu_A(x)\geq h\}$, and A and B are fuzzy numbers defined in Definition 1.

CONSTRAINT POSSIBILISTIC LINEAR REGRESSION MODEL

In the conventional regression analysis, deviations between the observed values and the estimated values are supposed to be due to observation errors. We assume, on the contrary, that these deviations depend on the indefiniteness of the system structure . Thus, we regard these deviations as the possibility of system parameters. Let us consider fuzzy outputs $Y_i=(y_i,e_i)_L$ although inputs x_i, i=1,...,N remain crisp. The problem is to decide fuzzy parameters $A_1^*,\dots, A_n^*$ such that

$$Y_i^* = A_1^*x_{i1} +\dots+ A_n^*x_{in} \supsetsim_h Y_i \quad ; i=1,..,N \quad (6)$$

In a possibilistic linear regression model, the following are assumed:

i) The data can be represented by a possibilistic linear model:

$$Y_i^* = A_1^*x_{i1} +\dots+ A_n^*x_{in} \triangleq \mathbf{A}^*\mathbf{x}_i \quad (7)$$

where $\mathbf{x}_i=(x_{i1},\dots, x_{in})^t$.

ii) The type of fuzzy parameter A_i is given by Definition 1.

iii) Given input-output relations $(\mathbf{x}_i,Y_i)$, i=1,...,N and a threshold h, it must hold that

$$[Y_i^*]_h \supset [Y_i]_h \ , \ i=1,\dots,N \quad (8)$$

iv) The experts knowledge is given as

$$\underline{\alpha}_i \leq \alpha_i \leq \overline{\alpha}_i \ , \ \underline{c}_i \leq c_i \leq \overline{c}_i \quad (9)$$

v) The index of fuzziness of the possibilistic linear model is

$$J(c) = \sum \mathbf{c}^t \cdot |\mathbf{x}_i| \quad (10)$$

where $\mathbf{c}^t|\mathbf{x}_i|$ is the width of Y_i.

Under the above assumptions, our problem is to obtain fuzzy parameters $A_j=(\alpha_j, c_j)_L$, $j=1,\dots,n$ that minimize $J(c)$ in (10) subject to the constraints (8) and (9). This problem can be solved as the following LP problem:

$$\min J(c) = \sum \mathbf{c}^t \cdot |x_i|$$

subject to

$$\begin{aligned} y_i &\leq |L^{-1}(h)|\mathbf{c}^t|\mathbf{x}_i| - |L^{-1}(h)|e_i + \mathbf{x}_i^t \cdot \boldsymbol{\alpha} \\ -y_i &\leq |L^{-1}(h)|\mathbf{c}^t|\mathbf{x}_i| - |L^{-1}(h)|e_i - \mathbf{x}_i^t \cdot \boldsymbol{\alpha} \\ \underline{\alpha}_i &\leq \alpha_i \leq \overline{\alpha}_i \ , \ \underline{c}_i \leq c_i \leq \overline{c}_i, \quad i=1,\dots,N \end{aligned} \quad (11)$$

APPLICATION TO MEASUREMENT OF OFFICE WORK

Measurement method of office work

Standard time is important not only to determine the number of workers, but also to increase the work efficiency. For instance, Taylor systems and other improved systems have resulted from the work measurement in time study and motion study. Also, the number of workers is determined based on the time study or motion study which estimates the time of each unit job. But it takes much cost to measure the time and occasionally it is difficult to precisely do so. The aim in this section is to propose the method of a constraint possibilistic linear regression for the estimation of office work.

a)Measurement problem of office work

Assume that the service of a certain section consists of n unit jobs, and let us denote the number and the time of each unit job as x_i and A_j, respectively. The total working time of the section, Y, is modeled as follows:

$$Y = \sum A_i \cdot x_i \quad (12)$$

In the estimation of the total time, we must take allowance for rest and personal need into account. In the case of office work about 20 % and about 80 % are taken for allowance and substancial working, respectively. Therefore, the total time is constructed as a fuzzy number Y_i with its central value y_i and the 10 % fuzziness in both sides. Then, our problem is to decide the fuzzy number $A_j(j=1,\dots,n)$ using the data in Table 1 where $Y_i(i=1.,,,.N)$ is a fuzzy number. Once we determine the fuzzy number A_j, we can estimate the time of unit job j for a month per person.

b)The estimation of the number of workers

When the number of unit jobs, x_1 to x_n, is given, using these estimated constants A_1 to A_n for a month, we can estimate the number of office workers required, N, as follows:

$$N = \sum A_j x_j \ / \ T \qquad (13)$$

where T denotes the working time for a month per person. This estimation N is a fuzzy number. Then, we can interprete N as possibilistic number. In other words, the number of office workers needed is included in the possibilistic interval value N.

Analysis of an office work in a bank

Let us analyze the office work for the job of computer operators in a bank. Office workers are required to perform three types of data input using a computer terminal and other miscellaneous works such as working on customers requirements which are hardly devided into several concrete unit jobs. Note that we define the time of the miscellaneous works as the proportion to the total time required in the section. Therefore, our model can be rewritten as

$$Y_i = A_1 \cdot x_{i1} + A_2 \cdot x_{i2} + A_3 \cdot x_{i3} + A_4 \cdot T \cdot N_i \ ; \ i=1,..,N \quad (14)$$

where Y_i is a possibilistic total time obtained for this section at the branch i, while x_{i1}, x_{i2} and x_{i3} denote data input of accounts, handling of certification deposit and transfer handling, respectively. N_i is the number of worker at the branch i. Table 1 illustrates data obtained in several branches of the bank in August, 1985.

Concerning the customer service of this bank it is previously known from experts that all values A_1 to A_4 must be satisfied the constraints given in Table 2. We employed h=0, because of the sufficient number of data. The constraint possibilistic regression model was applied to obtain the possibilistic values A_1 to A_4 shown in Table 3. Let us estimate the number of office workers required through this model in order to check the validity of these values A_1 to A_4. Table 4 illustrates the result which explains that all new samples except samples 9,14,16 and 28 are included in the estimated interval. Fig.1 shows estimated fuzzy numbers of workers at four branches.

REFERENCES

[1] Tanaka,H., S. Uejima and K. Asai : Linear regression analysis with fuzzy model, IEEE Trans. SMC-12 (1982), 903-907.

[2] Tanaka,H., T. Shimomura, J. Watada and K. Asai, : Fuzzy linear regression analysis of the number of staff in local goverment, FIP-84 at Kauai, Hawaii on July 22-26 (1984).

[3] Tanaka,H. : Fuzzy data analysis by possibilistic linear models, Int. J. of Fuzzy Sets and Systems (to appear).

Table 1 Data obtained in August

	Branch office	Number of unit jobs: Accounts handling x1	Certification deposit x2	Transfer handling x3	Total working time Y
1	Shinjuku	5076	666	156	$(37206\ ,\ 3721)_L$
2	Shibuya	4991	446	94	$(29972\ ,\ 2997)_L$
3	Jiyuugaoka	3343	139	23	$(21704\ ,\ 2170)_L$
4	Ueno	4160	307	88	$(25838\ ,\ 2584)_L$
5	Sapporo	8215	519	115	$(44440\ ,\ 4444)_L$
6	Sendai	6264	473	46	$(41340\ ,\ 4134)_L$
7	Takasaki	2521	334	38	$(19637\ ,\ 1964)_L$
8	Utsunomiya	3465	317	25	$(25838\ ,\ 2584)_L$
9	Chiba	4738	434	47	$(39273\ ,\ 3927)_L$
10	Funabashi	3412	202	22	$(19637\ ,\ 1964)_L$
11	Hiratsuka	5638	397	53	$(31005\ ,\ 3101)_L$
12	Gifu	2186	318	27	$(20670\ ,\ 2067)_L$
13	Shizuoka	4144	443	76	$(20670\ ,\ 2067)_L$
14	Osakaekimae	3972	620	150	$(36173\ ,\ 3617)_L$
15	Hankyuumeda	6011	485	40	$(42374\ ,\ 4237)_L$
16	Nanba	5133	535	132	$(41340\ ,\ 4134)_L$
17	Abenobashi	3085	397	92	$(17570\ ,\ 1757)_L$
18	Sakai	6303	523	77	$(33072\ ,\ 3307)_L$
19	Kyoto	4648	610	130	$(44441\ ,\ 4444)_L$
20	Wakayama	4678	577	87	$(28938\ ,\ 2894)_L$

Table 2 Constraints on parameters

Accounts handling A_1	Certification deposit A_2	Transfer handling A_3	Rate of miscellaneous work A_4
$1.0 \le \alpha_1 \le 15.0$	$1.0 \le \alpha_2 \le 15.0$	$1.0 \le \alpha_3 \le 15.0$	$0.5 \le \alpha_4 \le 0.7$
$1.0 \le c_1 \le 10.0$	$1.0 \le c_2 \le 10.0$	$1.0 \le c_3 \le 10.0$	$0 \le c_4$

Table 3 Estimated fuzzy parameters

Accounts handling	$A_1 = (1.083\ ,\ 1.00)_L$
Certification deposit	$A_2 = (10.21\ ,\ 1.00)_L$
Transfer handling	$A_3 = (\ 1.00\ ,\ 1.00)_L$
Rate of miscellaneous work	$A_4 = (\ 0.70\ ,\ 0.023)_L$

Table 4 Checking data and estimation of the number of office workers

	Branch office	Accounts handling	Certification deposit	Transfer handling	Estimated number	Number of workers
1	Honteneigyobu	22260	6675	1641	[19.0 , 43.4]	14.8
2	Ginza	4760	78	278	[0.3 , 4.0]	3.7
3	Nipponbashi	10948	1566	491	[4.6 , 14.4]	6.6
4	Toranomon	6581	634	235	[1.9 , 7.4]	4.6
5	Gotanba	2423	281	125	[0.8 , 3.0]	1.1
6	Shinjukunishi	3836	332	78	[1.0 , 4.1]	1.8
7	Ikebukuro	3696	496	92	[1.5 , 4.7]	2.9
8	Oizumi	2170	208	15	[0.6 , 2.4]	0.8
9	Mitaka	3286	3234	33	[9.3 , 15.4]	1.5
10	Chofu	877	109	5	[0.3 , 1.1]	0.5
11	Aomori	2504	249	27	[0.8 , 2.8]	0.9
12	Mito	3423	320	46	[1.0 , 3.8]	3.0
13	Omiya	5514	262	17	[0.9 , 5,1]	2.9
14	Kawagoe	485	61	3	[0.2 , 0.6]	0.9
15	Yokohama	9883	752	159	[2.4 , 10.2]	7.0
16	Totsuka	4197	888	108	[2.6 , 6.6]	1.5
17	Nagoya	9758	795	226	[2.4 , 10.4]	5.5
18	Imaike	3173	178	6	[0.6 , 3.0]	1.6
19	Osaka	18009	3861	1036	[11.1 , 28.9]	12.4
20	Ibaragi	1086	177	12	[0.5 , 1.5]	0.7
21	Higashiosaka	3557	399	20	[1.2 , 4.2]	3.0
22	Tsu	918	140	14	[0.4 , 1.2]	1.2
23	Nara	4346	388	27	[1.2 , 4.7]	2.5
24	Kobe	7474	1425	245	[4.1 , 11.2]	4.4
25	Himeji	6256	766	90	[2.3 , 7.6]	3.5
26	Nishinomiya	2281	334	8	[1.0 , 3.0]	2.2
27	Kakogawa	811	170	3	[0.5 , 1.3]	0.6
28	Tsukaguchi	700	99	3	[0.3 , 0.9]	1.1
29	Hiroshima	4353	622	64	[1.8 , 5.6]	3.0
30	Tokushima	2772	433	29	[1.3 , 3.7]	2.0
31	Kochi	2069	231	76	[0.7 , 2.5]	1.2
32	Fukuoka	4948	420	131	[1.3 , 5.3]	3.3
33	Kitakyushu	2646	348	31	[1.0 , 3.3]	2.4
34	Miyazaki	1380	90	2	[0.3 , 1.4]	0.8
35	Kagoshima	2677	238	15	[0.7 , 2.9]	0.9

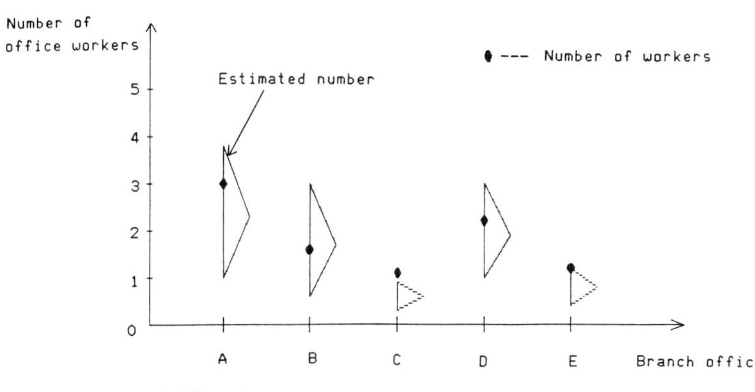

Fig.1 Estimation of office workers in a bank

ORGANIZATIONAL AND MANAGERIAL DECISION MAKING

Human Factors in Organizational Design and Management - II
O. Brown, Jr. and H.W. Hendrick (Editors)

AN EVALUATION OF THE UTILIZATION OF MULTI-ATTRIBUTE UTILITY (MAU) MODELS IN MANAGERIAL DECISION MAKING

Dan Zakay

Psychology Department
Tel-Aviv University Ramat-Aviv 69978
Israel

46 subjects participated in a decision-making experiment, in which their actual choices were predicted on the basis of various models, including a complete MAU model, an MAU model with equal weights, partial models in which only the most important attribute or the two most important attributes were considered, and a random model in which weights' values were determined randomly. It was found that the complete MAU model was the best predictor of actual choices. The significance of these results for the utilization of MAU models in managerial decision making is discussed.

Managerial decision making is very often done via the Multi-Attribute Utility (MAU) Model, a model which is very popular and studied in decision-making training courses. (Gardiner & Edwards, 1975; Swezey, 1979). According to the MAU model, each alternative is decomposed to a set of attributes and each attribute is assigned a utility value: (Uj) and an importance weight: (Wj). The overall utility of an alternative Ai is: (Wilkie & Pessemier, 1973)

1) $MAU_i = \sum_{j=1}^{N} W_j U_j$. Where J-is the attribute's index.

However, despite the appeal of the MAU model, it is not clear whether the benefit of assigning different weights to the attributes is significant as compared, for example, with an equal weighting method. This is an important question since the process of dealing with weights is time consuming. The purpose of the present study was to test this question.

METHOD

Subjects - Twenty males and twenty six females, all of them business administration students, participated in the experiment.

Experimental tasks

The rationale of the experiment was to predict actual decisions on the basis of MAU measurements. Two decision contents were used: Vocational choice and choice among university applicants (which is very similar to choosing candidates for work). These two contents were selected since they were of interest and relevance to the subjects. Each alternative in each content was composed of three attributes.

Measurement procedure

Two parameters were measured: 1) The subjective value of each attribute in each alternative, and 2) The subjective importance weight of each attribute. The subjective values of the attributes were measured on an ordinal scale through a process whereby the subject evaluated in each attribute an "ideal" level and then ranked other possible values in that attribute as compared to that "ideal" level. By this method it was also possible to define for each subject his/her "ideal" alternative (IA) for each content.

Importance weights were measured by a graphic method in which attributes were treated as weights and the subjects had to place them on a scale where the weight would tip the scale as a function of its distance from the midpoint. By following this procedure the subject is required to make ratio level judgements.

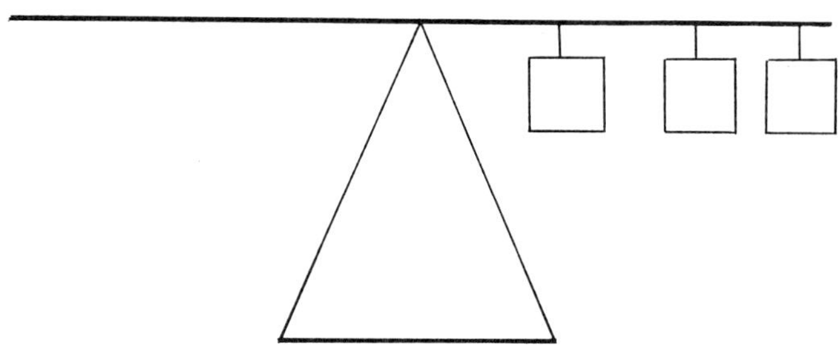

Figure 1

The Form of Balance and Weights Used for Weights' Measurement

The measurement was done on all of the relevant attributes simultaneously and thus possible interactions among attributes were taken into consideration, (Zeleny, 1976). The distance from the midpoint to each attribute was taken as its weight. The MAU of each alternative was then computed for each subject.

Procedure

The experiment consisted of two sessions. In the first session the importance weights and the attribute's values were measured and in the second session, which took place two days after the first one, the experimental task was administered. Each subject was tested on both applicant and vocational choices. 28 binary choices of each content were presented to SS' individually and it was asked to mark that alternative which was preffered most in each choice. It was allowed to choose both alternatives in case of indifference. The choice situations were administered in random order.

Statistical analysis and results

The hypothesis which was tested was that by using a complete MAU model, i.e. - by calculating the MAU of each alternative on the basis of weighting by subjective values of all three attributes, a significant prediction of actual choices made by each subject will be obtained. In order to test this, five different models were compared:

a) A complete model, using all the attribute's values and all the importance weights as measured.

b) A partial model using only the two most important attributes and its importance weights.

c) A partial model using only the most important attribute and its importance weight.

d) An equal weight model using all the attributes but ignoring its importance weights (i.e. assuming equal weights).

e) A random model using randomly selected numbers as the importance weight's values. (The random numbers were in the range of the measured importance weights). The distribution of choices of each subject was regarded as a binomial distribution with P=0.5 (Becker et al., 1963). For each subject a binomial test was carried out on the basis of the number of choices which fit the predicted choices according to a specified model. This was done for each one of the five models and for each decision content. The results of the binomial tests and confidence intervals for vocational and applicants choices are presented in tables 1 and 2.

Table 1

Percentage of subjects who were found to fit the various models significantly.

content	α level	A	B	C	D	R
AP+	0.10	0.0	0.0	0.0	2.0	2.1
	0.05	6.3	26.0	0.0	58.6	2.1
	0.01	93.4	54.3	0.0	2.0	0.0
	Total	99.7	80.3	0.0	62.6	4.2
J++	0.10	6.6	11.1	0.0	8.8	4.4
	0.05	4.4	8.8	0.0	17.7	2.2
	0.01	73.3	46.6	0.0	26.6	0.0
	Total	84.3	66.5	0.0	53.1	6.6

\+ Applicants ++ Jobs
A - complete model. B - two attributes. C - one attribute.
D - equal weights. R - random weights.

Table 2

Confidence intervals (α = 0.05) for proportions of subjects who were found to fit the various models significantly, according to decision content and type of model

Content / Model*	A Complete	B Two Dimensions	D Equal Weights
Applicants	0.92 - 1.00	0.66 - 0.89	0.49 - 0.75
Jobs	0.71 - 0.92	0.51 - 0.78	0.39 - 0.67

* Only the three best models are presented

In order to test whether the difference in predictive power between the various models was significant, a 2x3 analysis of variance was performed. One factor was decision content (C) with two levels. The second factor was type of model (M) with three levels (models A,B,C). These three models were selected since, in all of them, the measured weights were considered and the only difference was the number of attributes counted. Thus, factor (M) may be regarded as the number of attributes counted. Each subject (S) was tested on all the contents and models. Hence, the design was: C x M x S, with repeated measurements. The dependent variable was the standard score U for the proportion of "correct" choices (those choices which fit the prediction of the tested model) of each subject.

A significant main effect ($F(1,44) = 8.4$, $p < 0.01$) between decision contents was found, as well as a significant main effect between models ($F(2,88) = 44.81$, $p < 0.01$). No significant effect was found between subjects, and no significant interaction was found. The cell means of U scores are presented in table 3.

Table 3

Cell Means (U* Scores) of the ANOVA

Decision Task:	A: Three Dimensions	B: Two Dimensions	C: One Dimension	Total
Applicants	3.28	2.30	0.48	2.02
Jobs	2.76	1.82	-0.33	1.41
Total	3.02	2.06	0.07	1.71

* Mean U values for the proportion of correct predictions for each subject.

The proportion of predicted choices according to decision content and predicting models are presented in figure 2.

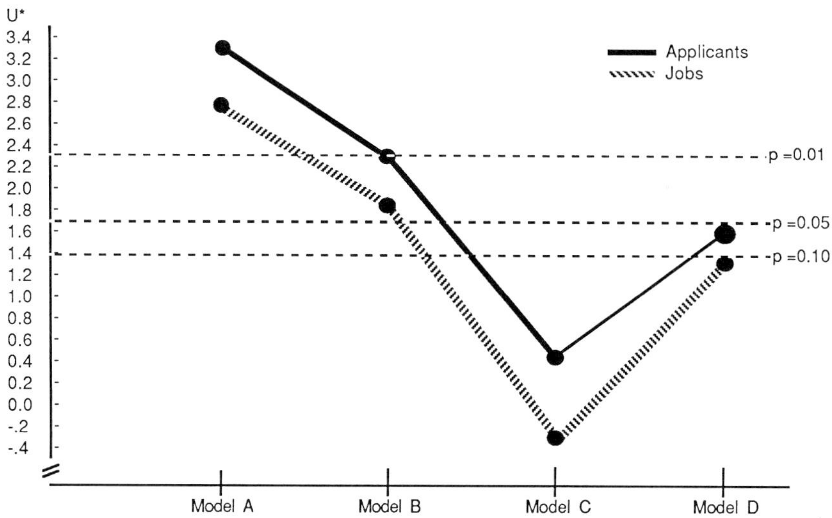

* Mean U values for the proportion of correct predictions for each subject

Figure 2

Proportion of Predicted Choices According to Models A, B, C, & D.

DISCUSSION

The results obtained clearly demonstrate the superiority of the complete model over all the others. This is true for both applicant and vocational choices. The results are indicating the possibility that SS' indeed utilized internally importance weights and that measured importance weights did well as predictors. It is highly reasonable to assume that subjects urilized mainly the complete MAU strategy as it was found that when only one or two attributes were considered, accuracy of predictions fell sharply. Another conclusion that can be drawn on the basis of the present study is that the presentation form of the alternatives is of importance and is influencing the ease of use of a specific decision strategy. In the experiment reported here Alternatives were presented so that attributes were emphasized and hence the utilization of an MAU strategy was probably facilitated. It is recommended that in any managerial decision a preliminary stage of structuring the decision problem and of decomposing the alternatives to its attributes should precede the MAU analysis. However, it should be remembered that binary choices with three attributes are not too complicated and hence subjects could do well in their intuitive decisions, but this is not the case when complicated managerial decisions are considered. In such cases intuitive judgements might be suboptimal (Hogarth, 1980). Since the actual utilization of an MAU model in managerial decisions is relatively simple and demands reasonable amounts of time and other resources, it seems that the use of a complete MAU model as a decision aid tool is justified and recommended.

However, in light of methodological problems which characterize decision making experiments (Ebbessen, Konecni, 1978), this conclusion must be tested in the case of real life managerial decisions.

References

Becker, G.M., Degrout, H.M., & Marschak, J. (1963). Stochastic models of choice behavior. Behavioral Science, 8, 41-55.

Ebbesen, E.B., & Konecni, V.J. (1978). On the external validity of decision-making research: What do we know about the decisions in the real world? In: Proceedings of the conference of cognitive processes in choice and decision behavior. Chapel Hill, N.C.: University of North Carolina.

Gardiner, P.C., & Edwards, W. (1975). Public values: MAU measurement for social D.M. In: Schwartz, S., & Kaplan, M. (Eds.), Human Judgement and decision processes. New York: Academic Press. 1-38.

Hogarth, R.M. (1980). Judgement and choice. New York: Wiley.

Swezey, P.W. (1979). An application of a multi-attribute utility model to training analysis. Human Factors, 21(2). 183-189.

Wilkie, W.L., & Pessemier, E.A. (1973). Issues in marketing use of multiattribute utility models. Journal of Marketing Research, 10. 428-441.

Zeleny, M. (1976). The Attribute Dynamic Attitude Model (ADAM). Management Sciences, 23, 12-26.

Human Factors in Organizational Design and Management - II
O. Brown, Jr. and H.W. Hendrick (Editors)

ON THE USE OF TIME: THE IRREPLACEABLE RESOURCE

P.A. Hancock

Department of Safety Science and
Human Factors Department
Institute of Safety and Systems Management
University of Southern California
Los Angeles, CA 90089-0021
U.S.A

ABSTRACT

This paper reviews differing aspects of the role of time in the operation of organizational systems. Such facets can range from the understanding of time as a unique functional resource of the organization to the personal use by key management individuals. It is this latter behavioral perspective that is the concern of this work. Our present operational perception of time as an immaleable and uncontrollable progression is questioned. Personal time utility analysis is addressed as a potential avenue through which to maximize temporal efficiency.

I. PREAMBLE: LAWS OF TIME

As $T \rightarrow \infty$, then $t \rightarrow > \infty$ ———————— (1)

$dT/dT = t^{(n)}$ ———————— (2)

$T \rightarrow T + 1$, but $T + 1 \nrightarrow T$ ———————— (3)

$t \propto 1/d$ ———————— (4)

The above expresses some behavioral laws of time that may be equated with the verbal descriptions below:

(1) No matter how much time there is available (T), there is never enough time (t) available.

(2) As an important deadline approaches (n), time goes more quickly. The more important the deadline the quicker the effective time (t) disappears.

(3) For behavioral purposes time (T) only flows in one direction.

(4) Effective time (t) goes down as distractions (d) go up.

This work was supported by a grant NCC 2-379 from NASA, Ames Research Center. Dr Michael Vidulich was the scientific monitor for the grant.

II. INTRODUCTION

It is strange to think of laws of time outside the realm of physics. Time is usually recognized as a component of equations of motion where for example velocity is the product of acceleration and time, or it is an issue in a theory of relativity where the speed of light remains constant and ostensibly stable phenomena such as space and time prove relative. It is from these insights that we derive our common conception of time as an inexhorable progression or flow which continues independent of our efforts or our perceptions.

Yet we recognize many occasions in which our behavior does not appear at all scaled by a regular progressive flow but is dictated by a variety of factors that relate to our own internal state and also the environment and tasks which surrounds us. So it is that simple laws of time which relate to the physical action of non-sentient matter do not capture the complexities which arise between perceivers and their environment. These observations are important for managers and operators who often appear constrained or limited by time as an indispensible and irreplaceable resource. The purpose of the present paper is to explore some aspects of time as they impact sucessful goal-oriented behavior. The paper is limited in that it does not consider time as a functional organizational resource, or as an emergent property of action in a temporal dimension. These aspects will be the topic of consideration at a later time.

III. THE INS AND OUTS OF TIME: ENDOGENOUS VS EXOGENOUS SOURCES OF INFLUENCE

There are numerous factors which potentially and actually influence the perception of temporal progress. These may be conveniently dichotomized into two distinct categories. These are exogenous or environemtnal sources and endogenous or internal influences. It is to these latter factors that we first turn.

III.1 Endogenous Influences: The Question Of An Internal Clock

There has been a long and mostly disappointing search for the human internal clock (see Doob, 1971; Hancock, 1983; Hetschel, 1976 for reviews). One potential reason for this failure in sentient human subjects is the naive perception of what would compose such an entity. For example, the search for an internal clock implies a single discrete element although it is probable that in accord with a variety of characteristics of human behavior, e.g., attention, that control of temporal behavior resides in several distributed loci. More importantly, the search has focused on processes that possess **clock-like** qualities. Or, more simply, we have tried to find a mechanical watch inside the active central nervous system. We have examined candidate processes that have simple and recognizable rhythmicity, with periodicities approaching that of a common second, or some simple multiple thereof. We wish such rhythmicity to be stable and yet not unalterable in that this would not jibe with current experimental knowledge. As might be expected, this search encompassing electroencephalographic (EEG) measures, heart rate, blood pressure and a plethora of other high frequency physiological fluctuations has rendered little of significance.

Rather than the above functions, which appear as reflections of energy utilization, the overall level of the major entraining bodily rhythm, i.e., temperature, has proved more influential. The notion of a simple "chemical clock" for controlling the perception of brief intervals has been the subject of controversy since Hoagland's (1933) initial postulation (see Bell, 1966; 1975; Fischer, 1966; Hancock, 1984; Hetschel, 1979). An examination of experimental evidence suggests that it is only when the thermal homeostatic mechanism is overcome by some external influence that temporal flow appears to vary under the impact temperature change. This concept is expressed in Figure 1 below.

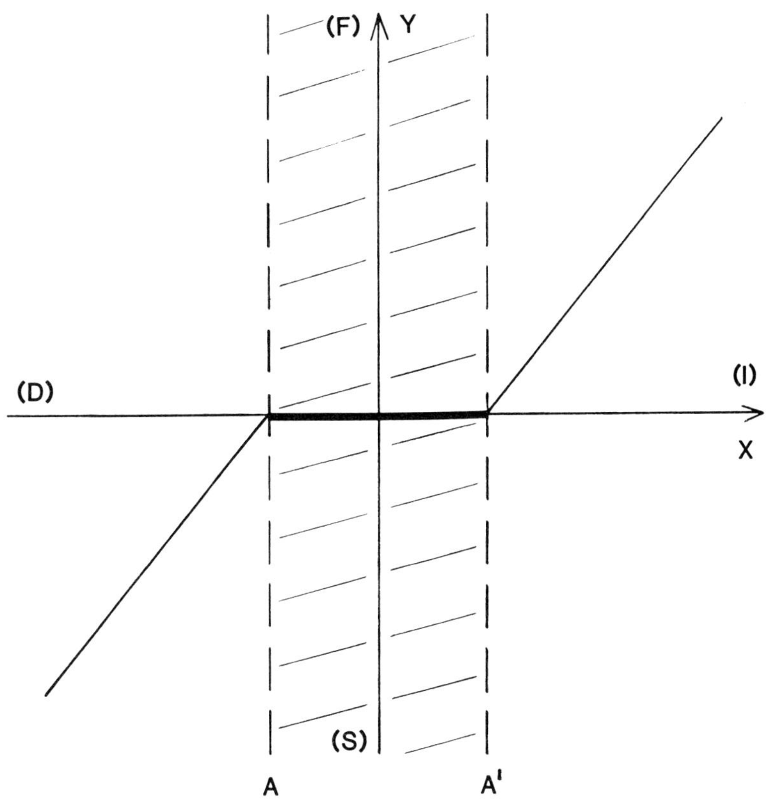

Figure 1: Temporal flow rate expressed as a function of change in body temperature. On the y axis, temporal flow rate is expressed as either faster (F) or slower (S) than a reference normative level. On the x axis, energy expressed in the form of temperature either increases (I) or decreases (D). The range of functional adaptability A to A' represents a region in which efficient compensation acts to retain temporal flow at a constant level.

The region of adaptability as indicated on the Figure is that in which neither the body temperature nor the temporal flow rate is affected. Outside this region, denoted as A to A' in the diagram, where external conditions are sufficient to superceed the ability to regulate against the perturbing environmental effect, then temporal flow covaries with the physiological reflection of energy such that less energy slows the system while excess energy speeds the perception of rate of progression. A note should be made concerning the monotonic functions presented outside the region of adaptability. Although these are presented as straight line functions it is posssible that they represent exponential relationships. Insufficent elucidating data are available to distinguish between these possibilities, although from some of our previous work on stress and attention (see Hancock & Chignell, 1985) we would favor the curvilinear function.

III.2 Endogenous Influences: A Summary

From the current perspective, endogenous influences exert relatively little effect on temporal perception under normal conditions. Clearly there are baseline state changes such as those associated with circadian function that affect temporal flow, but it is only under exceptional conditions that endogenous factors achieve supremacy over their environmental counterparts. Such occasions may be where exogenous influences are reduced by artificial isolation as in spacecraft or submarines, or on those occasions where stress intrinsic to an environment is sufficient to overcome the immediate dynamic adaptive capability of the operator. Such circumstances might occur in life-threat conditions such as battle, or under the driving stress of achieving a critical performance deadline in complex, time-limited, operational systems.

III.3 The Impact Of The Environment

Such is the immediacy of the environment it is not possible to discount the temporal information which imminently impacts the perceiver. One factor which has often been suggested to influence perceived temporal flow is the complexity of this information. Complexity is usually a term which eludes precise definition, or where an initial attempt has been made it reflects some simple physical characteristic of a series of nonsense stimuli. There are two key elements in terms of information content and these are first the reduced uncertainty as initially postulated in information theory (Shannon & Weaver, 1949). However, the second and more elusive element is the meaning which such information communicates to the perceiver. This meaning must be interpreted in terms of the utility of the information in reducing the "distance" between the perceiver and some desired goal-state. These two factors are contingent upon the physical rate of presentation (see Hancock, this volume).

The time-varying nature of this summed information may be approximated as a wave-like structure which presents an environmental "signature" in a manner directly comparable with the way in which we have represented stress previously (Hancock & Chignell, 1985). Indeed, summed information represents the stress that impinges upon the active information processing nervous system. Variation in temporal rate under the impact of the dynamic information "signature" can thus be represented by the same description intrinsic to Figure 1. In this representation, the excess or insufficiency of energy in the system represents the information signature, and while the ability to transduce this information to useful action remains within the range of adaptability then temporal flow is largely undisturbed or in effect controlled. Where such information strays to extremes, temporal flow is either speeded or slowed contingent upon the "signature" at hand.

In terms of behavioral adaptability, there is reason to suspect that the intrinsic ability to process information, mediated through attentional resource capacity may itself be subject to wave-like fluctuations (Jones, 1976). It is possible that the extent of the region of adaptability described may represent the outer threshold of the ability to integrate or synchronize external input with the intrinsic variation expressed as waves of attentional efficiency.

III.4 Exogenous Sources: A Summary

When we wish to know environmental time we look at a clock and this provides a most powerful external signal which allows immediate synchronization with collective social time. However, unless we wish to spend all productive time staring at an external timepiece, other factors must enter into the rate of perceived flow. Among the candidates that have been identified are stimulus numerosity (the absolute number of stimuli per event epoch) and stimulus complexity. However, these are interpretable qualities in that the individual perceiver attaches greater or lesser importance to them in response to the particular task at hand. The ability to process information, as related to the process of adaptive synchronization, would certainly suffer from repetitive switching from one source to another as will occur with progressive interuption. Such disturbances not only act to desynchronize the information assimilation process, they also serve to reduce the external time available for task completion. Little wonder texts on time management focus on the isolation of the manager from such disruptive influences (Mackenzie, 1972; Webber, 1972).

IV. TIME UTILITY ANALYSIS

There are a variety of characteristics concerning our collective temporal behavior that are common across individuals faced with managerial and operational constraints. Some are expressed in the initial section of this work but that is clearly not an exhaustive listing. From the perspective of the system manager and operator it is important to recognize which are actual and which are artificial constraints. It is the case that our perception of temporal flow governs the manner in which we structure both problems and their subsequent resolution. In such circumstances, it is the perception of time which is of greater importance than the recognition of some physicalistic and immaleable external flow (see Bergson, 1910).

What practical approaches can be used to maximize the utility of effective time. A first step is the individual identification of some of the characteristics of these periods. A necessity is to plot out the function of effective time, defined as goal-oriented work periods efficiently directed toward task problem resolution, in terms such as:

1) Time of Day (Occurence, Periodicity, Frequency)

2) Loci of Interruptions (Source, Frequency, Priority)

3) Actual Length of Effective Work Period (Start-Stop Thresholds)

4) Specific Occupation (Attitude Toward Work Task)

Such insights may give direction toward greater temporal efficiency as required by managers and operators.

V. SUMMARY

In the design and operation of specific systems, the operator or manager is faced with decisions concerning the allocation of available resources. All such resources are potentially renewable at differing unit costs. The exception to this rule is time. As such it is a unique and irreplaceable commodity whose usage should be most carefully considered. In this work a brief analysis into factors which affect an individuals temporal behavior indicates a number of potential degrees of freedom which allow greater control over temporal progression than has been formerly appreciated. Extension of these capabilities forms the focus of present research work.

VI. REFERENCES

Bell, C.R. (1966) Control of time estimation by a chemical clock. _Nature_, _210_, 1189-1190.

Bell, C.R. (1975) Effects of lowered temperature on time estimation. _Quarterly Journal of Experimental Psychology_, _27_, 531-538.

Bergson, H. (1910) _Time and free will_. New York: McMillan.

Doob, L.W. (1971) _The patterning of time_, Yale University Press: New Haven.

Fischer, R. (1966) Biological time. In: J.T. Fraser (Ed.). _The voices of time_, Braziller: New York.

Hancock. P.A. (1983) An endogenous metric for the perception of brief temporal intervals. Unpublished Doctoral Dissertation, University of Illinois.

Hancock, P.A. (1984) An endogenous metric for the control of perception of brief temporal intervals. _Annals of the New York Academy_, _423_, 594-596.

Hancock, P.A. (1986) Stress, information-flow, and adaptability in individuals and collective organizational systems. In: _Proceedings of the Second International Symposium on Organizational Design and Management_, H. Hendrick and O. Brown (Eds.). North Holland: Amsterdam.

Hancock, P.A., & Chignell, M.H. (1985) The principle of maximal adaptability in setting stress tolerance standards. In: R.E. Eberts and C. G. Eberts (Eds.). _Trends in Ergonomics/Human Factors II_, North Holland: Amsterdam.

Hetschel, R.F. (1979) Hyperthermia in normal human beings: effects on velocity of the visually evoked response and on perception of time.

Unpublished Doctoral Dissertation, University of Kansas.

Hoagland, H. (1933) The physiological control of judgments of duration: evidence for a chemical clock. Journal of General Psychology, 9, 267-287.

Jones, M.R. (1976) Time, our lost dimension: Toward a new theory of perception, attention, and memory. Psychological Review, 83, 323-355.

Mackenzie, R.A. (1972) The time trap: How to get more done in less time, McGraw-Hill: New York.

Shannon, C., & Weaver, W.A. (1949) The mathematical theory of communication, Urbana: University of Illinois Press.

Webber, R.A. (1972) Time and management, Van Nostrand-Reinhold: New York.

Human Factors in Organizational Design and Management - II
O. Brown, Jr. and H.W. Hendrick (Editors)

ANALYSING THE EFFICIENCY OF MULTIPLE SAMPLING PLANS OF MIL-STD-105 D UNDER SWITCHING RULES

A.M.A. Haridy*, A.Z. El Shabrawy**, M.F. Abdin***
* Associate Professor, Faculty of Eng., Ain Shams Univ., Cairo, Egypt.
** Consultant, Center of Organization and Job Analysis, Cairo, Egypt.
*** Assistant Professor, Faculty of Eng., Ain Shams Univ., Cairo, Egypt.

In this paper, the computer simulation technique is used to analyse the efficiency of multiple sampling plan of the MIL-STD-105 D under the switching rules of the standard. A suitable simulation model for the sampling process that actually happens in practice was formulated. A computer program was designed to solve this model taking into consideration all the inspection phases and the switching procedures of the standard.

INTRODUCTION

The MIL-STD-105 D [1] is one of the most widely used of all attribute sampling inspection schemes. It is a system of lot-by-lot acceptance sampling plans which are producer oriented. The consumer protection is provided through the use of the switching rules of the standard, which go into effect when a specified number of lots are rejected within a given sequence of lots. The standard contains single, double and multiple sampling plans for normal, tightened and reduced inspection. The inspection process begins with normal inspection, and the switching to reduced, tightened or discontinuation of inspection is done according to the switching rules of the standard.

Since the standard was adopted by the US government, efforts did not cease to evaluate its efficiency [2]. The Japanese were pioneers in this aspect. They endorced almost completely the details of the standard, they found out some deficiencies [3] and they therefore suggested some modifications on the standard in order to improve its dynamic characteristics and to simplify its use.

Dr. A.M.A. Haridy, one of the authors, has examined the efficiency of single and double sampling plans of the MIL-STD-105 D under the switching rules of the standard, using the computer simulation technique, and proved that there was an obvious deficiency in the switching mechanism of the standard when operating the single and double sampling plans. The results were published in 1982 [4] and 1985 [5].

This paper is devoted to testing the efficiency of the multiple sampling plans of the MIL-STD-105 D under the switching rules of the standard, using the computer simulation technique.

The plan of work is to:

(1) Study in detail the MIL-STD-105 D.

(2) Design a simulation model for the multiple sampling inspection process that actually happens in practice.
(3) Design a simulation computer program to solve the designed model as a whole.
(4) Analyze the obtained results.
(5) Attain the final conclusion concerning the efficiency of multiple sampling plan of the MIL-STD-105 D under the switching rules of standard.

DESIGN OF THE SIMULATION MODEL

The designed simulation model is a stochastic one, which simulates the multiple sampling inspection process that actually happens in practice. A matrix of dimension 10000 is used to represent the inspection population: Three lot sizes are used with the different sample sizes: lot size 100 for sample size code letters A,B,C & D; lot size 1000 for sample size code letters E,F,G,H & J; lot size 10000 for sample size code letters K,L,M,N,P,Q & R. In each case the lot may has a percent defective (P) which is constant during the sampling process to ensure that the binomial formula applies. For each vale of (AQL) we choose six values of (P) ranging from (P=0.5 AQL) to (P=1.75 AQL) with difference (0.25 AQL) between each two successive values. For each sample size code letter the inspection process is carried out at the different values of (AQL's); in such a way that; at each (AQL) value, 1000 samples are inspected for each value of (P). The samples are drawn randomly and inspected under the switching rules of the MIL-STD-105 D.

DESIGN OF THE SIMULATION COMPUTER PROGRAM

A special computer program is designed, which can easily solve the above mentioned simulation model as a whole, and gives a complete picture of the switching characteristics of the multiple sampling plans of the standard. This enabled us to judge the efficiency of the MIL-STD-105 D's multiple sampling plans when used in practice. The computer runs are carried out at the computer center of Moncton University, Canada.

RESULTS AND ANALYSIS OF RESULTS

Table 1 and 2 show the results obtained for sample size code letters F & K as an example of the results obtained from the computer.

Assuming that the standard is real efficient, it is supposed that, when using the multiple sampling plans in practice, the switching mechanism of the standard should have the features shown in the following figure:

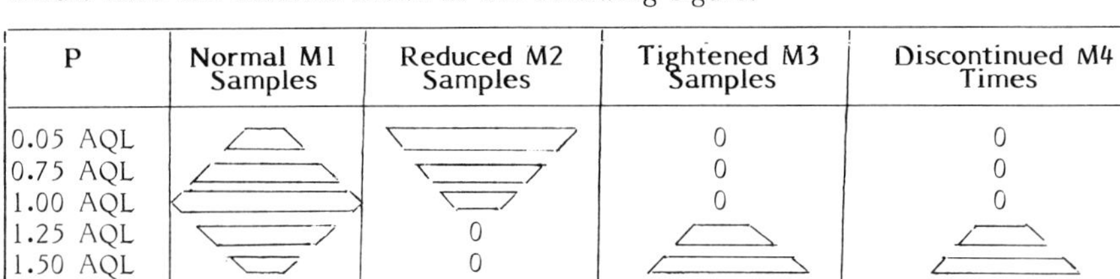

Ideal Switching Mechanism

TABLE 1- RESULTS FOR SAMPLE SIZE CODE LETTER F

NN = 1000
N = 5
N1 = 2

AQL = 0.02500

1. KCN = 0 KRN = 2 KCR = 0
KRR = 2 KCT = 0 KRT = 2

2. KCN = 0 KRN = 2 KCR = 0
KRR = 2 KCT = 0 KRT = 2

3. KCN = 0 KRN = 2 KCR = 0
KRR = 2 KCT = 0 KRT = 2

4. KCN = 0 KRN = 3 KCR = 0
KRR = 3 KCT = 0 KRT = 3

5. KCN = 1 KRN = 3 KCR = 0
KRR = 3 KCT = 1 KRT = 3

6. KCN = 1 KRN = 3 KCR = 0
KRR = 3 KCT = 1 KRT = 3

7. KCN = 2 KRN = 3 KCR = 1
KRR = 3 KCT = 2 KRT = 3

LV = 2

NSAM	P	M1	M2	M3	M4	MS1	MS2	MS3
1000	0.012500	70	930	0	0	7	7	0
1000	0.018750	99	901	0	0	9	9	0
1000	0.025000	125	870	5	0	12	11	1
1000	0.031250	179	821	0	0	14	14	0
1000	0.037500	133	862	5	0	9	8	1
1000	0.043750	325	656	19	0	26	23	3

AQL = 0.06500

1. KCN = 0 KRN = 3 KCR = 0
KRR = 3 KCT = 0 KRT = 2

2. KCN = 0 KRN = 3 KCR = 0
KRR = 3 KCT = 0 KRT = 3

3. KCN = 1 KRN = 4 KCR = 0
KRR = 4 KCT = 0 KRT = 3

4. KCN = 2 KRN = 5 KCR = 0
KRR = 5 KCT = 1 KRT = 4

5. KCN = 3 KRN = 6 KCR = 1
KRR = 6 KCT = 2 KRT = 4

6. KCN = 4 KRN = 6 KCR = 1
KRR = 6 KCT = 3 KRT = 5

7. KCN = 6 KRN = 7 KCR = 2
KRR = 7 KCT = 4 KRT = 5

LV = 8

NSAM	P	M1	M2	M3	M4	MS1	MS2	MS3
1000	0.032500	155	845	0	0	15	15	0
1000	0.048750	170	830	0	0	17	17	0
1000	0.065000	200	800	0	0	20	20	0
1000	0.081250	340	660	0	0	34	34	0
1000	0.097500	336	664	0	0	33	33	0
1000	0.113750	403	597	0	0	39	38	0

AQL = 0.04000

1. KCN = 0 KRN = 2 KCR = 0
KRR = 2 KCT = 0 KRT = 2

2. KCN = 0 KRN = 3 KCR = 0
KRR = 3 KCT = 0 KRT = 2

3. KCN = 0 KRN = 3 KCR = 0
KRR = 3 KCT = 0 KRT = 2

4. KCN = 1 KRN = 4 KCR = 0
KRR = 4 KCT = 0 KRT = 3

5. KCN = 2 KRN = 4 KCR = 0
KRR = 4 KCT = 1 KRT = 3

6. KCN = 3 KRN = 5 KCR = 1
KRR = 5 KCT = 1 KRT = 3

7. KCN = 4 KRN = 5 KCR = 1
KRR = 5 KCT = 2 KRT = 3

LV = 4

NSAM	P	M1	M2	M3	M4	MS1	MS2	MS3
1000	0.020000	121	879	0	0	10	10	0
1000	0.030000	158	842	0	0	13	13	0
1000	0.040000	216	784	0	0	15	14	0
1000	0.050000	274	683	43	1	24	16	7
1000	0.060000	325	638	37	0	28	21	7
1000	0.070000	362	617	21	0	26	21	4

AQL = 0.10000

1. KCN = 0 KRN = 4 KCR = 0
KRR = 3 KCT = 0 KRT = 3

2. KCN = 1 KRN = 5 KCR = 0
KRR = 4 KCT = 0 KRT = 3

3. KCN = 2 KRN = 6 KCR = 0
KRR = 5 KCT = 1 KRT = 4

4. KCN = 3 KRN = 7 KCR = 1
KRR = 6 KCT = 2 KRT = 5

5. KCN = 5 KRN = 8 KCR = 2
KRR = 7 KCT = 3 KRT = 6

6. KCN = 7 KRN = 9 KCR = 3
KRR = 7 KCT = 4 KRT = 6

7. KCN = 9 KRN = 10 KCR = 4
KRR = 8 KCT = 6 KRT = 7

LV = 14

NSAM	P	M1	M2	M3	M4	MS1	MS2	MS3
1000	0.050000	230	770	0	0	23	23	0
1000	0.075000	320	680	0	0	32	32	0
1000	0.100000	337	663	0	0	34	33	0
1000	0.125000	450	550	0	0	45	45	0
1000	0.150000	440	560	0	0	44	44	0
1000	0.175000	491	509	0	0	50	49	0

TABLE 2- RESULTS FOR SAMPLE SIZE CODE LETTER K

NN = 10000
N = 32
N1 = 13

AQL = 0.00400

1. KCN = 0 KRN = 2 KCR = 0
KRR = 2 KCT = 0 KRT = 2

2. KCN = 0 KRN = 2 KCR = 0
KRR = 2 KCT = 0 KRT = 2

3. KCN = 0 KRN = 2 KCR = 0
KRR = 2 KCT = 0 KRT = 2

4. KCN = 0 KRN = 3 KCR = 0
KRR = 3 KCT = 0 KRT = 3

5. KCN = 1 KRN = 3 KCR = 0
KRR = 3 KCT = 1 KRT = 3

6. KCN = 1 KRN = 3 KCR = 0
KRR = 3 KCT = 1 KRT = 3

7. KCN = 2 KRN = 3 KCR = 1
KRR = 3 KCT = 2 KRT = 3

LV = 2

NSAM	P	M1	M2	M3	M4	MS1	MS2	MS3
1000	0.002000	529	243	228	11	49	17	32
1000	0.003000	550	99	351	21	52	9	42
1000	0.004000	478	3	519	35	63	1	61
1000	0.005000	441	3	556	41	65	2	63
1000	0.006000	381	6	613	51	70	2	68
1000	0.007000	352	0	648	57	70	0	69

AQL = 0.01000

1. KCN = 0 KRN = 3 KCR = 0
KRR = 3 KCT = 0 KRT = 2

2. KCN = 0 KRN = 3 KCR = 0
KRR = 3 KCT = 0 KRT = 3

3. KCN = 1 KRN = 4 KCR = 0
KRR = 4 KCT = 0 KRT = 3

4. KCN = 2 KRN = 5 KCR = 0
KRR = 5 KCT = 1 KRT = 4

5. KCN = 3 KRN = 6 KCR = 1
KRR = 6 KCT = 2 KRT = 4

6. KCN = 4 KRN = 6 KCR = 1
KRR = 6 KCT = 3 KRT = 5

7. KCN = 6 KRN = 7 KCR = 2
KRR = 7 KCT = 4 KRT = 5

LV = 7

NSAM	P	M1	M2	M3	M4	MS1	MS2	MS3
1000	0.005000	598	367	35	3	50	45	4
1000	0.007500	677	137	186	17	55	35	19
1000	0.010000	607	47	346	33	48	12	36
1000	0.012500	450	15	535	52	58	3	55
1000	0.015000	404	18	578	57	63	4	59
1000	0.017500	362	2	636	63	66	2	64

AQL = 0.00650

1. KCN = 0 KRN = 2 KCR = 0
KRR = 2 KCT = 0 KRT = 2

2. KCN = 0 KRN = 3 KCR = 0
KRR = 3 KCT = 0 KRT = 2

3. KCN = 0 KCT = 3 KRT = 0
KRR = 3 KCT = 0 KRT = 2

4. KCN = 1 KRN = 4 KCR = 0
KRR = 4 KCT = 0 KRT = 3

5. KCN = 2 KRN = 3 KCR = 0
KRR = 4 KCT = 1 KRT = 3

6. KCN = 3 KRN = 5 KCR = 1
KRR = 5 KCT = 1 KRT = 3

7. KCN = 4 KRN = 5 KCR = 1
KRR = 5 KCT = 5 KRT = 3

LV = 4

NSAM	P	M1	M2	M3	M4	MS1	MS2	MS3
1000	0.003250	553	82	365	28	55	12	42
1000	0.004875	442	7	551	44	63	2	61
1000	0.006500	355	0	645	59	69	0	68
1000	0.008125	300	0	700	64	74	0	74
1000	0.009750	325	0	675	62	70	0	70
1000	0.011375	271	0	729	72	73	0	73

AQL = 0.01500

1. KCN = 0 KRN = 4 KCR = 0
KRR = 3 KCT = 0 KRT = 3

2. KCN = 1 KRN = 5 KCR = 0
KRR = 4 KCT = 0 KRT = 3

3. KCN = 2 KRN = 6 KCR = 0
KRR = 5 KCT = 1 KRT = 4

4. KCN = 3 KRN = 7 KCR = 1
KRR = 6 KCT = 2 KRT = 5

5. KCN = 5 KRN = 8 KCR = 2
KRR = 7 KCT = 3 KRT = 6

6. KCN = 7 KRN = 9 KCR = 3
KRR = 7 KCT = 4 KRT = 6

7. KCN = 9 KRN = 10 KRC = 4
KRR = 8 KCT = 6 KRT = 7

LV = 13

NSAM	P	M1	M2	M3	M4	MS1	MS2	MS3
1000	0.007500	672	323	5	0	59	57	1
1000	0.011250	758	172	70	4	66	56	9
1000	0.015000	756	67	177	13	53	33	20
1000	0.018750	595	38	367	34	55	16	38
1000	0.022500	459	9	532	48	62	6	56
1000	0.026250	390	4	606	60	64	2	61

AQL = 0.02500

1. KCR = 0 KRN = 4 KCR = 0
KRR = 4 KCT = 0 KRT = 4

2. KCN = 1 KRN = 6 KCR = 0
KRR = 5 KCT = 1 KRT = 5

3. KCN = 3 KRN = 8 KCR = 1
KRR = 6 KCT = 2 KRT = 6

4. KCN = 5 KRN = 10 KCR = 2
KRR = 7 KCT = 3 KRT = 7

5. KCN = 7 KRN = 11 KCR = 3
KRR = 8 KCT = 5 KRT = 8

6. KCN = 10 KRN = 12 KCR = 6
KRR = 9 KCT = 7 KRT = 9

7. KCN = 13 KRN = 14 KCR = 6
KRR = 10 KCT = 9 KRT = 10

LV = 24

NSAM	P	M1	M2	M3	M4	MS1	MS2	MS3
1000	0.012500	418	574	8	0	36	35	1
1000	0.018750	606	327	67	4	48	39	8
1000	0.025000	596	132	272	20	55	23	31
1000	0.031250	486	12	502	48	57	5	51
1000	0.037500	357	2	641	62	67	2	65
1000	0.043750	298	4	698	69	72	1	70

AQL = 0.06500

1. KCN = 1 KRN = 7 KCR = 0
KRR = 5 KCT = 0 KRT = 4

2. KCN = 4 KRN = 10 KCR = 1
KRR = 7 KCT = 3 KRT = 9

3. KCN = 8 KRN = 13 KCR = 3
KRR = 9 KCT = 7 KRT = 12

4. KCN = 12 KRN = 17 KCR = 5
KRR = 12 KCT = 10 KRT = 15

5. KCN = 17 KRN = 20 KCR = 7
KRR = 13 KCT = 14 KRT = 17

6. KCN = 21 KRN = 23 KCR = 10
KRR = 15 KCT = 18 KRT = 20

7. KCN = 25 KRN = 26 KCR = 13
KRR = 17 KCT = 21 KRT = 22

LV = 69

NSAM	P	M1	M2	M3	M4	MS1	MS2	MS3
1000	0.032500	199	801	0	0	19	19	0
1000	0.048750	404	566	30	1	36	30	5
1000	0.065000	604	145	251	17	54	26	28
1000	0.081250	421	1	578	54	61	1	60
1000	0.097500	287	0	713	70	73	0	73
1000	0.113750	220	0	780	77	78	0	78

AQL = 0.04000

1. KCN = 0 KRN = 5 KCR = 0
KRR = 4 KCT = 0 KRT = 4

2. KCN = 3 KRN = 8 KCR = 1
KRR = 6 KCT = 2 KRT = 7

3. KCN = 6 KRN = 10 KCR = 2
KRR = 8 KCT = 4 KRT = 9

4. KCN = 8 KRN = 13 KCR = 3
KRR = 10 KCT = 6 KRT = 11

5. KCN = 11 KRN = 15 KCR = 5
KRR = 11 KCT = 9 KRT = 12

6. KCN = 14 KRN = 17 KCR = 7
KRR = 12 KCT = 12 KRT = 14

7. KCN = 18 KRN = 19 KCR = 9
KRR = 14 KCT = 14 KRT = 15

LV = 40

NSAM	P	M1	M2	M3	M4	MS1	MS2	MS3
1000	0.020000	231	769	0	0	24	23	0
1000	0.030000	465	525	10	1	42	40	1
1000	0.040000	573	286	141	13	51	35	15
1000	0.050000	518	52	430	42	55	10	44
1000	0.060000	408	11	581	58	61	2	59
1000	0.070000	267	0	733	73	74	0	74

AQL = 0.10000

1. KCN = 2 KRN = 9 KCR = 0
KRR = 6 KCT = 1 KRT = 8

2. KCN = 7 KRN = 14 KCR = 3
KRR = 9 KCT = 6 KRT = 12

3. KCN = 13 KRN = 19 KCR = 6
KRR = 12 KCT = 11 KRT = 17

4. KCN = 19 KRN = 25 KCR = 8
KRR = 15 KCT = 16 KRT = 22

5. KCN = 25 KRN = 29 KCR = 11
KRR = 17 KCT = 22 KRT = 25

6. KCN = 31 KRN = 33 KCR = 14
KRR = 20 KCT = 27 KRT = 29

7. KCN = 37 KRN = 38 KCR = 18
KRR = 22 KCT = 32 KRT = 33

LV = 110

NSAM	P	M1	M2	M3	M4	MS1	MS2	MS3
1000	0.050000	70	930	0	0	7	7	0
1000	0.075000	338	652	10	0	30	28	2
1000	0.100000	583	300	117	9	43	30	13
1000	0.125000	446	6	548	50	61	3	58
1000	0.150000	263	0	737	73	74	0	79
1000	0.175000	210	0	790	79	80	0	79

While the results obtained reveal that in some cases this standard do not follow the ideal features shown above, but it behaves as follows:

(1) For small sample sizes that range from 2 to 50-normal inspection, code letters from A to J :

a. For values of $P \leqslant AQL$, the standard goes very close to the ideal

b. For values of $P > AQL$, too much misswitching to reduced inspection occur, while almost no switching to tightened inspection or discontinuation of inspection occurs.

(2) For large sample sizes that range from 80 to 1250-normal inspection (sample size code letters from K to R):

a. For vlaues of $P \leqslant AQL$, too much misswitching to tightened inspection especially for small values of AQL.

b. For values of $P > AQL$, switching either to reduced or tightened inspection tends to be very close to the ideal behaviour for all values of AQL.

CONCLUSION

From the above mentioned analysis it is quite evident that there is an obvious deficiency in the switching mechanism of the MIL-STD-105 D when operating the multiple sampling plans. This requires the introduction of some modifications or additions to the standard in order to bring the switching mechanism as close as possible to the expected ideal behaviour.

LIST OF SYMBOLS

NN = lot size
N = Sample size for normal & tightened inspection
N1 = Sample size for reduced inspection
AQL = Acceptable Quality Level
KCN = Acceptance number for normal inspection
KRN = Rejection number for normal inspection
KCR = Acceptance number for reduced inspection
KRR = Rejection number for reduced inspection
KCT = Acceptance number for tightened inspection
KRT = Rejection number for tightened inspection
LV = Limit number for reduced inspection
NSAM = Counter for total number of inspected samples
M1 = Counter for total number of samples inspected under normal conditions
M2 = Counter for total number of samples inspected under reduced conditions
M3 = Counter for total number of samples inspected under tightened conditions
M4 = Counter for number of times of discontinuation of inspection
MS1 = Counter for number of times going into normal inspection phase
MS2 = Counter for number of times going into reduced inspection phase
MS3 = Counter for number f times going into tightened inspection phase

The values (N, N1, KCN, KRN, KCR, KRR, KCT, KRT & LV) are obtained from the tables of the standard for each value of (AQL). The value (NSAM) is always 1000 samples.

REFERENCES:

[1] Department of defense, USA, "Sampling Procedures and Tables for Inspection by Attributes, MIL-STD-105 D".

[2] Schilling, E.G.; and Sheesley, J.H., "The Performance of MIL-STD-105 D under the switching Rules (Part 1: Evaluation)", JQT, V 10 n 2, Apr. 1978.

[3] Koyama, T., "Switching Characteristics Under MIL-STD-105 D", Technometrics, V 21 n 1 Feb. 1979.

[4] Haridy, A.M.; and Adel, M., "Analysing the Efficiency of the Switching Rules of MIL-STD-105 D", Proc. of 13th anu. conf. of Pittsburgh on Modeling & Simulation, Pittsubrgh, Pa, USA, V 13 Pt 1, 1982, pp 315-319.

[5] Haridy, A.M., "Examining The Efficiency of MIL-STD-105 D's Double Sampling Plans" Proc. of 16th annu. conf. of Pittsburgh, Pittsburgh, Pa, USA, 1985.

OFFICE AUTOMATION,
MANAGEMENT AND DESIGN

Human Factors in Organizational Design and Management - II
O. Brown, Jr. and H.W. Hendrick (Editors)

TRENDS IN THE AUTOMATION OF AMERICA'S OFFICES IMPLICATIONS FOR US PUBLIC POLICY

Benjamin C. Amick, III & Vary T. Coates

Congress of the United States
Office of Technology Assessment
Washington, D.C. 20510

This paper briefly discusses a report, Automation of America's Offices, prepared by the Office of Technololgy Assessment for the Education and Labor Committee of the US House of Representatives and the Labor and Human Resources Committee of the US Senate. The Office of Technology Assessment (OTA) is a nonpartisan analytical support agency serving the United States Congress. It provides objective analysis of major public policy issues related to scientific and technological change.

The Automation of America's Offices report is a broad assessment that focused on a variety of issues including: technologies of the office, employment, training and education, changing organizational environments, the quality of worklife and client confidentiality and data security. While the report did not comprehensively analyze the public policy issues brought out in the analysis, it suggested areas of consideration for Congress over the next decade and into the 21st century. OTA chose a 15 year time span because Congress will be less concerned with the emphemeral effects and transitional problems than with the long-range structural changes. Many of the structural changes are likely to become clearly visible only after office automation has been widely adopted and organizations learn to use its full capabilities.

DEFINING OFFICE AUTOMATION

Defining an office and office automation is difficult. Almost any place where information handling is the main activity is called an office, whether it is one person at a desk or a complex hierarchy of executives, professionals, and clerical workers. The office is wherever "office work" is done, and "office work" is the processing and use of information for the purposes of tracking, monitoring, recording, directing, and supporting complex human activities. One of the striking consequences of information and communication technologies is that together they make much "office work" independent of space and time. For the OTA report "office automation" broadly referred to the application of microelectronic information technology and communication technology to "office work".

TRENDS IN OFFICE AUTOMATION

The dominant trends in office automation, from 1985 to 2000, are likely to be:

- a continuing strong movement toward microcomputers and toward distributed data access and data handling, usually superimposed on rather than superceding centralized automatic data processing;
- more powerful, easier to use software;

- a strong trend toward linking and networking of microcomputers, minicomputers, mainframes, and periphal and supporting systems;
- increasing choice among technological options for accomplishing information handling objectives;
- more and more capture of data at the point of origin, decreasing the need for repeated keyboarding and centralized data entry, and;
- growing capability for communication, between devices, between organizations, and between locations.

By the mid-1990's nearly every office will have at least one computer, just as nearly all offices today have telephones. There will probably be a terminal of some kind for at least every two to three office workers. Since many organizations will by then have adapted their work process and work environment to the new technology - although restructuring and change will surely continue - many solutions to current transitional problems should be available. Office productivity should increase significantly.

EMPLOYMENT TRENDS

In the immediate future, the anticipated effects on employement are only possible or probable. There is little evidence that decline in growth in, or reduction of, the number of office jobs is already affecting large numbers of people. It is possible that there will be continuing strong demand for additional office workers. The most likely outcome, however, is slowing growth in the number of office jobs and even eventually an absolute reduction in the number of jobs in offices.

At present the Federal government is poorly equipped to detect or understand early signals of problems arising from structural changes in the economy related to technology. It is difficult to track and demonstrate even long established employment trends because the kinds of data collected by the Bureau of the Census, the Bureau of Labor Statistics, and other government agencies change over time. The way in which those data are aggregated or disaggregated, and the way they are labeled also change.

Concommitant with the capitalization of office work, there is occurring a shift in the status of employees from full-time to part-time. A large increase in the number of full-time jobs lost to part-time or temporary workers without benefits packages and coverage by worker protection mechanisms (perhaps occurring due to office automation) could ultimately burden taxpayers or lead to degradation in the level of social well-being. If Congress chooses to intervene, it could constrain coversion of employees to other status by legislative confirmation of recent IRS rulings that an independent contractor accepting work from only one organization is in fact an employee; a manadatory minimum ratio between full-time and part-time or temporary employment; or mandatory pro-rating of benefits packages and social security contributions. Alternatively, Congress could alleviate some of the problems of involuntary part-time employment by further actions to encourage or require Voluntary Reduced Work Time arrangements, making involuntary part-timers eligible for partial unemployment benefits.

If office automation reduces the number of office jobs, women and members of minority groups who have historically been disavantaged in the job market, are likely to be most negatively affected. Congress may then wish to consider a series of steps to alleviate their disadvantage. It could:

- o take explicit notice of these concerns and factors in discussion of bills concerning pay equity and comparable worth, which are now before committees;
- o provide incentives for maintaining or improving ratios of female to male employment in labor force reductions associated with automation;
- o provide incentives for maintaining the share of employment held by minority groups in office work; and
- o provide subsidized child care facilities or increased tax deductions for child care to increase the employment options open to working parents.

OFFICE AUTOMATION & WORKER HEALTH

Along with employment effects, the health effects of office automation have been a topic of political controversy. Visual and musculoskeletal problems currently can be alleviated through appropriate office and workstation design; stress-related illness or disease is more difficult to control and may emerge as the greatest public health problem among office workers in the future. Rewiewing the current information and ongoing research, there is at this time no good basis for fear of VDT effects on reproductive processes, but research should continue to be monitored for new findings.

In view of the concern about suspected hazards associated with computers and VDTs, options for Congress to consider include: 1) a public information program to inform employers and office employees about what is known about the effects of office work on health and well-being, and what can be done to reduce visual and musculoskeletal strain and psychological stress; and 2) directing the Occupational Safety and Health Administration to develop ergonomic advisories, guidelines, or standards for use in public and private sector office and in design of equipment and furniture used in office automation. If these actions and self-initiated organizational management actions to alleviate these concerns are inadequate, then Congress may wish to consider legislation such as limiting the number of hours that workers can be required to spend continuously in VDT use.

In any case, because scientific evidence is still equivocal, and significant risks involved in computerized office work would affect a large proportion of Americans, Congress could support the funding of research on the relation of stress to chronic disease and illness and on the possibility of reproductive hazards related to VDT use. It may also wish to establish national mechanisms to monitor the health status of American office workers.

Congress may wish to consider the desirability of educational or information programs for employers, employees, and their associations with regard to problems associated with office automation and strategies that organizations have successfully used for alleviating them. Other possibilities to be considered under some circumstances are clarification of worker rights under existing labor-management negotiations procedures, or changes in labor law to require worker involvement in technology-related discussions.

HOME-BASED WORK

The increased ability to do information processing work at home via electronic communication is another consequence of office automation. Home-based clerical work, with the work performed on a piece-rate or hourly basis, could increase significantly in the next few years. Already controversial, it would then become much more so.

Because this activity is already controversial, Congress may have to formulate policy regarding it within the next several years. Proposals have been made in Congress to provide tax credits for the purchase of computers for home-based work (and other nonrecreational uses). Other ways to encourage home-based work include legislative actions voiding IRS rulings and court decisions that make many home-based workers employees rather than independent contractors, and removal of other regulatory or tax barriers to home-based work.

However, Congress may instead wish to discourage home-based computer-mediated clerical work. It could then consider a ban on paid computer-related employment when the primary or sole work site is the worker's home, but would probably wish to distinguish this employment from similar activities designed to establish entrepreneurial small businesses. Congress may, instead, wish to clarify through legislation the conditions under which home-based workers are employees entitled to the normal worker protections afforded by law and equity, and to require pro-rated benefits for workers accepting work to be done in their homes.

Instead of banning or strongly discouraging home-based office work, Congess could insisit on rigorous application and enforcement of existing worker protection laws and regulation to home-based work, including wage and hour laws, occupational safety and health regulations, and all applicable reporting requirements. This strategy would require strengthening the resources and capabilities available to enforcement agencies.

Rather than banning or discouraging home-based work, Congress could adopt a strategy of increasing the range of employment options for workers through tax deductions or provision of day care centers; alternative means of caring for children, aged, or disabled dependents; and by further acting to increase mainstream employment opportunities for handicapped workers.

POLICY ISSUES FOR CONGRESS

A number of policy issues have been identified above, and OTA has concluded that:

- o Many of the concerns that have been raised with regard to office automation are transition problems that will be solved by market forces and by the common sense and ingenuity of users -- they do not require Federal action of any kind.
- o Some concerns -- most importantly, the possible effect on future white-collar employment levels -- are matters of long-term national interests. They merit watchful attention, and Congress should begin now to make sure that the nation is prepared to deal with them if and when conditions warrant action. This, however, will be

difficult to do because the capability to detect and understand emerging structural change in the economy is poor. Better economic and employment data and development of improved forecasting techniques are urgently needed.

- There are a few specific issues of immediate concern; some of them are already the subject of legislative proposals before Congress or State legislatures. These include, for example:
  - -documented or suspected health problems associated with office automation,
  - -proposals to encourage or to ban home-based office work, and
  - -largely unexamined relationships between pay equity and comparable worth questions and the changes associated with office automation.
- There are also specific issues of concern to Congress in its oversight of Federal agencies and activities, including:
  - -the implications of Federal office automation for procurement, personnel, and budgetary policies, and
  - -adequacy of present provisions to assure data security and confidentiality and continuation of Federal functions under emergency conditions.
- An emerging issue, which may become more important and more controversial in the near future, is the beginning trend in off-shore sourcing of data-entry work.
- There are several groups for whom the effect of office automation are particularly important and whose interests merit special attention:
  - -women in clerical, managerial, and professional office positions,
  - -minority workers, and
  - -small business firms.

References

U.S. Congress, Office of Technology Assessment, Automation of America's Offices (Washington, DC: U.S. Government Printing Office, OTA-CIT-287, December, 1985).

Human Factors in Organizational Design and Management - II
O. Brown, Jr. and H.W. Hendrick (Editors)
Elsevier Science Publishers B.V. (North-Holland), 1986

MANPRINT: AN APPROACH TO OFFICE AUTOMATION IN THE U.S. ARMY

Karl J. Gunzelman

Headquarters, Department of the Army
Office of the Deputy Chief of Staff for Research, Development and Acquisition
The Pentagon
Washington, D.C.

The Army's information management program requires a multi-specialized approach to information system design and management. Having identified human engineering problems with the current office automation of an Army agency, it became apparent that human factors engineering was required in the information system development process. The Manpower and Personnel Integration (MANPRINT) initiative, which integrates multiple domains into the materiel development process, is considered for use as an Information Systems design methodology.

OVERVIEW OF THE ARMY INFORMATION MANAGEMENT PROGRAM

Realizing that information management is essential to organizational success the United States Army has taken steps to improve the effectiveness in utilizing its information resources. In May, 1984, the Chief of Staff of the Army, General John A. Wickham Jr. approved the establishment of the Information Mission Area (IMA) (Doyle and Craven, 1985). The Information Mission Area (IMA) was established in recognition that the lines separating traditional automation (includes office automation), telecommunications, visual information, records management and printing and publishing were rapidly disappearing and the need for their management integration was imperative. The office, Assistant Chief of Staff for Information Management (ACSIM) was established to organizationally focus Army management attention on the integration of those subdisciplines throughout the Army and to supervise information system planning and implementation.

Information systems are to be developed through a systematized, time-phased plan transitioning the Army from the present to the future taking into account guidance from Department of the Army Staff, resource constraints, and an ability to absorb technology. The entire purpose of Army information management is to improve the quality and flow of information available to the Army. It focuses on the expressed information needs of the Army's decision makers, at all levels, by supporting a variety of functions, with finite resources. Figure 1 shows the components of the IMA, its environment, and the management activities.

The IMA consists of three major groups along a continuum: Sustaining Base, Strategic, and Theater/Tactical (U.S. Department of the Army, 1985a). Each of these three groups include the above mentioned subdisciplines, with the associated information systems developed and managed through an architectural approach (Kolesar, 1985). The Army has adopted the concept of a three-tier information system architecture based on distributing processing power to three specific hierarchial levels within the Army Structure. These levels include: Regional, Installation and End-User (Mabius, 1985). Data stored at the Regional and Installation level would constitute the major portion of the Army's Corporate Data Base and be available to the end-user. Analysis of the end-users needs are necessary to determine appropriate interfaces to most productively interact with the automation and organizational environments.

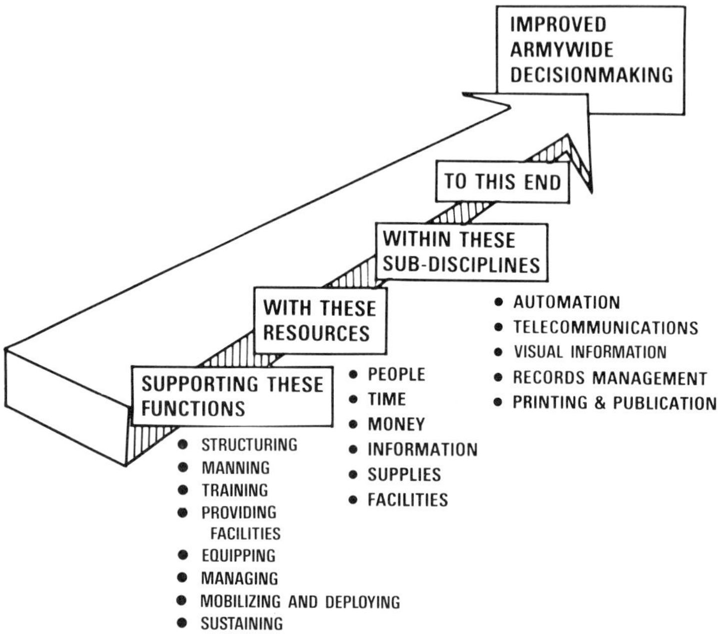

Figure 1. The Purpose of Army Information Management

INFORMATION SYSTEMS PLANNING

The information management process is predicated upon an Information Systems Plan (ISP), which examines the organization's information needs. It provides a structure for planning the near- and far-term information needs of the Army's decision-makers. The plan is designed to support an organization's need for improved information systems. The plan (U.S. Department of the Army, 1984) is based on five basic concepts:

1. The long-range plan must support the goals and objectives of the organization thus implying a "top down" approach.
2. The plan must address information needs at all management levels.
3. The plan must provide consistent information implying that data should be managed as a shared resource for use by decision-makers.
4. By designing information systems to suport the basic activities of the organization, the plan should survive organizational and management changes.
5. The plan calls for implementation by subsystem; while planning is top-down, implementation must be bottom-up.

The methodology employed to improve information systems is a formal requirements study that identifies information requirements and relationships of the organization. It focuses on a top-down senior decision-maker view of the organizational processes (a group of logically related decisions and activities required to manage the resources of the organization) performed and information classes (logical groupings of information used or created in the organization) needed (Department of the Army, 1985b). The outcome of the study, elaborated in the ISP, addressing the functional processes and the data classes and systems that support those process are displayed in an information model. The information model (figure 2) provides matrices displaying: 1) organizational processes against information classes, and 2) organizational elements against the processes. A weakness of this approach is that it fails to address the personnel subsystem, and the interaction it has with the organization (structure & environment) and the information system's automation devices (engineering & technology).

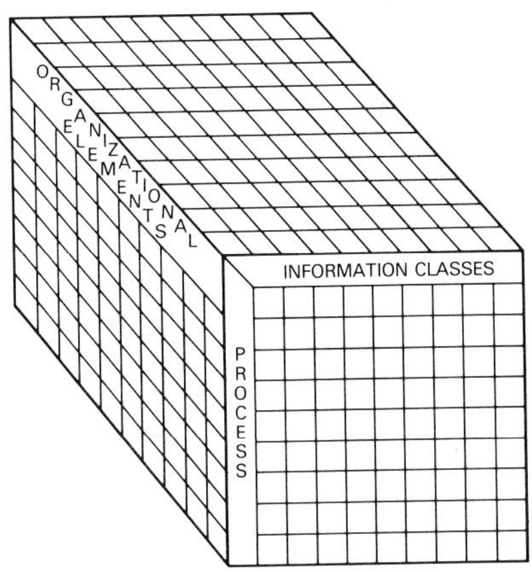

Figure 2. The Information Model

MANPOWER AND PERSONNEL INTEGRATION (MANPRINT)

The Manpower and Personnel Integration (MANPRINT) initiative is an umbrella concept that identifies and integrates human factors engineering, manpower (allocations), personnel (qualifications), training, system safety and health hazard considerations prior to, during and after the Materiel Acquisition Process (MAP). Each of these MANPRINT domains are analyzed in association with their common threads of cost, performance, and workload and as they relate to the underlying issues displayed in figure 3. MANPRINT is oriented on constraining the design of materiel systems and associated support requirements so that acquired systems can be operated and maintained efficiently, safely and within the manpower structure, personnel skill, and training resource constraints of the Army. The engineering design philosophy of MANPRINT is focused on optimum system performance on the battlefield (user environment) which includes consideration of both soldier and equipment capability (U.S. Department of the Army, 1986a). The MANPRINT initiative is iterative; subsequently it is integrated throughout the materiel acquisition process and during all phases of the life cycle for those materiel systems falling within Army Regulation 70-1 (see U.S. Department of the Army, 1986b).

MANPRINT emphasizes front-end analyses and planning of soldier-materiel system design. The ultimate goal is for better total system performance by optimizing the equipment and human subsystems' interface. It is understood that the interaction of the subsystems is important, with the total system performance equal to the product of the equipment and human subsystems' performance. This relationship can be expressed in the following equation form:

Equipment Performance (PE)	X	Human Performance (PH)	=	System Performance (PS)

MANPRINT is an option oriented process as opposed to an objective oriented process. Because analyses occur early, decision-makers are provided more useful information upon which to make tradeoffs in areas such as quality and numbers of people, training, technology, conditions and standards. Rather than concentrating specifically on the equipment performance parameters, of the traditional development process, decision makers are afforded the opportunity to pursue a more balanced approach to materiel acquisition.

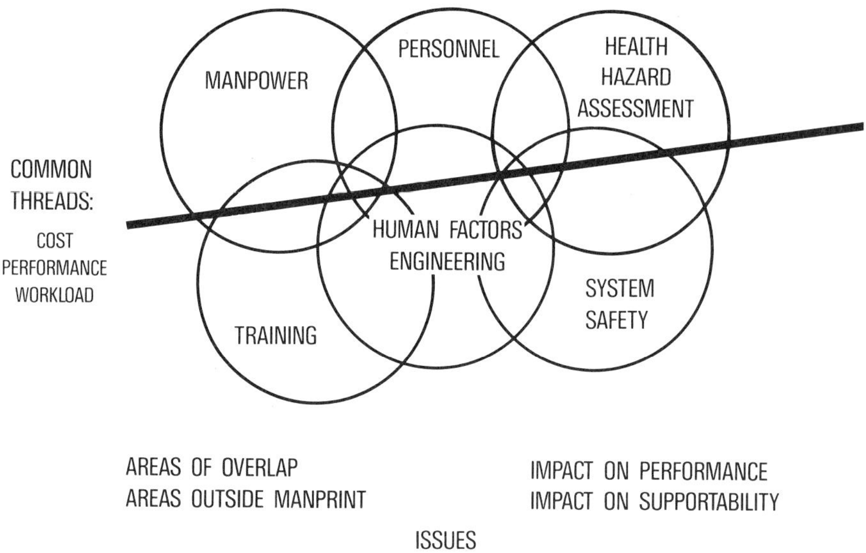

Figure 3. The MANPRINT Domains

MANPRINT AND OFFICE AUTOMATION

An initiative to come out of the ISP for a major Army Agency on the Department of the Army Staff, was the need to provide each Action Officer a management information system workstation (U.S. Department of the Army, 1985c). The initial task for developing the prototype workstation, a turn-key system consisting as a minimum of a keyboard, a VDT screen, and a central processor, with each designed into a mapped, local area network (LAN), was assigned to an software engineer with Action Officer experience. Lacking in the original design were ergonomic considerations in designing the workstations, researching human factors issues associated with software design, and examining the sociotechnical implications of automation. That agencies ISP findings acknowledged human factors related problems to current office automation devices but no solutions, nor any plan to solve them:

> 1. Some people are afraid of learning to use the system or, if required to learn the mechanics of the system operations, will not discipline themselves to use the facilities available. This issue is seen as a psychological phenomenon.
>
> 2. Some people perceive automation output as dull and without "pizzazz." Certainly there are categories of information which can be best transmitted orally because the inflection of the voice is more effectively presented as a horseblanket (presentation of large amounts of information in a dense and analogous format).
>
> 3. Some computer output requires clarification through eyeball-to-eyeball expansion regardless of the accuracy of the written information displayed. This instant clarification may never be replaced by automation (U.S. Department of the Army, 1984, p 28).

MANPRINT was then suggested to the information system management team, by the recently appointed human factors engineer, as the design methodology for designing and implementing the agencies information system. Because MANPRINT is concerned with total system performance, it considers the integrated system. As shown in figure 4, an understanding of that system is provided by an analysis of the organization (structure & environment), the machine (engineering & technology), and the man (people) subsystems, as well as the interactions that occur.

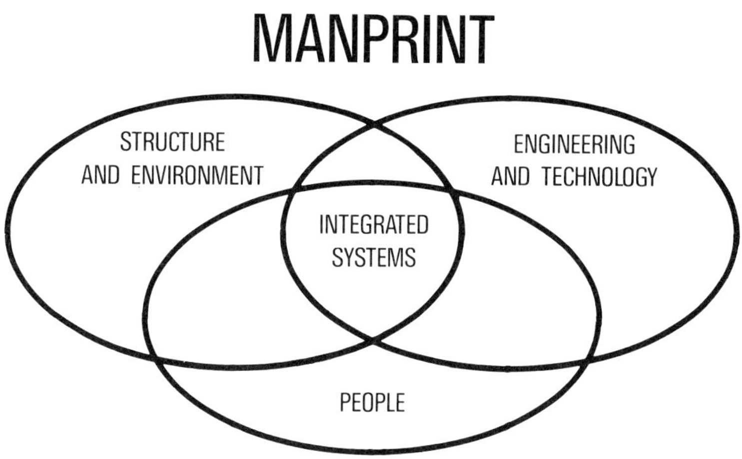

Figure 4. MANPRINT for Integrated Information Systems

This model of MANPRINT leaves the design team the flexibility to use any number of approaches in integrated systems analysis and measurement. Consideration had been given to the macroergonomics approach to system design pursued by Hendrick (1985) which focuses around establishing the design of the integrated system's organizational structure; then the design proceeds with the microergonomic design of jobs, work groups, and the human-machine interface. A focus on redesigning the Department of the Army hierarchial structure first, however, would prove an ominus task.

If the senior leadership could see a connection between office automation and productivity, as it pertains to the lessons learned about simple organizational form, lean staff with few administrative layers and few people at the top (Peters and Waterman, 1982) perhaps the organizational structure would change; consider this a case for "top-down" planning and "bottom-up" implementation stated in the ISP. Another approach considered, was a sociotechnical process (Pava, 1983) that views the office organization as an open system that interacts with a complex environment; it exhibits systematic complexity; and the effective office organization, in turn, generates a level of variety that matches the level of flexibility required by its mission and environment. While the implementation of an information system based on the analyses of a technical subsystem (tools & techniques) and its social system (division of labor & methods of coordination) counterpart would prove very interesting and profitable, the process is unfamiliar to the information management team. The process also resembles, too closely, the Army's recently terminated Organizational Effectiveness program.

The design methodology accepted to integrate the domains of MANPRINT into the information system architecture has been expounded upon in great detail by Meister (1971, 1982, and 1985), and utilized by Baily (1982), and more recently integrated into a IBM training program (Chapanis & Shafer, 1986). The process begins with a mission analysis. The mission area analysis identifies the functional requirements of a new system necessary to meet some future need(s). The method for determining the user's needs within the agencies information system will consist of an function analysis and subsequent allocation of functions based on user input and the structure of the organization.* Function allocation and verification consist of three major steps:

1. Determine system requirements
2. Determine system functions
3. Allocate functions between men and machines:
   a. specify alternate man-machine configurations and functions;
   b. verify that the human can perform the functions tentatively assigned to him, and that his performance will satisfy system requirements.
   c. select the most effective man-machine concept by comparing alternate designs.

Once the future system is identified an analysis of the present (baseline) system will provide the data necessary to determine the systems' delta. Once again, the present information system is determined by measuring variables within the MANPRINT domain taxonomy and as they apply to information system functions. Additional steps, beyond function allocation and verification, must then be accomplished. These steps will further refine the baseline analysis process and provide more specific data toward the information system design:

Task Description and Analysis

1. Describe and analyze tasks
2. Determine equipment requirements implied by the tasks.
3. Analyze tasks in terms of their demands.

Identify Man-Machine Interfaces

1. Select appropriate interface components.
2. Properly arrange control-display components.

The delta will provide the data necessary to update the Information Management Plan. The subsequent IMP will be implemented by applying a body of knowledge surrounding the domains of MANPRINT (e.g., Smith, 1983 & Helander 1985) as they apply to information systems. Figure 5 summarizes the entire focus of MANPRINT on information systems design.

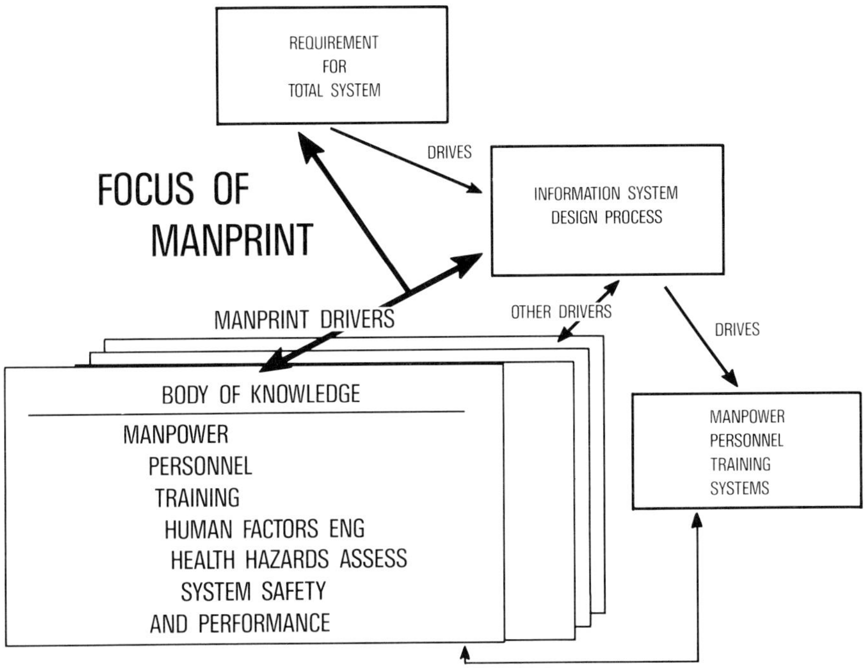

Figure 5. Focus of MANPRINT

CONCLUSION

The results of the analyses are not yet available for presentation, but they will make for interesting discussion when completed. It is the intention of this applied research that the planning for new information systems will inevitably lead to a more productive Army headquarters, by incorporating human factors considerations into the systems engineering approach.

MANPRINT, a design process originally intended for acquiring materiel systems has been tailored to meet the idiosyncratic requirements of information systems. Having considered a number of alternative approaches for integrating human factor considerations into information systems design, the paradign professed by Meister (1971, 1985), and utilized by Baily (1982) is accepted to assess behavioral issues in a multi-specialized system's approach.

* While the processes suggested by Pava (1983) and Hendrick (1985) were not selected for implementation, organizational considerations contained therein will be a part of the total system analysis.

NOTE: The Manpower and Personnel Integration (MANPRINT) initiative is an accepted design process by the Department of the Army. Its application to information systems design, however, are the opinions and assertions of the author. They should not be construed, in any way, as reflecting the views of the Department of the Army.

REFERENCES

Baily, R.W. (1982). Human Performance Engineering: A guide for system designers. Englewood Cliffs, NJ: Prentice-Hall.

Chapanis, A. and Shafer, J.B. (1986). Factoring Humans into FSD Systems. FSD Technical Directions, 12, (1), 15-22.

Doyle, D.K. and Craven R.E. (1985, July). Misson: To Bring Order To Information Revolution. ARMY, 30-33.

Helander, M.G. (1985). Emerging Office Automation Systems, Human Factors, 27, 3-20.

Hendrick, H.W. (1985). Wagging the tail with the dog: Organizational design considerations in ergonomics. ODAM, 3, (4) 5-16.

Kolesar, G.F. (1985, July). The Architecture Goal: Systems that 'Talk'. ARMY, 36-39.

Mabius, L.J. (1985, October). Office Automation in the Army: Planning for the Users. Government Executive, 20-24.

Meister, D. (1971). Human Factors: Theory and Practice, New York: John Wiley & Sons.

Meister, D. (1982). The role of Human Factors in system development. Applied Ergonomics, 13. 119-124.

Meister, D. (1985). Behavioral Analysis and Measurement Methods, New York: John Wiley & Sons.

Peters, T.J. and Waterman, R.H. (1984). In Search of Excellance. New York: Harper and Row.

Pava, C.H.P. (1983). Managing New Office Technology: an organizational strategy. New York: Free Press.

Smith, M.J. (1983). Occupational Stress due to Computerized Office Technology. National Institute of Organizational Safety and Health, Cincinnati, Ohio.

U.S. Department of the Army. (1984). Information System Plan. Report for the Office of the Deputy Chief of Staff for Research, Development and Acquisition. Washington, DC: U.S. Government Printing Office.

U.S. Department of the Army. (1985, a). AR 25-1: The Army Information Management Program. Washington, DC: U.S. Government Printing Office.

U.S. Department of the Army. (1985, b). AR 25-5: Information Management for the Sustaining Base. Washington, DC: U.S. Government Printing Office.

U.S. Department of the Army. (1985, c). Information Management Plan. Report for the Office of the Deputy Chief of Staff for Research, Development and Acquisition. Washington, DC: U.S. Government Printing Office.

U.S. Department of the Army. (1986, a). AR 602-2: Manpower and Personnel Integration (MANPRINT). Washington, DC: U.S. Government Printing Office.

U.S. Department of the Army. (1986, b). AR 70-1: Systems Acquisition Policy and Procedure. Washington, DC: U.S. Government Printing Office.

Human Factors in Organizational Design and Management - II
O. Brown, Jr. and H.W. Hendrick (Editors)

OFFICE ERGONOMICS: THE ROLE OF MANAGEMENT

Marilyn S. Joyce
The Joyce Institute:
A Professional Training Corporation
1313 Plaza 600 Building
Seattle, Washington 98101
U.S.A

Although human factors and ergonomic considerations continue to achieve ever-increasing importance in office environmental design, job design and productivity, management, except in a few leading edge corporations remain unconvinced or uninformed of the critical nature of these considerations. This paper reports on a study done in the Northwest to determine awareness levels of management. The focus is on examining those benefits of ergonomic implementation, specifically of ergonomics training, which will have value to management and heighten their awareness about the importance ergonomics.

INTRODUCTION

Over the past 20 years, implementation of automation has been delegated by top management to the manufacturers of the technology, to data processing managers in their own companies, to mid-management who install computers to improve productivity. Consequently, implementation has been "technology driven." In fact, even now, many managers, HRD professionals and safety professionals are unaware of the fact that they could be applying ergonomic principles as an integral part of their jobs and consequently having a significant impact on the productivity and well-being of workers.

Our work with many major U.S., Canadian and Australian corporations, as well as our knowledge of the literature, has led us to some conclusions about what the role of management should be, what the obstacles to implementing human factors solutions are, and how we can, as human factors professionals, communicate the benefits of our work to management.

Therefore, the purpose of this paper is:

1) to examine management's awareness of ergonomic issues;
2) to suggest the benefits of ergonomic application to management.

MANAGEMENT'S AWARENESS OF ERGONOMICS

Obstacles to Implementation

The most basic obstacle to the widespread implementation of ergonomics in office environments is that management does not understand the significance of the issues and does not realize the benefits to be gained. Recent studies have demonstrated the lack of awareness or commitment to the human issues.

Dr. Alan Westin, after his extensive study which included visits to 110 organizational sites using VDTs and interviews with 1,100 employees, plus 650 supervisors, managers and executives, concluded that: "Few organizations have developed centralized, master OA plans to bring together all the players and provide a framework for integrating both the technical and human dimensions of office automation" (Westin, 1985).

In a 1984 survey of management in over 100 midwest firms, of whom approximately two-thirds had direct responsibility for purchasing, over 50 percent reported they depended primarily on "magazines, trade publications or vendors and manufacturers for ergonomic information...about two-thirds reported they had difficulty in obtaining needed information" (Sauter, 1984).

Other obstacles to the implementation of ergonomics center on organizational inadequacies, namely, a lack of communication among the various departments within organizations. Often, medical directors, human resource development (HRD) managers and information-system managers never communicate with each other nor with purchasing agents. In addition, the often confrontational nature of labor relations leads to the perception of management that ergonomics is a labor demand and consequently impedes implementation.

Northwest Study

In order to determine the awareness levels of management regarding ergonomics from a training and consulting perspective, The Joyce Institute: A Professional Training Corporation, had a study conducted by an independent research firm, Hebert Research, Inc. The elements of the 1985 study that relate to this paper are summarized below.

Methodology

A total of 100 businesses in the Northwest were interviewed by trained Hebert Research data collection staff. The businesses were selected through probability sampling methods by Standard Industrial Classifications.

Industries examined included the following:

Industry	Percent of Total Sample
Insurance	32.0%
Financial Institutions	18.0%
Manufacturing	18.0%
Legal Services	10.0%
Medical Services	6.0%
Government	6.0%
Oil/Utilities	6.0%
Communications	4.0%

The data was analyzed using generally accepted univariate measures of central tendency and dispersion. The multivariate analyses involved Chi square, analysis of variance, and other appropriate tests of independence and association. Multivariate analysis was conducted on three specific variables. The variables relevant to this report are: respondents' job titles, and the likelihood to use ergonomics awareness training. The breakdowns are as follows:

Respondents' Job Titles Classifications	% of Total Sample
President/Vice President Category	28%
General Managers/Department Heads	34%
Data Processing and Personnel Directors	38%

The statistical procedures utilized were applied with a .95 confidence level for estimating values and/or providing significant inferences. A .05 significance level was used as the criterion to test hypotheses. The statistical analysis included in this market research was conducted in May, 1985. Marketing variables may change over time as a result of competitive and management practices. Hebert Research recommends that the research analysis be regularly reviewed and updated as required by The Joyce Institute.

Analysis

A significant finding emerged as it affects human factors professionals: 58% of the sample reported that they were not familiar with the term "ergonomics", yet, budget approvals were dominated by President/Executive (19%), Partner/Owner (11%) and Vice President of Administration (14%). Therefore since 44% of all budget approvals are from top management, it is critical that they, as well as General Managers, Personnel Directors and others understand the issues. In fact, budgets for training and consulting were even proposed by top management in 29% of the sample.

COMMUNICATION OF BENEFITS TO EMPLOYERS

Describing Ergonomics to Management

As part of the Hebert Research study, the term "ergonomics" was described as:

> "the relationship between person and machine. In general, it means providing a good office environment, good job design and good training in response to automation in the work place."

When respondents were asked to evaluate the importance of ergonomics based on that definition, they responded positively. Evaluations were made using a 0 to 10 scale, with 10 meaning extremely important. The overall average [mean] rating was determined to be 7.14. An estimated 48.9% of the respondents gave a rating of 8, 9, or 10, reflecting very high importance evaluations, with approximately 47% giving a moderate importance evaluation of 4, 5, 6, or 7. The mode, or most frequently given evaluation was 8, with more than 20% giving that response.

Evaluations of importance or benefit were conducted, using a 0 to 10 scale on five executive management objectives for training and consulting. The following table reports the average [mean] and the proportions reporting a high (8, 9, 10) importance or benefit evaluation.

Objectives	Average [mean]	% High Evaluation
Increased productivity	8.26 *	72.8%
Improved employee satisfaction	7.98	61.0%
Decreased work-related fatigue	7.48	52.0%
Familiarizing executive staff with a new field	6.24 **	31.9%

* A statistically significant difference (.0424) was determined to exist among three job title groups on the importance of increased productivity. The three groups are more thoroughly defined in the methodology. The average [mean] ratings are reported below.

Job Title Category	Average [mean]
Data Processing or Personnel Directors	8.68
General Managers	8.38
Presidents/Vice Presidents	7.57

** Statistically significant differences were found, with respondents who had given a high likelihood of using ergonomic awareness training giving higher ratings: 7.57 for familiarizing executives (.0155)

Benefits to Employers

Respondents in the study were asked to evaluate their likelihood of using training or consulting services that emphasized certain topics. The highest single category was "Productivity Improvement through Ergonomics" (30%), followed by "Legal Responsibilities and Liabilities (29%). "Redefining Personnel/Organizational Needs" and "Safety in the Workplace" rated a 19%.

Training Results

Training has been demonstrated to have some measurable results, particularly in terms of productivity, personnel needs, and safety.

- Based on client measures, we have received reports of various productivity increases. (See Figure 1.)

- Training can enhance employees' perception of the employer and therefore improve morale. A Boeing Aerospace manager said: "There was less fatigue involved. We had fewer complaints after the training of headaches and problems with the use of the computer terminal...I guess one of the side benefits that we hadn't really anticipated was that some of the people indicated, 'Gee, management's really concerned about us.' they appreciate it. It was something we weren't expecting".

Figure 1
On-The-Job Productivity

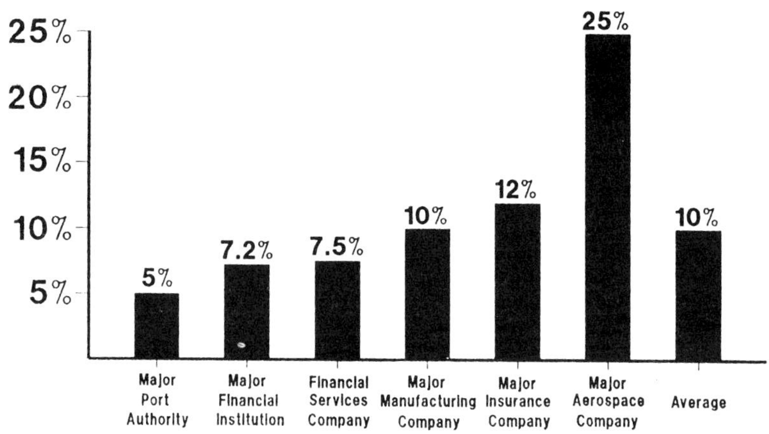

- Training is a preventative measure -- it can help participants counteract occupationally induced stress symptoms. Our results indicate that training, in fact, does help workers feel better and work better. Our survey results corraborate the studies done by researchers and indicate positive results from training. Survey results from pre- and post-training of 480 participants yield the following results (See Figure 2).

Figure 2
Effects of VDT Work

PRE		POST
81%	Back Pain	70%
88%	Painful Neck/Shoulders	80%
53%	Painful Arms/Legs	52%
86%	Irritated Eyes	85%
62%	Blurred Vision	60%
88%	Eyestrain	87%

In addition, we have conducted Management Surveys after implementation and have found that 86% reported that ergonomic training "helped them deal effectively with worker concerns about automation".

CONCLUSION

Role of Management

What then, is the role of management? We think it includes the following:

- to understand technology from the point of view of its impact on workers; that is, to recognize that the new technology is creating not just new ways of doing old jobs, but new jobs -- and a new culture;

- to see the "big picture" by recognizing that the technology is only a small part of the office automation issue. Ergonomics, with all its aspects must be central to all decisions. Not just data processing or facilities people, but line managers, HRD professionals and safety professionals should be included in the decision process so that hardware, software, furniture and training are considered integral and equally important parts of the whole. Training must include everyone -- managers, supervisors, programmers and end-users at all levels in an organization, especially people performing high production tasks such as word processing or customer service;

- to recognize that the health complaints (eyestrain, fatigue, etc.) are real, whether physiological or psychological, and that they will not go away by being ignored. Employee awareness has been heightened by their own experiences with discomfort on the job and reinforced by reports in the press and crusades by worker organizations.

However, there is still a long way to go. Once respondents understood the value, 64% gave a high or medium rating when asked "the likelihood of using ergonomic awareness training". Our response has been to design a Managing Office Ergonomics program to provide management with knowledge that will allow them to make informed purchase and organizational decisions and to promote communication among various departments within the organization whose decisions affect the productivity as well as the health and safety of employees.

* Datahealth is a federally registered service mark of The Joyce Institute: A Professional Training Corporation in Seattle, Washington.

REFERENCES:

[1] Bureau of National Affairs, "VDTs in the Workplace: A Study of the Effects on Employment," Washington, D.C. 1984.

[2] Galitz, Wilbert O., The Office Environment: Automation's Impact on Tomorrow's Workplace, (Administrative Management Society Foundation, Willow Grove, Pennsylvania, 1984).

[3] Hebert, James D., Hebert Research, Bellevue, Washington, (1985).

[4] Hendrick, Hal W., "Cognitive Complexity, Conceptual Systems and Organizational Design and Management: Review and Ergonomic Implications" in: Hendrick, H.W., and Brown, O. Jr. (eds.) Human Factors in Organizational Design and Management: Proceedings of First Symposium held in Honolulu, Hawaii, August 1984 (North-Holland, Amsterdam, 1984).

[5] Johansson, Gunn and Gunnar Aronsson, "Stress Reactions in Computerized Administrative Work," Journal of Occupational Behaviour, Vol. 5, (John Wiley & Sons, Ltd., New York, 1984).

[6] Joyce, Marilyn S., "Ergonomic Skills Training: Discussion of Methods Used To Improve Skill and Comfort Levels of Workers Using Video Display Terminals," in: Proceedings of the Human Factor's Society, 28th Annual Meeting, (San Antonio, 1984).

[7] Sauter, Steven L., L. J. Chapman and S. J. Knutson, "Improving VDT Work: Causes and Control of Health Concerns in VDT Use," (Dept. of Preventive Medicine, University of Wisconsin, 1984).

[8] Smith, Michael J., "Ergonomic Aspects of Health Programs in VDT Operators," in: Matthews, M. and Webb, R.D.G. (eds.), Proceedings of the 1984 International Conference on Occupational Ergonomics, (Toronto, May 1984).

[9] Westin, Alan F., Heather A. Schweder, Michael A. Baker and Sheila Lehman, The Changing Workplace: A Guide to Managing the People, Organizational and Regulatory Aspects of Office Technology, (Knowledge Industry Publications, Inc., White Plains, N.Y., 1985).

[10] "Why Business Must Face Up To The Screen Safety Issue," Management Technology, Vol. 2, No. 9, (January 1985).

Human Factors in Organizational Design and Management - II
O. Brown, Jr. and H.W. Hendrick (Editors)

OFFICE DESIGN AS DONE BY AN INTERIOR DESIGNER

Lucile B. Rowe, ASID and William E. Hartung, HFS

Rowe Design Associates
1710 Mitchell Road
Silver Spring, MD 20903-1230
USA

While there is only a tenuous connection between the fields of Interior Design and Human Factors, there is an overlap between these fields concerning the design of office work spaces. A brief summary of the elements and procedures used by Interior Designers is presented in the hope that this will expand the awareness of members of the Human Factors community, engender ideas for new avenues of research, and create new links between fields.

INTRODUCTION

The goal of the workshop on OFFICE DESIGN AS DONE BY AN INTERIOR DESIGNER is to expand the awareness of professionals in the field of Human Factors by exposing them to the processes used by a professional Interior Designer when designing offices. Professionals in the field of Human Factors are aware of many facets of the impact of environment on the person, but the designer considers other factors which, as yet, lie outside the field of Human Factors and which are important to creating an enjoyable work place. For example, throughout their work on the tasks of office design, designers attempt to establish a balance between aesthetics and functionality.

CLIENT OBJECTIVES AND LIMITATIONS

Client Objectives

Designers begin by helping their clients clarify their objectives; for example, image and status differentiation are objectives which are commonly held but not often articulated. Designers point out to the client that these are valid and important objectives. Worker comfort, accessibility, privacy, and productivity are objectives which are more clearly shared by Interior Designers and the Human Factors professionals.

Limitations

Both professions must deal with a variety of limitations and constraints in the course of a project. Common constraints include time, money, space, and organizational structure. Additional limitations on the designer's work include architectural features, existing furniture, essential equipment, and the client's personal aesthetic taste.

FACTORS AND TOOLS

Function

When Interior Designers use the term Function they mean the degree to which a thing does what it is designed to do. The scale of functionality varies from pure function (e.g., a wrench) to pure aesthetic (e.g., a painting). Interior Designers work to establish a balance betwen the functional and aesthetic qualitites of each element in the total office environment. A goal is to weight the balance in favor of function while appearing to favor aesthetics.

Research

Anthroprometrics, Productivity, and other research performed by Human Factors professionals are used by the Interior Designer in formulating the design of an office. In addition, the designer's own research gathers data on the client's goals, structure, personnel, equipment, etc. This client specific research is primarily assembled through surveys and interviews.

Space Planning

Space Planning is the first expression of the developing design concept. It provides the basic skeleton of the finished office and it is crucial to good function. In developing a space plan, beyond the usual limitation of the physical space, the Interior Designer must account for other factors such as status of each employee in the office, the communication links which exist or are needed, and privacy requirements to name a few.

Design Elements

The term Design Elements covers the more subjective qualities that the Designer uses in creating a design concept. Included under Design Elements are unity, color, line, volume, balance, rhythm, and texture.

Unity, which holds the design together, has a scale from boring to chaotic. While variety is needed for stimulation, human comfort requires a balance closer to boredom than to chaos. Achieving proper balance requires that special attention be given to the degree of unity in the design.

Color can be the most stimulating and satisfying element for individualizing and humanizing a work space. Neutral colors, such as grey and beige, provide unity. Stimulation and excitement is gained by introducing contrasting colors or colors of different values. Some colors have particular uses because they produce specific reactions in most people. For example, orange and red stimulate activity and might be used in a cafeteria to stimulate people to eat faster and move on. The image or personality of a business can be expressed in the choice of color scheme. A conservative and dignified image is perceived when almost any color is blended with grey to acheive a subtle hue. Pure primary colors against a white background produce a striking and dramatic effect which can be seen as young, bright, and informal. In a work area with high levels of light and glare, dark colored walls will soften the harshness of the environment and reduce glare. In short, used properly, color can work magic.

The controlled use of line, volume, and form is subtle and harder to recognize. A band of color in a carpet can be used to lead people from one space into another. The perceived shape of a wall can be changed by painting it with bands of color. Thus form is influenced by the use of lines. The volume or visual weight of any object varies directly with the degree of contrast it has in relation to its background.

Balance and rhythm are two more subtle elements used by Interior Designers. When all elements of a space are symetrical, it is called formal balance which conveys a sense of rest. Asymetry, on the other hand, conveys a sense of movement and flow. Repeated shapes (i.e., lines, volumes, colors, and forms) produce a visual rhythm which is sensed at a deep psychological level. A subtle form of rhythm is produced by the light and shadow contrasts which occur when a row of down lights are placed in the ceiling along a wall. Balance and rhythm influence whether a work space is perceived as organized or chaotic.

Texture is the most under-used tool for humanizing a workspace. A typical office has many hard surfaces and tends to feel cold, barren, and unfriendly. Fabric panels, textured wall coverings, natural wood, and fibre art are texture elements which can be added to humanize the space.

The elements discussed above are the tools of the Interior Designer which are least understood by the rest of the world. Therefore, when they are artfully used the results seem most magical.

Architectural Features

For purposes of this discussion, utilities, decorative detailing, partitions, acoustics, and cabinetwork are all considered under the heading of architectural features.

Utilities tend to be at the purely functional end of the spectrum of functionality, but their artful integration requires aesthetic manipulation. Computers in the modern electronic office too often generate an unsightly and unsafe confusion of wires; but, good planning can make them nearly invisible. Raised floors, flat cables and wire chases in the furniture are means for eliminating the mess of wires while maintaining flexibility.

Decorative detailing, such as columns and mouldings, helps to set the style and personalize the space. The choice of mouldings, for instance, can express the style and status of the company or even of an individual. The materials used in a space also help to deliniate the boundaries of areas which should be perceived as separate from the normal work areas.

Worker comfort is closely linked to the visual and acoustical privacy afforded by partitions. Moveable partitions are used to create individual work spaces in an open office environment. However, they should be chosen and used with care. The Buffalo Organization for Social Technological Innovation studied the effects of work environment on worker comfort [1]. They found that to satisfy a worker's need for privacy in an open office plan required three full walls extending above standing eye level.

Sound levels in office environments often reach levels at which most workers will experience discomfort and fatigue. Full height partitions and sound absorbing materials should be included in the basic plan for any large office space. Conversations nearby are a particularly distracting to most workers.

The proper design of built in cabinets for a work space achieves functional and aesthetic goals simultaneously. Aesthetically, cabinets aid in expressing the design concept while functionally they can make storage more efficient and accessible.

Some of the architectural elements discussed above are less visible than furnishings; and yet they absorb a significant portion of the budget and fulfill a significant part of the functional requirements. The average cost for interior construction and furnishings of an office is $54 per square foot. Of this total, $21. is for furnishings and $33. is for construction [2].

Furnishings
The term furnishings normally refers to those items which are moveable. Selection of the style of furniture used in an office is a basic expression of the design concept. Factors to be considered in this choice include the work being performed and special equipment needs. There is a plethora of furniture which is customized to accomodate computer equipment so that little compromise of style need be made. While ergonomic seating is also readily available, users often do not understand what seating position is correct for them. Further, when they have an ergonmically designed piece of furniture they usually don't know how to adjust it for their own best comfort. Choosing the style and quantity of cabinets for file storage is often neglected or badly underestimated. Storage needs, if not addressed, can take over the whole office like a creeping paralysis.

Lighting
Lighting is another area of the design concept where conflicts may develop between function and aesthetics. The most common way this conflict occurs is when a decision maker chooses the building standard "two by four" fluorescent ceiling fixtures, usually for reasons of cost. The result is a harshly lit area with glare and reflection which cause fatigue and strain. This outcome is unsatisfactory both functionally and aesthetically. While there are no perfect solutions to most lighting problems, careful planning can avoid such outcomes. By using a variety of fixtures and by varying the light levels according to work needs, glare and fatigue can be reduced and a more interesting and pleasing environment can be created.

Wall coverings
Plain painted walls are both barren and perishable. Textured wall coverings, on the other hand, help to humanize the space, are less perishable, and may have valuable acoustical properties. Wall coverings are second only to floor coverings in the square footage they occupy in a visual scene. Because they occupy so much of the visual field, wall coverings must be chosen with care and attention to the overall design concept.

Floor coverings
Floor coverings occupy the largest portion of the visual field, are subject to the greatest wear, yet they are often selected solely on the basis of the lowest price. The purchase price, in fact, should be only one factor in the choice; ease of maintenance, static electricity, durability, color, the need for texture, and fire/safety considerations are also important. Although its initial cost is higher than roll carpet, carpet tile may prove more cost effective in the long run, because the individual tiles can be

removed and replaced as necessary. With carpet tile it becomes easy and inexpensive to change the wiring or to replace stained areas. Hard rubber, vinyl, marble, and wood each have properties that indicate their use in specific circumstances. Ease of maintenance often dictates the final choice, because a stained or worn surface is neither attractive nor safe.

Window treatments
Most office workers place great importance on knowing what is happening outside and windows provide their only contact with the natural world. Window treatments, therefore, need to provide control of light and glare without closing out the view. This can be done with blinds, semi-opaque draperies, bronze coatings, or vertical blinds. Additional possibilities include lattice work and sliding panels if there is a generous budget.

Accessories
A design project remains incomplete until the accessories have been installed. They express the image desired and personalize individual spaces. Their drama and visual focus provide the essential stimulation that prevents a unified scheme from becoming a boring scheme. Signage supplies the necessary cues to make workers and guests feel secure and comfortable.

Summary
The tools and factors discussed above are used to develop a design concept. Surveys and interviews together with research reports provide the data. Space planning is the skeleton around which the design takes place. Design elements such as color, texture, and rhythm are humanizers, and unity holds it together. Architectural details develop the style; furnishings make it work; lighting makes it visible; wallcoverings provide texture and reduce maintenance; floorcoverings support the system; windows admit the world; accessories wrap it up. A completed design concept is a whole far greater than the sum of its parts. It is through its parts, however, that the whole is expressed and is experienced.

DESIGN PROCESS

The process described here is the one used by one of the authors, and is individual. Essential to this process is an ability to accept inputs from a client on at least two levels; the level at which they can verbalize their needs, and the level at which they reveal their desires by their non-verbal reactions to individual elements and to whole situations. This process also requires the ability to mentally image, in color, the space being designed and to fit elements into this mental image.

Starting the process
The design process starts with input of the client's needs and the limitations and constraints of the job. These data are sorted and examined for internal incompatabilities, such as unrealistic space requirements or budget. Sketching out a preliminary space plan is next. During this early phase of development the designer is beginning to form a mental image of the finished space. The image at this point is still indefinite yet it has sufficient substance to serve as a model against which potential elements can be tested.

Fleshing out the design
When a reasonably solid space plan is developed, the designer proceeds with the selection of elements. When selecting each element, the designer constantly is checking how it fits into the original mental image and is comparing this with the ongoing inputs from the client. As each new piece is fitted into place the image as a whole may undergo a change which necessitates a rethinking of the earlier selections. Thus the development of the design concept is both additive and evolutionary. In most cases there are many alternatives for each element and the selection of the "right" item is subject to availability and technical limits. Also, the designer wishes to retain some alternatives for each selection in case they are needed later.

Continued evolution of the design
During the design process and even after the implementation has begun, the client will have changes in needs and desires. These changes must be evaluated for long term implications; it is not unusual for a client to overrule the space plan and try another arrangement, only to return to the plan after a brief experiment. Ideally, a space plan should allow for a certain amount of flexibility so that such experimenting can be carried out without incurring additional cost in furnishings or construction.

SUMMARY

There is no final conclusion to any design process, for it is always changed as new needs arise. The goal itself is evolutionary in that its realization lies in an enriched environment that continues to work well throughout the inevitable changes.

A basic frustration for design professionals is the lack of feedback from the occupants of a completed project. Funds are rarely provided for studies after occupancy; yet without this data designers are limited to whatever anecdotal comments they can collect. Designers are sometimes left with a nagging suspicion that some of their design decisions may produce outcomes which are contrary to their objectives and the best interests of their clients.

It is the authors hope that this exposure to the tools and methods of Interior Design will stimulate the interest of Human Factors professionals. We further hope that this stimulation will be sufficient to engender some new directions for research as well as better communications between these fields.

REFERENCES

[1] Buffalo Organization for Social Technological Innovation, Report: The Impact of Office Environment on Productivity and Quality of Working Life, (Westinghouse), 1984.

[2] Loebelson, A., Giants 2, The Second 100 Interior Design Giants, Interior Design, August 1984.

SYSTEM SIMULATION, MODELING AND ANALYSIS

Human Factors in Organizational Design and Management - II
O. Brown, Jr. and H.W. Hendrick (Editors)

Incorporating a Systems Analysis Model in Addressing the Issues of Office Automation

Michelle M. Robertson
Hal W. Hendrick
Steve Rosenberg

Institute of Safety and Systems Management
University of Southern California
Los Angeles, California
U.S.A.

Utilizing data gathered from several case studies, the utility of a seven step systems analysis model is demonstrated. The problem addressed was that of a decrease in productivity caused by occupational stress from video display terminals. This applciation also illustrates the benefit of integrating ergonomic and organizational approaches to organization problems.

INTRODUCTION

The introduction of new technology often creates a new set of problems for management. Failure to adequately address these problems may result in a failure to meet the initial set of goals. In some cases, both productivity and quality of work life may be even less then before the new technology was introduced. This was demonstrated by the classic Tavistock studies of the introduction of the long wall coal mining method in Great Britian (De Greene, 1984).

As the trend towards office automation continues, the comfort, productivity, and safety of the worker represents a set of problems of significant importance to organizations. As a result of initial office automation experiences, organizations are beginnning to recognize that successful introduction of automation requires an optimum interface between the worker and the new technology. The dramatic increase in the computerization of office work has been universally experienced as presenting a wide spectrum of worker/automation interaction concerns (e.g., physiological, behavioral, cognitive, comfort and normative adaptability of the worker). Some of these problems manifest themselves in the form of decreased productivity, occupational stress, and a decreased quality of work life. Since many of these office automation organizational problems are extremely complex and multi-causal, many of the underlying factors may not be identified and assessed. Consequently, only the more obvious symptoms of the problem will be analyzed and thus result in a superficial, narrow and non-effective solution (Robertson & Hendrick, 1984).

In order to assure a more thorough inter-disciplinary and comprehensive approach to organizational problems, such as those described above, addressing the impact of automated office systems on workers requries an integrated approach that incorporates the methodological framework of a systems analysis.

The purpose of this paper is to demonstrate a generalized seven step systems analysis model (Mosard, 1982). Representational case studies and other relvant data sources pertaining to office automation are utilized to illustrate the model.

THE SEVEN STEPS OF THE SYSTEMS ANALYSIS MODEL

The proposed system analysis model consists of seven steps. These are 1. defining the problem; 2. setting the objectives and developing an evaluation criteria table; 3. developing alternatives; 4. modeling alternatives; 5. evaluating alternatives; 6. selecting an alternative; and 7. planning for implementation. This model is a modification of one proposed by Mosard (1982) based on earlier work in systems engineering by A.D. Hall (1969). Each of these steps is represented by a graphical model or appropriate table. This model is viewed in context of dealing with improving the workers' relationship within contemporary office environments.

Defining the Problem

Step I in the systems analysis approach is defining the problem. With large-scale problems and issues, a fairly high level of complexity is generally inherent which in turn necessitates a reasonably formal, systematic analytical approach to the problem definition (Mosard, 1892). Warfield (1972) proposed an analytical tool that has proven useful in defining large, complex problems, called a "problem factor tree." Developing the problem factor tree is accomplished by identifying the major problem, sub-problems, and causal factors including their interrelationships. The major problem would be sectionalized by identifying problem elements, including needs, constraints and the perscns/groups involved.

In order to produce a graphical "model", these problems and problem factors should be precisely stated and carefully linked together by an iterative process (Mosard, 1982). The lower level problem factors in the model contribute to the major problem. Feedback loops also may be shown. The problem factor tree then depicts a hierarchal, logical structure of the encompassing problem elements.

Figure 1, depicts the defined problem involving a increase in absenteeism and turnover, and a decrease in productivity and performance of clerical workers caused by occupational stress from Video Display Terminals (VDT's). Two sub-problems are defined along with many causal problem factors.

Setting the Objectives and Developing Evaluation Criteria

The second step of the systems analysis methodology is setting the objectives and developing evaluation criteria for selecting the best alternative for achieving the objectives. An objectives tree is a hierarchal, graphical structure of objectives that addresses the problems that have been identified and structurally analyzed (Mosard, 1983). This objectives tree can be created by first identifying the major needs, goals, objectives and sub-objectives. The objective tree for the proposed problem is depicted in the upper half of Figure 2. The logic of the objectives tree is that the lower level objectives contribute to the

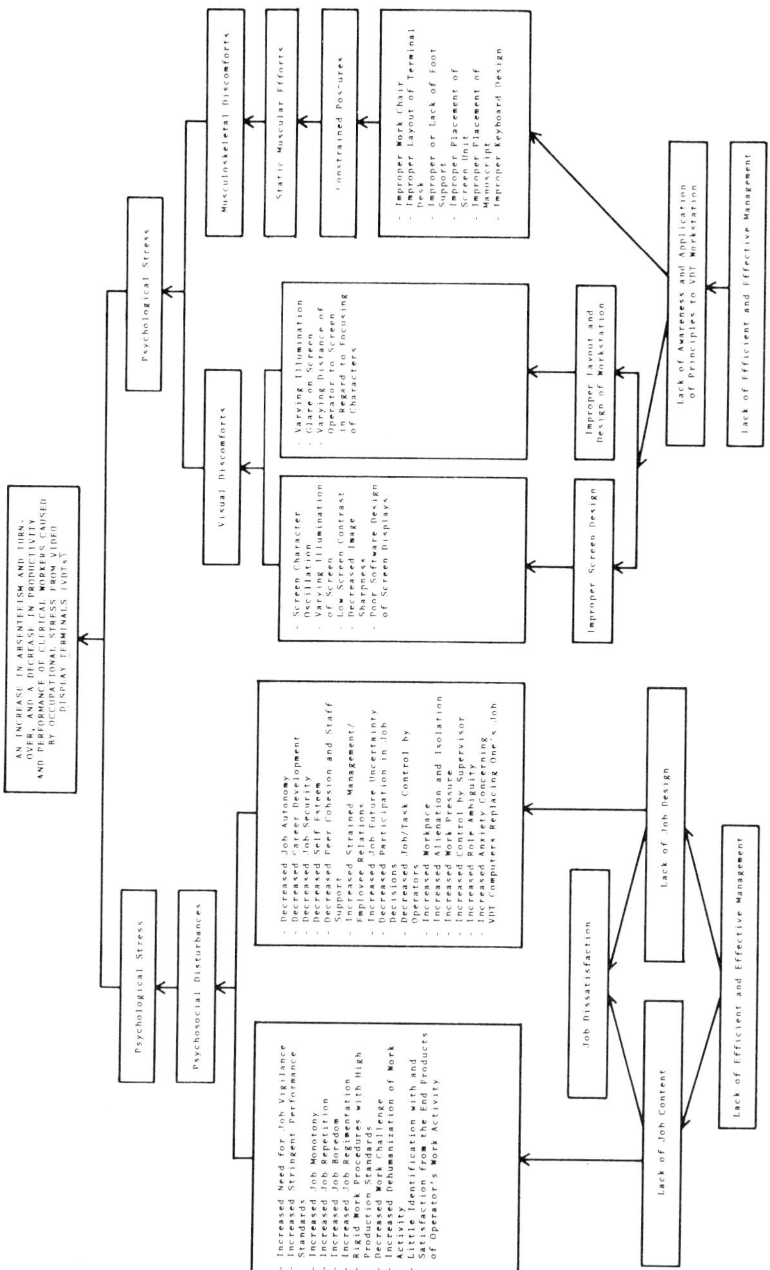

PROBLEM FACTORS TREE

FIGURE 1

attainment of the middle level objectives which contribute to the upper level needs or goals (Warfield, 1972). Analyzing the degree of interaction between objectives, constraints, and the persons/groups that are involved in the process, facilitates the development of the Objectives Tree. (Mosard, 1982). After the objectives are set, and precisely defined in terms of specific objective measures, a preliminary criteria table may be developed. This criteria table will be utilized to evaluate alternative ways of accomplishing the objectives. Decision criteria usually will include various risks, costs, benefits, and measures of effectiveness (Mosard, 1983).

Developing Alternatives

Step III of the systems analysis approach requires identifying alternative ways of attaining the objectives and sub-objectives. An alternative can be defined as a specified set of activities or tasks designed to accomplish an objective (Mosard, 1983). Quade (1975) recommends that a wide range of alternatives be developed by first identifying major classes of alternatives and then developing distinct alternatives within each class. In figure 2, these alternative sets of activities are portrayed below the Objectives Tree. Subsequently, this creates an Objectives/Activities Tree. Hybrid alternatives also may be created which incorporate the best features of several of the initially identified alternatives. The Objectives/Activities Tree in Figure 2 depicts the major goal of decreasing absenteeism, and turnover, and increasing productivity and performance by alleviating the occupational stress from VDT's. For each alternative the activities also are exhibited. The four hybrid alternatives (D-F), represent only four of the many possible combinations of common activities that could be specified.

The preliminary decision criteria table is presented in figure 3. This will be utilized to evaluate alternatives later in this study. The criteria are based on a long term perspective. The cost criteria are based on both short term and long term perspectives.

Modeling the Alternatives

The fourth step of the systems analysis approach is modeling the alternatives. Modeling, in this context, is defined by Mosard (1982) as the development of either a descriptive or predictive model representing each alternative set of activities (or representing the entire system) in such a way as to allow alternative configurations of the system to be analyzed. The system element interrelationship and/or gross resoure requirments should be depicted in order to aid in determing the effectiveness and other consequences of each alternative set of activities (Mosard, 1983). There are many possibilities of modeling techniques that may be utilized, such as flow charts, simulation and systems dynamics modeling.

The model utilized in this illustration was that of an input-output flow diagram. This is a simplified modeling technique that allows alternative configurations of the system to be analyzed. In utilizing the input-output flow diagram, the inputs consist of people, resources and information, and the outputs are the results and products. These outputs can, in themselves, become the sources of inputs to other sub-systems, and thus extend the diagram to fully represent the entire system being analyzed.

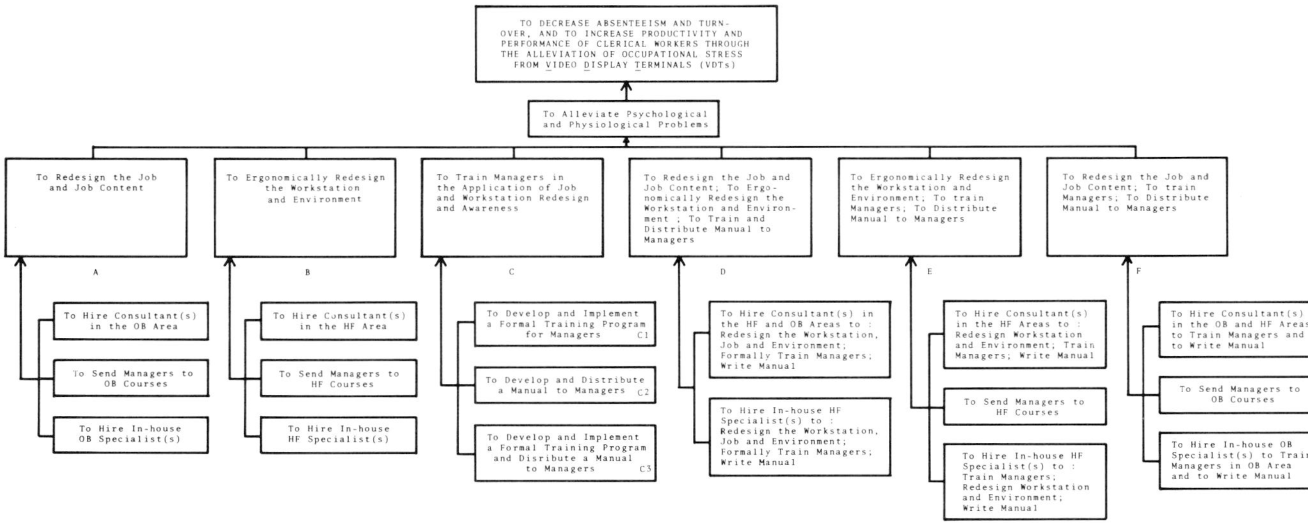

OBJECTIVES/ACTIVITIES TREE

FIGURE 2

SCOPE	RISK OF FAILURE	COSTS (short/long term)	BENEFITS / EFFECTIVENESS	
1. Will Help The Entire Organization Internally/ Externally?	1. Employees Resistance To Change	1. Materials	1. Increase in Morale	5. Decrease in Turnover
2. Long Term Program Effectiveness	2. Employees Learning New Skills	2. Resources	2. Decrease in Absenteeism	6. Decrease in Job Stress
	3. Management Acceptance	3. Loss Of Productivity	3. Decrease in Workman's Compensation Cases	7. Decrease in Number of Litigation Cases
		4. Training Programs	4. Increase in Product Quality	8. Decrease in Insurance Premiums
		5. Decrease in Product Quality		9. Increase in Productivity

Preliminary Decision Criteria

(Long Term)

Figure 3

Figure 4 exhibits only one alternative, D, in the simplified model of the input-output flow diagaram. There are two phases of this model that are depicted, the redesign phase and the operations phase.

Evaluating Alternatives

Step V of the systems analysis methodology is evaluating alternatives. This is accomplished by evaluating and comparing each alternative set of activities utilizing the major decision criteria. These criteria include effectiveness, risk, cost, and benefit for all appropriate future conditions. An evaluation criteria scorecard is developed for use in evaluating and comparing alternatives.

Figure 5 depicts an evaluation criteria scorecard for the identified problem and its respective six alternatives. For alternative C, each activity is separately evaluated (C1, C2, & C3). The scorecard is based on a long term perspective and incorporates the preliminary decision criteria table (figure 3). Since this problem is not based on a specific case study, quantitative figures were developed only for the cost and benefit criteria. The basic cost figures were derived from studies done by Danoiff (1982), Bailey (1984), 9 to 5 (1984), and from the experience of one of the authors in conducting contractural training programs and performing job re-design. Other costs items that were not included in the figures were the indirect costs of poor performance and product quality. In order for these to be accurately assessed, they must be defined within each specific organization where an established baseline can be made. For the benefit figures, percentages of the proposed increase in productivity were developed (Dainoff, 1982; Bailey, 1984). Additional benefits also can be included, which are in many cases more substantial than just productivity percentages. These include the number of workman's compensation and ligitation cases, insurance premiums, and the dissemenation of accurate knowledge concerning VDT use throughout the organization. For the other two criteria, (effectiveness and risk of failure) qualitative rankings were developed using a five point scale. In actually applying this systems

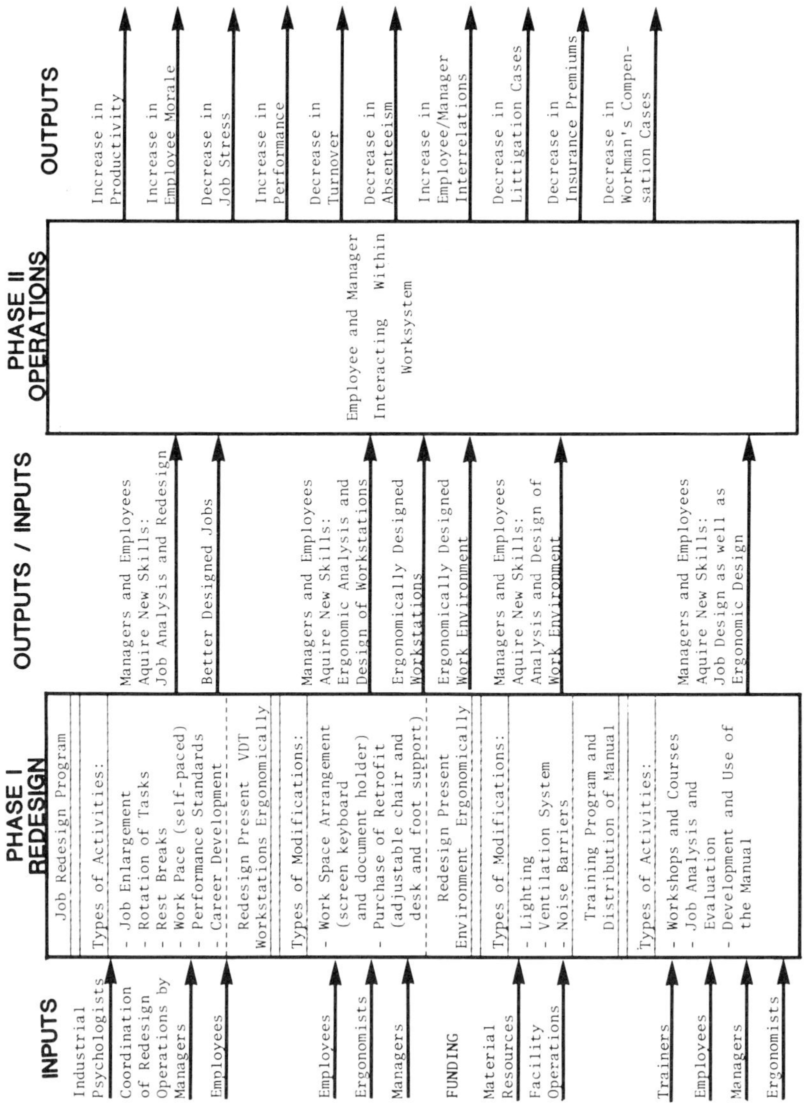

INPUT /OUTPUT MODEL ALTERNATIVE 'D'

FIGURE 4

analysis approach within the context of a specified, well-defined organization and its encompassing environment, the individual "score" for each criteria for each alternative could be quantitative and/or qualitative in nature. These scores would be derived by considerable data gathering, analysis, and interpretation by resource teams possibly consisting of Ergonomists, Organizational Psychologists, Systems Analysts, Human Resources Personnel, Employees and Managers.

ALTERNATIVES	EFFECTIVENESS	RISK OF FAILURE	COSTS	BENEFITS
ALTERNATIVE A: Redesign Job/ Job Content	moderate	low	moderate ($125,000)	moderate (10%)
ALTERNATIVE B: Ergonomically Redesign Workstation and Environment	moderate	low-moderate	moderate ($145,000)	moderate (10%)
ALTERNATIVE C: Train Managers in Job & Workstation Redesign & Awareness	C1 low-moderate C2 low-moderate C3 moderate	C1 moderate-high C2 moderate-high C3 low-moderate	C1 low ($40,000) C2 low ($20,000) C3 low-moderate ($60,000)	C1 low-moderate (7%) C2 low (5%) C3 moderate (10%)
ALTERNATIVE D: Redesign Job/ Job Content/ Ergonomically Redesign Workstation & Environment/ Train & Distribute Manual	high	low	high ($330,000)	high (26%)
ALTERNATIVE E: Ergonomically Redesign Workstation & Environment/ Train Managers & Distribute Manual	moderate	low	moderate-high ($205,000)	moderate-high (15%)
ALTERNATIVE F: Redesign Job/ Job Content/ Train & Distribute Manual	moderate	moderate-low	moderate ($185,000)	moderate (15%)

EVALUATION CRITERIA "SCORECARD"
Rating Scale: Very Low, Acceptable, Moderate, High, Very High
(costs based on 100 employees)
(% of productivity increase)
FIGURE 5

Selecting an Alternative

This process involves establishing the importance or weight of each evaluation critieria. This may be accomplished by either objective or subjective measures. A decision table is utilized in combining and structuring of the evaluation information. The final step is the selection of the desired alternative set of activities. Figure 6 exhibits a decision table for the six alternatives of the identified problem. The future conditions were based on the assumption that the proposed project has not yet been funded. Therefore, three levels of potential funding have been identified. The six alternatives thus can be evaluated in terms of each level of funding. For illustrative purposes, an arbitrary probability for each future condition was assigned. In this example, the alternative set of activities that should be selected, and that was actually selected by one organization (Hall, 1985) is alternative D. This large multinational corporation has experienced a large number of Carpal Tunnel Syndrome (CTS) cases arising from the use of VDT's. Because of such a high number of CTS cases reported in such a short period, the company's insurance rate increased by 300%, they had a half million dollars in ligitation and workman compensation cases, and hysteria had spread throughout the organization. The organziation immediately responded with the set of activities that are identified in alternative D. One year later, the organization had experienced the benefits listed in figure 3. (Hall, 1985). Thus, it is judged and has been demonstrated that alternative D is an excellant alternative in terms of the combination of effectiveness, low risk, and high benefit for moderate to high costs.

FUTURE CONDITIONS	HIGH LEVEL OF FUNDING	MODERATE LEVEL OF FUNDING	LOW LEVEL OF FUNDING
Probability of Future Condition	.25	.50	.25
ALTERNATIVE A: Redesign Job/ Job Content	1	2	3
ALTERNATIVE B: Ergonomically Redesign Workstation	2	2	3
ALTERNATIVE C: Train Managers in Job & Workstation Redesign & Awareness	1	1	3
ALTERNATIVE D: Redesign Job/ Job Content/ Ergonomically Redesign Workstation & Environment/ Train & Distribute Manual	2	3	1
ALTERNATIVE E: Ergonomically Redesign Workstation & Environment/ Train Managers	2	3	3
ALTERNATIVE F: Redesign Job/ Job Content/ Train & Distribute Manual	1	2	3

DECISION TABLE

Rating Scale: 1-relatively low probability of success
2-relatively moderate probability of success
3-relatively high probability of success

FIGURE 6

Planning of for Implementation

Step VII in this systems analysis methodology is planning for implementation. This is accomplished by developing network plans for initiating and implementing the selected alternative set of activities. Also, a schedule and a sequence of tasks, responsibilities and requirements should be developed for the implementation activities. This schedule might include a contingency plan with scheduled decision points and decision responsibilities. There are several scheduling techniques that are available for implementation planning, including PERT/CPM charts, the Gantt chart and the DELTA chart. These may be used either alone or in conjunction with one another. Figure 7 presents a DELTA chart for the implementation of alternative D. This chart provides an overview of plans for initiating and implementing the selected activities.

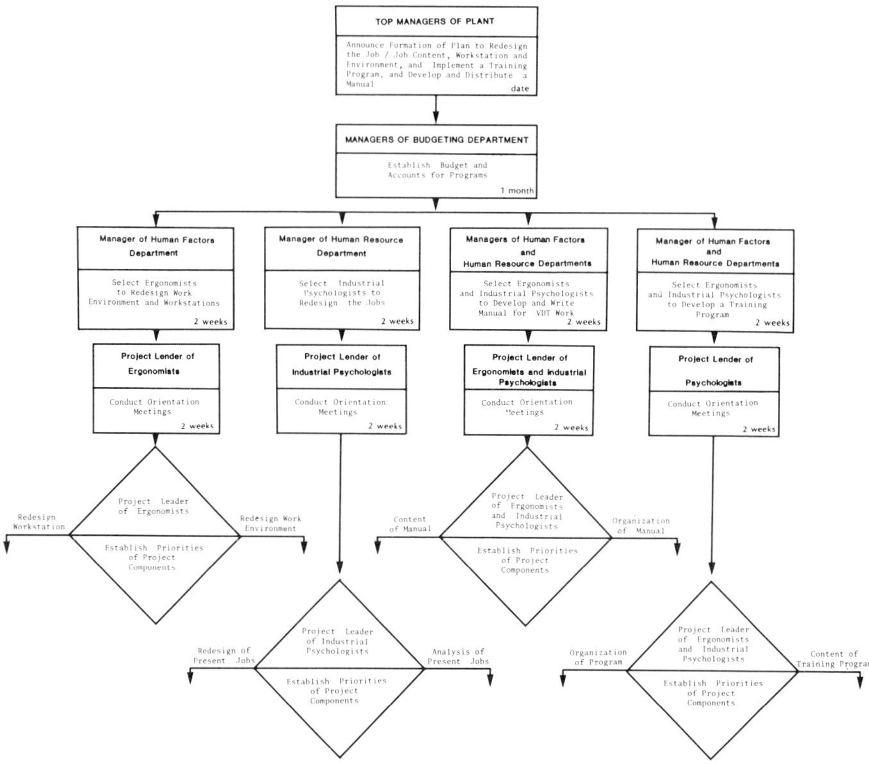

DELTA CHART
ALTERNATIVE "D"

FIGURE 7

CONCLUSION

It is often typical of organizations that address the issue of occupational stressors caused by VDT's to select a less optimal, atomisitic solution. For example, focusing only on physical and environmental conditions and neglecting overall job design. A benefit of the systems analysis model is the integration of ergonomic and organizational behavior approaches to organizational problems, at the macro as well as the micro systems level. Organizations are now at a point where they are willing to allocate various types of organizational resources in order to adequately address office automation problems. It is proposed that if organizations would work through their office automation problems via a systems analysis model, such as the one put forth in this study, more effective solutions will arise along with an increase in benefits. Given the increasing complexity and rapidly changing technology occurring in organizations today, a systems analysis approach is likely to make an even greater contribution to improving the worker/automation interface in the future than it has in the past.

REFERENCES

[1] Bailey, R.W., Is ergonomics worth the investment?, in: Ericsson Information Systems, (ed), Proceedings of the World Conference on Ergonomics in Computer Systems (Ericsson Communications, Garden Grove, 1984).

[2] Brill, M., Margulis, S., & Konar, E., Using office design to increase productivity. (Buffalo Organization for Social and Technological, Buffalo, 1984).

[3] Cakir, A., Hart, D.J., & Stewart, T.F., Visual display terminals: a manual covering ergonomics, workplace design, health and safety, task organization (John Wiley and Sons, Winchester, United Kingdom, 1980).

[4] Campaign for VDT safety, the economics of implementing VDT legislation, position paper, (National Association of Working Women, Washington, D.C., 1985).

[5] Dainoff, M.J., Fraser L., & Taylor, B.J., Visual, musculoskeletal, and performance differences between good and poor workplaces, in: Proceedings of the 26th Annual Human Factors Society, (Human Factors Society, Santa Monica, 1982).

[6] DeGreene, K., Sociotechnical systems, Factors in analysis design and management, (Prentice-Hall, Inc., Englewood Cliffs, New Jersey 1973).

[7] Hall, A.D., A three dimensional morphology of systems engineering,IEEE Transactions on Systems Science Cybernetics., SSC-5 (1969) 156-160.

[8] Hall, P., RSI - will court action be the next step?, RSI and Workplace Health, (W.D. Scott & Co. Pty. Ltd., North Sydney 1980).

[9] Helander, M.G., Billingsley, P.A., & Schurick, J.M., An evaluation of human factors research on visual display terminals in the workplace, in: Muckler, F.A., (ed), Human Factors Review (Human Factors Society, Santa Monica 1984).

[10] Louis Harris & Associates, The steelcase national study of office environments: do they work? No. II, Comfort and productivity in the office of the 80's, (1980).

[11] Mosard, G., A generalized framework and methodology for systems analysis, IEEE Transactions on Engineering Management, EM-29 (1982) 81-87.

[12] Mosard, G., Problem definition: tasks and techniques, Journal of Systems Management, 34 (1983) 16-21.

[13] Quade, E.S., Analysis for public decisions (American Elsevier, New York, 1975).

[14] Robertson, M.M., Hendrick, H.W., A systems analysis apporach to integrating OD and ergonomic solutions to organizaitonal problems, in: Hendrick, H.W., Brown, O., (eds) Human Factors in Organizaitonal Design and Management (North Holland, Amsterdam 1984) 33-43.

[15] Smith, M.J., Cohen, G.F., Stammerjohn, Jr., L.J., & Happ, A., An investigation of health complaints and job stress in video display operations, Human Factors Journal 23 (1981) 387-400.

[16] Springer, T.J., Visual display terminal workstations: a comparative evaluation of alternatives. Paper prepared for State Farm Mutual Auto Insurance Co, Bloomington, Illinois (1982).

[17] Stewart, T.F., Practical experiences in solving vdu ergonomic problems, in: Grandjean, E., & Vigliani (eds), Ergonomic Aspects of Visual Display Terminals (Taylor & Francis Ltd, London, 1980).

[18] Warfield, J.N., Toward interpretation of complex structural models, IEEE Transaction on Systems Management Cybernetics, SMC-2 (1972).

Human Factors in Organizational Design and Management - II
O. Brown, Jr. and H.W. Hendrick (Editors)

A MODEL FOR MANAGING EVOLUTIONARY SYSTEMS

Robert F. Bachert
Armstrong Aerospace Medical Research Laboratory
Wright Patterson Air Force Base, OH 45433

Glen R. Gallaway; Battelle Columbus Division
1637 Forestdale Avenue; Beavercreek, Ohio 45432

A business in today's climate of rapidly changing technology must face and deal with the challenge of continual organizational and managerial adaptation and evolution to achieve goals. The alternative is usually failure to progress, inability to compete, and the possibility of going out of business. It is easy to talk about dealing with change but difficult to accomplish. A practical method to describe and evaluate business goals and structure and deal with change is introduced in this paper.

INTRODUCTION

We often assume that a successful business has an extensive understanding of the details of its operation. In reality, this is not a valid assumption. Many companies are not aware (in detail) of the structure, effectiveness, and cost of their internal operation. They may even question why they should be concerned about knowing this type of detail since things appear to be going fine without that understanding. However, the business world is constantly evolving and changing. Technologies such as factory and office automation ensure that very little stays constant. What makes a business successful today may not be the same for tomorrow. So as the successful business's environment and market change that business must learn to change with it (and in a successful way). Once a business realizes that it is important to better understand how it operates and what its effectiveness is, the next step is to analyze its operation. This is done in order to build a database of information that can later be used to develop a picture (model) of how (and how well) business is functioning. Examples of information that is needed are presented below:

1. PRESENT BUSINESS OPERATION - A business must have an understanding of present business goals, operation, the management process, organizational structure, effectiveness and cost of each business function, and how all of these factors interact to produce the product or service.
2. ALTERNATIVE OPERATIONS INFORMATION - Data must also be gathered on alternative ways of doing business as a basis for evaluating the effects of changing structure. This knowledge must be constantly evaluated to determine if the data continues to represent a continually evolving environment.

3. TECHNOLOGY IMPACTS - Each of these individual functions (i. e. personnel resource management, engineering, finance, etc.) have their own technologies that can impact a business's profitability as much as that of a product and its product technology (e.g. use of computers to do accounting and engineering and personnel records, etc.).
4. ORGANIZATION/MANAGEMENT - The organizational structure and management style that are best for this business.
5. MARKETPLACE - Customers want to know how the product or service best satisfies their needs . A business needs to know how to interface with their customers and provide the product or service required.

For many businesses, the data collection and analysis step is not easy to accomplish. It is quickly apparent that many of the parameters are constantly changing (evolving) and that many deal with personnel issues. These are both factors with which businesses are often unwilling (or unable) to deal with effectively. Even if a business is able to collect the appropriate data they often don't know how to synthesize the data into a good functional model of the business.

The above statements seem to imply that the process to develop understanding and provide guidance to an organization's operation is only a job for the big business that has the resources to pursue the problem. Not so. We are suggesting that there are tools and techniques that can help most companies evaluate themselves and develop better ways to manage their future.

DEALING WITH CHANGE

We mentioned above that technology, market conditions, products requirements, and management and the organization needs are constant drivers of change. These types of changes separately and together indicate that evaluation, direction, and control of a business must be managed by a process that considers the normal change process. We will look at business operations in terms of an evolutionary process (accounting for change; dynamic modeling). The approach will be to integrate the business knowledge and the evolutionary process into terms of a comprehensive business systems model. This evolutionary systems model can then be used to look at and evaluate the current business's structure and operation and to evaluate the impacts that planned or unplanned changes could have on the business.

In the broad sense, a business must understand and selectively control all of the functions that management is responsible for. A business must also continually adapt the organization's structure to effectively perform all these functions that guide, direct, and control the business.

Computer technology will be used as an example of the change forces that are working on corporations. Computers and data processing are a pervasive technologies impacting most departments in a business. Robotics, management information systems, and office automation have an will continue to change operations procedures and organizational structures. Artificial

Intelligence is also a pervasive technology rapidly being incorporated into most enterprises and eventually will drastically change the way business is performed. These challenges must be addressed.

THE SYSTEMS CONCEPT

It might be argued that all systems are evolutionary in nature. Here, systems are defined to be one or more humans aided by tools or machines, working to solve a problem or achieve a goal. [These are Human Systems and the phase "Human Systems Science" can be used to describe the scientific study of such systems]. The study of systems and the system development methodologies yield approaches to accomplish effective and efficient evaluation.

Systems can be studied and analyzed by individuals outside (i.e. consultants, researchers) the system or those that are a part of the system being evaluated. This paper examines how systems can incorporate internal components (includes personnel) to evaluate and redesign the system as part of a continuing or evolutionary process. A generic model has been developed as a study and communication vehicle to accomplish these objectives.

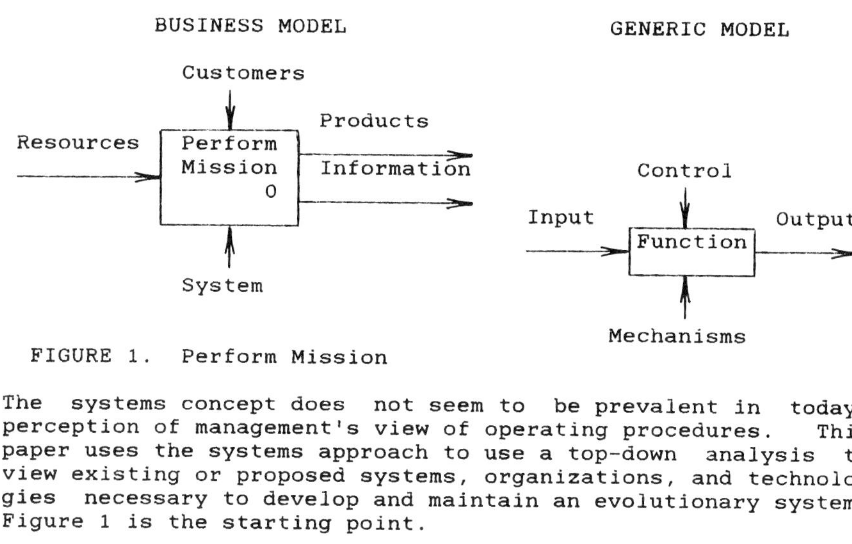

FIGURE 1. Perform Mission

The systems concept does not seem to be prevalent in todays perception of management's view of operating procedures. This paper uses the systems approach to use a top-down analysis to view existing or proposed systems, organizations, and technologies necessary to develop and maintain an evolutionary system. Figure 1 is the starting point.

The box in Figure 1 represents the function of a business or person(s), i.e., the function being performed by a system to produce an output using resources given as input and controlled by external influences. If this system is contained within a larger system, which is often the case, then other controls, such as corporate rules and regulations, would exist and should be indicated in the model.

A functional model is developed by decomposing the block shown in Figure 1. This decomposition is given in Figure 2.

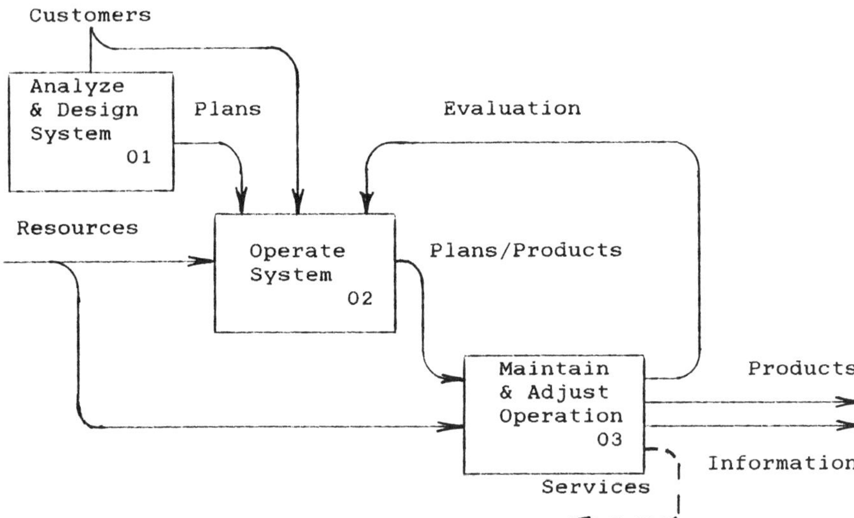

FIGURE 2. Perform Mission Diagram

The functions shown in the first level decomposition indicate a model from an evolutionary systems approach. These are the generic functions that are necessary to enable components of a system to design or redesign the system or subsystems. The term "components" is used here because it has not yet been established whether management or labor (or both) will perform the task. Note that the mechanism, i.e., system, shown as performing the function in Figure 1, has been omitted in Figure 2. Services, shown as a dashed arrow are system internal and will be expanded and discussed below.

An overview of the decomposition of each of the three functions in Figure 2 is given by the hierarchic function description and corresponding definitions to the functions given (box numbers) in Table 1.

Table 1. Function Description and Definitions of Figure 1.

#	Function Description
0	**Perform Mission**: This is the mission or activity (function) of the business or organization (system).
01	**Analyze and Design System**: The functions carried out to develop requirements and synthesize the new system.
11	**Perform Strategic Planning**: Development of long term goals, such as corporate planning, R&D, organization development, manpower planning, policy making
12	**Perform Tactical Planning**: Development of short term goals, usually one to five year plans. Daily decision making using Management Information Systems.

- 13 **Design and Integrate Operational System**: Design and evolve to new system structure and environment using requirements and plans developed. Industrial engineering, standards, etc.
- 02 **Operate System**: This includes all activities necessary to manufacture or produce to systems product other than services and management/administration, etc.
  - 21 **Manage Contracts and Budgets**: Finance, accounting, budgeting, and programming.
  - 22 **Manage Logistics**: Purchasing, inventory, equipment, and building maintenance.
  - 23 **Produce Products**: The problem solving, project, or product producing component.
- 03 **Maintain and Adjust System**: Activities necessary to keep the system operating in an efficient and stable or evolving environment.
  - 31 **Market and Transition Product**: The interface with the customer or user of the system.
  - 32 **Monitor, Review Evaluate**: Management functions for quality control of all system components including products.
  - 33 **Administer**: All administration not included in other functions, e.g. public relations, security, safety, etc.
  - 34 **Maintain, Improve Competency**: Personnel functions, training, resource evaluations, real property and environment monitoring and control.

The mechanisms or resources necessary to accomplish the functions include: 1) The organizational structure of people, and 2) The tools, equipment, and physical environment used. These resources are generated and maintained by the functions indicated in Table 1 and in shown in Figure 2 by the dashed arrow labeled services. In designing a system, the functional model is usually developed first and the mechanisms last. Human systems allocation decisions and technology issues and decisions are made during mechanism design and assignment. Training and physical environment follows.

The model presented is generic in nature and should incorporate all the tasks/activities done by various systems. The organizational structure varies and buildings may look different but the basic list of functions performed to produce products or solve problems is common to all business systems. The products range from goods to service, transportation, communication, agriculture, education, banking, or government but the system uses subsets of the generic list of functions to produce its product.

This model can best be used as a guide for planning, designing, monitoring and controlling the evolution of a total system. For example:

A business realizes that it must evolve to remain competitive. Using the model as a guide, present departments are shown in the models as the mechanisms necessary to accomplish the functions. Where mechanisms do not exist,they must be created, or where existing ones are inadequate, adjustments are made. All new functions are implemented. This represents the present system architecture and is the starting point for the transition to a new system architecture.

ANNOTATED BIBLIOGRAPHY

Harrington, J. J. Understanding the manufacturing process: Key to successful CAD/CAM implementation. (Manufacturing engineering and material).

This book uses the IDEF methodology developed by the Air Force to model and discuss the manufacturing process or functions of an enterprise. Starting with the function "conduct a manufacturing enterprise" and a brief discussion of corporate management the book concentrates on the "manufacture product" function. A comprehensive model is given. The perspective is product oriented and the life cycle of the enterprise as a system is absent, i.e., the viewpoint of an evolutionary enterprise is not given.

Evolutionary Information Systems, edited by John Hawgood (North-Holland, Amsterdam, 1982). Proceedings of the IFIP TC8 Working Conference on Evolutionary Information Systems, Budapest, Hungary, 1-3 September, 1981.

This book represents the results of a conference sponsored by the International Federation of Information Processing (IFIP) TC 8 working group. The purpose of the conference was "to explore the role of information systems in a changing world, including functional aspects such as environmental and human factors which impose requirements and constraints on development as well as structural aspects, such as how formal design techniques can make information systems adaptable to changing requirements." Many concepts, and techniques, and discussions presented should be amenable to systems development in general.

Integrated Computer-Aided Manufacturing (ICAM) Architecture Part III, Volume I - Architecture Part III Accomplishments, AFWAL-TR-82-4063, AFWAL/ML, Wright Patterson AFB, Ohio.

This report presents an overview of a project initiated to maintain and update existing manufacturing architecture and develop training courses to assist in the dissemination of IDEF applications, concepts, and procedures. Although the evolutionary perspective is absent, the methodology discussed is applicable to such an approach.

ERGONOMIC FACTORS IN HUMAN–COMPUTER INTERACTION

Human Factors in Organizational Design and Management - II
O. Brown, Jr. and H.W. Hendrick (Editors)

SADT ™: AN EFFECTIVE TOOL FOR KNOWLEDGE ACQUISITION

Ken Evers
Constance Hoyland

SofTech, Inc.

Robert F. Bachert

Armstrong Aerospace Medical Research Laboratory

The primary bottleneck in the process of building knowledge-based systems is usually the acquisition of the appropriate problem solving knowledge. Therefore, if an effective knowledge acquisition tool were used, the development cycle of an expert system could be shortened and the validity of the knowledge base could be improved. SADT is a technique with a proven background for extracting and integrating systems and problem solvlng knowedge from experts. SADT contains the features necessary to scope the expert system of interest, to collect the knowledge, and to validate that information without placing an undue burden on the experts while providing the information in a form translatable into the expert system computer program. This paper outlines the capabilities of SADT and the utilization of these capabilities to fulfill the requirements for knowledge acquisition.

INTRODUCTION

Expert systems are software systems that perform a function that normally takes humans a relatively long time to complete. Knowledge engineering is the name that has been given to the activity responsible for acquiring knowledge and building the expert systems. At least eleven generic categories of knowledge engineering applications and the problems they address can be examined [1]. The categories are interpretation, prediction, diagnosis, design, planning, monitoring, debugging, repair, instruction, and control. The important thing to note is that problems are being addressed, i.e., the expert systems to be built will be problem solvers.

™ SADT is a trademark of SofTech, Inc.

Even though expert systems are designed to replace humans, they are also used to augment human performance, either procedural and/or cognitive. New techniques are needed to identify functions in systems performed by humans that can be replaced by expert systems. "The design of management information technology for military systems should probably begin with a task analysis that subserves the requirements of the knowledge engineering porocess; i.e., a model of the human knowledges and performances to be augmented or replaced" [2].

Knowledge acquisition is the process of acquiring the expertise for problem solution or the process of modeling the task or function performed by the expert. The latter is an area of system development or systems analysis. SADT has been successfully used for problem solving, systems development modeling, and human-machine systems analysis [3]. It provides the methodologies for knowledge acquisition and the framework for knowledge engineering. This paper provides a short overview of SADT and discusses how it can be used as a tool to perform the knowledge engineering tasks.

SADT

SADT is best noted as a requirements engineering tool utilized for large-scale software development, where project management demands partitioning work into logical units that can be assigned to team members. SADT is used to obtain a well-documented system description which links the functions performed within a system together by means of the information transferred among the functions. By decomposing a system's functions into progressively finer detail, one ultimately arrives at a set of primitive actions. In this manner, SADT is an extremely powerful technique for working out a clear-cut understanding of an at-first obscure and nebulous subject, documenting the understanding, and then communicating that understanding to others. It is this concept within SADT which makes it an effective tool for knowledge acquisition. To accomplish its goal, SADT uses a graphic representation, consisting of boxes and arrows, to describe the system in a top-down manner such that it gradually exposes more and more detail about the system.

The process of developing an SADT system model is accomplished through an author-readership cycle and involves an individual knowledgeable in SADT and one or more individuals knowledgeable in the system. During the authorship portion of model construction, the author interviews the system experts in order to obtain pertinent system information and to transform that information into the SADT diagrams.

Once the system has been defined, or bounded, the knowledge engineer interviews the system experts to collect the pertinent information and to represent that information via the SADT model. The knowledge engineering process is accomplished in a top-down, controlled manner. At a general level, an expert is capable of expressing the primary stimuli and information that he uses and his obvious responses. As he decomposes his action into greater levels of detail, he may not be able to express, and possibly not realize, the information or guidelines he uses all of the time because they have become so commonplace to him that he is no longer aware of them. Through the SADT procedure, the knowledge engineer is able to identify and categorize not only what the expert says but is also able to identify holes and inconsistencies in the information so that additional searching can be done.

This evaluation of information is expecially critical if there exists multi-sources for the information, or pockets of knowledge. Under these conditions, SADT provides the knowledge engineer with the structure necessary to integrate the various information and check for consistency and completeness among expert sources.

In order to be the most effective, the interviewing process must not place any undo restrictions, guidance, or burden on the expert. The process must provide a comfortable environment for the expert as well as make effective use of time. The framework of SADT consists simply of boxes and arrows, the use of which is known by the knowledge engineer but does not have to be known by the expert. Therefore, the expert describes his processing activities in his own terminology and the knowledge engineer then builds the model using that terminology within the SADT framework rather than translating the expert's knowledge into a specialized language.

This ability to retain the terminology of the expert within the SADT model becomes important when verifying that what the author has respresented in the model was indeed what the expert said. This verification is accomplished through the readership process of the SADT methodology. During this process the expert reviews his information as it is presented in the model and informs the author of necessary corrections to the information. This information verification is efficient because the expert reviews information within his area of expertise and in the terminology familiar to him.

As the knowledge base advances, any manual record keeping becomes unwieldy and project management becomes difficult. The knowledge engineering process must be computerized so that the information can easily be input, accessed, modified, and displayed. As the size of the knowledge base increases, the approach must provide management insight into the total program, the current state of the program, and the magnitude of the work remaining. Also, as the size of the knowledge base increases, the more likely it is that multiple informational sources will exist. Therefore, the approach must be relied upon as a communication and documentation vehicle as well as an information collection vehicle. SADT has been implemented on a VAX 11/780 and on a Macintosh computer. Therefore, all collected information can be represented and modified through a computerized capability.

CONCLUSION

The knowledge engineering requirements stated thus far are by no means complete, but they provide a preliminary insight into what capabilities the knowledge engineering method must posess. SADT, the basis for the knowledge engineering tool, has been demonstrated to provide these capabilities plus many more.

SADT is a top-down, hierarchical functional decomposition approach for defining and understanding a system. By decomposing an expert's view of the system and by associating his knowledge of the system with these views, it is possible to gradually identify and collect the information required from the system.

The SADT notation is composed of boxes and arrows which graphically illustrate the functions the expert considers, the information that he considers for each function, and the informational relationships among the functions. The graphical notation provides a good communication tool, provides an exceptionally good managemant tool, and encourages a team effort for the project. The SADT methodology, therefore, provides an effective tool for knowledge engineering.

[1] Hayes-Roth, F., Waterman, D.A., Lavat, D.B., Building Expert Systems, (Addison-Wesley Publishing Co. Inc., 1983)

[2] Harris, S.D. and Owens, J.M., Some Critical Factors That Limit The Effectiveness of Machine Intelligence Technology in Military Systems Applications, Special Report 84-2, Naval Aerospace Medical Research Laboratory, Pensacola, Florida.

[3] Ross, Douglas T., SADT Today: A Retrospective on an Idea, SofTech, Inc., 1985 Special Issue of IEEE COMPUTER Magazine of Requirements Engineering.

[4] SofTech, Inc., ICAM Architecture, Part II, AFWAL-TR-81-4023, Volume I - Volume IV, Air Force Wright Aeronautical Laboratories, Wright-Patterson Air Force Base, Ohio, (1981).

Human Factors in Organizational Design and Management - II
O. Brown, Jr. and H.W. Hendrick (Editors)
Elsevier Science Publishers B.V. (North-Holland), 1986

EFFECTS OF AN AUTOMATED SURVEY SYSTEM UPON RESPONSES[1]

Linda M. Doherty
Navy Personnel Research and Development Center
San Diego, CA 92152-6800

Marie D. Thomas
College of Mount St. Vincent
Riverdale, NY 10471

Recent developments in computer capabilities and communications are enabling quality surveys to be designed, administered and analyzed efficiently and accurately. This paper describes (1) components of one automated survey system, (2) field results comparing computerized and paper and pencil surveys, and (3) future research issues on the effects of automated surveys upon individuals and responses.

INTRODUCTION

Survey research in the past has primarily used computer technology as a tool to statistically analyze questionnaire responses. While this is a necessary function, computers now facilitate surveys to be designed, administered, analyzed and reported efficiently and in a timely manner. Existing computer technology is sufficiently inexpensive so that most computers can communicate with remote terminals to collect and analyze atttitudinal information, and integrate that information with other data bases. Currently, the private sector uses computer capabilities and communications to conduct surveys quickly and efficiently on a broad range of topics that include elections, Nielson television ratings, and consumer product evaluation. The Census bureau is testing one such automated survey system, Computer Assisted Telephone Interviewing (CATI) system, where interviewers interact with a computer to make contact with respondents by telephone, and ask them branching, in-depth questions that are recorded immediately into the computer, updating existing files (Nicholls, 1983).

Besides the obvious speed of conducting surveys on computers, thereby eliminating the traditional paper and pencil instruments themselves, there are several unique aspects to conducting surveys using an automated system that involve and affect the respondent. Questionnaires may now be adapted to individuals, since the computer has the ability to easily and quickly skip blocks of questions and probe issues in depth through extensive branching of questions. Scales could be constructed that would be sensitive to capture differences in individuals' attitudes, and then using the flexibility of the computer, adapt questions and scales to more precisely measure individuals' attitudes. Second, using the computer will decrease errors at all phases of questionnaire design, administration, data collection, coding and analysis. Individuals at terminals will find it all but impossible to enter incorrect responses, or leave items blank. And third, an automated sysstem has the capability of collecting sensitive information from individuals with more confidentiality than traditional methods.

PURPOSE

Since computer administration of surveys is a relatively new phenomenon, little research has been conducted comparing this method with other data collection methods in terms of efficiency and quality of data. While the automation function is straightforward, the effect of the technology on survey responses is yet to be determined. This paper focuses on (1) briefly describing the components and capabilities of one automated survey system being developed for the Navy, (2) outlining some field study results comparing an automated survey with a traditional paper and pencil survey, and (3) delineating some future research issues that impact on individuals and their responses, that may now be easily addressed with automation.

AUTOMATED SURVEY SYSTEM COMPONENTS

The Navy Personnel Research and Development Center is developing an automated survey system to assess the attitudes of the Navy civilian workforce. CENSUS (Computerized Executive Networking Survey System) consists of using microcomputer technology, in particular IBM ATs as host computers. Software specifically designed to allow simultaneous access by several participants in the survey was developed for CENSUS. Surveys are presented to respondents in remote sites by means of Northern Telecom Displayphones that serve as both the telecommunications and terminal input devices. Data are transferred during the time of survey administration to the IBM AT using commercial phone lines. Results are obtained by applying computer programs that combine survey information with existing demographics, providing almost immediate feedback to policy makers about the attitudes of individual segments of the workforce.

The computer technology is being exploited to develop a complete stand-alone survey system capable of being used independently by managers and policy makers, enabling them to interact with menu-driven computer programs and access information much like a decision support system. While each of the survey components is being developed independently as part of a research effort, they will be integrated to form a complete system. The specific components consist of (1) automated survey administration and data collection presently developed as the prototype CENSUS system, (2) automated data analyses on longitudinal survey data bases that have been previously integrated with existing demographic data bases, (3) computerized sampling strategies to identify any segment of the workforce for querrying, (4) a computerized reporting system with management reports, tables and graphs automatically produced, and (5) a computerized authoring system that aids managers unsophisticated in survey design to develop high quality, timely, and useful surveys.

RESULTS OF FIELD TESTS

A prototype CENSUS survey system has been field tested on three separate occasions (twice in San Diego and once in the Washington, DC area) with a sample of 600 identified each time, located in 13 to 16 separate Navy installations. Each survey was conducted during business hours for one week, and focused on civilian personnel issues (e.g., training needs and the retirement system). Response rates were high, ranging from 60% to 85%, depending upon the level of commitment by management to participating in the the survey. Respondents' attitudes toward the survey and the computer were overwhelmingly positive. Virtually no one had difficulty using the terminal, obtaining a phone line, and most anticipated future computerized surveys. Almost no one refused to use their social security number as an identifier when told that their survey responses would be linked with information in their personnel data bases--even though they were told that participation was voluntary. In addition, participants during the second San Diego survey were asked if they wanted feedback on the results from the first survey. Over 90% chose to view bar graphs and textual material representing selected overall results from the previous survey.

Useful results were presented to policy makers within a week of survey completion. Results demonstrated the feasibility of conducting automated surveys using a microcomputer. In terms of the advantages of timeliness, administration costs, and logistics, automated surveys are an improvement over traditional mail-out paper and pencil surveys.

To test the effects of the automated survey system on individuals' responses, CENSUS was compared to a paper and pencil version of a survey consisting of 60 questions on attitudes toward proposed changes to the civil service retirement system. Previous evidence exists (McBrien, 1985) that respondents view computerized surveys more favorably, find the task easier, and more interesting than paper and pencil surveys. Also, Kiesler & Sproull (1985) and Erdman, Klein, & Geist (1983) found that when comparing electronic surveys to a paper mail surveys, there were fewer completion mistakes, fewer blank items, and fewer refusals to answer questions in the electronic survey. These issues were addressed in the comparison between the two versions of the retirement survey.

The paper and pencil survey on retirement was administered to 46 individuals at a Navy research laboratory located in Maryland during the last week in August, 1985, while an automated version was being administered to a sample of employees in the Washington, DC area in 16 activities (N = 307). Comparisons between a comparable sub-sample of the automated group (N = 27) and paper and pencil group were made to assess the error rate of the respondents on the paper version (e.g., multiple responses and inappropriate branching). All respondents were also encouraged to complete a written comment sheet regarding how they liked/disliked the questions, and for the automated group, how they liked/disliked the computer. Forty-four from the paper and pencil group and 18 from the comparable computer group responded.

Results indicated that 3% of the total responses were in error for the paper and pencil survey, with half of those errors representing incorrect branching, i.e., individuals who did not branch properly when instructed to do so. There were essentially no errors in recorded responses or hardware, software or communications malfunctions for the computerized survey group. Also tested was the reliability of items by repeating three questions at the beginning and end of the questionnaire. Both groups responded essentially the same on the repetition of these items, even though paper and pencil respondents were able to refer back to the initial items later in the survey, and would presumably respond more consistently than the computer respondents. Comments from the computerized survey were overwhelmingly positive toward taking a future survey by computer rather than by paper and pencil. Over 89% of the overall computer group would prefer an automated survey over a traditionally administered survey. The computer group also expressed many written comments regarding the usefulness, importance and belief that their opinions would be considered. One explanation is that people believed the survey was important and valid attitude assessment instrument because it was presented on the computer.

A major aspect of how the technology differentially affects survey responses was in the comments individuals provided as to what they liked about the computerized and paper and pencil versions of the survey. In written responses about the automated version, many individuals responded that they found the survey to be engaging and believed the results would be used by policy makers. Many respondents who answered that they had little knowledge of the retirement system expected the computer to present them with a tutorial concerning retirement benefits, while they paper and pencil respondents had no such expectation. This result highlights the potential of computerized surveys to be used as educational tools, and hence, influence future attitudes.

RESEARCH ISSUES

In general, the research issues focus on exploiting the technology to improve the quality and usefulness of surveys, while developing an understanding of the effects that automation has on individuals and their responses. Two important potential areas of research applying computerized surveys are as follows.

Computerized administration of surveys can make a substantial contribution in collecting responses to sensitive questions. Previously, the method of randomized responses (Warner, 1966; Campbell & Joiner, 1973) has been used to elicit answers to sensitive questions, particularly in situations where people have suspicions about their anonymity. This method requires the individual to respond to a randomly presented question that is either sensitive or not, without the experimenter's knowledge of the specific question being asked. By knowing probabilities associated with answers to known questions, an estimate of the number in the sample who agree with the sensitive questions may be calculated. Some evidence indicates that using computers may elicit fewer socially desirable responses (Kiesler & Sproull, 1985; Kiesler, et al., 1985), and presumably more honest responses. While the content of the retirement surveys reported did not directly ask sensitive questions, there is some evidence that the respondents were willing to disclose their identity, (using social security numbers), and did not feel that doing so changed their responses. Respondents may perceive that completing a survey on an automated system is more anonymous that other methods, and, hence, they may be more willing to disclose personal and sensitive information.

Some advantages of obtaining answers to sensitive questions by computer over the randomized response method are that (1) there are fewer ethical questions when asking individuals questions directly, rather than indirectly, (2) frequency of responses do not have to be estimated, and (3) the computerized system is far less cumbersome and time comsuming to use. This is a particularly potential important application area in military research, where selection of highly trusted personnel to work on classified projects is a top priority, and being able to have confidence in individuals' responses to sensitive issues is essential.

A second area important where the technology of survey systems may have an effect on responses of individuals is in feedback systems within organizations. Historically, surveys have been a device for tapping attitudes and opinions of employees in organizations. However, there was more concern for the measurement issues of the instrument itself than for the informational value or feedback aspects of the system. Currently, as popular management books advocate, there is an emphasis on employee involvement in decisison making at all levels in organizations. This could lead to a resurgence in the use of surveys, as managers need to better understand the opinions of the employees prior to making decisions that affect their working lives (Hinrichs, 1985). By integrating computerized survey technology into computerized management information systems, comprehensive feedback may be provided that would improve organizational communication, an essential ingredient in the tools of modern management.

Employee involvement in the CENSUS system is evidenced by the interest in receiving feedback and the anticipation for future surveys by the field test respondents. By automating reports of results, policy makers, managers and employees alike will be able to have meaningful and timely feedback on the attitudes of the workforce. Involvement in the organization by individuals at all levels should improve by incorporating the technology of presentation and communication methods. These feedback methods would include improved information displays, such as the enhanced use of graphics (Fienberg, 1979). In addition, a true feedback system will be established when managers become involved in developing questionnaires and accessing the data base, interactively through easy to use menu-driven computer programs. By evaluating the needs of managers, more effective feedback mechanisms will be developed to provide input to future surveys, and will be the final component in the automated survey system.

REFERENCES

(1) Campbell, K. & Joiner, B.L. How to get the answer without being sure you've asked the question. Amer. Statistician, 27(1973) 229-31.

(2) Erdman, H., Klein, M.H., & Greist, J.H. The reliability of a computer interview for drug use/abuse information. Behavior Research Methods & Instrumentation, 15(1983) 66-68.

(3) Fienberg, S.E. Graphical methods in statistics. Amer. Statistician, 33(1979) 165-178.

(4) Hinrichs, J.R. Employee involvement through survey feedback. The Industrial-Organizational Psychologist, 23(1985) 16-18.

(5) Kiesler, S. et al. Affect in computer-mediated communication: an experiment in synchronous terminal-to-terminal discussion. Human-Computer Interaction, 1(1985) 77-103.

(6) Kiesler, S. & Sproull, L.S. Response Effects in the Electronic Survey, 1985 (unpublished paper).

(7) McBrien, B.D. The Role of the Personal Computer in Data Collection, Data Analysis, and Data Presentation: a Case Study, 1985 (unpublished paper).

(8) Nicolls, W.L., II. CATI research and development at the Census Bureau. Sociological Methods & Research, 12 (1983) 191-197.

[1] The opinions expressed in this article are those of the authors, are not official, and do not necessarily reflect the views of the Navy Department.

Human Factors in Organizational Design and Management - II
O. Brown, Jr. and H.W. Hendrick (Editors)

A COMPUTER-AIDED EVALUATION OF FLIGHT TEST SUPPORT

Robert C. Schwalm
Michael Ellison

Fort Worth, Texas
U.S.A.

This paper discusses the development of an approach to simplify the process of organizing and addressing reported deficiencies and anomalies during aircraft development and flight testing. The rational taxonomy which forms the framework of this data base management concept is structured as a hierarchy designed to meet the end user's needs for sorting and accesing system reports. The requirements for automating this tool are also addressed.

INTRODUCTION

The purpose of this paper is to discuss an approach being developed which is intended to simplify and eventually automate the process of organizing and addressing reported deficiencies and anomalies during aircraft development and flight testing. This data base management concept, referred to as Computer-Aided Evaluation of Flight Test Support (CEFTS), is the result of two independent attempts to develop taxonomies which classify system deficiencies and anomalies. Each taxonomy had its own weaknesses, and the CEFTS concept which developed from those efforts is intended to rectify those weaknesses while maintaining the strengths of each.

The objectives of CEFTS are five-fold:

a) to develop a taxonomy for classification and sorting of reported system deficiencies and anomalies both for operators (in this case, pilots) and/or maintenance personnel

b) to develop an analysis procedure or decision tool to enter reports into the taxonomy

c) to develop a semi- or fully-automated system to facilitate entry of reports into the taxonomy

d) to develop a means for readily searching for particular types of information across levels of the data base

e) to develop a feedback system to identify relevant materials (e.g., military standards) for the user.

The result will be a computer-aided system which is capable of determining the nature of a reported deficiency or anomaly, where it belongs in an overall data base of reports, and how the user might address such a deficiency.

The advantages of such a system are numerous, and the savings in manpower and costs is significant. Advantages for data base management include the ability to sort and search large numbers of deficiency reports easily, the ability to organize both operator and maintainer reports either separately or together, an institutional memory of deficiencies and anomalies, and a "lessons learned" feedback and correction system. Personnel and costs savings include a continuity in data collection across personnel and time, consistency in interpretation of deficiencies, reductions in the number of on-site personnel, and more efficient use of on-site personnel.

Background

The concept being discussed represents largely the strengths of two previous attempts to develop taxonomies for classifying deficiencies and anomalies. The two efforts were initiated independently, specifically in recognition of the need to be able to organize the many watch items (WITs) and service reports (SRs) which are generated during the flight test, modification, and routine operation and maintenance of aircraft systems. Common to both efforts was the recognition that there must be some way of sorting and searching the large number of WITs and SRs which are generated by any system, that that organization should be based on the nature or source of the reported deficiency or anomaly, and that there should be some capability for tracking those reports over time. Implied by each taxonomy was the need to provide a source of information for how those deficiencies might be corrected once they have been identified. Each of the two efforts is described in greater detail in the paragraphs which follow.

The first approach attempted to classify operator/aircraft deficiencies in terms of the standards addressed by MIL-STD-1472C, "Human Engineering Design Criteria for Military Systems, Equipment, and Facilities." The result was a standards-based taxonomy with eight broad categories drawn directly from MIL-STD-1472C. Category headings included:

a) Control/Display Complexity

b) Task Complexity

c) Display Symbology

d) Error Control

e) Operator Feedback

f) Information Content

g) Head-Up Displays

h) Illumination

Based on the frequency of reported deficiencies in each category, solutions to the more common types of problems could then be recommended.

In the second approach, a problems-based taxonomy was generated as part of a reverse engineering or "lessons learned" analysis of an aircraft program. The intent of the effort was to classify reported operator/system deficiencies by nature and source so that the deficiencies could be tracked and used to suggest improvements for subsequent aircraft modification. (The logic of this classification process resides in the need to identify classes of behavior, as argued by Fleishman (1967) and as implemented by Fleishman and by others (e.g., Parker, 1967.) The three highest-level categories of this taxonomy were Controls, Displays, and Equipment. Second-, third-, and fourth-level subcategories specified, respectfully, the source of the deficiency or anomaly (e.g., design), the general problem area (e.g., anthropometrics), and finally the specific problem type (e.g., strength). An ad hoc decision tree could then be used to sort the reports into lowest-level categories where they could be counted, trends identified, and needs suggested.

Neither effort at categorizing system deficiencies was fully successful. Most of the problems with each had to do with category size and overlap (i.e., not totally inclusive or mutually exclusive). (Other problems with the development and use of rational taxonomies are described in detail in Bennett, 1971.)

APPROACH

Requirements

A review of the strengths and weaknesses of the two preliminary taxonomies concluded that an effective taxonomy must:

a) be tied directly to standards

b) be developed independently of specific problems

c) have mutually exclusive and totally inclusive categories

d) have categories of equal weight

e) have categories which can be specifically defined

f) exhibit a continuity between higher and lower levels of the hierarchy

g) be amenable to automation.

These, then, are the requirements to be addressed during development of the CEFTS taxonomy.

Method

The approach for development of the CEFTS taxonomy and the data base management system of which it is a part was relatively simple. The first step was development of the taxonomy itself. A rational approach was used to define the levels of the taxonomy and to determine the categories of which it is comprised. All candidates were subjected to an informal test to determine whether they meet the requirements specified above.

The second major step will be development of a decision tree. To accomplish this will require a review of the literature to determine the nature and features of the decision tree most applicable to the CEFTS concept. A modification of the decision flow diagram process used by Mallamad, Levine, and Fleishman (1980) offers promise as a decision aid. Some sort of expert system might also be effective as part of the decision tree. Given the variety and complexity of the many deficiency and anomaly reports which arise, an expert system may, in fact, be required to help the user sort through those reports.

The use of an expert system coincides with the third major step, that of automating the data base management system to the greatest extext possible. Automation assures a reduction in labor and costs and is, therefore, essential. Although the data collection and data entry procedures on-site will involve minimal automation, the remainder of the procedure--from input locally, to classification and recommended solutions--can eventually be automated to a large extent.

Methodology

Once the requirements specified above have been translated into the components of a data base management system, the framework for organizing and analyzing system deficiencies will be ready for implementation. The Computer-Aided Evaluation of Flight Test Support network would work as follows. Briefly, after watch items, service reports, or deficiency reports have been generated as the result of pilot debriefings, those reports are transmitted over ARPANET (or similar phone line) to the local computer system supporting CEFTS. At that point each report is sorted and organized according to the composite taxonomy. This sorting is accomplished manually using an ad hoc decision tree until the automated capability comes online. A more sophisticated decision tree--perhaps an expert system--will be developed for use with the automated system. Output from the classification process includes an organized data base, a tie-in to the status of each report, and a reference to any relevant texts, standards, guidelines, or resource documents.

OCCUPATIONAL IMPLICATIONS

The importance of this system lies in the organization it imposes on a large and unwieldy data base and in the potential for automation it offers. The CEFTS system will sort system deficiency or anomaly reports and classify them according to a rational taxonomy that is founded in both the requirements for human factors (e.g., MIL-STD-1472C) and the nature of the problems being addressed. The use of a data base management system/taxonomy also provides a greater institutional memory than any individual could offer. Thus, both operator and maintainer reports can be incorporated in the same data base and a history of system reports across time and across sequential production blocks will also be available. Accessing this institutional memory is therefore a more cost-effective and time-efficient process than otherwise possible.

The CEFTS system will also reduce personnel requirements and labor costs since it will allow more effective use of on-site personnel and is relatively impervious to the disruptive effects of personnel turnovers, changing support requirements, and reduced funding. Finally, although CEFTS will be applied almost exclusively to human factors problems, the methodology itself could be easily modified for use with a variety of other areas.

The ultimate benefit to be derived from the Computer-Aided Evaluation of Flight Test Support is that it provides an objective and readily available data base from which design/redesign candidates can be determined. At the very least, the taxonomy provides an immediate counting of aircraft deficiencies, thereby suggesting a priority for addressing them. In this way, CEFTS is not just a data base management tool; it is also a design tool to assist the design engineer and a decision aid to support the program management decision process.

REFERENCES

(1) Bennett, C. A., Toward empirical, practicable, comprehensive task taxonomy, Human Factors 13 (1971) 229-235.

(2) Department of Defense, Human engineering design criteria for military systems, equipment and facilities (MIL-STD-1472C) (U.S. Government Printing Office, 1983).

(3) Fleishman, E. A., Performance assessment based on an empirically derived task taxonomy, Human Factors 9 (1967) 349-366.

(4) Mallamad, S. M., Levine, J. M., and Fleishman, E. A., Identifying ability requirements by decision flow diagrams, Human Factors 22 (1980) 57-68.

(5) Parker, J. F., Jr., Identification of performance dimensions through factor analysis, Human Factors 9 (1967) 367-373.

Human Factors in Organizational Design and Management - II
O. Brown, Jr. and H.W. Hendrick (Editors)

AUTOMATION IN RETAIL:
Effects On Consumer and Store Personnel

Mark S. Hoffman

Human Factors
NCR Corporation
Cambridge, Ohio
U.S.A

Automation in retail business equipment has provided various managment tools for measuring the company's performance. The introduction of computer based Point-Of-Sale terminal systems has created large data bases of sales information; these have often been too voluminous for in-store personnel to use. Next-generation computer architecture will have the capability to control user interfaces to communicate valuable complex interrelationships using the daily sales productivity data. Advancements in artificial intelligence and microelectronics will provide many challenging opportunities for the human factors professional/ergonomics specialist to explore the techniques for introducing this new equipment into the marketplace.

INTRODUCTION

Retail business is one of the oldest industries in existence. Practices and behavioral roles of the retailer and customer have remained constant until the industrial revolution; the art of selling mechandise and services was done primarily in a one-on-one relationship, without formal social structures and business regulations shaping either role. The salesmanship process of bargaining was based on the ability of the business persons (usually the sole proprietor), and his abilility to obtain the best price for the goods. This form of retail business is still flourishing in today's society among the large retail corporations; these smaller businesses are commonly referred to as Specialty Stores.

Three technological revolutions in retail business have occurred within modern society. These changed the methods of conducting business transactions, the organizational structure and division of responsibilities within the store, and the behavioral roles of the business person and the customer. Under today's competitive business climate, the store management is required to deal with large volumes of sales productivity and inventory management information, and make fast decisions to meet new marketing opportunities without negatively affecting the current base of retail operations.

Advancements in intelligent microcomputer based systems and artificial intelligence have the potential to improve the business person's ability to make expedient decisions. These new technologies when introduced into the retail marketplace will affect users in all levels of the organization. The goal is to further enhance their abilities to perform their roles by providing more flexible methods for using information gathered during the sales transaction. However, considerable human factors research into the organizational structure is required to ensure that the integration of this new technology into retail business achieves the desired goals.

The purpose of this paper is to explore the history of the organizational structure of retail businesses, review the impact of introducing transaction processing equipment into the organization and explore how it has affected the roles and responsibilities of the users. Areas for research on the impact of future technological introductions into the retail community and its effect on organizational design are identified.

CASH REGISTER ERA

Specialty stores dominated the world markets until the until the 1870's. These stores were typically proprietorships offering a limited line of merchandise or services i.e. groceries, hardware, restaurants, etc. to a specific market. The store manager, usually the owner, manually maintained ledgers. Calculation errors were made frequently. These stores served a limited customer base. Competition was minimal since the inventory capacity and distribution capabilities limited the market. The ability to sell goods was primarily based on the availability of the merchandise. In the United States, the specialty store thrived because mass distribution of goods was difficult; regional distribution channels were limited to railway system and horses.

The Department Store evolved in Europe with Bouciicaut in Stockholm (Ref 1). Bouciicaut has been commonly referred to as the "Father of the Department Store". He began using a fixed high mark-up on dry goods and eliminated the flexibility of universal bargaining. He also created the showroom concept for customers to browse, and initiated the practice of accepting merchandise returns. The competitive edge for this new concept in retailing was the limited capacity of the specialty store to display and maintain a large selection of merchandise.

Department stores began appearing in the large cities throughout Europe and the U.S. during the period of 1880-1900. The major U.S. department store chains, of which most are current members of Federated Department Stores (Lord & Taylor, Macy's, Stewarts, Lazarus, Marshall Fields, and May Company) were founded during this time period. The gross profit margin was approximately 20% (Ref 2). Typical weekly wages for sales personnel was $5.00. Selling was done on a one-on-one basis, since the majority of the customers were affluent. The concepts of in-store engineering and the design of workstations for processing sales transactions were virtually nonexistent; sales transactions were conducted over countertops, where the sales clerk's only responsibility was to manually log the purchases into a ledger and complete the tilling operation with a cash box.

The invention of the cash register, pneumatic tubes, and the telephone represented the first technological revolution in retail businesses. The first cash register with a paper journal, the Paper-roll Register, appeared on the market in the 1880's (Ref 3). This provided sales accountability for each transaction. It facilitated the division of labor allowing management to separate itself from the selling operation, maintain control over the store, and concentrate on managing inventory turn-over and capital assets. It also provided a cost effective solution for maintaining low wages. Pneumatic tubes first adopted by Marshall Fields (Ref 1) permitted centralization of cash flow, having all transactions settled through one department which was located away from the selling floor. The telephone provided the first system of centralized customer credit approvals.

Department stores competed with specialty stores by merchandising goods to rural areas through catalogue sales. These items were typically clothing and hardware which the local specialty stores could not afford to stock. Sears, Wards, and Penneys expanded this segment of the retail market.

The popularity and widespread use of the automobile fostered many changes in retailing during 1900-1920. Department stores flourished with tremendous growth and expansion in major cities through the U.S.. Trade unions began to appear in retail workplaces because of the wide separation in wages and responsibilities between management and sales clerks. During this time period, four distinct types of retail businesses emerged: department stores; specialty stores; food distribution; and food services. These types of retail business differed in marketing requirements, organizational structure, inventory turnover cycles, and distribution methods. Mass transportation provided the means for retail management to centrally control several stores within an urban area.

Many specialty stores failed during this time period because of the enactment of the federal income tax in 1913, the inability to manage and plan long-term capital acquisitions, and the inexistence of a formal organizational structure and division of responsibilities. Large retailers merchandised clothing fashions using the influence of motion pictures, and newpaper promotions. These promotional mediums fostered the growth in consumerism enjoyed by today's retail business.

Organizational structures for these large retail companies included separate departments for accounting, merchandising, purchasing, sales, and customer services. Technological innovations in cash registers included electronically controlled cash drawers and three printers (journal, receipt, and slip). An example of this is the NCR Series 500 Cash Register. These machines provided gross sales and department totals using mechanical counters. These sources of printed sales history were used by accounting departments as quantitative estimates to measure inventory turnover.

ELECTRO-MECHANICAL ERA

This era began after the Great Depression of the '30s. Retail data processing equipment was characterized with the invention of the mechanical accounting machine. These were used to manually transpose sales totals from cash registers and validate daily sales productivity using hardcopy output from journal printers. Departmentalization and segregation of job responsibilities continued to expand. The creation of "middle management" occurred because of the regional expansion of large retail companies. During this period, the department store had to refocus its public image as a supplier of products to provider of customer services; this was because of inexpensive cost of capital, overhead costs, and increased competition (Ref 3).

During this period, Marks & Spencer, a U.K. based retailer began an new type of retail business similar to those referred to in today's market as mass merchandise stores (Ref 1). This concept was to sell merchandise at 50% gross profit margin with a limited selection of goods in each department. They were one of the first retailers concerned with the design of workstations for processing customer transactions. Sales transactions were processed using the front-end workstations which were more typical of the side sequential workstation used throughout the food distribution industry. Customer behaviors were changed through forced queuing. The workstation design gained significant attention because traditional countertop designs used in department stores could not accommodate the larger transaction sizes and volumes of customers created by this new form of retailing.

Four factors following WWII in the US changed the direction of retail business practices more than any other occurrence prior to that time. These were the commissioning of the interstate highway system in 1956, the post-war population explosion, the proliferation of television in most homes, and the massive relocation of population centers from the urban centers to suburbia. The large centralized retailers were forced to build shopping facilities in suburbia. The expansion of the shopping centers occurred during the late '50s. These usually housed a department store, a supermarket, and several specialty stores.

The interstate highway system allowed large retail corporations to locate distribution warehouses for maintaining regional inventories for stores around major metropolitan urban areas. Televison became the dominant medium used by retailers used for promoting merchandise; these changes in transportation and mass communication positioned the modern day consumer as a significant source of economic stability and growth in terms of today's market. Prior to that time the primary sources of economic growth were construction, basic manufacturing, and agriculture.

Expansion of large retail corporations created many new positions in middle managment and sales. However, the methods for measuring sales productivity, merchandising goods, controlling inventory and distribution of goods within each region had not changed. Processing data was still done using mechanical adding machines by large accounting departments; and the retail floor manager only used sales totals by departments. The growth of retailing during the post-war era seemed limited by the available data processing technology.

POINT-OF-SALE ERA

Point-Of-Sale (POS) transaction processing by today's standards had its origins in the mid-60's, with the invention of the electronic cash registers, e.g. the NCR Class V. This type of equipment had the capability to consolidate daily transaction sales information into one register and generate detailed audit reports. Store level consolidation provided the store manager with concise information for directing the merchandise sales in the store. This in-store network of cash registers represented a significant technological advancement.

POS concepts were fully enhanced with the minicomputer based terminal systems. Food distribution and general merchandise (department stores, specialty stores, etc.) developed merchandise marking systems to take advantage of the optical reader technologies introduced in the late '70s. Food distribution adopted the Uniform Product Code (UPC) and the general merchandise industries adopted Optical Character Reader code (OCR).

Two different computer architectures were developed for retail industries, based on the transaction speed requirements, to provide sufficient customer service and inventory turnover rates. Processor based systems were designed for UPC scanning systems because updating files in realtime was needed for daily replenishment of goods. Intelligent terminals were used in the non-food industry with in-store distributed networks. This architecture was chosen because the terminals were considerable distances apart, and inventory turnover was seasonal. Sales information did not have to be applied to the central inventory file in realtime. Daily inventory updates were not required in the department store; if the network included a region of several stores, this process was usually performed once per week.

Large scale integration of POS into retailing occurred first in the US, then in Europe and Japan. These new types of terminal systems provided sales control, inventory management, employee productivity, and other types of information valuable to the store management for monitoring daily activities in the store. It also provided a means for testing different merchandising techniques. This information was available to the user by invoking a simple key code at the terminal. During this era of technological transition, retail companies developed data processing personnel to control the distribution of these massive data bases created from the in-store systems. These departments were also responsible for channeling the different types of information to the appropriate departments through the organization.

The introduction of POS computer based systems fostered many movements within the four types of retail industries and across nations. Management, sales clerks, and customer roles have evolved in several directions. Social trends in the food distribution industry reflect the widest range of cultural attitudes within the major retail markets.

Bar code scanning has become a standard mode of data capture for supermarkets in the US. Japan and Europe have recently begun to scan items; however, the lag in adapting this technology has been partially affected by the labor unions and the differences in customer shopping behaviors. Worker unions resisted scanning because of the fear of reductions in the workforce and the loss of self-esteem. Prior to the introduction of scanning, many checkers took pride in their ability to quickly and accurately process customer transactions using key entry.

The introduction of scanning POS systems created movements in designing more effective transaction processing workstations. Criteria for acceptable customer queues were adapted by retailers to promote images of customer service. There has been essentially no significant research in customer queuing behavior reported in the formal literature. A few general queuing models for predicting a store's ability to process customer transactions have been developed for specific applications (Ref 4,5). Queuing behavior represents a significant void in behavioral research.

Cultural conventions have affected the success of POS. For example in the U.S., customers have their items bagged by store personnel; the customers bag their purchases in Japan and Europe. The influence of the ergonomics community in the design of transaction processing workstation is evident in Europe and Japan. Checkers operate terminals in seated workstations in Europe. Modular terminals in Europe and Japan are required for most retail environments.

Adaptation of POS in the retail industries has also created some concerns from the users viewpoint. Some of these are as follows:

(A) The most commonly monitored performance parameter by management indicates the rate of itemization by checker; this is usually expressed as items per minute. The fear of management misusing this information to eliminate slow performers caused problems.

(B) The extended capability to capture product information at the POS provides data which is often too voluminous; the time required to interpret this data competes with the many other daily tasks of the crowded time schedule of the store manager.

(C) Customer acceptance of the new technology has sometimes been slow. Many customers showed their apprehension toward this new technology, concluding that key entry is a fairer and more accurate method of data entry. Many suspicions have been ill-confirmed because store management failed to maintain current shelf prices in the computer look-up file.

NEXT GENERATION MICROCOMPUTER ARCHITECTURE

Trends in the microcomputer industry are demonstrating tremendous computing speed and capacity. ROM storage is continually becoming miniaturized and less expensive. Advancements in intelligent computers and artificial intelligence offer the potential to provide the retailer with synthesized information and decision alternatives for satisfying short-term problems. However, methods and technologies for designing user interfaces are crude, at best. These need to be developed so that flexible interactive user interfaces can demonstrate the benefits of the technology to users in the retail business community.

Self-service transaction processing systems in the four retail industries will be introduced as a means to provide additional convenience without affecting the existing floor space. Self-service terminals are attractive to the retailer with high labor expenses and training costs. This new form of purchasing products allows the sales person to service several customers within a given time period.

The widespread use and acceptability of personal computers has already influenced the market in terms of flexibility to be used in virtually any application and environment. The next generation computer architecture with multi-dimensional graphics capability will provide the opportunity to design user interfaces which can visually communicate multi-dimensional complex relationships in the performance data to the actual store manager. Specialists are currently needed to interpret much of the sales data being gathered at the store.

Workstation design and ergonomics will play a greater role in the retail environment because the experiences and knowledge of the problems adapting new POS equipment in checkers workstations and the general awareness and appreciation of ergonomically designed products because of the popularity of microcomputers.

Robotics will play a major role in inventory retrieval and reshelving products. A Japanese chain, Seiyu-Nokenda, already has an experimental supermarket where all shelving and unloading of merchandise from the trucks is done by robots.

The design of the retail business organization and management practices will change. More expedient inventory replenishment methods will be used to control the continually increasing cost of inventory. 12-15% of the workforce in the U.S. is employed in the retail industry (Ref 6). This is a large segment, considering that these values do not include the service industries, e.g. lodging, petroleum, etc. As technology continues to change, the growth rate in retail employment will continue to climb; it has been growing by 2.4% during the past fifteen years.

The future holds significant challenges for the human factors/ergonomics professional in the retail industries. The introduction of POS has only been partially successful from the human factors viewpoint because most store management personnel are not trained or equipped to fully utilize the performance data base already available. Likewise, the responsibilities and tasks associated with processing customer transactions are becoming easier and require virtually no formal training. The success of introducing new systems into the retail markets which utilize more powerful computing power will depend upon the willingness of management to train employees and provide relatively stable employment within the organization.

References

1. Pasdermadjian, H. *The Department Store: Its Orgin, Evolution, And Economics*. London: Newman Books, Ltd., 1954.

2. McNair, M.P.& May, E.G. "The American Department Store 1920-1960", Harvard Graduate School of Business Administration, Publication No 166.

3. "The Cash Register Era", *Celebrating The Future*, NCR Publication, 1984.

4. Hoffman, M.S. & Cramer, M.L. "Workstation Design Through A Simulation Model", *1981 Annual Proceedings Of The Human Factors Society*, October 1981.

5. Warren, E.H., Jr. "Estimating Waiting Time In A Queuing System", *Society of Decision Sciences*, Vol 12, January 1981.

6. *Statistical Abstracts of the US 1985*, US Dept of Commerce, 1985.

Human Factors in Organizational Design and Management - II
O. Brown, Jr. and H.W. Hendrick (Editors)

Integrated Software: The Promise, the Product, and the Problems

James A. Carter Jr.
Department of Computational Science
University of Saskatchewan
Saskatoon, Saskatchewan, CANADA
S7N 0W0

Although personal computers have been commercially avaialble for a decade, it is only in the last two years that they have become highly respectable and highly pervasive within organizations of all kinds. IBM is often credited with this new respectability due to its introduction of a product into the market in 1981. The widespread use of microcomputers, however, only caught on two years later with the introduction of the promice of their being a major productivity tool for various levels of workers throughout an organization. This promice, although facilitated by the hardware produced by IBM and others, was largely supported by the introduction of a new generation of integrated software packages. These software packages typically stressed their general purpose nature and their ease of use by the average office worker. In this way they provided the promice to do a variety of tasks for a variety of individuals and to be able to combine the results produced by individuals for the good of the overall organization. This paper analyzes this promice, the products which claim to meet this promice, the problems with these products, and future directions towards better realizing this promice.

THE PROMICE

The range of potential users of computers in an organization is as broad as the range of the individuals in an organization. These users may range from senior management to technical workers to clerical workers and even unskilled laborers (1). This range of individual responsibilities and skills poses a major problem in designing systems that must meet the needs of a number of individuals in an organization. The design is further complicated by the various levels of computing experience that end users in today's society may possess. This variation in experience results from the difference in the amounts and the kinds of use each individual makes of computers both within and outside the organization.

Integrated software packages meet the challenge of this diverse user body by concentrating on a few general purpose applications that each can be used for performing a variety of fundamental business tasks. The typical applications provided by an integrated application package include: word processing, electronic spreadsheet, business graphics, data base, and communications. Managers can use these applications for planning, while technical and clerical workers can use them for producing various types of products on a day to day basis. Each type of worker can use these applications at whatever level of sophistication best suits them. All workers can exchange information between applications and, with the help of the communications application, between individuals in the organization.

Levels of Integration

The promice can be further anlayzed into a number of integrating features. These features have been grouped into six levels of integration, by Carter (2), based on the structure of the tasks performed by and for the user. Tasks performed by a user are based on FUNCTIONS of DATA that are obtained via an INTERFACE to PROGRAMS or TASKS within a unifying program. These Capitalized words identify five types of possible integration of user tasks above the level of total NON-INTEGRATION.

With NON-INTEGRATION each application is developed as a totally distinct program. Although the program may grow in complexity, the lack of integration, in its original design, limits its extent of use. Without integration of some form, the user must re-enter data into programs to fulfill other application needs. Further, similar functions may behave very differently between the various programs used.

DATA INTEGRATED programs are separate programs which share common data files or use some type of data base maqnagement system. Information generated by one program can be used by another in some manner. USER INTERFACE INTEGRATION makes separate programs, with separate data, interact with the user in the same way. The result of combining both DATA INTEGRATION and USER INTERFACE INTEGRATION is PROGRAM INTEGRATION. The barriers of accessing only one application at a time and of requiring the user to interface between programs with the computer's operating system are overcome with TASK INTEGRATION. TASK INTEGRATED systems include the various applications as tasks within a single multi-purpose program. FUNCTION INTEGRATION is a proposed level of integration beyond TASK INTEGRATION in which a user could define any number of applications as virtual tasks within a single system of shared functions.

THE PRODUCTS

The new generation of integrated software packages is only a new generation in terms of the promices it makes and not in terms of the technology. This new generation is actually composed of three different technical approaches: program environments, families of integrated programs, and all-in-one packages. Further analysis will show that these approaches are not new ones but only refinements of existing technologies.

Program Environments

Program environments, such as IBM's *Top View* and Microsoft's *Windows*, provide basic frameworks for potential integration of other application programs. As such they are more like extensions to operating systems rather than actual end user applications in their own right.

Program environments provide primarily data level integration. They do this by providing translation services between various sources of data and various applications wishing to input that data. They allow blocks of data from one application to be used as input to other applications. In some instances, they may also provide the capabilities of reading various types of formatted data files as input into an application that would not otherwise be able to read the data.

Although program environments can provide a common user interface for their super operating system like functions, the primary interfaces between the user and the applications are those provided by the application progarms running under the environment. These differences in application programs leave the area of user interface integration largely a matter of chance rather than design.

Program environments currently provide more of a promice than a product for running multiple applications at one time. This is due to both limitations in the amount of memory and the type of features available to the applications that run under the environments. Once the operating system and program environment are loaded into a typical personal computer there is seldom room available for more than two average size application programs. Currently a number of extended memory management standards have been proposed that would get around this limitation, but only for programs written to conform to these standards. Similarily, to be able to utilize fully the features of a program environment, an application program must be "well behaved". To be well behaved is to follow further standards, which most successful applications have found necessary to circumvent in order to provide adaquate performance for the end user. Thus, while program environments may provide easy access to multiple applications in the future, they provide only very limited access at present.

Families of Integratable Programs

Families of integratable programs have been around for over a decade, primarily in the area of accounting applications. The main contribution of the new families of integrated applications, such as Innovative Software Inc.'s *Smart System*, is their integration of general purpose applications. By utilizing separate programs for each major application, families of integrated programs tend to concentrate on the quality of the individual applications before the quality of the integration, although they may provide high qualities of both.

Families of integratable programs provide primarily program level integration. They tend to maximize both data and user interface integration while keeping individual applications separate from one another. Thus where windowing is allowed, the multiple windows are only multiple views of data from within a single application and do not provide the user multiple applications.

All-in-one Packages

All-in-one packages, such as Ashton-Tate's *Framework* and Lotus' *Symphony*, provide the greatest emphasis on integration. All-in-one packages utilize task level integration to provide the user a number of general purpose applications. Data can be readily moved between applications or shared by them. The user interface design and functions are highly consistent throughout all applications. Multiple applications can be accessed simultaneously through separate windows which can be linked together, where necessary. To facilitate the high level of interaction and of data processing possible with all-in-one packages, they often allow the user a high level command language for defining and storing particular combinations of functions, applications, and data.

Although their aim is to provide a multi-purpose tool, all-in-one packages usually appear to be extensions of one focal application that is more highly developed than the others. This can be accomplished by adding other complimentary applications to a well developed central application, such as a word processor or a spreadsheet. This focus is evidenced in the user metaphor or conceptual model of the particular all-in-one package. Thus *Framework's* frames appears to be more word oriented while *Symphony's* cells appears more number oriented.

It has been suggested that while an all-in-one package may meet most of the needs of one kind of worker, it may not meet the needs of all kinds of workers. Recent new releases of these packages have emphasized both the continued advancement of their focal application and the strengthening of the other applications, thus expanding their potential markets.

THE PROBLEMS

By the end of 1985, various software reviewers started to ask what happened to the promices of integrated software packages. Sales of packages had leveled off, as if the market were saturated, long before everyone has converted to integrated software. New product announcements recently have been generally only enhancements to existing products rather than dramatic improvements such as experienced in the previous few years. The promice, therefore, appears unfulfilled.

To analyse the causes of this disallusionment and to evaluate whether or not they are well founded, it is necessary to analyze how the products have or have not lived up to the needs of the organizations which were to be their benificiaries. If integrated packages have failed to be all things to all people, is it the fault of the applications, the technology, the people, the organizations, or all of the above? Based on the answer to this question, one should also ask if improvements can be made to better meet the needs of organizations and to better fulfill the promice.

Applications

The benefits of the integration of applications may be offset by a variety of limitations they impose upon their adopters. The more integrated the product, the more limitations that must be accepted in order to use it. These limitations include the number and types of applications, the state of the art of the applications, and the extent of integration of the applications. They may both restrict the flexibility of the user and increase the complexity of learning and using the package.

The set of applications integrated in all-in-one packages or families of integratable programs has remained relatively limited and fixed (generally: word processing, electronic spreadsheet, business graphics, data base, and communications). While these applications may be very powerful tools, their general purpose design (that makes them powerful) also requires the user to have a sophisticated understanding of how to apply them to specific problems. While a data base or a spreadsheet can be used to plan a schedule and then business graphics can be used to illustrate it, a scheduling application would do this for the user in a much simpler manner. Thus, in order to set up a schedule within an integrated package where no scheduling application exists, the user must become a developer as well as a user.

A variety of solutions exist to the problem of having only general purpose applications provided as the basics in an integrated system. There has developed a widespread availability of books of "spreadsheet templates" which are basically programs for the user to type into a general purpose program to produce a specific purpose application. Unfortunately, the results of using different "templates" to produce specific applications also usually results in the loss of integration between these applications. In these circumstances, using a lower level of integration, such as a program environment to integrate separate special purpose programs, may provide just as good integration with less trouble for the user. A compromise solution is to provide a program environment within an all-in-one package, such as *Framework's* DOS Window, and a set of guidelines to the user as to when it is better to use a specialty package rather than trying to develop your own.

Developers of integrated packages have the dual demands of designing for integration as well as the various applications contained in them. Often, the state of the art of individual applications has progressed beyond that of comparable applications in integrated packages, leaving the integrated packages needing to catch up. Thus, the adopting of an integrated package may be seen to tie the user to capabilities that may always be trailing the leaders.

In word processing, the state of the art has grown to include spelling checking, mail merging, outlining, variable styles and sizes of print fonts, and primitive graphics. Only outlining was available first from an integrated package (*Framework*). Many of these other features are only starting to appear in new releases of integrated packages. Specialized decision support systems and knowledge based expert systems are being marketed to meet needs, first met by spreadsheets and data bases, in easier and more powerful manners. Distributed systems and local area networks are superseding the style of occasional communication supported by most integrated packages, providing an additional dimension of data integration. "Fourth Generation Languages" are gaining in popularity over traditional command languages for the ease they provide users in developing specific applications.

Along with each of the advances in the state of the art, there comes a period of exploring the advantages and disadvantages of the advances. Thus, working at the state of the art is most advantageous for those users who are either very sophisticated or very much in need of the new features which it provides. For the majority of users, working with slightly older but proven and tested versions of applications may fulfill all of their needs with less trauma. Thus, the lack of state of the art applications is not really a problem for most users. A compromise solution would again be a feature similar to a DOS Window, to be used only where absolutely necessary.

The amount of integration between each application in integrated packages, although greater than in other cases, has remained relatively low. For example, few word processors can comfortably incorporate graphics with text without considerable effort because of the separateness of the word processing and graphics applications. Use of alternative sources of applications, such as with program environments or DOS windows, often causes even greater concerns in integration. Decisions regarding this problem can only be made based on the need for integration and the current state of the art.

Technology

The state-of-the-art of the technology has also limited the promice of integrated packages. Hardware capacity, software complexity and the costs associated with both have forced developers to make tradeoffs between the level of integration and the quality of applications within their packages. Conflicting conceptual models and user metaphors, designed to help the user, have added to the reluctance of users to adopt integrated packages.

Many desirable integrating features can only be achieved with an increase in the hardware capabilities beyond that of current typical personal computers. High processor speed and large main memory are required to support high resolution bit mapped graphics and the concurrency of multiple functions, applications, or users. Local area networks with external communication capability are necessary to support integration in the form of distributed systems.

Further developments in software technology will also be necessary to achieve the full promice of integrated applications. The development of functionally integrated systems can provide a technology which overcomes many of the current limitations and also allows for easy addition of other applications to an integrated package. The development of knowledge bases will allow customization of systems to individual user's levels of expertise and need while maintaining compatibility and consistency between users where necessary.

Cost acts as a limiting factor both in what is available and in what the user is willing to pay. Hardware advances are continually providing better computing power for the money that users are willing to spend. Costs, however, will remain a limitation in the area of software. Typical users will only pay the equivalent price of two individual packages for an integrated package, since in most cases they will only make an equivalent amount of use. To be successful, developers must include much more than the equivalent of two individual packages. Thus, developers must rely on much higher volume sales to make up for their greater development costs. Recent sales trends have not justified this expectation. The similarily greater amount of effort in developing updates to integrated packages was recently reflected in the relatively high prices charged for upgrades to some of the poppular integrated packages. Further complicating the cost situation is the expectation of organizations to be able to buy site licenses or to get volume discounts for software while the developers wish to sell only individual copies at fixed unit prices. Unless the economics of software costs can be worked out to the mutual satisfaction of users and developers, advances in integrated packages may be few and far between.

The complexities of design and marketing of integrated packages have led to the introduction of a number of conceptual models and user metaphors to explain them. Systems may be called electronic desktops, outlines, slideshows and many more names. Individual applications may reside in windows, frames, libraries, spreadsheets, and many more structures. Each of these models or metaphors is designed to show how universal and easy to use the system is to at least a target group of users. Each shows certain biases or viewpoints of the developers and certain stereotypes of the intended users. Unfortunately, familiar as all of these concepts are to most users, they may be perceived differently both qualitatively and attitudinally. Either individually or in a small consistant group they do a good job in helping explain the purpose and workings of an integrated package. The confusion comes from the users being exposed to conflicting terms and concepts from a variety of sources besides the particular package being used. Further confusion results in the importance placed on these analogies in explaining the system to all users, even though not all users are likely to be so intuitive as to think in terms of metaphors.

People and Organizations

Difficulties with people and organizations realizing the promice may be attributed to the concept of the personal computer. In many cases personal computers are thought of only as a tool for an individual to use to accomplish an individual task. Neither the power of the tool nor the value of the information, which is really a major corporate asset, is understood. Further, since it is individuals who are using personal computers individually, many organizations provide little or no support to them, leaving everything to the individual's inniative and descretion. Strong efforts at providing organizatinal support, such as organizational information centers and new methods of managing semi-independent workers, are necessary to transform isolated individual users of integrated software packages into an integrated team of information workers.

It must be recognized that integrated packages can only do some things for some people. Since typical users are limited in the number of applications they will use and since some individual packages are more powerful than their counterparts in typical integrated packages, some users will choose to use one or two specialized packages rather than one integrated package. Various users within an organization, therefore, may make individual cases for specific packages for thier own use.

While such a choice may be in the interests of the individual it often overlooks the value of standardization and synergy within an organization. To overcome this difficulty, integrated packages need to not only inprove the quality of their individual applications, but also to allow for the easy addition of other applications within the main integrated package.

The co-ordination and integration of personal computers into organizations requires more than just individual initiatives. Although individual initiative is often the instigator, once the number of users and uses grows, some centralized organization is required to ensure that this growth is productive and efficient. One potential structure for providing such an organization is the concept of an information center.

Information centers are often an outgrowth of an organization's data processing department trying to provide quality user support for non-traditional data processing activities. They encourage and assist the end user to learn about data processing methodologies and to develop small or one time applications for themselves. Information centers, therefore, have a similar purpose to that of most personal computers in allowing the development and use of applications that would not otherwise have been undertaken by the data processing department.

Information centers typically provide a pool of specialists centrally located that support large numbers of users in different departments. This support may include: educating users in the various aspects of data processing; providing on-site resolution of problems; providing access to reference materials; providing access to hardware and software for evaluation and development puropses; providing assistance in developing small or one time applications; providing assistance in selecting and purchasing hardware and software for end user use; and providing and controlling access to corporate networks and data bases.

The information center approach is desirable since it uses "a carrot rather than a stick". It still leaves the user in control of applications and expendatures, while encouraging the user to standardize and integrate in order to obtain the various benefits of information center services. Those types of hardware and softwware that the information center supports and which it may even make available at very attractive prices become the de facto standards for the organization. Users not following these standards do so on their own and at potentially much higher costs.

As the use of personal computers grows, even within an integrated manner, so also grows the potential independence of the worker in the methods chosen to achieve the work. This affects both sides of the worker - manager relationship. More and more tasks can be performed directly by the worker on the computer without going through any intermediary management approval or supervision. With advances in communications technology it even will be possible for many workers to perform thier work on a personal computer at home, many miles from the manager. This increasing autonomy of workers gives rise to a number of potential problems involving the co-ordination of tasks to ensure that their results integrate appropriately without having missing parts or wasteful redundancies. It changes the emphasis of management from assigning work to be done in a certain way and supervising the work to assigning work the results of which must interface in a certain way and synthesizing the results. This change parallels changes in software technology by concentrating on what needs to be done rather than on how it is to be done. Until managers become more involved in goal setting, arbitrating and evaluating and less involved with managing, the human link in integrating computing within an organization will remain weak and the integration of applications will not evolve beyond the level at which it exists in the software being used by a number of individuals in the organization.

THE FUTURE

To date, most integrated packages appear to be built around meeting the perceived needs of one or more applications rather than around meeting the actual needs of the user. This approach has maximized the understanding of software developers along with the current state of hardware technology. Advances in applications have been largely tied to the available size of computer memory and the speed in processing. These advances have resulted in more data and functions in applications, and a more sophisticated user interface to these applications. The set and the structure of individual applications, however, has remained relatively fixed, and only now is starting to slowly expand.

To change the focus to the needs of individual users, a new approach to the basis of software integration is necessary, such as that of function integration. The state-of-the-art, however, has not yet reached function level integration. To do so will require both a new understanding of the user and a new state of technology to implement this understanding. To change the focus further to the needs of individual users within an organizational context requires both advances in computing and advances in the organization of the workplace. These organizational changes in the workplace are not just due to the adoption of integrated software, but have been developed and discussed in other settings. Where such methods of organization exist already, higher levels of integration may also exist.

RECOMMENDATIONS

The following are a set of recommendations for organizations wishing to achieve the promice of integrated software:

1. The decision to adopt and standardize on integrated software should be made at the organizational level. Once such a decision is made it should be communicated to all individuals in the organization in such a manner as to show them why it was a good decision and to encourage them to comply with it.

2. All-in-one packages provide the highest current level of integration and the greatest potential for achieving the promice of integration. The adoption of an all-in-one package as the basis for integration within an organization should only be made, however, if it is supported by a suitable cost/benefit analysis that proves this generalization.

3. The need for additional application programs beyond those contained in an integrated package needs to be recognized and allowed for in any attempt to standardize on an integrated package. The use of such additional packages, however, should be limited to those cases where the benefits to the organization are greater than the costs of deviating from the standard integrated package.

4. Where possible, additional application programs should be accessable and should be accessed from within the standard integrated package.

5. Consideration needs be given to how the results of integrated software packages can be integrated both with each other and with the needs of the organization.

6. Where conceptual models and user metaphors of integrated systems are used they must be pertanent to the people in the organization who will use them if they are going to be of any use.

7. There is a need for information center support for individual users of integrated software in an organization in order to educate the users and to assist them to achieve the fullness of the benefits of the software.

8. There is a need for a goal oriented and integrating management approach to the use of integrated software in an organization if the fullness of the promice of integration is to be realized for the organization.

9. Since the state of the art of integrated software is evolving due to advances in technology and understanding, organization should be prepared to change in order to achieve new benefits as they become available, rather than expecting to adopt a single package that will solve all problems both now and in the future.

REFERENCES

[1] Carter, J.A., User-Oriented Structured Design of Data Processing Applications, in: Hendrick, H.W., and Brown, O. Jr. (eds.), Human Factors in Organizational Design and Management, (Elsevier Publishers B.V., Amsterdam, 1984), 259-268.

[2] Carter, J.A., Human Factors Concerns For Integrated Software, in the Proceedings of the 18th Annual Congress of the Human Factors Association of Canada, (Hull, Quebec, 1985).

Human Factors in Organizational Design and Management - II
O. Brown, Jr. and H.W. Hendrick (Editors)

USER PREFERENCE IN MENU BASED INTERFACE DESIGN

R. Glenn Walker
James. A. Carter Jr.

Department of Computational Science
University of Saskatchewan
Saskatoon, Saskatchewan
Canada

1. Abstract

This study develops a list of suggested command names for use in building menu trees for applications. It is based on the taxonomy of user-oriented functions by Carter(1985). It minnimizes conflict between function names and proposes a methodology that can be used to standardize design of menu interfaces. The problems of organization and form are not addressed by these results.

2. Menu Design

2.1. Introduction

This paper deals with the design of menu systems from the perspective of the naming of functions utilized by the designer. The goal is to bring user preference into the picture and thus involve the user in the design of the system.

Traditionally designers have neglected the user and by doing so, many have met with disaster [Thimbleby (1982)]. By accomodating user preferences, the designer can make the use of the system more straight forward to the user and thus assure a more successful design [Good, Whiteside, Wixon, & Jones (1984)]

2.2. Traditional Design Philosophy

Good, Whiteside, Wixon, and Jones (1984) stated that there are two basic strategies for the design of systems: i) adapt the user to the system. ii) adapt the system to the user. Traditional design philosophy tends to adopt the first method as opposed the the second. This reflects the usual tendency to ignore the user during the design phase.

Design methodologies that take into account the user will be more likely to be accepted. The user who feels neglected will feel resentment towards the system. This can get in the way of successful integration of the system into the work environment. The use of user input into the design of the system can enhance the acceptance of the system.

2.3. Importance of a Good Design Methodology

Whiteside, Jones, Levy, & Wixon (1985) showed in their study on user performance vs interaction style that interaction style does not affect performance, but the design of the chosen interaction style will. This makes the importance of a good design methodology tantamount to the success of any system design.

Thus it is important to study how interface design can take into account user preference. Regardless of interface style it is important that the user has a say in how he/she will interact with the system. This study attempts to meet these needs dealing with menu interface design. A related study by Jean Hantke, similarily deals with these needs for command language interfaces. Future work will combine the results of these two studies, and others if necessary, to obtain results which can be used for all styles of interfaces.

2.4. Issues in Menu Design

Tullis(1985) states two issues for menu design. Both have to do with structure of the menu. The first was the concern over the organization of the menu and how the logical relationships among the many functions can be presented to the user in a coherent way. This is important as it aids the user in successfully navigating the menu structure with fewer wrong turns when trying to locate a function. Tullis attacked this problem by using a sorting technique to find out what the perceived logical relationships were among functions. He found that the majority of users perceived the relationships among the functions in an object-oriented fashion.

Tullis(1985) also identified another issue. Granted that the logical relationships talked about above have been discovered, how does one best represent these functions in a hierarchy? This question deals with the issue of whether depth is more preferable than breadth. Depth refers to the number of levels employed in a menu hierarchy. Breadth refers to the width of the menu hierarchy. Both are affected greatly by the number of options which appear at any node in the menu hierarchy. Tullis tested users to determine which was more desireable in terms of fewer errors and less effort to complete a task. He found that breadth was more desireable. This was due mainly to the fact that users found that in a menu where great depth was employed they had difficulty in determining which option they should follow to find the function they wanted. This was due mainly to the level of generality which had to be employed in higher level choices in order to group options at lower levels.

Tullis, however, ignored an issue which he could have gotten input from the user on. In both of the above cases, Tullis used user preference to determine what the best solution to each issue was. However, he ignored the aspect of what function names would the users prefer. Instead Tullis used the already determined function names that were part of an existing operating system. By doing this, Tullis limited the portability of his results to other systems. The current study addresses this overlooked aspect. By looking at user preference to determine what function names users would prefer we are allowing the user to define his own work environment. As well, by not tying it to a particular system the results obtained can be applyed in general accross a wide variety of systems. This implies that a standard can be derived that is totally based on user preference.

2.5. The Need for Design Standardization

The need for standardization becomes clear because of the possible confusion between systems. However Treu(1982) points out that this type of standardization would be difficult to achieve due to the vast variety of systems that exist. Despite the difficulties, any move towards standards can only be beneficial to the end user of the designs.

Benbasalt (1978) points out that any design standard must incorporate the characteristics and needs of the user to be successful. This is where surveys of user preference come in as an important tool for determining a standard.

2.6. Earlier Work Related to This Study

This paper is an extension of the work being carried on by Jim Carter et al at the University of Saskatchewan. Carter and Everett (1984) developed a user-oriented hierarchy of functions. This hierarchy consisted of common user-oriented functions that were found to exist in a number of systems surveyed. These systems varied in scope from operating systems to application packages. The goal was to come up with a list of the functions necessary for a user to accomplish any common tasks. The current hierarchy consists of 89 different terminal nodes each representing some homogeneous function.

Once the hierarchy was developed, Carter (1985) developed a set of operational definitions and set about conducting surveys of users. These surveys were given as an open ended questionaire allowing the users to respond with their own word choices. This was done to find out what users' preferences were for command names for each of the functions in the hierarchy. The goal was to supply the basis for further research into command-based and menu-based systems.

3. Methodology of This Study

3.1. Introduction

3.1.1. Purpose

The purpose of this methodology is to propose a method by which user preference can be incorporated into the design of menu interfaces. A method is proposed by which surveys are used to measure user preference; and through simple analysis procedures, determine what the best function names are according to what the user prefers. The results obtained are not tied to any particular system. Therefore, the results obtained using this methodology should be transportable to any type of system. This is important since it forms the basis of a standard for menu interfaces across the many types of systems that exist. The measurement of user preference is vital to successful design standardization procedures as was stated earlier.

3.1.2. Definitions

In order to proceed with the discussion of a proposed methodology, it is first necessary to define a few of the basic terms and structures that were used by this study.

The concept of node refers to one of the terminal nodes in the user-oriented hierarchy of functions mentioned earlier. Each node consists of operational definitions and associated suggested function names along with data collected for each suggested function name. There were 89 such nodes in the hierarchy at the beginning of this study.

An operational definition refers to a context free definition of a function in the hierarchy. Each node had three such definitions included so that a wider variety of surveys could be made up for the number of people being surveyed.

Suggested function names refer to a group of descriptive words which the user is presented with in the course of the survey and asked to make a choice as to which he/she prefers as the name for the particular function that the suggested function name is associated with in the hierarchy. Lists of associated function names for each node were taken from Carter's 1985 study as a basis for this study. For examples of function names, operational definitions, and nodes, see Appendix B: Detailed User Survey Results.

Goodness, conflict, and object orientation are also terms which are important to this study. Goodness refers to the suitability of a function name. This suitability is determined by user preference. Obviously, the more users prefer a function name the more suitable it becomes. Conflict refers to the occurrence of multiple uses of the same suggested function name by users in naming functions in the hierarchy. This can be due to the close relationship between the functions (ie. they are next to each other in the hierarchy) or confusion between functions that are not closely related (ie. functions are separated from each other in the hierarchy). Object Orientation refers to whether people tend to think of the object as the primary concern when making choices from the suggested command names. Put another way, does the user consistently choose verbs over nouns or nouns over verbs.

3.2. Surveys

The surveys were formed from the operational definitions and suggested function names. The operational definitions and suggested function names were largely derived from those used by Carter (1985). In Carter's study, however, only one operational definition per node was employed. For this study, two additional operational definitions were employed with each node to allow greater variety in makeup of surveys. The total number of unique operational definitions employed were 264 with 88 lists of suggested function names. Each survey consisted of 30 operational definitions along with a list of suggested function names. There were 10 surveys made up. This made it necessary to duplicate some of the operational definitions. No one survey showed the same definition twice.

The user was presented with a series of thirty screens each consisting of an operational definition and an associated list of suggested funtion names. The user was asked to use the cursor keys to highlight his/her choice and then to select it by pressing return. The software developed allowed multi-user access to the results database. The equipment used was a local area network at the University of Saskatchewan consisting of an Ethernet and 50 Zenith 150 microcomputers. All software was developed in Turbo Pascal. An example survey screen is included below with an operational definition and associated suggested function names:

Your students do assignments in files on the computer. The students can do anything they want to with their files until the due date. Then you tell the computer not to let the students do anything else to the files - only you can see them so you can mark them. What command would you use to tell the computer not to let them see the file?

access accessible available change code password permission permit revise

The subjects of this survey were all students in a computer literacy course. Their opinions represent the view of naive users. These people represented a broad spectrum of experience ranging from the sciences to the humanities. The surveys were administered to 10 lab sections of approx 40 people each. Each node had at least three lab sections replying to it for a typical coverage of at least 100 people. Some technical problems caused problems with some of the nodes where coverage was less than 100. These nodes will have to be resurveyed at a later time.

The survey was administered to the students as part of their first session in the lab. They were told that it was a part of an experiment being carried on and that their performance was not being graded. The results were collected in a common database. The software mentioned above allowed multi-user access to the database. A frequency count was kept broken down by operational definition for later analysis.

3.3. Data Analysis Methodology

3.3.1. Goodness

Rank ordering was used to determine the goodness of any suggested function name.Words were ordered from best to worst based on the total of the observations for all three definitions included with each node.

The initial best set of function names was indicated by taking the most prefered words until a 60% coverage of preference was reached or the top 6 function names were reached. This represented a starting point from which further analysis could be done. Specifically, looking at the conflicts in function name usage that occured.

3.3.2. Conflict

In the conflict analysis, the problem was to minimize the conflicts that occured between nodes. This was done by following a series logical steps.

The first was to look at a conflict and see if any of the nodes involved in the conflict were closely related (ie. side by side in the hierarchy). If this was the case, then an attempt was made to minimize the conflict by combining the nodes according to the methodology employed by Carter (1985). If such a combination could be done it was carried out and the resulting combination of the data replaced the previously separate nodes. This was done to take care of conflicts that arose out of confusion between related functions which was understandable.

After the combination step, if a conflict still existed then it was solved by looking at each node, except the one with the highest relative frequency, and trying to substitute other suitable terms. The substitutions took place by substituting less prefered words that would not cause further conflict with other nodes. The goodness rule of 60% or 6 function names had to be maintained in order for a substitution to be successful. Also, no function name for which fewer than 5% of the users surveyed showed preference was used. If no successful substitution within a node could be made then the conflict was left as unresolved for now.

3.3.3. Noun/Verb Selection

Noun/Verb orientation was determined by looking at the frequency of nouns and verbs that were chosen by users taking the survey. Each node was analyzed to determine the percentage of nouns and verbs that appeared at each level of preference. This gives some indication of how much importance users place on using nouns and verbs for function naming. This will show an indication of how users are action or object oriented in their choices of function names.

3.4. Use and Limitations of the Methodology

This methodology can be used by system designers to design menu interfaces that will accomodate their particular user population. The methodology could also be extended by seeking the preference of a suitably large sample of users and developing a standard for use in developing a wide range of systems.

It can also aid in the evaluation of existing packages. The hierarchy and associated command names can be used to evaluate the effectiveness of any particular system in meeting the preferences of users. It can also serve as a basis of comparison between systems. This can be invaluable information when looking at several systems that all appear to be similar.

This methodology is not without limitation. The results obtained are clearly only as good as the data fed into it. One must be careful to recognize that the results are directly related to the population surveyed. For example, the results used here represent the preferences of naive users. The methodology does nothing to adjust for other types of users not surveyed and leaves the full burden on the user to determine the suitability of generalizing the results.

This methodology does not offer a way to resolve conflicts between nodes that are far apart in the hierarchy. This is something that must be done independently through some other method. In other words, the logical relationships between data are not explicitly analyzed by this methodology. The relationships among the nodes examined are taken as given and analysis proceeds from that point.

4. Findings

4.1. Introduction

The detailed results of the application of the methodology are too long to be presented here. They are included in the full technical report [Walker and Carter (1986)]. The summary findings of the application of the methodology are given here.

4.2. Goodness

The goodness of the data was analyzed by looking at the average level of user preference that was satisfied at each of the first six levels. The table below displays what the averages were before conflict resolution took place.

Preference Table

Preference Level	Coverage of User Preference	Percentage Increase
1	33.85%	33.85%
2	53.35%	19.50%
3	67.07%	13.72%
4	76.86%	9.79%
5	80.13%	3.27%
6	82.87%	2.74%

The table clearly shows that the user preference climbs sharply from a low of 33.85% user preference coverage on average for the first ranked suggested function name to 82.87% user preference coverage on average by the time you take into account the top six suggested function names. If an average preference level of 60% was required, the table shows that to obtain this we would on average be required to look at the top three user prefered words. The methodology used above stressed a level of 60% preference coverage or a maximum of six function names. These results would indicate that the maximum of six function names is too high. This accounts for the fact that in applying the goodness criterion of six function names in the results obtained in the survey, the maximum of six function names was never enforced. The sixty percent preference criterion was the limiting boundary in choosing the intial set of user prefered function names.

4.3. Conflicts

The original results obtained from the raw survey results yielded 53 conflicts involving 143 function names.

After applying the methodology discussed. The number of conflicts was reduced to 32 involving 78 function names. This was done mostly through the substitution of lower ranking alternatives. There were six conflicts which could be lessened by the combination of the nodes. These six combinations involved 13 of the previous nodes in the hierarchy. The result was the 40% reduction in the number of conflicts through the removal of 65 conflicting entries.

Three reasons why conflict resolution could not be fully accomplished were discovered while applying the methodology.

Insufficient response was identified in four cases as a possible factor in preventing combinations that would have reduced conflict. These nodes would require resurveying before any conclusive decision can be made about their final disposition.

There were twelve cases where combinations of sibling nodes were not possible since they did not meet the criterion set out by Carter(1985) for agreement among combinable siblings. Four of the twelve cases involved the insufficient response problem mentioned above. In the other eight, the problem was simply that the criterion for combination could not be met with the current results. In order to accomodate the combination of these nodes the standards for the combination of a node would have to be lowered. The determination of whether this is desirable will have to be made in further studies.

The lack of viable alternatives for a function name in conflict was the most prevalent cause occuring in 49 cases. The characteristic way in which the user preference was highly concentrated in the top six levels left a number of alternative that fell below the threshold level of 5% preference and as outlined in the methodology were not used. This was especially evident where the words in conflict represented a large block of user preference. A good example of this was the conflict over "password". In each case the user preference for this function name ranged from 50% to 63.50% making substitution impossible.

Among the 32 nodes that remain in conflict, there are several observations that can be made about the causes of these conflicts. One example involves the function name "password". It occurs in strongly prefered roles in three different parts of the hierarchy:

1.1.3.5.2	Assign a Password	63.50%
1.2.7.1	Change a Password	50.00%
1.2.7.2	Changed Access	22.73%
2.3	Change Signon Password	51.69%

Because of this strong preference it is impossible using the methodology to replace it and still maintain a 60% goodness criterion. This might indicate that the user sees a strong similarity among these functions. In other words the user is showing how he/she perceives the logical relationship among the four different functions. If this is the case then a rearrangement of the hierarchy is in order. In this case, a rearrangement would involve the combination of the nodes. This type of refinement represents an ongoing process which must occur until the logical relationships shown in the hierarchy reflect those perceived by the majority of users.

Another relationship among nodes in conflict could lead to reduction in the level of conflict. These are those conflicts where the nodes in conflict represent possible aliases for each other. An example of this is the conflict as a result of the use of the function name "continue":

2.5.4	Continue Processing	32.81%
3.1.4.1	Display Next Help Screen	39.86%

A strong argument can be made that the displaying of the next help screen is the act of continuing processing help. The second use of "continue" represents a more object specific alias of the first use. In this case the help process is a specific instance of processing in general. This points to the possibility of functions existing in the hierarchy that do in fact represent the same function. The user is making us aware of this through his clear use of preference to indicate that he/she views both these functions in a similar way.

Some undetermined amount of confusion may be accounted for by the potentially leading nature (bias) of the definition being given. This can lead to otherwise unexplainable conflicts. The definition may inadvertantly suggest an answer to each individual. This is a problem which must be looked at after each iteration of the survey so improvements can be made where possible.

4.4. Noun/Verb Selection

When looking at the noun/verb orientation we find that the verbs represent 75% of all the function names prefered and nouns account for the rest. This displays a large preference for verbs in describing functions. This seems consistant with the results obtained by Tullis (1984) where he found that the majority of people in his study grouped actions by the object they affected.

However, a 25% preference for nouns represents a large chunk of preference that cannot be ignored. If we look at the noun/verb usage at each level of preference we see some interesting trends. The following table shows what the levels of useage are for nouns and verbs for the first six levels of user preference:

Preference Level	Nouns	Verbs
1	30.00%	70.00%
2	20.00%	80.00%
3	35.00%	65.00%
4	30.00%	70.00%
5	35.00%	65.00%
6	50.00%	50.00%

Nouns seem to appear consistently across the top levels of preference. If nouns only existed at the lower levels then their elimination could be possible without upsetting the balance of user preference. But, users seem to prefer nouns in certain circumstances. This shows that although users prefer verbs more, they still insist on using nouns to describe certain functions. This indicates a hybrid of action/object orientation is necessary to satisfy user preference completely.

References

Benbasalt, I. (1978), The Impact of Cognitive Styles on Information System Design, MIS Quarterly, 2(2).

Carter, J. A. (1985), A Taxonomy of User-Oriented Functions, Technical Report 85-1, Universtity of Saskatchewan, Department of Computational Science.

Carter, J. A. and Everett, H. J. M. (1984), A Hierarchy of User-Oriented Functions in Gotlieb, L. (ed.) Proceedings of the Canadian Information Processing Society Session '84, Calgary, Alberta.

Good, M., Whiteside, J., Wixon, D. and Jones, S. (1984), Building a User-Derived Interface in Communications of the ACM, 27(10).

Thimbleby, H. (1982), Interactive Systems Design: A Personal View in IEEE International Conference on Man-Machine Systems, July (1982), London, England.

Tullis, T. S. (1985), Designing a Menu-based Interface to an Operating System in Human Factors in Computing Systems, Proceedings of the CHI '85 Conference, San Francisco.

Whiteside, J., Jones, S., Levy, P.S., and Wixon, D. (1985), User Performance with Command, Menu, and Iconic Interfaces in Human Factors in Computing Systems, Proceedings of the CHI '85 Conference, San Francisco.

Human Factors in Organizational Design and Management - II
O. Brown, Jr. and H.W. Hendrick (Editors)

Behavioral Analysis of Software Developers for Increment of Software Productivity

Akinori Komatsubara, Toshiyuki Suzuki and Yoshimi Yokomizo
Department of Industrial Engineering and Management
Waseda Universiy, Tokyo 160, Japan

Environmental modification in software factories is one of the effective methods to increase software productivity. The environmental model which indicates the relation between the environment and element tasks of software development is proposed. To utilize this model, the method to divide the software development into element tasks is also discussed.

1. Introduction

As software crisis threatens our recent technetronic age, to increase the productivity of software products is strongly required. As there are several methods to increase the productivity of software products, environmental modification is said to be one of the effective methods (B. W. Boehm 1981). We have also experienced several cases that the software productivity is increased by the environmental modification of software factories (A.Komatsubara 1983). However it seems that there are few studies which discuss the theory of environmental modification. Therefore the modification sometimes falls into so-called "patch-type improvement". In this study, accordingly, environmental model which relates the productivity with environment of software factories is proposed. The application of this model will be also discussed.

The purpose of environmental modification to increase the software productivity are summarized as the following two points.

1) To produce the environment where the software developers can devote themselves to software development.

2) To maintain the motivation of software developers, as we know that the software developers' motivation will be lowered and sometimes they will make job hopping from the software factories where the enviroment is too poor. As the software developers are usually highly educated, they cannot satisfy their pride in so poor environment, and lower productivity.

The latter is related to industrial psychology. In this study we treat the former point.

2. Environmental Model

The following model is proposed in this study.

Suppose that the series of software development work to develop one unit software product can be divided into element tasks of number of N and that the following formula is proposed.

$$T = \sum^{n} U(i) \tag{1}$$

here T: Total required period of time to develop one unit software product
U(i): the actually used period of time to develop the #i element task.

Suppose that the each element task requires its specific environment to ensure the highest productivity.

The enviromental factor "a" is defined for each element task, which indicates the satisfaction level of the each environment of the each element task. The following formula is proposed.

$$U(i) = a(i) \cdot t(i) \quad \text{for any } i \tag{2}$$

here t(i): the theoretically required period of time to cope with the #i element task.

a(i): environmental factor which indicate the satisfaction level of the environment for #i element task. $a(i) >= 1$

This formula indicate that "a(i)" will take 1 when the environment is best for the element task, and the used period of time to cope with the #i element task, U(i), will be equal to the theoretically required period of time, t(i). It also indicates that the "a(i)" will be greater if the environment for #i element task will be poorer. This leads the longer U(i).

Moreover this formula shows that the a(i) is the ratio of the theoretically required period of time and actual used period of time for #i element task.

The following formula is obtained with the formula (1) and (2).

$$T = \sum^{n} a(i) \cdot t(i) \tag{3}$$

This formula indicate the relation of environment and element tasks and we name this environmental model.

3. Application of Environment Model

The following two points become important to utilize the environmental model.

1) the method how to devide software development work into element tasks

2) the meaning of element task

3-1 Devide of element task

We propose the environmental structure or environmental level in software factories showin in Fig.1. The size of element task shall differ according to the level where the environmental modification is aimed.

3-2 Meaning of the element task

As the software development tasks are intellectual task and it sometimes progresses without motions or visible activities, the activities or motions are not always related with the element task. For example, when a worker stop his motion and keep quiet, traditional Industrial Engineering will regard this as "rest". However in the software development, it may be "thinking", not "rest". Therefore in software development, same activity of task may have different meaning, and in such cases different environmental conditions are required. Accordingly in software development, element task must be divided with the meaning of it.

4. Case study of Behavioral Analysis

To improve the environment of individual working zone, activities of system engineers were observed. As the result, eleven element tasks shown in table 1 were observed. However if we consider the meaning of these element tasks for the software development, some kinds of element tasks were divided into several different element tasks. That is,

"talking" and "hearing" ---> relax talking and hearing,
or communication talking and hearing

"reading" -----------------> relax reading or
information aquirement reading

"keeping silent" ----------> rest or thinking

Therefore we know that the software development task at individual working zone consists of 15 element tasks basically. Based on these element tasks, we modified the environment of individual working zone and obtained increment of software productivity (K. Honda 1985).

5. Conclusion

As software development is completely different from those that the traditional IE treated, to obtaine satisfactory fruits in environmental modification requires new concept and method and the proposed model seems to be effective for understanding environmental modification in software factories.

Reference

B.W.Boehm, Improving Software Productivity. Proc. of COMPCON, Fall 1981, 2-16, 1981

A.Komatsubara et al, Increasing Productivity in Computer Software Development by Environmental Modification, Proc. of HFS 27th Annual Meeting, 746-750, 1983

K.Honda et al, Research on Work Environment for Software Productivity Improvement, Proc. of COMPSAC. Fall 1985, 241-248, 1985

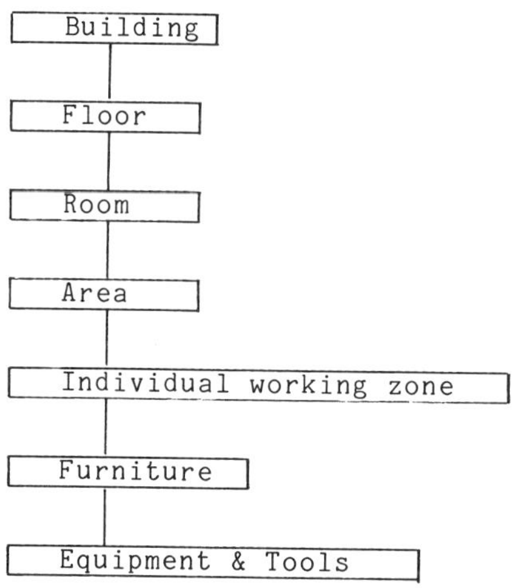

Fig.1 Level of Environment in Software Factory

Table 1 element tasks obtained by observation

1. Keybording	2. reading	3. talking
4. taking and moving articles		5. looking
6. writing	7. hearing	8. seeking
9. checking	10. keeping silent	11. arranging

Human Factors in Organizational Design and Management - II
O. Brown, Jr. and H.W. Hendrick (Editors)

ASSURING EFFICIENT CUSTOMER SERVICE - TOOLS TO MAKE IT FEASIBLE

AN ODAM WORKSHOP

James W. Suzansky, Ph.D.
and
Frederick A. Piumelli

Work Systems Associates, Inc.
65 West Valley Brook Road
Long Valley, New Jersey 07853

High quality customer service requires promptness and reliability from an organization's employees. Whether the service is sales, account inquiry, or equipment installation and repair, scheduling the proper number of employees at the times they are needed to meet customer demands is essential to the organization's continued success.

This workshop reviews the processes of "**Workforce Management**" described in a companion paper (Suzansky, 1986). These processes enable long- and short-range employee scheduling to assure that the proper number of employees are available to efficiently meet customer needs.

Recent innovations in personal computer capabilities have resulted in software which supports cost-effective use of **Workforce Management** in most organizations. The workshop uses a case study with hands-on demonstrations of the software and manual procedures available to assist this increasingly essential component of sound management practices.

Reference

[1] Suzansky, J. W., But I can never reach you... Assuring customer service through workforce planning and management, in Brown, O. and Hendrick, H.W. (eds.), Second International Symposium on Human Factors in Organizational Design and Management (North-Holland, Amsterdam, 1986).

Human Factors in Organizational Design and Management - II
O. Brown, Jr. and H.W. Hendrick (Editors)

Relationship Between Symbol-Background Color Differences and Legibility of a CRT Display

Isamu YOROIZAWA and Masayuki INOUE
NTT Electrical Communications Laboratories
1-2356 Take Yokosuka
JAPAN

Legibility of a CRT display is subjectively evaluated against symbol-background color difference. Results show that legibility becomes greater by increasing the difference in luminance and chromaticity. Moreover they show that the difference in luminance has a greater effect than that in chromaticity. Consequently, legibility can be defined by an equation derived from quantitative analysis. The results can be used as a standard indicator for selecting colors when creating color frames. In addition, this indicator can be applied to improvement in legibility for visual display terminal operators in an office.

1. INTRODUCTION

It is generally accepted that a color display is superior to a monochrome one in discrimination, visibility, readability, and attractiveness,[1] and that it even reduces user reaction time and misreading[2],[3]. It is obvious that suitable colors must be chosen to make the most of these advantages.

Several studies have already been made of the relation between color and legibility of characters[4]-[7]. However, no systematic investigations are found on the relationship between legibility and the differences in chromaticity and luminance. Thus the legibility of color displayed characters has never been quantitatively defined.

In order to obtain a standardized indicator of the legibility of various color combinations used for displayed characters and background, subjective evaluation tests were conducted from the following three viewpoints: (1) relation between character color and legibility on a colorless background (Experiment-1), (2)relation between legibility and difference in chromaticity between characters and background (Experiment-2), and (3)degree of influence on legibility by differences in chromaticity and differences in luminance (Experiment-3). In addition, in order to quantitatively define legibility, a legibility equation was derived using the results of Experiment-3.

2. TEST CONDITIONS

2. 1. DISPLAY EQUIPMENT USED FOR THE EXPERIMENT

For displaying colored text, a 20-inch high-definition color CRT display was used along with a video memory having a spatial address of 1280(horizontal) by 1024(vertical) pixels. Use of this display equipment allows the character color and background color to be varied independently.

2.2. DISPLAYED CHARACTERS

Since a text format can be considered a basic type of character display, a passage containing 85 Japanese characters was displayed. Characters consisted of 64X64 dots each. There were 16 pixels between each character, 48 pixels between lines, and 16 characters in each line.

2.3. SUBJECTS, SCALE, TEST PROCEDURES, AND VIEWING CONDITIONS

The subjects were ten students with normal vision. A scale of 1 to 5 shown in Table 1 was used to evaluate legibility. For each of the experiments, one frame of the text was displayed on the screen for 20 seconds. Within this time the subjects were asked to evaluate legibility of the characters. Viewing conditions were as shown in Table 2.

Table 1. Legibility evaluation scale

5	:	very legible
4	:	legible
3	:	ordinary
2	:	less legible
1	:	least legible

Table 2. Viewing conditions

CRT surface reflection luminance	15cd/m2 (max. in center of CRT surface)
Room illumination	180 lx on the CRT surface 500 lx at the subject
Viewing distance	2m

3. PRELIMINARY EXPERIMENT

In general the sensory properties of color are described as hue, saturation, and lightness, and for a standard quantitative expression of these the Munsell color system[8] is used. In order to use both physical quantitative expressions and these sensory expressions, the following preliminary experiment was performed. Samples of the 13 colors listed in Table 3 were lined up on the display, and the subjects were asked to judge brightness of the colors on a scale of 1 to 10. The judged value of brightness was taken as the equivalent of lightness in the Munsell color system, and the hue and saturation of each color were determined using the corresponding CIE chromaticity diagram, also as shown in Table 3.

Table 3. Colors used in preliminary experiment, Experiment-1, and Experiment-3

No.	Luminance [cd/m2]	Chromaticity x , y	Lightness JSC(C. L.)*	Hue/ Saturation
1	43.6	0.60, 0.34	5.4(0.88)	7.5R / 16
2	51.7	0.41, 0.23	4.8(0.59)	5 RP/ 16
3	80.2	0.26, 0.14	6.4(0.72)	5 P / 24
4	39.0	0.18, 0.09	4.2(0.48)	7.5PB/ 24
5	29.2	0.15, 0.08	4.5(1.09)	5 PB/ 22
6	187.0	0.20, 0.26	7.9(0.97)	5 B / 10
7	140.0	0.30, 0.58	7.5(0.88)	10 GY/ 17
8	157.0	0.33, 0.56	8.2(0.48)	7.5GY/ 15
9	208.0	0.40, 0.50	8.1(0.57)	2.5GY/ 11
10	79.2	0.51, 0.41	5.0(0.79)	5 YR/ 10
11	227.0	0.27, 0.28	10.0(-)	white
12	0.0	- , -	1.0(-)	black
13	26.	0.27, 0.28	3.7(0.88)	gray

*JSC means lightness score judged by subjects on a scale of 1 to 10. C.L. means 95% confidence limit of JSC.

4. EXPERIMENT-1 (RELATION BETWEEN CHARACTER COLOR AND LEGIBILITY)

In order to determine the legibility of individual colors, an evaluation was made of colored characters against a colorless background.

4. 1. CHARACTER COLOR AND BACKGROUND CONDITION

For the character colors, the 13 colors shown in Table 3 were used. Color of the background was black (luminance of 0 cd/m2 measured in direct contact with the screen) or gray (13.2 cd/m2).

4. 2. EXPERIMENTAL RESULTS

The relation between the average of the opinions (MOS: Mean Opinion Score) in this experiment and luminance and chromaticity are shown in Figures 1 and 2 respectively. The result of a gray background was similar to that of a black background. These results indicate the following: (1)higher luminance tended to result in greater legibility, and (2)if the CIE chromaticity diagram is bisected by a line running from the base point to white, colors distributed in the upper left half were more legible than those in the lower right half. In other words, hues in the spectrum bluish green--green--yellowish green--yellow--orange were judged more legible than red--purplish red--purple--violet--blue.

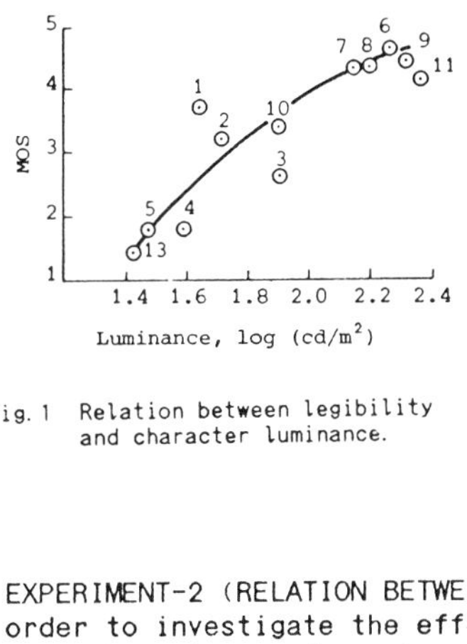

Fig. 1 Relation between legibility and character luminance.

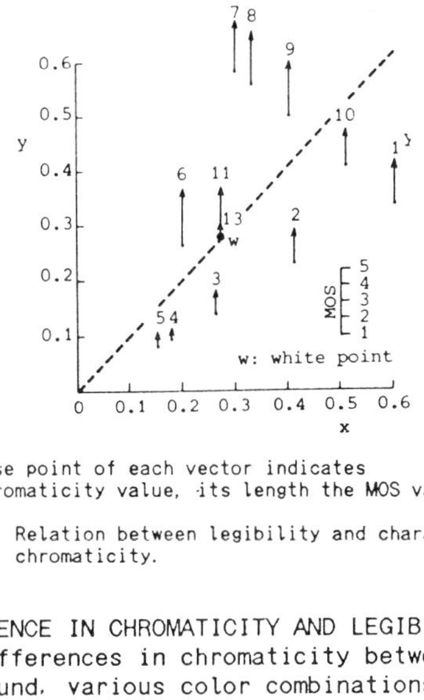

Base point of each vector indicates chromaticity value, its length the MOS value.

Fig. 2 Relation between legibility and character chromaticity.

5. EXPERIMENT-2 (RELATION BETWEEN DIFFERENCE IN CHROMATICITY AND LEGIBILITY)

In order to investigate the effect of differences in chromaticity between colored characters and a colored background, various color combinations were visually evaluated at a fixed level of lightness.

5. 1. COMBINATIONS OF CHARACTER COLOR AND BACKGROUND COLOR

At a lightness equivalent to 5 on the Munsell scale, 56 different combinations of color pairings of 8 colors shown in Table 4 were evaluated.

Table 4. Colors used in Experiment-2

No.	Luminance [cd/m2]	Chromaticity x, y	Chromaticity u', v'	Hue/ Saturation
1	50.0	0.52, 0.32	0.36, 0.50	5R / 12
2	47.5	0.47, 0.27	0.35, 0.46	10RP / 14
3	49.3	0.33, 0.19	0.29, 0.37	10P / 16
4	55.7	0.26, 0.18	0.22, 0.35	2.5P / 14
5	46.8	0.20, 0.18	0.17, 0.34	7.5PB/ 13
6	42.1	0.26, 0.42	0.14, 0.50	5 G / 8
7	41.4	0.30, 0.51	0.14, 0.54	10GY / 10
8	50.4	0.43, 0.40	0.25, 0.52	10YR / 6

5.2. EXPERIMENTAL RESULTS

The test results are shown in Figure 3. It can be seen that legibility improves as the distance on the chromaticity chart between the character color and the background color increases. Influence of hue of the character color and background color is shown in Figure 4, revealing that the greater the difference between the two hues, the greater the judged legibility. These points are also concluded in the case of lightness 6 from the results of additional experiments.

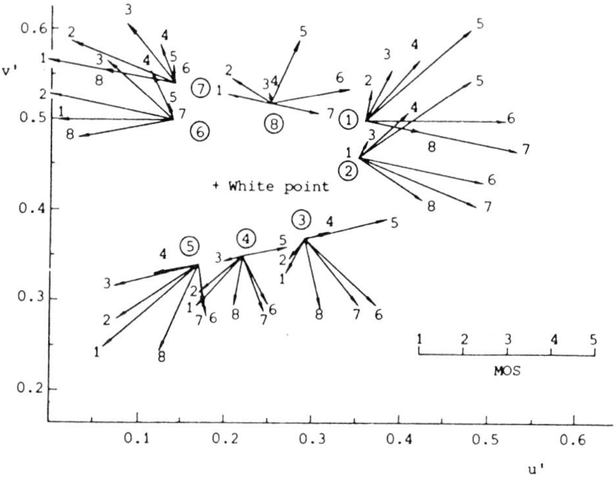

Circled numbers located at the vector base points indicate character color, and numbers located at vector endpoints indicate background color. The base point of each vector indicates character chromaticity value, and its length the MOS value. In addition, the direction of each vector was determined by combining the character's point and background's point on the CIE 1976 UCS chromaticity chart.

Fig. 3 Relation between legibility and character and background chromaticity.

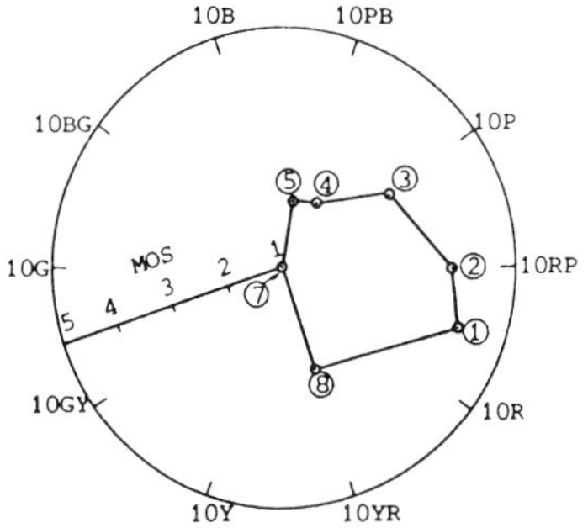

Lightness 5, background color 6. Circled numbers indicate character color.

Fig. 4 Samples of results showing the relation between legibility and character hue and background hue.

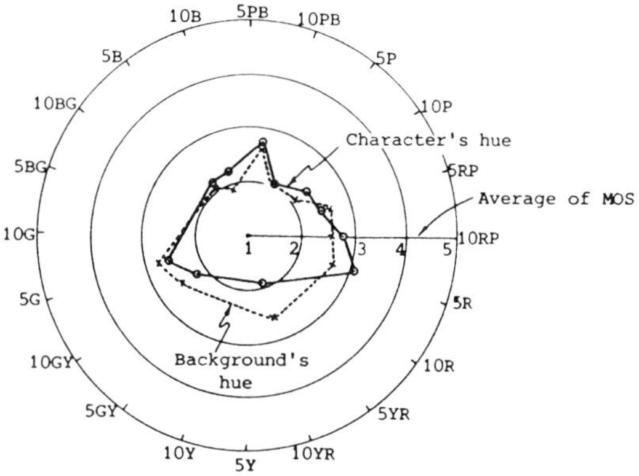

Fig. 5 Relation between legibility and character hue and background hue.

5.3. QUANTITATIVE EXPRESSION OF CHROMATICITY DIFFERENCE CORRESPONDING TO LEGIBILITY

In order to determine a quantitative expression of chromaticity difference corresponding to legibility of displayed characters, the following correlation coefficients(CC) were calculated. These CCs were between the four types of physical measures and the MOS values: (1)distance apart on the CIE chromaticity diagram: CC=0.48, (2)distance apart on the CIE 1976 UCS chromaticity diagram: CC=0.66, (3)distance apart on the Munsell hue circle that shows saturation in terms of distance from the center: CC=0.50, (4)the angle formed by the two colors on the Munsell hue circle: CC=0.47.

The second CC has the closest relationship with judged legibility. Therefore, distance apart on the CIE 1976 UCS chromaticity diagram was chosen for expression of chromaticity difference after this experiment.

5.4. RELATION BETWEEN HUE AND LEGIBILITY

From the results at lightnesses 5 and 6, average MOS values for each of the colors used for characters and background were obtained. The results, given in Figure 5, indicate the following two points: (1)colors which are highly legible for characters also tend to result in high legibility when used as background color, and (2)colors in the range green--yellowish green--yellow--orange were judged more legible than those in the purple--violet--blue range, which agrees closely with the pattern observed in the case of colored characters on a colorless background.

6. EXPERIMENT-3 (DEGREE OF INFLUENCE ON LEGIBILITY BY DIFFERENCES IN CHROMATICITY AND DIFFERENCES IN LUMINANCE)

In order to determine the extent of the effect on legibility by differences in chromaticity, and likewise by differences in luminance, a visual evaluation test was conducted in which the various colors were paired in terms of different chromaticity and different luminance.

6.1. COMBINATIONS OF CHARACTER COLOR AND BACKGROUND COLOR

For the character color all 13 of the colors listed in Table 3 were used, and for the background six colors (Nos. 1, 3, 5, 7, 9, 13 in Table 3) were chosen. Of the 78 possible combinations, eliminating those of the same color, 72 different pairings were evaluated.

6.2. EXPERIMENTAL RESULTS

The MOS values for each of the combinations are given in Table 5. Based on the results of section 4 and 5 above, differences in luminance and differences in chromaticity can be cited as physical factors governing legibility. Here the correlation between legibility and difference in luminance (correlation coefficient 0.79) is stronger than that of differences in chromaticity (0.10). This indicates that differences in luminance have a greater degree of influence on legibility.

Table 5. MOS values for various color combinations (Results of Experiment-3)

Char. / Background	1 7.5R	2 5RP	3 5P	4 7.5PB	5 5PB	6 5B	7 10GY	8 7.5GY	9 2.5GY	10 5YR	11 white	12 black	13 gray
1	—	1.0	1.6	1.7	1.8	3.7	3.4	3.3	4.2	1.0	4.0	3.9	2.6
3	1.6	1.6	—	1.8	2.6	2.6	2.2	2.1	3.2	1.3	3.4	3.2	2.3
5	2.2	2.7	3.2	1.0	—	3.4	4.1	3.7	4.5	2.8	4.6	1.6	1.1
7	3.0	3.0	2.1	2.3	3.6	1.4	—	1.0	1.9	2.2	2.4	3.7	3.2
9	3.4	3.6	2.6	3.1	3.9	1.7	2.0	1.2	—	3.1	1.0	3.7	3.6
13	2.2	2.6	3.0	1.6	1.8	4.5	4.4	4.3	4.6	3.3	4.0	3.3	—

7. AN EQUATION FOR DEFINING LEGIBILITY

To obtain an equation to define legibility, linear multiple regression analysis was carried out using the MOS values obtained in Experiment-3 as the criterion variable. The differences in luminance C (defined as absolute[log (BL/BB)], where BL is character luminance and BB is background luminance) and the differences in chromaticity U (distance apart on the CIE 1976 UCS chromaticity diagram) were used as predictor variables. As a result the following multiple regression equation was obtained.

$$MOS = 1.29 + 2.46C + 1.33U \qquad (1)$$

The multiple correlation is 0.80. Thus legibility can be defined to a certain extent by using this equation.

8. CONCLUSION

As a result of a series of subjective evaluations on the legibility of various combinations of character color and background color on a CRT display, the following conclusions are made: (1)the greater the difference in luminance and in chromaticity between the character color and background color, the greater the judged legibility, (2)of these, the difference in luminance had a greater effect than the difference in chromaticity, (3)for both character color and background color, colors in the range green--yellowish green--yellow--orange tended to have greater legibility than those in the purple--violet--blue range. Moreover, legibility of color text on a CRT display can be defined by using equation (1). The results given here can be used as a standard indicator for selecting colors for characters and background when creating color video frames. In addition, this indicator can be applied to improvement in legibility for visual display terminal operators in an office.

REFERENCES

[1] Hatada and Sakata:"Psychological visual perception and display"(in Japanese), J. of Institute of Television Engineers of Japan, Vol. 31, No. 4, pp245-255, (1977)

[2] Keister, R. S. :"Data entry performance on color versus monochromatic display", Proc. Hum. Factors Soc. 25th Annu. Meet., pp736-740, (1981)

[3] Kopala, C. J. :"The use of color-coded symbols in a highly dense situation display", Proc. Hum. Factors Soc. 23rd Annu. Meet., pp397-401, (1979)

[4] Institute of Color Science of Japan :"Color science hand-book"(in Japanese), Tokyo Univ. Press, (1980)

[5] Nishimura and Morimoto:"Legibility of color CRT display"(in Japanese), Tech. Report of Inst. Television Engineers of Japan, VVI35-1, (1980)

[6] Ookoshi, Yabiki and Uenishi:"Picture quality evaluation of character-graphic display and a consideration on color encoding method"(in Japanese), IECE Japan annu. meet., 1189, (1981)

[7] Welbank, M. :"An investigation of prestel coloured texts on coloured background", 10th International Symp. on Hum. Factors in Telecommunications, (1983)

[8] Wyszecki, G. and Stiles W. S. :"Color science", John Wiley & Sons, Inc., (1967)

Human Factors in Organizational Design and Management - II
O. Brown, Jr. and H.W. Hendrick (Editors)

Human Factors in Voice-Input Applications (II)

Rinzou Ebukuro

C&C Sensing Systems Development Div.
The NEC Corporation
Minatoku Shiba, Tokyo
Japan

Three mora place name word set requires little training in acquisition of voice input skill was presented in former conference. Further error analysis study have been performed on three mora word set in succeeding study and presented here. Word phonemic structure will give effect on the voice input performance. Definition of the disparity of word pair is given and experimental result of relationship between voice input performance and word pair disparity is demonstrated in this paper.

INTRODUCTION

Although the technology development in voice input application is actively beeing carried out and many successful application have been developed, little achievement in the study of performance analysis and comparison method has been attained. The importance of the study has been exclaimed to bring more successful application development of voice input system.[1]

This paper claims the importance of word set design and error word pair analysis and proposes new procedure of estimation method as one of the solutions for voice recognition estimation and comparison method and also the basic tool for the human factors analysis.

ASSUMPTIONS

Voice input performance will also be affected by word set property (assumption 1). If the word set is composed of similar word pairs, no applicable performance for practical use will be expected. However, the similar word pairs are always reminded by the voice input operators, the performance will supposedly be improved (assumption 2). If the word set is composed of different words, better performance can be expected (assumption 3).

DISPARITY ANALYSIS[2]

Disparity D between two words is defined as follows:

$$D = (J, K) \qquad \text{.......... (1)}$$

Where J: Phonemic number difference between two words.
K: Phoneme difference between two words.

Example: Disparity [TOMATO / TOBACO]

No phonemic number difference between the words and two phoneme differences are found on the word pair, thus the disparity D becomes (0, 2).

Disparity index I is also defined as follows:

$$I = J + K \qquad \text{..........} \quad (2)$$

Disparity index I for the above example becomes,

$$I = 0 + 2 = 2$$

Disparity coordinate D(J, K) is also defined. The above word pair example must be placed in the coordinate D(0, 2) as shown in Figure 1.

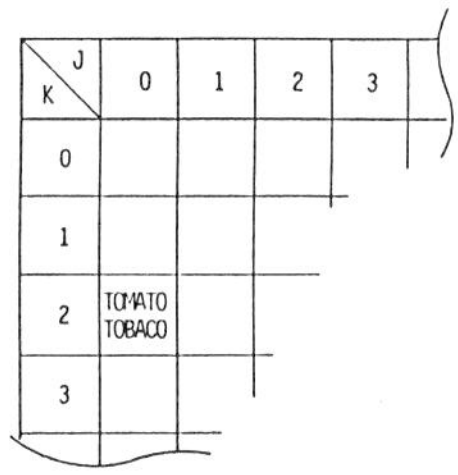

Figure 1
Disparity Coordinate

PREPARED WORD SET[3]

Three types of prepared word sets are presented in Table 1. G1 is composed of the word pairs having isolation index I ≥ 3 except two word pairs. G2 includes word pairs on I = 0 ∿ 6, and G3 includes I = 1 ∿ 6, but word pairs on I ≤ 2 are arranged in adjacent.

Disparity analysis result of the prepared word set is shown in Figure 2, and word pairs number placed in disparity coordinates are given in Figure 3.

METHOD

Japanese native speakers six male subjects ranges 24 ∿ 51 years old engineers participated to this examination.

All the vocalizations are recorded on the cassette tape in eleven passes for each word set. The third vocalization was used for standard pattern registration and fourth to tenth, total seven vocalization passes, were applied for data collection.

NEC DP-200 voice input unit programmed for isolated speech recognition with Shure SM-12A noise cancelling microphone were used for the test. Recorded vocalizations were input to DP-200 through its remote control terminal. All the recognition result were printed out on NEC NA-200 M20 printer.

Table 1 Word Set Groups

G1				G2				G3			
No.	WORDS	No.	WORDS	No.	WORDS	No.	WORDS	No.	WORDS	No.	WORDS
1	·A·OBA	11	·O·OTA	1	·ITAYE	11	HIRATA	1	·ITAYA	11	MAGARI
2	KUSABA	12	HI·OKA	2	TAKASE	12	YOSIMA	2	·IWAYA	12	FUNA·I
3	TAKI·O	13	·U·EDA	3	HIRANO	13	MAGAYA	3	TAKASE	13	FUNAKI
4	MIYOSI	14	NAKANO	4	YOSINO	14	·O·OTE	4	TAKA·E	14	NAKANO
5	TAKASE	15	TAKASE	5	MAGARI	15	YOSINO	5	HIRANO	15	·OKANO
6	SE·IKE	16	·OKA·O	6	FUNA·I	16	·OKANO	6	HIRATA	16	·OKA·O
7	TOYOMI	17	TOYOMI	7	·OKA·O	17	TAKA·E	7	YOSINO	17	·O·OMI
8	·ITAYA	18	KANO·O	8	·IWAYA	18	NAKANO	8	YOSIMA	18	·O·OTE
9	NAGATA	19	FUNA·I	9	NAGATA	19	FUNAKI	9	NAGATA		—
10	YAMAZU	20	YOSINO	10	TAKA·E	20	·O·OMI	10	MAGAYA		—

G1

```
 1 ·A·OBA
 2 KUSABA  22
 3 TAKI·O  32 15
 4 MIYOSI  23 06 15
 5 TAKASE  23 05 12 05
 6 SE·IKE  14 15 23 15 14
 7 TOYOMI  23 06 14 04 05 15
 8 ·ITAYA  13 13 24 14 14 24 15
 9 NAGATA  22 04 14 06 04 15 06 13
10 YAMAZU  23 05 14 06 04 15 06 14 04
12 ·O·OTE  03 24 33 23 23 13 22 14 24 24
13 HI·OKA  12 14 24 13 15 04 14 22 14 15 13
14 ·U·EDA  03 22 33 24 24 14 24 13 23 24 04 13
15 NAKANO  23 05 12 06 03 15 06 14 03 04 24 15 24
16 TAKASE  23 05 12 05 00 14 05 14 04 04 23 15 24 03
17 ·OKA·O  23 23 12 24 22 33 23 13 23 23 22 33 23 21 22
18 TOYOMI  23 06 14 04 05 15 00 15 06 06 22 14 24 06 05 23
11 KANO·O  31 14 03 14 14 24 14 24 14 14 32 23 33 13 14 13 14
19 FUNA·I  33 13 05 14 14 24 14 23 14 14 33 24 32 14 14 13 14 04
20 YOSINO  24 05 13 06 06 14 05 15 06 05 23 15 24 04 06 22 05 14 15
```

G2

```
 1 ·ITAYA
 2 TAKASE  14
 3 HIRANO  13 05
 4 YOSINO  15 06 04
 5 MAGARI  14 04 05 06
 6 FUNA·I  23 14 14 15 13
 7 ·OKA·O  13 22 22 22 23 13
 8 ·IWAYA  01 14 13 15 14 23 13
 9 NAGATA  13 04 05 06 03 14 23 13
10 TAKA·E  23 10 14 15 13 04 12 23 13
11 HIRATA  12 05 02 06 05 14 23 12 03 14
12 YOSIMA  14 06 06 02 06 15 23 14 05 15 05
13 MAGAYA  12 04 05 06 02 14 23 12 02 13 04 05
14 ·O·OTE  14 23 24 23 24 33 22 14 23 32 23 23 24
15 YOSINO  15 06 04 00 06 15 22 15 06 15 06 02 06 23
16 ·OKANO  04 13 12 12 14 23 10 04 14 22 14 14 14 13 12
17 TAKA·E  23 10 14 15 13 04 12 23 13 00 14 15 13 32 15 22
18 NAKANO  14 03 03 04 04 14 21 14 03 12 05 06 04 24 04 11 12
19 FUNAKI  14 05 05 06 04 10 23 14 05 14 05 06 05 24 06 14 14 05
20 ·O·OMI  14 24 24 23 23 32 22 14 24 33 24 23 24 02 23 13 33 24 23
```

G3

```
 1 ·ITAYA
 2 ·IWAYA  01
 3 TAKASE  14 14
 4 TAKA·E  23 23 10
 5 HIRANO  13 13 05 14
 6 HIRATA  12 12 05 14 02
 7 YOSINO  15 15 06 15 04 06
 8 YOSIMA  14 14 06 15 06 05 02
 9 NAGATA  13 13 04 13 05 04 06 05
10 MAGAYA  12 12 04 13 05 04 06 05 02
11 MAGARI  14 14 04 13 05 05 06 06 03 02
12 FUNA·I  23 23 14 04 14 14 15 15 14 14 13
13 FUNAKI  14 14 05 14 05 05 06 06 05 05 04 10
14 NAKANO  14 14 03 12 03 05 04 06 03 04 04 14 05
15 ·OKANO  04 04 13 22 12 14 12 14 14 14 14 23 14 11
16 ·OKA·O  13 13 22 12 22 23 22 23 23 23 23 13 23 21 10
17 ·O·OMI  14 14 24 33 24 24 23 22 24 24 23 32 23 24 13 22
18 ·O·OTE  14 14 24 32 24 23 23 23 23 24 24 33 24 24 13 22 02
```

Figure 2
Disparity Analysis Result for Table 1 Word Sets

K \ J	0	1	2	3
0	2	0	0	0
1	0	0	1	0
2	0	5	10	3
3	6	15	22	7
4	13	37	19	-
5	15	18	-	-
6	16	-	-	-

(1) G1

K \ J	0	1	2	3
0	2	4	0	0
1	1	1	1	0
2	6	11	8	3
3	5	17	24	3
4	15	32	11	-
5	16	13	-	-
6	17	-	-	-

(2) G2

K \ J	0	1	2	3
0	0	3	0	0
1	1	1	1	0
2	5	8	7	2
3	4	14	18	2
4	13	29	12	-
5	16	6	-	-
6	11	-	-	-

(3) G3

Figure 3
Word Pairs Distribution

RESULT

No substitution error was found in G1 word set test among six operators in seven passes repeated vocalizations.

Figure 4 (1) gives all substitution error word pairs for G2 word set vocalizations, and Figure 4 (2) shows all substitution error word pairs for G3 word set vocalizations by the six subjects.

Table 2 shows summary of total error E and rejection R in the tests.

Table 2 Total Error and Rejection Rate in Each Test

G1		G2		G3	
E	R	E	R	E	R
0	36	11	28	5	11
0%	4.3%	1.3%	3.3%	0.7%	1.5%

Relationship between disparity index I and error rate to the total utterance is shown in Figure 5. Error rate on disparity index I is calculated in following manner; for example, in Figure 4 (1) G2, total two substitution error word pairs are found in the coordinates D(10) and D(01), i.e. at I = 1, and then, total utterance in this test is 6 subjects x 7 passes x 20 words = 840, thus the error rate ε becomes, $\varepsilon = 2 \div 840 \times 100 = 0.23\%$.

K \ J	0	1
0	- -	F [FUNA.I / FUNAKI]
1	D [ITAYA / IWAYA]	B [NAKANO / ·OKANO] D [·OKANO / NAKANO] D [·OKANO / NAKANO]
2	D [ITAYA / IWAYA] E [YOSINO / YOSIMA] [YOSIMA / YOSINO] F [HIRATA / HIRANO]	F [·IWAYA / NAGAYA]
3	A [NAGATA / HIRATA] (A — Participant A)	0

(1) Error word pairs for G2

K \ J	0	1
0	-	D [FUNA.I / FUNAKI] D [FUNAKI / FUNA.I]
1	0	D [·OKANO / NAKANO]
2	0	B [·ITAYA / NAGAYA]
3	B [NAGATA / HIRATA]	0

(2) Error word pairs for G3

Figure 4
Error Word Pairs

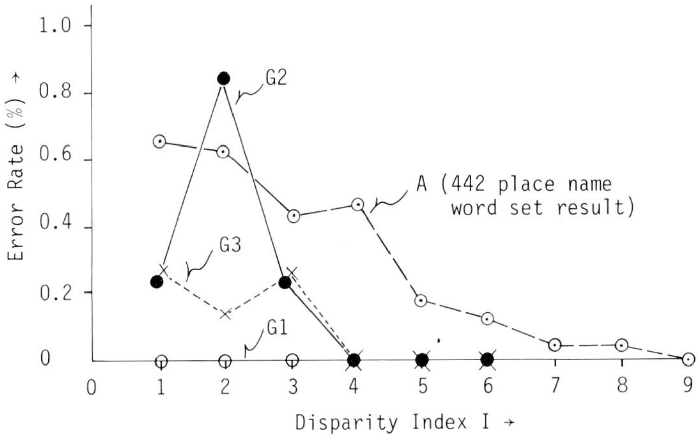

Figure 5
Relationship Between Disparity Index and Error Rate

Line A in Figure 5 is one of the other test result examples, put for further confirmation. The test was performed by four male subjects on 442 place name word list in which two to eight mora words were randomly listed in two pass vocalizations.

DISCUSSION

The scale of this experiment is not enough to give concrete result. However, following apparent tendency is found in the test result.

(1) No error word pairs are found in G1 word set test in which word pairs on $I \leq 2$ are excluded except two word pairs (result by these word pairs on $I = 0$ have been neglected). Assumption 3 will be explained with this result.

(2) Error rate in G2 word set test is larger than G3 test. This result will support the assumption 2.

(3) Thus the assumption 1 would be supported by those which above mentioned result.

(4) Error rate on the disparity index decreases as the index increase. This result is backed with the other larger scale test result, line A in Figure 5.

From these test result similar word pairs might be explained by the disparity index figure. In this case similar word pairs will be the word pairs on $I \leq 2$ or 3, and in the case of line A, $I \leq 3$, or 4.

CONCLUSION

(1) Word pair disparity analysis is presented and procedure for word set design and error analysis are given here. This disparity method will suggest one of the basic procedures for voice input performance estimation and comparison, and further the experimental result shows that the word set property will give some effect on human behavior especially in the acquisition of operation skill.

(2) Effect of the word set property which is characterized by the word phonemic structure in the word set on the voice input performance is also found in this experimental result and the concept of similar word pair from the phonemic word structure is presented and some rough figures are indicated from the experimental result.

(3) The relationship between word length and acquisition of voice input operation skills were discussed in the former paper [4]. The experimental result of this paper indicates that the phonemic structure of the words is also one of the factors on the acquisition of operation skills in addition to the word length.

REFERENCES

[1] James L. Franagan et. al., Automatic Speech Recognition in Severe Environments, Comittee on Computerized Speech Recognition, Comission on Engineering and technical Systems, National Reserch Council. National Academy Press. (1984)

[2] R. EBUKURO, Discussion on the Voice Input Word Set Estimation Method, proceedings of Academic Meeting of Tohoku Chapter of the Society of Instrument and Control Engineers (SICE) No. 203 (1984)

[3] R. EBUKURO, Estimation of Voice Input Performance with Estimated Word Set, Proceedings of Academic Meeting of Tohoku Chapter of SICE No. 204 (1984)

[4] R. EBUKURO, Human Factors in Voice-input Applications, Proceedings of First Symposium on Human Factors in ODAM, North-Holland, pp 199-203 (1984)

Human Factors in Organizational Design and Management - II
O. Brown, Jr. and H.W. Hendrick (Editors)

A Methodology for Concept Communication Support in Business Presentation Using Animated Graphics

Yutaka KASAHARA and Toshiaki MIYASHITA

C&C Systems Res. Labs., NEC Corporation
Miyazaki 4-1-1, Miyamae-ku, Kawasaki-shi,
Kanagawa, 213 JAPAN

A methodology for supporting concept communication, in business presentation using animated graphics, is proposed. Two important issues in the business presentation using animated graphics are: 1) Concept communication gap between a presenter and his audience. 2) Unskillful utilization of animated graphics. To cope with these problems, the methodology introduces: 1) Factors greatly influencing audience's way of thinking. 2) Factor control method with user friendly interface. The effectiveness of the methodology is discussed through presentation material creating experiments, using a prototype system implementing the methodology.

INTRODUCTION

The most important objective for a presenter in business presentation is to convey his or her ideas and intentions to the audience. Animated graphics not only assist a human in understanding the content of information and grasping a tendency, but also stimulates audience's insight and enhances their cognition[Thalmann et.al 1985]. Then, potential usefulness of business presentation using animated graphics seems to be widely recognized[Dworkin 1975, Thalmann et.al 1985].

However, business presentation using animated graphics is not so widely realized, despite the popularity of animated graphics in the fields of commercial film making. [Brooth et.al 1983, Reynolds 1982] These circumstances can be explained by the fact that utilization of animated graphics has not been fully explored in the business presentation field. Namely, it has not been clearly recognized how and in what situation to use animated graphics.

Then, the purposes of the proposed methodology are:
- To provide tools for skillful representation including animated graphics.
- To support precise communication of ideas and intentions between human beings by using animated graphics.
- Totally, to support making a consensus in the presentation environment.

CONCEPT OF BUSINESS PRESENTATION USING ANIMATED GRAPHICS

Features of Business Presentation Using Animated Graphics

Animated graphics can represent time variant information and spatially moving items in motion picture format. Business presentation using animated graphics has the following desirable features.

(1) Representation of "PLUS 1" information dimension
Animated graphics can be used to represent one more information dimension, such as infomation change along time and artistic sense or feelings, in addition to 3-D geometric dimensions and other attribute dimensions like color.

(2) Impressive information representation for human beings
Movement is probably one of the most dominant visual forces in human experiences.[Dondis 1973] Based on human experiences regarding visual susceptibility to motion, a motion picture can represent visual illusion and psychologically impressive images. Thus, it becomes possible to represent feelings and/or imaginative concepts, which would require many words to explain or might even be impossible to explain using words only.

(3) Representation for thought changing
Animated graphics can represent thought changing and idea flow, as well as time-variant information and spatially moving items, in a motion picture. This way of utilizing animated graphics is one of the most important characteristics in business presentation.

These characteristics, if appropriately utilized, assist the presenter in building up his own concept into a visually attractive image and communicating the concept to the audience.

Problems in the Business Presentation Using Animated Graphics

First problem is concept communication gap between presenters and audience. Figure 1 shows information flow from a presenter to his or her audience. Original information is naturally or sometimes intentionally transformed as its flow brings the data to the audience. In other words, original facts and data are changed by adding a specific intention, changing media according to the individual differences in the manners and skills for presenters.(see Fig. 1) Under these circumstances, the information determined by the audience often has a certain difference from the information which a presenter wishes to convey.

Second problem is unskillful utilization of animated graphics. Because the animated graphics have extraordinary expression power, if enhanced presentation materials are inappropriately created, the possibility of a misunderstanding arising is much higher than in other ordinal presentations.

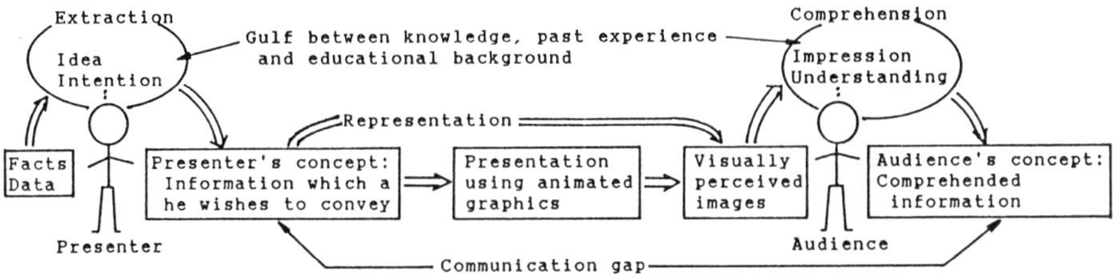

Figure 1. Information flow and concept communication

A METHODOLOGY FOR PRECISELY COMMUNICATING PRESENTER'S CONCEPT

A methodology, coping with the above mentioned problems, is to provide a mechanism for logically bridging the concept communication gap. The following are the two phases to accomplish the methodology.

- Analyzing key factors for business presentation.
- Providing users factor controlling mechanism with friendly interface.

Figure 2 shows the concept of the methodology.

Analyzing Key Factors for Business Presentation

In this phase, analysis proceeds according to the following steps based on information flow(see Figure 1) in the business presentation.

- Structure a business presentation.
- Extract factors for reducing the concept communication gap.
- Assign the factors at the related hierarchy in the structure.

Figure 3 shows a scenario model for a presentation using animated graphics. The model reflects the presenter's scenario writing process and has a hierarchical structure. In the model, the story should be coherent throughout the entire presentation. A story is composed of various scenes and their combinations. A scene is a presentation period representing one logically continuing concept. In other words, motion in a scene has continuity in time, location and thought dimension. A scene is composed of background, moving objects, text strings for explanations and their combinations.

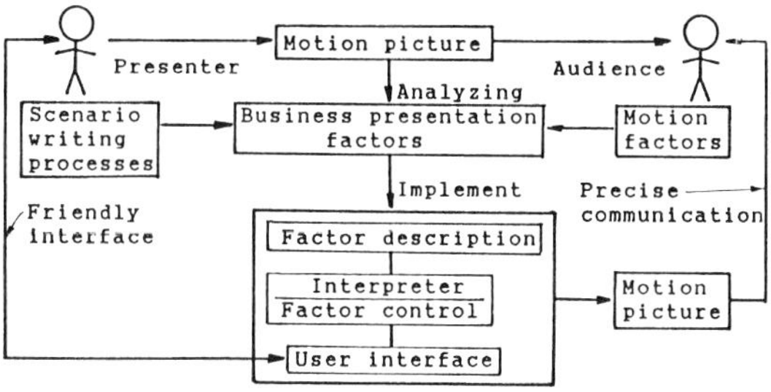

Figure 2. Concept of the methodology

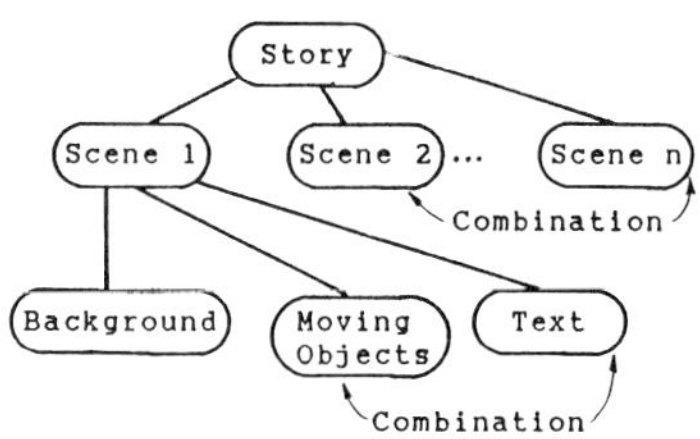

Figure 3. A scenario model structure

Business presentation factors

As previously described, The most important thing in a business presentation is to precisely represent the presenter's intention and feelings. Through the analysis of human scenario writing process and human motion perception, factors greatly influencing audience and factors reflecting the presenter's way of thinking are extracted. For example, the camera work factor relates intention appeal. Text enhancement such as blinking, reversing and their timings relate main points appeal. Table 1 shows the factors for business presentation and the corresponding human aspects according to the related hierarchy in the scenario model. The factors described in the Table 1 assist presenters in creating perceptible presentation materials for audiences. Consequently, these factors can support reducing the concept communication gap between a presenter and audience.

Table 1. Factors for effective business presentations

Level	Factors	Human ability for cognition and representation
Story	-Time for entire presentation -Scene combination Scene sequence Scene change Inter-scene effect -Scene layout Superimposing Multi-viewing	-Desired time -Logic progress Thought flow Topics change Attention draw -Paralell imaging Image overlapping Image concurrency
Scene	-Camera position -Camera work Viewing direction Viewing angle Camera motion -Combination of components Relative motion Motion timings Text synchronization	-Easy to create and attractive to the eye -Intention appeal Appeal a specific item Spatial imaging range Artistic sense or realistic feelings -Feeling Spatial Encompassment Feelings and artistic sense Thought progress
Component	-Object motion Speed Acceleration Smoothness -Text enhancement Highlighting, Blinking, Reversing, Coloring -Background motion	-Object oriented thinking Understanding speed Strong appeal Syncronization with the way of thinking -Mainpoints appeal Appeal specific text strings -Audience movement

Factor Control Mechanism and User Interface Design

The proposed mechanism is to provide methods for controlling important factors shown in Table. 1, as well as ordinal parameter description for creating a motion picture such as object shape and motion path.(see Fig. 4) Also, the factor specifying process for the presentation should be functionally easy to handle. Following described are features of user interfaces design.

(a) Introducing know-how of professional users
It is difficult for novice users to create pertinent presentation, even if a friendly interface were provided. To cope with this human ability limitation to encompass spatial imaginations, the following mechanism is provided.
- Database for professional user's know-how regarding the factors such as camera work and artistic sense.
- Providing selection and combination tools for these know-how.

(b) Standard value setting for the factors
Regarding the motion factors such as the speed, acceleration and text blinking and its timing, user specified values cannot often create pertinent motion. This is because users cannot imagine the relationships between the factors and the resulting motion. To cope with the problem, the system provide default values for these factors based on the psychological experiments, which extract the standard factor values for assisting human understanding of the resulting presentation.

(c) Factor modification using relative expression
Modifying the factor values is also difficult because of the same reason for the factor specifying process. To cope with the problem, a selection mechanism of relative expression, such as a little more smooth and faster than the previous version, is introduced.

(d) Timing chart
Timing chart is an important interface at the scene level. Figure 5 shows a timing chart design, which can specify object motion start and stop time, relative motion, text synchronization with the object motion and more. By graphically editing the timing chart, almost all motion combinations or relations can be controlled.

Other considerations for the user interface design are:
- Using selection method rather than commands specifying.
- Adopting mouse based graphic interface as many as possible.
- Using user terminology rather than using technical jargons.

These features for specifying the factors enable even novice users to create pertinent presentation materials for precise concept communication.

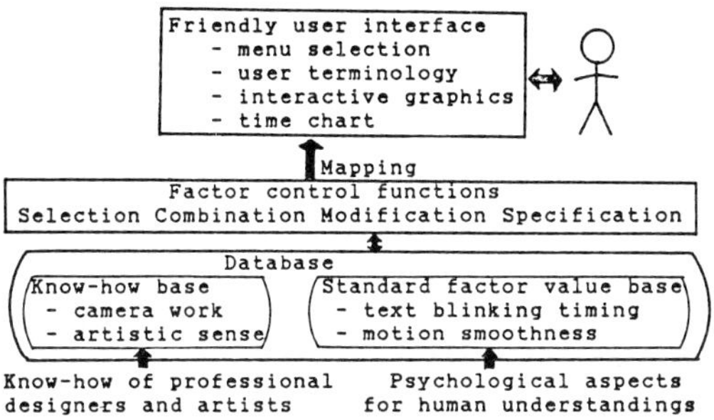

Figure 4. Factor control mechanism

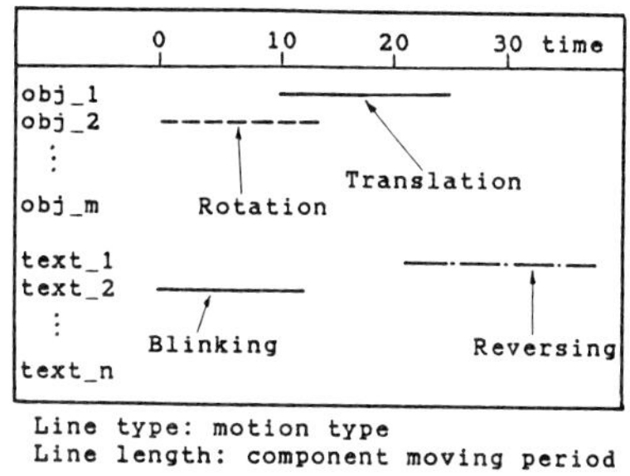

Figure 5. Timing chart design

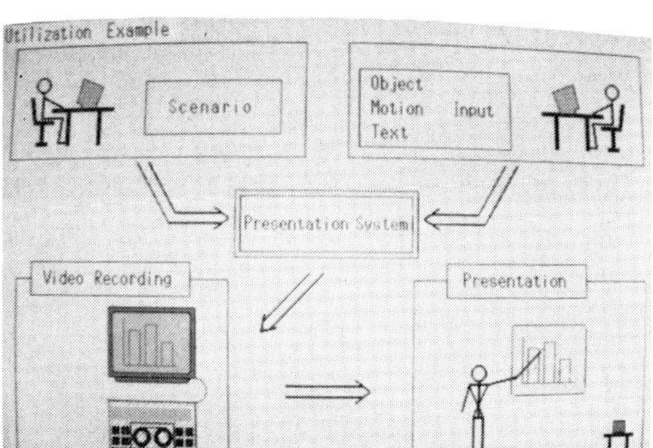

Figure 6. A system utilization image

Figure 7. Title presentation

EXAMPLES

A personal computer based prototype system, based on the methodology is still under development. However, some experiments for creating business presentation materials including motion picture were performed, by partly using the functions for the prototype system. Figure 6 shows a system utilization image created by using the functions of prototype system itself. 2D drawing function, combination of text strings and objects, scene layout are effectively used in enhancing the presentation. Figure 7 shows a frame in the title presentation. Title lettering data are input in 2D. Then, the system generates 3D shapes by sweeping. In regard to specifying the pertinent motion speed and smoothness, motion factors in the know how base and relative expression for factor modification are effectively used.

The examples shown here are all too simple to evaluate the usefulness of the methodology. However, through the motion picture creation processes, it has been confirmed that the consideration of the factors decides whether the resulting presentation can enhance the audience's impression and understanding.

CONCLUSION

A methodology for concept communication support in business presentation was proposed. Two important issues in the business presentation using animated graphics are: 1) Concept communication gap between a presenter and his audience. 2) Unskillful utilization of animated graphics. To cope with these problems, it was suggested that the methodology introduces: 1) Factors greatly influencing audience's way of thinking. 2) Factor control mechanism with user friendly interface.

In recent years, with the advent of personal computers and other office equipments, the opportunities for communicating through information media are more and more increasing in the office and educational environments. Then, the methodology, to reduce concept communication gap and to support making a consensus, should be effective in the business environment. It is also believed that developing the methodology is a first step toward improving human idea representation and concept communication using animated graphics, although the extraction of factors closely relating to human cognition is still a matter for debate.

ACKNOWLEDGMENT

The authors would like to thank Mr. M. Managaki for his encouragement and Mr. T. Tsurutani for discussion and suggestions. They would also like to thank members involved in the prototype system development.

REFERENCES

[1] Brooth, K. S., Kochanek, D. H. and Wein, M., Computers Animate Film and Video, IEEE Spectrum, Feb. 1983, 44-51.

[2] Dondis, D. A., A Primer of Visual Literacy, 1973, SAIENSU-Sha,1982(translated into Japanese).

[3] Dworkin, s., Technical Communication with Motion Picture, A Guide for Better Technical Presentation, IEEE PRESS, 1975, 200-203.

[4] Kasahara, Y., Miyashita, T., Kitakaze, S. and Tsurutani, T., A Presentation Support System Concept, Proc. IPSJ(Information Processing Society of Japan) 31st. Conf., Sept. 1985(in Japanese).

[5] Kasahara,Y., Miyashita,T. and Tsurutani,T., An Animated Things Model for Bridging the Communication Gap, IPSJ, Technical group memo on Graphics and CAD, July 1985(in Japanese).

[6] Reynolds, C. W., Computer Animation with Scripts and Actors, Computer Graphics Vol.16, No.3, 1982, 289-296.

[7] Thalmann, N. M. and Thalmann, D., A New Tool for Business Graphics: The Animated Data System, Proc. Trends & Applications using Computer Graphics, IEEE, May 1985, 116-122.

TRAINING SYSTEMS

Human Factors in Organizational Design and Management - II
O. Brown, Jr. and H.W. Hendrick (Editors)
Elsevier Science Publishers B.V. (North-Holland), 1986

A Hand-Held Training Aid In A Military Environment: Description and Proposed Evaluation[1]

Carol Ann Robinson Ph.D.

Navy Personnel Research and Development Center
San Diego, CA 92152

ABSTRACT

This paper describes a portable, low-cost, computerized, hand-held training aid that can supplement classroom training and provide off-site training and practice opportunities in a variety of Army, Air Force, and Navy technical specialties. The engineering design, software and courseware incorporated the latest in electronic, human factors, and training technologies. Software and courseware were designed to provide intrinsic motivation and enhance the effectiveness of training. The evaluation plan that describes the research questions and associated measures is also presented.

BACKGROUND

Recent advances in semiconductor technology have made possible the development of low-cost, hand-held, computerized training devices. These devices offer many of the features of larger, traditional, computer tutoring systems, as well as the ability to bring military training to nontraditional learning environments, such as study halls or barracks or in the field.

The U.S. Army Research Institute for the Behavioral and Social Sciences (ARI) commissioned the development of a small, low-cost, battery-operated, hand-held computer with a one-line video display screen and voice/sound simulation capability that can be used with a booklet for introducing new military terminology. The prototype device, called TUTOR, responds to differing knowledge and motivational levels and rates of learning. It provides frequent corrective feedback and a variety of instructional routines that include gaming features to provide intrinsic motivation and enhance the effectiveness of training.

Field tests showed that students used the TUTOR widely and, with equivalent amounts of training time, they retained material presented by TUTOR more effectively than material presented by conventional media (Wisher, 1985).

The success of TUTOR encouraged wider applications for this device. A joint military service effort to develop a training aid that can be used by all branches of the U.S. military has produced the computerized hand-held instructional prototype or CHIP. CHIP is not limited to job-related vocabulary training but can be used to also teach job-related procedural tasks. This upgraded device includes expanded gaming features, record-keeping, progress-checking, and courseware authoring capabilities.

DESCRIPTION

Like TUTOR, the CHIP technology is designed around a familiar and useful medium--the book. This book or booklet contains graphics, instructional text, and test questions that are used with CHIP's prompts and feedback. Separate plug-in cartridges and different accompanying booklets will enable CHIP to supplement classroom training for various technical training schools or to provide remedial training, on-the-job refresher, or sustainment training.

Hardware Description

CHIP (Figure 1) is about the shape and size of a three-ring binder and weighs, including batteries but excluding the booklet, about 5 pounds.

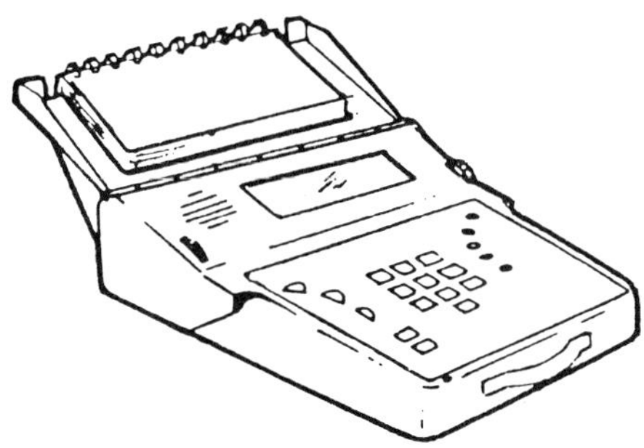

Figure 1. CHIP.

CHIP was designed to be easy to carry, set up, and operate under a variety of conditions. Its case, display, keyboard, and audio and video features incorporate the latest in human factors technology. The lid snaps open and folds back to support the booklet, which places the booklet in the same visual plane as the display screen. The pages of the booklet flip up over the top rather than to the side, minimizing the likelihood of the wind accidently turning the pages when CHIP is used in a field environment.

The audiovisual area slopes up from the keyboard area for an optimum viewing angle of approximately 22 degrees. The speaker and display screen are side by side to make audiovisual outputs emanate from the same general area. A liquid crystal display screen capable of displaying eight 40-character lines allows simultaneous display of several lines of text and arcade type gaming graphics.

The audio and video controls and the most frequently used input keys are arranged so that they can be easily operated by the thumb. This feature allows for the student to devote more attention to the visual information on the screen and in the booklet and less attention to finding the right keys.

Keys A,B,C,D, and E (Figure 2), which are used extensively in responding to multiple-choice questions, are laid out in an arc on the right side. An embossed ridge bisecting the key C, as well as an embossed dimple, help the students find the center key with their thumb. The five YES, NO, GO, ERASE, and SAY special function keys are on the left side with the Go key identified by an embossed dimple and ridge. The less frequently used numerical keys are arranged in a 3x3x1 pattern in the center of CHIP.

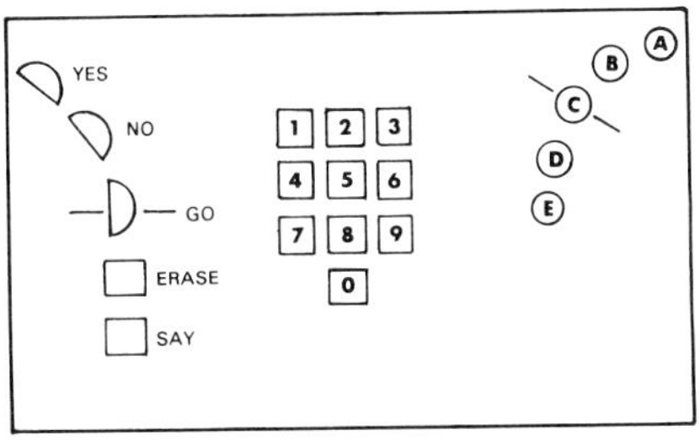

Figure 2. Keyboard Layout.

The CHIP's main printed circuit board contains a microcontroller, memory, logic, speech, and driver circuits. The removable courseware cartridge contains a smaller printed circuit board, which is enclosed in a plastic housing for protection.

CHIP is powered by four or eight removable D-size batteries. With eight nickel-cadmium batteries, CHIP will operate for about 65 hours without recharging.

Software/Courseware Design

The basic philosophy behind the development of CHIP is that learning will be more likely to occur if it is convenient and interesting. CHIP makes learning more convenient than conventional methods by being easily transportable. CHIP makes learning more interesting by including several instructional routines based on training principles from cognitive psychology and video gaming techniques to improve intrinsic motivation and training effectiveness.

Four independent but mutually supportive instructional routines that can be used for any training application were developed. These are:

1. Pregame. This routine has two modes: warmup and explanation. The warmup consists of questions in the booklet. The students respond to each question by pushing the appropriate key on the keyboard. After the students have finished all the questions, voiced feedback tells them the number of correct answers. The explanation mode which follows, consists of tutorial material containing embedded questions with immediate feedback. The pregame does not include any gaming features because they might distract students during the initial explanation process (Malone, 1981).

2. Roll Call. This drill and practice gaming technique requires the students to match a list of ten words with their definitions, or vice versa. Three choices of answers per word or definition continue to cycle on the screen, one at a time, until the student selects an answer. The goal of the game is to create on the screen a formation of 10 soldiers, as at morning muster. For each correct answer selected, CHIP says "Here, Sir" and adds a soldier to the formation on the screen; for each incorrect answer, CHIP responds "Missing"

and displays a blank slot in the formation. If the student scores 100 percent, the CHIP says "All present and accounted for, Sir."

3. Target Practice. This routine may be used to augment vocabulary training or present general multiple-choice or matching questions. Typically, CHIP asks questions about a picture in the booklet. The gaming technique in this exercise features a "friendly" and an "unfriendly" projectile at either end of the screen. Each correct answer selected on CHIP moves the friendly projectile along a trajectory toward a target; an incorrect answer moves the unfriendly trajectory. A new projectile is fired upon each response, its distance depending on the number of correct and incorrect responses to date.

4. Mine Field. This instructional routine reviews and drills in step-by-step procedures. There are two principal modes: A job-step review that displays the steps of a job one at a time in the order they should be performed and a game that provides drills about the sequence in which these steps are performed. The main objective of this game mode is for students to successfully progress through a mine field by correct responses to procedural steps.

The other three routines can also be used to supplement the job-step routine in training procedures. To accommodate individual differences in learning and also to provide challenge to the student, each game has a basic and an advanced level.

Courseware is typically divided into 30 or 40 lesson segments. Each lesson segment uses the routines that are applicable to its subject matter. Students control the order in which they work on lesson segments.

CHIP includes software to keep and retrieve student records including student number, the number of instructional units completed, and student performance data. These records can be uploaded to a desktop microprocessor for detailed analysis by training personnel. An authoring aid allows training personnel to create new instructional bases.

PROPOSED EVALUATION

CHIP will be evaluated in at least five applications. Each service will select the appropriate test environment (e.g., classroom or field) and use (e.g., classroom training or refresher training).

The performance of students trained by CHIP will be compared with that of students trained by conventional methods. The significant research questions and associated measures to be analyzed (listed below) will be adapted as needed to particular test applications:

1. The effectiveness of CHIP for training will be determined by final exam games, learning gains as reflected in the difference in pre- and posttest scores, time-to-completion of course segments, academic setbacks, attrition rates, hands-on tests, and transfer of training.

2. The usage patterns for CHIP will be analyzed by determining the time to complete given segments and the human factors features that affect use (e.g., readability, sound, ease of using booklet, key placement, and so on).

3. Student and instructor attitudes toward CHIP, computer-assisted instruction, traditional instruction, and related factors will be assessed by a questionnaire.

4. The background and ability of the students using CHIP will be determined by an assessment of ASVAB scores, background knowledge, age, and educational history.

5. How usage patterns, attitudes, and student ability and background relate to training effectiveness will be determined.

Human factors and logistical data such as storage and distribution requirements, reliability of devices, and transportability will also be collected.

Evaluation findings from all applications will be integrated and collapsed for the preparation of a cost and training effective analysis (CTEA). Findings will include data on training effectiveness, usage patterns, reliability, availability and maintainability.

Based on TUTOR's evaluation students using CHIP are expected to show better retention of materials. Research suggest that military student attitudes are positive toward computer-aided instruction (Robinson, Tomblin, & Houston, 1981 and Robinson, Smith, & Thode, in preparation). Preliminary results suggest that expanded gaming features result in greater usage, and retention of instructional materials.

[1] The views expressed in this article are those of the author, are not official, and do not necessarily reflect the views of the Navy Department.

REFERENCES

[1] Malone, T.W. Toward a theory of intrinsically motivating instruction. Cognitive Science, 4, (1981) 333-369.

[2] Robinson, C.A., Tomblin, E.A. and Houston, A. Computer-managed instruction in Navy technical training: an attitudinal survey (NPRDC Tech. Rep. 82-19) (Navy Personnel Research and Development Center, San Diego, December 1981).

[3] Robinson, C.A., Smith W.H., and Thode W.F. Microprocessor-based on-site training for passive acoustic analysts: II. Lesson development and evaluation. (Navy Personnel Research and Development Center, San Diego, in preparation).

[4] Wisher, R.A. The development and test of a hand-held computerized training aid. Paper presented at NATO Symposium on Training Technology (Brussels, Belgium, April 1985).

Human Factors in Organizational Design and Management - II
O. Brown, Jr. and H.W. Hendrick (Editors)

Ergonomic Knowledge in a U.S. Automotive Plant: Correlates of Practical and Theoretical Job Analysis Capabilities

Jeffrey K. Liker and Bradley S. Joseph

University of Michigan
Industrial and Operations Engineering
Center for Ergonomics
Ann Arbor, MI 48109

Ergonomic knowledge is typically applied in industry by employees with relatively little formal education in the field. Many companies are sending employees to short training courses to improve ergonomic knowledge and skills. Yet little is known about the competence of these personnel to effectively analyze operator stresses before or after short training courses. This paper examines the distribution of ergonomic knowledge across a broad range of personnel in a large automotive parts plant, and assesses the effects of training on ergonomic knowledge and skills.

INTRODUCTION

Ergonomic research and development on the relationship between job/workplace design and physical stresses on operators has yielded a plethora of insights and practical tools for designing ergonomically sound workplaces. In moving from theory to industrial application, a variety of technology transfer mechanisms are possible. These include direct application of analysis and design principles by university faculty and their students (e.g., as consultants or full-time staff members), training sessions to transfer knowledge to existing personnel involved in job and workplace design who typically lack formal education on ergonomics, and the use of ergonomic taskforces in industrial plants in which ergonomic experts become a "resource" to the task force (Liker et al.,1984; Joseph,1986).

The use of training in ergonomic principles and the ergonomic task force approach appear to be on the rise in industry. For example, Ford Motor Company recently let a major multi-year contract with the U.M. Center for Ergonomics in which a primary deliverable has been training of Ford's engineers, safety professionals, supervisors, and in some cases even shop-floor employees. In a contract with the UAW in 1984, a fund based on employee and company contributions was established for major U.S. automakers to improve health and saftey in the workplace. Part of this multi-million dollar fund is earmarked for training in ergonomics. Regardless of whether these events in the auto industry represent a statistically reliable trend toward greater ergonomic awareness for industry in general, it is clear that many companies are relying on in-house personnel to analyze and redesign jobs, and the main knowledge base of these "ergonomic practitioners" is often a combination of personal experience and the knowledge gained in relatively brief (i.e., several days at the most) ergonomics courses.

Such company practices raise obvious questions about the competence of these personnel to effectively analyze operator stresses. This paper presents preliminary data bearing on the following issues: To what degree can existing industrial personnel industrial personnel accurately diagnose operator stresses associated with job and workplace design? Do training sessions in ergonomics actually improve the analytic abilities of industrial personnel? Does industrial experience provide insights that are consistent with scientific knowledge on operator stresses? Does general education help equip plant personnel to analyze operator stresses? Do hourly workers have special insights because of their direct experience on the job?

THE STUDY

Early in 1983 over forty industrial personnel in an automotive transmission parts plant were selected to participate in one of several task forces established to analyze and redesign jobs and workplaces to reduce operator stresses (Joseph et al.,1986). Membership in these groups was heterogeneous in terms of education, years of industrial experience, and job responsibilities -- including engineers, managers, skilled maintenance workers, and job setters. In preparation for their role in the ergonomics program, all participants attended a four- or eight-hour training session conducted by ergonomics experts from the U.M.'s Center for Ergonomics. The training was performed by Thomas Armstrong, a well known authority on ergonomics, using traditional lecture methods with extensive visual aids (sophisticated slides and models of the human anatomy). Both prior to and after this training, participants were tested on their ability to recognize work related stresses on the upper extremity (i.e., shoulder, elbow, and wrist). The tests included true-false questions and analysis of actual jobs shown on slides. Participants' analyses of the degree of stress on the shoulder, elbow, and wrist on operators shown in the slides (using a five-point scale ranging from no stress to high stress) was compared to analysis by experts in the Center for Ergonomics. The five experts who analyzed the jobs were in high agreement on the degree of stress. Details on the scoring procedures and measures of agreement among experts can be found in Joseph (1986: Appendix A).

The two testing methods appear to test two types of knowledge. The true/false questions test what Ceci and Liker (1986a) refer to as "academic knowledge," which is more traditional in scope and concentrates on the assimilation and memorization of factual information. By contrast, analysis of jobs depicted in slides is a better simulation of the actual mental processes involved in analyzing real jobs. Ceci and Liker refer to this type of reasoning process as "non-academic knowledge." For example, in their studies of expertise in racetrack handicapping (Ceci and Liker, 1986b) demonstrate that "experts" use highly sophisticated cognitive processes compared to "non-experts;" yet there is no correlation between groups in measured I.Q. or education (I.Q. and education are very highly correlated). Other cognitive-anthropological studies have found examples of persons who perform low on I.Q. tests yet regularly engage in what appear to be sophisticated reasoning in the context of their jobs (Super,1980; Fahrmeier,1984). This context-specific reasoning is more likely to be reflected in analysis of job stress in the slides.

These studies led us to hypothesize that individuals with greater formal education would score higher on the true-false questions compared to their less educated counterparts, but industrial experience would play a greater role in analysis of actual jobs depicted in slides. We also expected the training to be more effective for the more highly educated as it was conducted in a traditional lecture format. Fortunately, there was virtually no correlation between education and years of industrial experience so collinearity between these independent variables did not need to be addressed in the analysis.

RESULTS

Pre- and Post-Training Comparison

Table 1 shows the average ergonomic quiz scores for all participants who took the training and completed the pre- and post- training quizzes. Scores are presented separately for the assessment of stress in jobs depicted in slides and factual knowledge questions. For each slide, the subject was directed to rate the stress on the shoulder, wrist, or elbow. Therefore, Table 1 presents separate scores for slides emphasizing each specific joint (approximately 8 slides/joint).

There was a general tendency for all scores to increase immediately after the training indicating that the training generally increased participants' factual knowledge of ergonomics and their ability to detect stressful work situations. Statistically significant differences (.05 level) between the pre- and post-scores were found for the wrist slides and the true/false-multiple choice questions. Note also that participants tended to be particularly bad at rating stress on the wrist prior to training. Stresses to the wrist might be less intuitive than stresses on the shoulder and elbow. The ability to spot these subtle stresses to the wrist better distinguishes ergonomic abilities across subgroups (discussed below) and is also more sensitive to training compared with the more obvious stresses of the elbow and shoulder. Moreover, the five experts were less likely to be in agreement on the wrist.[1]

TABLE 1
Percent Correct on Pre- and Post-Training Quiz Scores by Question Type (all persons who completed training and both quizzes, N=43)

Question Type	Pre-Training Scores	Post-Training Scores	Difference
Shoulder Slides	77.6	78.9	1.3
Wrist Slides	66.5	73.4	6.9*
Elbow Slides	68.2	68.5	0.3
True/False	81.7	90.7	9.0*

*P-value less than 0.05

It should also be noted that the pre-training test scores, in particular, the true/false-multiple choice questions, averaged 73.5 percent. This demonstrates considerable intuitive knowledge from on-the-job experience. (No one in the sample had prior formal education in ergonomics.) Another explanation for this may be that the test was too "easy" (i.e., obvious). Because of time constraints, a proper pre-test was not run to determine if this were true.

[1] Agreement was measured by correlations between ratings of slides across pairs of experts. There was stronger agreement among the experts on the degree of stress on the shoulder (average r=.87) and elbow (average r=.76) compared with the wrist (average r=.61).

While participants scored higher on quiz scores immediately after the training, did they retain this knowledge for later use? If the knowledge gained by the training is lost in a relatively short period of time, we cannot expect the training to improve the overall effectiveness of an ergonomics program. To test retention of knowledge, a sub-sample of participants who took the post-training quiz immediately after the training took the quiz again two to three weeks after the training. Overall, there was no significant loss or gain in ergonomic knowledge.

The Role of Education and Experience

The high pre-training scores indicate a substantial amount of intuitive ergonomic knowledge, yet these mean scores mask considerable variation in individual scores. Why did some individuals score higher than others? The correlates of ergonomic knowledge provide insights into the process by which this knowledge was acquired and the types of training that might be targeted to particular subgroups. Table 2 summarizes correlations between education and experience with pre-training quiz scores.

We expected education to be correlated with factual knowledge (i.e., true/false and multiple-choice factual questions) but not necessarily with the knowledge needed to analyze job stress (slide questions). We also expected the better educated to learn more from the training as they had more experience learning from lectures (a peculiarly academic form of knowledge aquisition).

TABLE 2
Correlations of Pre-Training Quiz Scores with Years of Education and Industrial Experience (N=49)

Questions	Education (Years)	Industrial Experience (Years)
Shoulder Slides	-.145	.176
Wrist Slides	.156	.358*
Elbow Slides	-.042	.066
True/False	.311*	.126

*P-value less than 0.05

Data presented in Table 2 supports the hypotheses on education and pre-training differences. There is a significant positive correlation (.05 level) between scores on the true/false and multiple choice questions and education for the pre-training quizzes. However, the slide scores did not show a similar relationship, indicating that educational level was not related to participants having the cognitive ability to analyze jobs prior to the training. We also examined the relationship between educational attainment and *changes* in quiz scores (Joseph,1986) and found no relationship. Generally, for the slides and true/false questions, those with a high school education and less learned as much or more than those with higher educational attainment. This finding suggests ergonomic training can benefit persons at varying education levels.

The correlations between years of industrial experience and pre-training quiz scores also support our hypotheses. Subjects with greater industrial experience were better able to diagnose stresses on the workers shown in the slides, particularly stresses on the wrist (r=.36). However, industrial experience did not correlate with the true/false-multiple choice questions, suggesting that experienced persons did not draw on abstract factual knowledge to analyze jobs, but rather used a more intuitive approach. Finally, there was no relationship between industrial experience and the amount participants learned (i.e., changes in quiz scores).

In conclusion, factual ergonomic knowledge was higher among those with more education; however, the practical ability to detect stresses on operators was greater among those with more years of industrial experience. We also examined the relationship between job status and quiz scores; however, job status was highly correlated with education and moderately correlated with industrial experience so it was not possible to isolate independent contributions. It is noteworthy that blue-collar job setters did not score consistently better or worse than their higher status counterparts on the slide portions of the quiz (though they did score relatively low on the wrist slides), but they performed significantly lower on the factual knowledge questions (Joseph,1986).

CONCLUSIONS

If ergonomic knowledge is to be effectively incorporated into actual workplace designs, dissemination of ergonomic principles must at least move beyond the walls of universities and research and development laboratories. Indeed, successful applications of ergonomics might require the active participation and involvement of shop floor workers. Certainly it is well established that commitment comes with involvement, but beyond commitment, shop-floor workers have intimate knowledge of their workplaces and how the work is actually done. We argue that this specific knowledge must be added to general knowledge of ergonomic principles to achieve effective workplace design and implementation (Liker er al.,1984).

The study reported here provides insights into the distribution of knowledge already existing in a large automotive parts plant, as well as the effectiveness of the traditional lecture as a means of knowledge dissemination. We expected variation in the way different people in plants think about and acquire knowledge of ergonomics, as captured in the distinction between "academic knowledge" and "non-academic knowledge" (Ceci and Liker, 1986a). The results showed that subjects with more years of formal education scored higher on that portion of an ergonomics quiz that reflected factual knowledge of abstract ergonomic concepts (i.e., "academic knowledge"), but education did not provide an advantage in analyzing stresses on actual workers shown in slides (i.e., "non-academic knowledge"). On the other hand, subjects with more years of industrial experience performed better than their less experienced counterparts on the slide portion of the quiz, but not on the factual knowledge questions.

Overall, lecture-based training improved ergonomic expertise as reflected in the ability to analyze the slides and answer factual knowledge questions. Counter to our expectation that individuals with more formal education would learn more effectively from the lectures, there was no clear relationship between the degree of improvement and education or industrial experience.

This preliminary evidence suggests that there is a considerable "ergonomic" knowledge base already existing in plants. Even before any formal ergonomic training, plant employees, white collar and blue collar, performed quite well on the ergonomic quiz. Moreover, different individuals were strong on different tasks depending on their education and experience. Hence, organizational processes (such as ergonomic task forces) which tap this extensive base of knowledge could go a long way toward improving work station design in plants. On the other hand, even a relatively short series of lectures improved the capabilities of the participants to detect stresses on operators, particularly in certain areas that are not highly intuitive (i.e., stresses on the wrist).

More research is clearly needed to further identify how different types of ergonomic expertise are distributed in the workforce, where training can make its biggest impact, and perhaps even ways of targeting different training formats and content to different subgroups in the workforce. The research reported here is clearly preliminary. Possible future research directions include:

1. Larger samples distributed across more differentiated categories of personnel, e.g., machine operators, job setters, skilled trades, line supervision, technical staff, and safety professionals.

2. A more careful delineation of stresses represented in slides. This would include a range of variation from no stress to moderate to extreme stress, as well as some subtle forms of stress that may not be obvious to the untrained eye. Also included should be interactions between force, duration, and frequency of exposure. We suspect that these interactions are less obvious than observing whether or not stress is present at a point in time.

3. Variation in the degree to which jobs depicted are of manufacturing processes familiar to subjects. These would range from jobs in the plant which subjects have seen or even worked, to jobs from similar processes in different plants, to jobs from entirely different manufacturing processes. The use of videotapes might provide important contextual data that is missing from slides. This would provide insights into the degree to which ergonomic knowledge is "coded" by subjects in context-specific terms or in abstract terms that can be transferred to different contexts.

4. Experimental work which varies training modes from lecture based to hands on.

5. Experimental work which varies the mix of personnel involved in actual identification of stresses on jobs. For example, does involvement of machine operators add to the quality of decisions compared to leaving the analysis to industrial engineers?

REFERENCES

Ceci, S.J. and J.K. Liker (1986a) "Academic and Non-Academic Intelligence: An Experimental Separation." In R.J. Sternberg and R.K. Wagner (eds.) *Practical Intelligence: Origins of Competence in the Everyday World*, Cambridge: Cambridge U. Press.

Ceci, S.J. and J.K. Liker (1986b) "A Day at the Races: A Study of I.Q., Expertise, and Cognitive Complexity." Journal of Experimental Psychology: General, In Press.

Fahrmeier, E. (1984) "Taking Inventory: Counting as Problem-Solving." In S. Scribner, (guest ed.) *Quarterly Newsletter of the Laboratory of Comparative Human Cognition,* Vol. 6: 1, San Diego: U. of Cal. Center for Human Information Processing.

Joseph, B.S. (1986) *A Participative Ergonomic Control Program in a U.S. Automotive Plant: Evaluation and Implications.* PhD dissertation, Ann Arbor: University of Michigan.

Joseph, B.S., J.K. Liker, and T.J. Armstrong (1986) "Group Decision-Making in an Ergonomics Program for a U.S. Automotive Plant: Correlates of Successful Meetings" in H.W. Hendrick and O. Brown,Jr. (eds.) *Human Factors in Organizational Design and Management, Vol 2.* Amsterdam: North-Holland, 1986.

Liker, J.K., B.S. Joseph, and T.J. Armstrong. (1984) "From Ergonomic Theory to Practice: Organizational Factors Affecting the Utilization of Ergonomic Knowledge." in H.W. Hendrick and O. Brown,Jr. (eds.) *Human Factors in Organizational Design and Management.* Amsterdam: North-Holland.

Super, C.M. (1980) "Cognitive Development: Looking Across at Growing Up." In C. Super and S. Harkness (eds.) *New Directions for Child Development: Anthropological Perspectives on Child Development*, 8, 59-69.

Human Factors in Organizational Design and Management - II
O. Brown, Jr. and H.W. Hendrick (Editors)

A Rationale For Computer Based Team Studies

Douglas L. Miller
Psychology Department
Old Dominion University
Norfolk, VA 23508
U.S.A.

Lori Bartels
Psychology Department
Old Dominion University
Norfolk, VA 23508
U.S.A.

Organizations today face an increasingly complex and technologically sophisticated work environment in which production is completed by teams of individuals, and specific problems are often addressed by cross-departmental groups or teams. The importance generally attributed to such team performance is reflected in the rapid growth in the last fifteen years of research on team training, team performance, and leadership effectiveness in teams.

Prior to a series of studies conducted by the late George E. Briggs and his associates at the Ohio State University (Briggs & Johnston, 1966a & b; Briggs & Naylor, 1965; Johnston, 1966; Johnston & Briggs, 1968; Naylor & Briggs, 1965) most training research involved only the acquisition of individual operator skills. Over the following decade research on team training increased (cf. Collins, 1977; Hall & Rizzo, 1975), and reviews of the topic in the mid to late 1970's were able to cite as many as 100 relevent studies, although not all of them dealt directly with team skill training (cf. Wagner, Hibbits, Rosenblatt & Schultz, 1977). Almost without exception the need for research and development in team training and team performance was recognized in those reviews.

Reviews in the mid 1970's show budding awareness of the value and necessity of team training independent of individual training (see, for example, Hall & Rizzo, 1975). With the growing recognition that a team is more than the sum of its individual parts (that team training cannot be conducted as merely an add-on to individual training; cf. Eggemeier & Cream, 1978), it has been increasingly recognized that there is a dual goal in team training; namely, (a) task skill development and maintenance, and (b) team skill development and maintenance. As the two become focused and converge, team performance improves and ultimately is maintained. Since the mid to late 1970's, research on team training and team performance has grown considerably, and some good reviews of the state-of-the-art have recently been published (see, for example, Dyer, 1984).

Nearly all of the recent research on teams has involved either (a) identification of the status or state-of-the-art of team training in the various military branches (Alluisi, 1977; Baum, Modrick, & Hollingsworth, 1981; Cooper, Shiflett, Korotkin, & Fleishman, 1984), (b) attempts to identify dimensions and variables that are unique to, and impact on, teams (Collins, 1977; Henriksen, Jones, Hannaman, Wylie, Shriver, Hamill, & Sulzen, 1980), or (c) development of highly situationally specific training technologies (Main, Rivenes, Steinemann, & Chadbourne, 1983). Rather than clarify our scientific understanding of teams, this research has served largely to increase awareness among experts of the variety of problems and issues that have yet to be adequately resolved in the areas of team training and team performance. At the same time, it has been recognized that misplaced faith in the natural evolution of team skills represents a waste of time and resources. For example, several recent military sponsored reports have placed a high priority on increasing research and development in the area of team training, maintenance, and evaluation (Alluisi, 1977; Baum et al., 1981; Denson, 1981; Dyer, 1984).

Despite this growing body of research, few of the results appear to be applicable to the solution of real organizational team problems (Thorndyke & Weiner, 1980). The available literature provides too little information concerning the most effective and efficient procedures for enhancing team skills (cf. Bass, 1982; Collins, 1977; Dyer, 1984; Hackman, 1982). Many theoretical and practical problems have beset the research related to team training and team performance, and important gaps in the literature remain. Furthermore, in several instances where data have been available, interpretations are ambiguous if not contradictory (cf. Egerman, Klaus, & Glaser, 1962; Horrocks, Herman, & Krug, 1961). Also, almost without exception the incremental time-dependent process involved in the acquisition of team skills has not been investigated. There is a gap in understanding how various patterns of team member interaction develop, change, and impact performance over the life cycle of a team. Thus, there are obvious needs for research in this area.

One area of particular importance in effective team performance, that of leadership, has been extensively researched. Many theories of leadership permeate the literature (Fiedler, 1967; House, 1971; Vroom & Yetton, 1973) providing the foundation for a large number of empirical studies, as indicated in *Stogdill's Handbook of Leadership* (Bass, 1981). The list of leadership variables studied include leader personality traits, physical traits, social skills, leader-follower interactions, democratic versus autocratic leadership, participative versus directive leadership, and the situational aspects of leadership (Bass, 1981). Although researchers have focused on many facets of leadership, little of this research has examined these issues directly in both evolving teams and established teams.

A Proposed Research Approach

This relative dearth of information seems to have resulted, in part, from the fact that there have been no standardized methods for the measurement and evaluation of the performances of teams during both training and asymtotic performance (see, for example, Dyer, 1984; Hackman, 1982; Larson, Sanders, & Steinemann, 1974; Nieva, Fleishman, & Rieck, 1978; Obermayer, Vreuls, Muckler, Conway, & Fitzgerald, 1974; Wagner et al., 1977). In their reviews, Nieva et al. (1978) and Thorndyke et al. (1980) specifically emphasized the need to develop standardized relevent team performance tasks that can be manipulated along a variety of dimensions (e.g., . the degree of team member interdependence) in a controlled setting.

Wagner and his colleagues (1977) cite the distinction made by Boguslaw and Porter (1962) between "established" and "emergent" task situations. They point out that in an "established" task situation (i.e., one in which the task is completely specifiable) team performance can generally be defined as the sum of the individual performances, and individual skill training is usually more effective (e.g., as indicated by the findings of Briggs and Johnston, 1967, and Hall et al., 1975). On the other hand, in an "emergent" task situation (i.e., one in which the task solution is not completely specifiable, thus allowing for unpredictable outcomes), individual team members must coordinate and communicate their activities (i.e., use "team skills"); in such cases, team skill training is usually more effective than individual training (cf. Eggemeier et al., 1978; McRae, 1968). Thus, in training groups of individuals to perform tasks that primarily involve individual skills, individual training should be employed. On the other hand, where the tasks primarily involve team skills, team training should be employed, and this becomes increasingly important as the task situation approaches more "emergent" rather than "established" types.

One of the major methodological problems involved in the study of team training (cf. Glanzer, 1962; Meister, 1976; Wagner et al., 1977) has been the development of emergent team tasks that are suitably flexible, yet reliable, and organizationally relevent. Yet, it is generally recognized that today's current complex and sophisticated organizations require team performance in settings characterized by emergent task situations as well as established task situations (see, for example, Eggemeier et al., 1978). The initial goal of the proposed research will be to develop and standardize a team training and performance assessment battery (TTAB) that can be used to investigate a variety of still unanswered questions concerning team training and team performance maintenance in both established and emergent situations.

Generally speaking, all approaches to laboratory performance assessment that exist in the literature can be placed on a continuum ranging from full-scale simulation to performance of specific laboratory tasks (Alluisi, 1967). The full-scale simulation technique represents the highest degree of "real" world fidelity in a controlled environment (Grodsky, Mandour, Roberts, & Woodward, 1966). While this simulation technique maximizes face validity for a specific task situation, it also limits generalizations to the specific environment to be simulated. At the other end of the continuum, measuring performance using specific tasks selected to represent common features in a wide variety of jobs, the findings may be generalizable to a wide variety of situations. Unfortunately, there is little or no face validity in simple laboratory tasks (Alluisi, 1969).

The TTAB represents a third approach to performance assessment, falling somewhere in the middle of the continuum. It is an attempt to maximize the advantages of full-scale simulation (high face validity) and the specific-task techniques (high generalizability). The TTAB utilizes a synthetic work situation in which a battery of organizationally relevent, yet generic, team and individual tasks are combined in the laboratory to form a work environment with specific and realistic jobs for each individual. Thus, reasonably high face validity is obtained, while strict laboratory control and generic tasks allow for reasonable generalizability (Alluisi, 1967). The control allowed by the TTAB along with its fairly high fidelity also give it obvious advantages over field research. Thus, this technique will provide researchers with operator performance that can be assessed, possesses a high degree of generalizability to other task situations, and provides the ability to assess both group performances and individual performances simultaneously in a controlled laboratory setting.

The TTAB would prove particularly useful in the study of teams as they develop, evolve or mature, an area currently being researched by the Center for Applied Psychological Studies at Old Dominion University. The device developed will allow for the controlled, systematic study and evaluation of the theory of team evolution and maturation currently under development, as well as the examination of the efficacy of other theories of team training, development, and behavior; the current lack of adequate theory in the area of team research is well documented (Dyer, 1984; McGrath and Altman, 1966). Laboratory research of this sort will provide a valuable complement to the field research that is currently underway. It will make possible systematic tests of hypotheses that can be derived from field observation.

In addition to leadership issues, several research topics seem to be uniquely suited to study with a standardized and generalizable synthetic-work team training assessment battery. One such issue concerns sequencing of individual and team training in problem solving behavior to optimize the acquisition and retention of those skills. As previously noted by Wagner and his colleagues (1977; see also Baum et al., 1981), the available literature provides little information concerning the with task situations.

Another vital element of team training and team performance in need of research that could be broached with the TTAB is the area of performance measurement and criteria development. The need for the development of improved criteria for improved performance measurement has been widely recognized (Baum et al., 1981; Dyer, 1984). The proposed TTAB is unique in its ability to define explicitly, quantify, and evaluate team performance across several team types and environmental conditions, while controlling vital performance features.

A final example of a potential area for research with the TTAB regards communications networks and protocols, and feedback (Baum et al., 1981; Hendricksen et al., 1980; Thorndyke et al., 1980). Different types of communication networks such as the circle, chain, Y, wheel, star, as identified by Leavitt (1951) may be studied to determine their effects on team effectiveness and efficiency, and team member satisfaction. Feedback issues that could be explored include (a) the impact and effectiveness of machine based, leader based, and other member based feedback, (b) individual versus group directed feedback, (c) timing features, (d) and other key content features (ratio of negative to positive feedback). Techniques for improving team communication, and thus team effectiveness (see, for example, the time sharing concept of Zimbardo, Linsenmeier, Kabot, & Smith, 1981) could then be compared, contrasted, and developed.

Plan of Research

The methods and procedures of data collection for the proposed research will be based upon those used for more than twenty years to assess (measure and evaluate) the performance of persons at work (cf. Alluisi, 1969; Alluisi, Beisel, Bartelloni, & Coates, 1973; Alluisi, Coates, & Morgan, 1977; Beisel, Morgan, Bartelloni, Coates, DeRubertis, & Alluisi, 1974; Chiles, Alluisi, & Adams, 1968; Morgan, 1974). These standard procedures will be modified and used in the management of a computer controlled and displayed battery of individual- and group-performance tasks termed the TTAB.

It is envisioned that this TTAB will be conceptually and functionally similar to the Multiple-Task Performance Battery (MTPB) used for nearly 20 years by Alluisi and his colleagues (Alluisi, 1967, 1969; Morgan & Alluisi, 1972; Morgan, Coates, Kirby, & Alluisi, 1984) to assess the performances of 5-person groups of subjects. However, the TTAB will differ from the MTPB in that (a) the TTAB will function primarily to provide measurements of the integrated, interactive performances of teams rather than just individuals, (b) the TTAB will consist of tasks designed to ensure that several of the major dimensions of organizational team tasks are represented in terms of their complexity, interdependencies, requirements for communication and coordination, etc., and (c) the TTAB will be designed to incorporate the features of a state-of-the-art interactive network of microprocessors. The TTAB will be structured around team performance tasks designed to provide considerable flexibility along the dimensions that are most important to the study of team training and performance. That is, the tasks will be configured so that manipulations can be introduced in variables such as team-member interdependencies, group cohesion, team workload, training schedules and procedures, etc. In addition, the team tasks will be designed so that they can be configured to different points along the dimension anchored at one end by "established" performance parameters and on the other end by "emergent" performance requirements (cf. Alexander & Cooperband, 1965).

In addition to the team tasks, the TTAB will contain two or three individual performance tasks. These tasks will be added to the team tasks, singly or in combination, in order to vary the workload of the team and in order to introduce realism associated with the fact that most operational situations require team members to work together on some tasks while also performing some duties independently. As with the team tasks, these tasks will be selected to represent different dimensions of individual performance that are judged to be important for successful performance in the organizational environment.

The synthesized battery will be capable of providing measurements of team performances during acquisition of a skill (training) as well as during later periods of asymtotic performance (work). These measurements will provide behavioral assessments of the efficiency of training and of the attainment of levels of skilled work performance both for individual team members as well as the entire work group. In addition, transfer conditions can be simulated in order to research a variety of issues related to the transfer of team training. The flexibility provided by this combination of state-of-the-art technology with the procedures of the synthetic work methodology will provide a powerful team training research tool.

Footnote

1. This paper was developed from a proposal prepared under the guidance of Dr. Ben B. Morgan, Jr. and Dr. Raymond H. Kirby at the Center for Applied Psychological Studies, Old Dominion University, and submitted to the United States Office of Naval Research.

REFERENCES

Alexander, L.T., & Cooperband, A.S. (August 1965). System training in team behavior (Technical Memorandum 2581). System Development Corporation.

Alluisi, E.A. (1967). Methodology in the use of synthetic tasks to assess complex performance. Human Factors, 9, 375-384.

Alluisi, E.A. (1969). In E.A. and I. Mc.D. Bilodeau (Eds.), Principles of skill acquisition. New York: Academic Press, pp. 59-101.

Alluisi, E. A. (1977). Lessons from the study of defense training technology. Journal of Educational Technology Systems, 5, 57-76.

Alluisi, E.A., Beisel, W.R., Bartelloni, P.J., & Coates, G.D. (1973). Behavioral effects of tularemia and sandfly fever in man. Journal of Infectious Diseases, 128, 710-717.

Alluisi, E.A., Coates, G.D., & Morgan, B.B., Jr. (1977). Effects of temporal stressors on vigilance and information processing. In R.R. Mackie (Ed.), Vigilance: Theory, operational performance, and physiological correlates. New York: Plenum Press, 1977.

Bass, B.M. (1981). Stogdill's Handbook of Leadership. New York: The Free Press.

Bass, B. (1982). Individual capability, team performance and productivity, In E. A. Fleishman (Ed.), Human Performance and Productivity (volume 1 of 3). Hillsdale, New Jersey: Lawrence Erbaum Associates.

Baum, D. R., Modrick, J. A., & Hollingsworth, S. R. (1981). The status of Air Force team training for command and control systems (Final Report vol. 1,3,5; Contract No. F33615-79-C-0025). Minneapolis, MN: Honeywell Systems and Research Center.

Beisel, W.R., Morgan, B.B. Jr., Bartelloni, P.J., Coates, G.D., DeRubertis, F.R., & Alluisi, E.A. (1974). Symptomatic therapy in viral illnesses; a control study of effects on work performance. Journal of the American Medical Association, 228, 581-584.

Boguslaw, R., & Porter E. H. (1962). Team functions and training. In R. M. Gagne (Ed.), Psychological Principles in System Development. New York: Holt, Rinehart, & Winston.

Briggs, G.E., & Jonston, W.A. (1966a). Stimulus and response fidelity in team training. Journal of Applied Psychology, 50 (2), 114-117.

Briggs, G.E., & Johnston, W.A. (1966b). Influence of change in system criteria on team performance. Journal of Applied Psychology, 50 (6), 467-472.

Briggs, G.E., & Johnston, W.A. (1967). Team training: final report, February 1966 - February 1967 (Report No. 1327-4). Orlando, FL: United States Naval Training Device Center.

Briggs, G. E., & Naylor, J. C. (1965). Team versus individual training, training task fidelity and task organization effects on transfer performance by three-man teams. Journal of Applied Psychology, 49, 387-392.

Chiles, W.D., Alluisi, E.A., & Adams, O.S. (1968). Work schedules and performance during confinemnet. Human Factors, 10, 145-196.

Collins, J. J. (1977). A study of potential contributions of small group behavior research to team training technology development. (Contract No. 00014-76-C-1076, NR 170-834). Alexandria, VA: Essex Corporation.

Cooper, M. A., Shiflett, S., Korotkin, A. L., & Fleishman, E. A. (1984). Command and control teams: Techniques for assessing team performance (Contract No. F 33615-81-C-0017). Bethesda, MD: Advanced Research Resources Organization.

Denson, R. W. (1981). Team training: Literature review and annotated bibliography (AFHRL-TR-80-40). Wright Patterson Air Force Base, OH: Logistics and Technical Training Division, Air Force Human Research Laboratory.

Dyer, J. L. (1984). Team research and training: A state-of-the-art review. Human Factors Review, 285-323.

Egerman, K., Klaus, D.J., & Glaser, R. (June 1962). Increasing team proficiency through training. 3: Decremental effects of reinforcement in teams with redundant members (Technical Report AIR B64-6162). American Institutes of Research.

Eggemeier, F. T., & Cream, B. W. (1978). Some considerations in the development of team training devices. Paper presented at the meeting of the American Psycological Association.

Fiedler, F. (1967). A Theory of Leadership Effectiveness. New York, McGraw-Hill.

Glanzer, M. (1962). Experimental study of team training and team functioning. In R. Glaser (Ed.), Training research and education. Pittsburgh: University of Pittsburgh Press, 379-408.

Grodsky, M.A., Mandour, J.A., Roberts, D.L., & Woodward, D.P. (1966). Crew performance studies for manned space flight. ER 14141. Baltimore: Martin Co.

Hackman, J. R. (1982). A set of methods for research on work teams (Technical Report No. 1; ONR). Yale School of Organization and Management.

Hall, E. R., & Rizzo, W. A. (1975). An assessment of U. S. Navy tactical team training: Focus on the trained man (TAEG Report No. 18). Training Analysis and Evaluations Group.

Henricksen, K. F., Jones, D. R., Jr., Hannaman, D. L., Wylie, P. B., Shriver, E. L., Hamill, B. W, & Sulzen, R. H. (1980). Identification of combat unit leader skills and leader-group interaction processes (Contract No. MDA903-78-C-2005). Alexandria, VA: Kinton Inc.

Horrocks, J., Herman, E., & Krug, R. E. (1961). Team training III: An approach to optimum methods and procedures. NAUTRADEUCEN 198-3. Port Washington, NY: U. S. Naval Training Device Center.

House, R.J. (1971). A path-goal theory of leader effectiveness. Administrative Science Quarterly, 16, 321-28.

Johnston, W.A. (1966). Transfer of team skills as a function of type of training. Journal of Applied Psychology, 50, 102-108.

Johnston, W.A., & Briggs, G.E. (1968). Team performance as a function of team arrangement and work load. Journal of Applied Psychology, 52, 89-94.

Larson, O. A., Sanders, S. I., & Steimemann, J. H. (1974). Survey of unit performance effectiveness measures (Technical Report No. 74-11). San Diego, CA: Navy Personnel Research and Development Center.

Leavitt, H.J. (1951). Some effects of certain communication patterns on group performance. Journal of Abnormal and Social Psychology, 46, 388-50.

Main, R. E., Rivenes, I. J., Steinimann, J. H., & Chadbourne, J. G. (1983). Tailoring shipboard training to fleet performance needs: V. design and production of training materials (Report No. 14-82-18). San Diego, CA: Navy Personnel Research and Development Center.

McRae, A.V. (June 1966). Interaction content and team effectiveness (Technical Report 66-10). Alexandria, VA: Human Resources Research Office.

McGrath, J. E., & Altman, I. (1966). Small group research: A synthesis and critique of the field. New York: Holt, Rinehart, & Winston.

Meister, D. (1976). Behavioral foundations of system development. New York, John Wiley and Sons.

Morgan, B.B.,Jr. (1974). Effects of continuous work and sleep loss in the reduction and recovery of work efficiency. American Industrial Hygiene Association Journal, 35, 13-30.

Morgan, B.B., Jr., & Alluisi, E.A. (1972). Synthetic work: methodology for the assessment of human performance. Perceptual and Motor Skills, 35, 835-845.

Morgan, B. B., Jr., Coates, G. D., Kirby, R. H., & Alluisi, E. A. (1984). Individual and team performances as functions of the team-training load. Human Factors, 26 (2), 127-142.

Naylor, J.C., & Briggs, G.E. (1965). Team-training effectiveness under various conditions. Journal of Applied Psychology, 49, 223-229.

Nieva, V. F., Fleishman, E. A., & Rieck, A. (1978). Team dimensions: Their identity, their measurement, and their relationships. Washington, DC: Advanced Research Resources Organization.

Obermayer, R. W., Vreuls, D., Muckler, F. A., Conway, E. J. & Fitzgerald, J. A. (1974). Combat-ready crew performance measurement system (AFHRL Final Report R-74-108 (I)). Brooks Air Force Base, TX: Air Force Systems Command (DTIC No. AD B005517).

Thorndyke, R. W., & Weiner, M. G. (1980). Improving training and performance of Navy teams: A design for a research program (Report No. R-2607-ONR). Santa Monica, CA: Rand Corporation.

Vroom, V.H., & Yetton, P. (1973). Leadership and Decision Making. Pittsburgh: University of Pittsburgh Press.

Wagner, H., Hibbits, N., Rosenblatt, R. D., & Schultz, R. (1977). Team training and evaluation strategies: State-of-the-art (HumRRO Technical Report 77-1). Alexandria, VA: Human Resources Research Organization.

Zimbardo, P. G., Linsenmeier, J., Kabot, L., & Smith, P. (1981). Improving team performance and participation via computer-mediated turn taking and informational prompts (ONR Contract No. N00014-78-C-0425). Stanford, CA: Office of Naval Research.

Human Factors in Organizational Design and Management - II
O. Brown, Jr. and H.W. Hendrick (Editors)

EFFECTIVENESS OF TEACHING METHODS: COMPUTER LITERACY OF END-USERS[1]

Urs E. Gattiker
Richard Williams
Dan Paulson
The University of Lethbridge

Computer literacy has been identified as one of the most important factors for the effective use of computer-based technology at the workplace. Managers need to know the most efficient methods available to teach computer skills to their workers in a short time. Such methods need to be suitable for all employees, whether academically gifted or weak. This paper investigates several approaches used to teach business students about micro-computers. The results show that hands-on teaching is superior to traditional classroom lectures, especially for academically less inclined students. Directions for future research and implications of these findings for management are discussed.

It is expected that by the end of this decade the majority of clerical as well as managerial employees in North America will use computers in their daily work (Giuliano, 1982; Gutek, 1983). The impact of computerization upon job design and work structure has already been substantial, making it an area of considerable interest for human resource management. Most of the relevant research has dealt with such areas as human factors engineering, financing and organizational design (e.g., Carter,1984; Doswell, 1983; Lieberman, Selig & Walsh, 1982, chap. 1 & 2). Unfortunately, while policy and strategy issues related to computer technology have taken most of the limelight in organizational literature (Megaw & Lloyd, 1984; Panko, 1984), education and training aspects have largely been ignored even though additional research in this area is necessary (Hebenstreit, 1985; Kahn, 1981). For example, a comparison of computer-assisted instruction (CAI) versus traditional classroom teaching methods would be useful when trying to determine the effectiveness of these approaches (Amarel, 1983). The amount of technical knowledge (e.g., programming) which may be required when striving for computer literacy has not been determined yet (Hebenstreit, 1983). It might well prove to be minimal. Academic background and individual learning capabilities may also be of consequence (Lepper, 1985). Finally, the type of equipment ultimately utilized (mainframe terminal or intelligent workstation) may dictate certain methods.

This paper investigates the effectiveness of the methods used in teaching computer literacy to business students at a Western Canadian University. Most of the students have previous work experiences, and many continue to work full-time while attending the university part-time. They acquire computer skills with the help of lectures as well as interactive CAI, which is provided in a micro-computer lab setting. The former teaches technical specifics about computers (hardware) and programming languages, while the latter concentrates on the equipment's use for such job tasks as word-processing, spreadsheets, accounting, database management and programming. The micro-computer course is a requirement for a management degree.

Training of the Office Worker

Much of the organizational literature stresses that only the effective use of technology really warrants its huge financial investment. An assessment of computer literacy in a company's workforce thus becomes the first step in achieving such effectiveness. However, a recent study of employees from several organizations across the Western U.S. showed that

[1]We would like to thank Eldon Gardner for his valuable advice during the data analyses. Please address all questions and comments to: Urs E. Gattiker, Technology Research Group, School of Management, The University of Lethbridge, Lethbridge, Alberta, T1K 3M4, CANADA.

hands-on experience and frequent computer use for certain job tasks does not necessarily constitute individual computer knowledge, let alone computer literacy (Gattiker & Larwood, 1986).

Contrary to popular belief, successful use of automated systems is not primarily dependent on technological advances and know-how, but rather on user acceptance of the electronic office (Dierkes & Von Thienen, 1984). Employees need to feel at ease working with the technology. People who interface directly with the computer and other electronic systems must have thorough knowledge of how to use their equipment most effectively. Therefore, organizations should provide sufficient instruction for the end-users within their workforce (Menashian, 1985).

Computer literacy. This is an important construct when assessing possible resistance to computerization (Gattiker, 1984; Dierkes & Von Thienen, 1984). Computer literacy has been defined in different ways for specific groups of employees (Bjorn-Anderson, 1983; Wynne, 1983). In this context, it includes, but is not limited to, the individual's capability to work with computer terminals or intelligent workstations. Such knowledge was usually related to technical aspects, as exhibited by computer scientists and similar professionals who often knew programming languages. However, today's proliferation of computer-based technology has brought about a radical change of the term computer literacy (Hebenstreit, 1983). In offices, it now describes the worker's ability to use appropriate software programs for certain clerical tasks, including business correspondence, report writing and spreadsheets (Card, Moran & Newell, 1984; Morgall, 1983). Computer knowledge may also mean some familiarity with the data processing capabilities of standard programs. This aspect is reflected more and more in the computer education of future managers (Gattiker, 1986).

Assessing the Effectiveness of Training Alternatives

Traditional computer training methods aimed at achieving computer literacy have included programming skills and knowledge about hardware (Lepper, 1985). Most of today's computer science majors are still taught in this way (Campbell & McCabe, 1984). Yet recent research shows a broader conception of the computer's role in education. For example, Taylor (1980) has described three uses of computers in schools: tutor, tool and tutee. In a tutoring capacity, the computer presents material, evaluates student responses, determines what to present next and keeps records of student progress. The computer serves as a tool when it provides statistical analysis, calculation, or word-processing. For students, the technology may become a calculator in statistics classes, a text editor for writing assignments, an aid to doing spreadsheets when solving accounting/finance problems, or a database management system. Finally, the computer assumes the role of tutee if students communicate in programming languages, such as BASIC and Pascal, which allow manipulation of the "capabilities" of the equipment (Kulik, Kulik & Bangert-Drowns, 1985).

Most of the studies assessing the efficiency of instructing individuals with CAI have been done with grade school students (Kulik, Kulik & Bangert-Drowns, 1985). CAI can provide frequent and highly relevant feedback directly from the student's work (Bork, 1985). Responses can then be carefully analyzed within the common everyday language of the student. However, the question of whether CAI is the best method for adults as well remains unanswered. Furthermore, the training needs for office workers may be quite diverse, depending on whether intelligent workstations, mainframe terminals or computer-aided design systems are used. Training methods for office personnel to acquire the necessary computer skills so far have included CAI and traditional classroom activities, such as lectures and self-study (Menashian, 1985). Clearly, researchers need to evaluate a wider range of educational approaches and their results than ever before (Kulik, Kulik & Bangert-Drowns, 1985).

Evaluation of training results. Obviously, CAI can help individuals overcome resistance to change by providing hands-on experience with the new technology, even if traditional lecturing about computers leaves the potential user awed and somewhat frightened (Hebenstreit, 1985). For instance, some U.S. research indicates that university freshmen who achieved high SAT scores and were near the top of their high school class performed better in computer science courses than their lesser ranked peers (Campbell & McCabe, 1984). Another study by Butcher and Muth (1985) concluded that students with a high grade point average (GPA) fared better in computer science courses than those with lower GPA's. However, since many employees today are being challenged to acquire computer skills, it is important for management to identify those teaching methods which successfully lead to adult computer literacy, regardless of academic competence (Brown, Sawyer & Hitchcock, 1984; Gattiker, 1986; Lepper, 1985).

It would be interesting to know how a student's past academic record explains his/her performance in a micro-computer training course. Would he/she do equally well in traditional classroom tests and exams versus non-traditional "take-home" assignments requiring actual use of the equipment? Answers to these issues and questions would help educators and managers alike to determine the appropriate teaching methods leading to computer literacy for employees who are confronted with technological innovations at the workplace (cf. Bjorn-Anderson, 1983; Kulik, Kulik & Bangert-Drowns, 1985; Menashian, 1985).

Research Issues

In this study, we tried to assess student performance in the acquisition of micro-computer literacy. We hoped to identify the specific techniques which would teach end-users the most useful skills for office settings, that is, facilitate the best man-machine interface (Menashian, 1985). The following hypotheses were developed:

Hypothesis 1a. Student performance in the CAI section of a micro-computer course is predicted by his/her overall academic performance.
Hypothesis 1b. Student performance in the CAI section of a micro-computer course is affected by his/her grade in a previously completed mainframe computing course teaching BASIC.
Hypothesis 1c. Student performance in the CAI section of a micro-computer course is affected by both his/her overall academic record and successful completion of a mainframe computing course in BASIC.

The models for CAI performance to be estimated here were in the form of linear regression equations:

Hypothesis 1a. CAI = constant + GPA
Hypothesis 1b. CAI = constant + GPBASIC
Hypothesis 1c. CAI = constant + GPA + CRBASIC

The variable CAI is the percentage grade that the student received in the CAI section, which is composed of ten assignments (20% of CAI mark total), two practical exams (20% each of CAI mark total), and two multiple choice exams (also 20% each of CAI mark total). The student's overall academic performance is quantified by his/her 'a priori' GPA and is designated as such. The university presently employs a four-point grade scheme with 4.0 (exceptional performance) resulting in an "A" grade, 3.0 being equivalent to "B", and so on. Below 1.0 ("D"), a failing grade is given with the rationale that the student did not comprehend a sufficient amount of the course material. If the student took the mainframe computing course as an option, the grade received in that course has been coded GPBASIC (i.e., A=4,B=3,C=2,D=1,F=0). A "D" or better in the BASIC course is coded as CRBASIC=1; failure in the course is coded as absence of mainframe experience (CRBASIC=0).

Hypothesis 2a. Student performance in the classroom segment of a micro-computer course is predicted by his/her overall academic performance.
Hypothesis 2b. Student performance in the classroom segment of a micro-computer course is affected by his/her grade in a previously completed mainframe computing course in BASIC.
Hypothesis 2c. Student performance in the classroom segment of a micro-computer course is affected by both his/her overall academic record and successful completion of a mainframe computing course in BASIC.

The models for CLASS overall performance can be expressed using the following linear regression equations:

Hypothesis 2a. CLASS = constant + GPA
Hypothesis 2b. CLASS = constant + GPBASIC
Hypothesis 2c. CLASS = constant + GPA + CRBASIC

The class mark (CLASS in percent) was made up of two exams. One assessed the student's general expertise regarding micro-computer functions, the other tested his/her knowledge about their use in the business world.

Hypothesis 3a. Overall student performance in a micro-computer course is predicted by his/her overall academic record.

Hypothesis 3b. Overall student performance in a micro-computer course is predicted by his/her grade in a mainframe course in BASIC.

Hypothesis 3c. Overall student performance in a micro-computer course is affected by both his/her overall academic record and successful completion of a mainframe computing course in BASIC.

The regression models used to evaluate overall performance in the micro-computing course were:

Hypothesis 3a. GPMICRO = constant + GPA
Hypothesis 3b. GPMICRO = constant + GPBASIC
Hypothesis 3c. GPMICRO = constant + GPA + CRBASIC

The student's overall grade in the micro-computer course is expressed in terms of the grade point that they received (A=4, etc.), and for simplicity is referred to as GPMICRO throughout this study.

Hypothesis 4. Students with superior performance in a micro-computing course are randomly distributed amongst 'a priori' GPA's.

Based on the above hypothesis, we were also interested in testing if "A" students would fare better with one particular teaching method for acquiring computer literacy than their academically inferior peers. Furthermore, would knowledge about mainframe computers and BASIC programming help academically less inclined students to perform better? The following hypotheses were made:

Hypothesis 5a. Student performance in the CAI section of a micro-computer course is predictable by means of his/her previous academic record and successful completion of a mainframe computing course in BASIC.

Hypothesis 5b. Student performance in the classroom segment of a micro-computer course is predictable by means of his/her previous academic record and successful completion of a mainframe computing course in BASIC.

Hypothesis 5c. Overall student performance in a micro-computing course is predictable by means of his/her previous academic record and successful completion of a mainframe computing course in BASIC.

The models for student performance using different GPA levels were also estimated via linear regression:

Hypothesis 5a. CAI = constant + GPAA + GPAB + GPAC + ACRBASIC + BCRBASIC + CCRBASIC

Hypothesis 5b. CLASS = constant + GPAA + GPAB + GPAC + ACRBASIC + BCRBASIC + CCRBASIC

Hypothesis 5c. GPMICRO = constant + GPAA + GPAB + GPAC + ACRBASIC + BCRBASIC + CCRBASIC

The letter following the acronym GPA refers to the student's grade point average group (note that if a student has a GPA of 3.8 then GPAA=3.8, GPAB=0 and GPAC=0). The prefix of CRBASIC operates in the same manner as for the GPA groupings. "D" grades for the micro-computer course appeared too infrequently to be of consequence here.

Take-home assignments may represent a fairly close parallel to office settings where employees are often asked to perform computer tasks under minimal supervision (Hebenstreit, 1983; Menashian, 1985; Samson, Graue, Weinstein & Walberg, 1984). Consequently, we also looked specifically at assignment performance as a representative variable of the individual's potential work situation.

Hypothesis 6. A student's practical (assignment) performance is predictable by means of his/her previous academic record and successful completion of a mainframe computing course in BASIC. To model ASSIGNMENT performance, we used the following regression:

Hypothesis 6. ASSIGNMENT = constant + GPAA + GPAB + GPAC + ACRBASIC + BCRBASIC + CCRBASIC

The variable ASSIGNMENT is the student's overall score resulting from the combined total of ten assignments done throughout the micro-computing course.

Method

Subjects and Design

Undergraduate management students must complete a micro-computing course to fulfill one of their graduation requirements. A BASIC programming course on the university's mainframe computer is also available to meet student demand for computer literacy, but may not be substituted for the micro course. Of the 128 students who participated in this project relating to micro-computers, about one half had already taken the mainframe computer class.

The micro-computer course is an attempt to impart computer literacy to management students, enabling them to work effectively with intelligent workstations in an office setting. Classroom methods consist of traditional lectures as well as CAI. Homework includes several assignments on micro-computers requiring the use of different software packages (e.g., Lotus, dBase and Word Perfect). Grading for this course is based on a traditional pencil test for the lecture part, and problem-solving tasks in a hands-on fashion for the CAI portion. In addition, two multiple choice exams relating to both parts are administered. All testing is done with time limits imposed.

Measures

Student grades were determined for the micro-computing course by considering overall performance, laboratory (CAI) performance, classroom performance and take-home micro assignments. The student's 'a priori' GPA and his/her grade achieved in the BASIC programming course (GPBASIC) were also obtained.

In this study, only students who received a passing grade (1.0 or better) in the optional mainframe computing course have been coded for the variable CRBASIC. Grouping variables were used to see if historically high GPA students may achieve GPMICRO=4. Students with 'a priori' GPA's of 3.50 and above were designated as "A" students, those with overall GPA's between 2.50 and 3.49 were classed as "B" students, and those ranging between 1.50 and 2.49 were labeled "C" students. Since no meaningful regressions can be run using such highly correlated variables as 'a priori' GPA and the GPBASIC (i.e., the latter being a part of the former), the use of CRBASIC is necessary. One might expect that students who do well overall would be more likely to have taken the mainframe computing course, but this was not so.

Statistics

All of the analyses for this project were accomplished on micro-computers with the statistical package SYSTAT. Multiple regression was used to determine not only the significance of factors but also the magnitude of the effect they might have on the dependent variable. To see if _k_ independent samples are from different populations, the Kruskal-Wallis one-way analysis of variance by rank was used (Siegel, 1956, pp.184-193). Sample values almost always differ to some degree, and the question is whether the variations among the samples signify genuine population differences or merely represent expected chance variations in random samples from the same population.

Results

In order to facilitate a logical presentation of our results, they will be divided into five sections, according to the relevant hypotheses.

Predicting student performance using CAI. Hypotheses 1a through 1c were supported as follows: A student's 'a priori' GPA indeed proved to be a significant determinant ($p < 0.001$) of his/her performance in the CAI section of the micro-computer course, and it also explained 31% of the total observable variance in his/her laboratory mark. GPBASIC accounted for 23.3% of the population variance in the CAI mark obtained (significance of the independent variable = $p < 0.01$). When taking into account both the student's historical academic performance and also the fact that he/she successfully completed the mainframe computer course (CRBASIC), over 36% of the population variance in the CAI mark can be explained.

Predicting student performance in the classroom. CLASS mark was determined almost as well with Hypotheses 2a and 2b as was CAI performance with Hypotheses 1a and 1b. The only important difference between the two groups is that although GPA is a significant predictor of class mark ($p < 0.001$), the other independent variable (CRBASIC) is not acceptable as significant in conjunction with the student's historical performance. The results show that H2a and H2b are supported by the data, whereas H2c can only be accepted with reservations.

Table 1

Regression Results for Student Performance: GPA and BASIC Course Grade

Dependent Variable		Constant		Independent Variables[1]			d.f.	Adj.R^2
CAI	=	42.459 (8.55)***	+	13.763 * GPA (7.60)***			126	0.309
	=	70.116 (23.50)***	+	5.267 * GPBASIC (4.45)***			61	0.233
	=	39.963 (8.28)***	+	13.720 * GPA (7.89)***	+	5.486 * CRBASIC (3.42)**	125	0.363
CLASS	=	36.726 (7.95)***	+	12.738 * GPA (7.57)***			126	0.307
	=	60.139 (19.70)***	+	5.227 * GPBASIC (4.32)***			61	0.221
	=	35.484 (7.66)***	+	12.716 * GPA (7.62)***	+	2.728 * CRBASIC (1.77)	125	0.319
GPMICRO	=	-0.347 (-1.01)	+	1.202 * GPA (9.63)***			126	0.419
	=	2.255 (9.17)***	+	0.369 * GPBASIC (3.79)***			61	0.177
	=	-0.553 (-1.70)	+	1.199 * GPA (10.21)***	+	0.453 * CRBASIC (4.17)***	125	0.486

[1] t ratios are in parentheses beneath estimated coefficients
*p< 0.05 ; **p< 0.01 ; ***p< 0.001

Predicting a student's overall performance in the micro-computer course. Hypothesis 3a stated that a strong relationship would exist between the grade point received by a student in the micro-computing course and his/her 'a priori' GPA, which was indeed confirmed. The results also supported Hypothesis 3b, that is, a student's grade achieved in a BASIC course predicts his/her overall performance in a subsequent microcomputer course. Hypothesis 3c held that both a student's historical academic performance and whether he/she had completed a mainframe computing course would be significant indicators of his/her overall performance in the micro-computing course. As shown in Table 1, not only are both of them significant (p<.001), but 48.6% of the variance in the final grade received by students is determinable with these two variables. Thus Hypothesis 3c is confirmed as well.

Academic consistency and its effect. With Hypothesis 4, we intended to determine whether or not it was appropriate to treat all students as coming from the same population. Specifically, we tested if a student's rank based on historical academic performance versus the grade received in the micro-computer course was randomly distributed. Table 2 indicates that the rank 'a priori' GPA's of students are not random. In other words, students receiving high grades in the micro-computer course originate from a different population of GPA students than those who received lower grades, refuting Hypothesis 4.

Predicting a student's performance with GPA and "traditional" computer literacy. The following three hypotheses are in reality a re-formulation of Hypothesis 3c, taking into account the fact that students with high previous academic performance will tend to perform differently academically weaker peers. As a result, the model to test the following hypotheses treats "A", "B" and "C" students as different groups than was shown to be true using a Kruskal-Wallis test previously.

Hypotheses 5a and 5b stated that a student's performance in the CAI section of the micro-computer course and final class grade are predictable, using the student's historical GPA and the mainframe course grade. However, Table 3 shows that having credit in a mainframe BASIC course is significant to performance in the CAI segment for "B" and "C" students only. For the CLASS section of the micro course, the mainframe BASIC course was not a significant predictor. Therefore, Hypotheses 5a and 5b are only partially supported.

Table 2

One-Way Analysis of Variance for 128 Cases

Dependent Variable GPA
Grouping Variable GPMICRO

Group	Count	Rank Sum
0	1	1.0
1	3	33.5
2	38	1476.0
3	51	3531.0
4	35	3214.5

Kruskal-Wallis test statistic 47.17
$p<.001$ assuming chi-squared distribution with 4 d.f.

According to hypothesis 5c, the predictive model of overall student performance in the micro-computer course (GPMICRO) would be predicted by having completed a mainframe BASIC course and the student's historical GPA. The results indicate that an academically low achiever benefits significantly from a mainframe BASIC course in terms of his/her overall grade in the micro-computer course, while their "brighter" counterparts do not derive much benefit (cf. Table 3). Thus hypothesis 5c can be accepted with reservations only.

Table 3
Regression Results for Student Performance: Different GPA and BASIC Course Grade Populations

Dependent Variable	Constant	Independent Variables[1]							
		GPAA	GPAB	GPAC	ACRBASIC	BCRBASIC	CCRBASIC	d.f.	adj $\underline{R}^2$
CAI =	39.275 (6.06)***	+ 14.381 * GPAA (6.39)***	+ 14.262 * GPAB (6.28)***	+ 13.545 * GPAC (4.62)***	- 2.229 * ACRBASIC (-0.36)	+ 4.611 * BCRBASIC (2.15)*	+ 7.949 * CCRBASIC (3.05)**	121	0.388
CLASS =	39.816 (6.11)***	+ 11.445 * GPAA (5.06)***	+ 11.092 * GPAB (4.86)***	+ 11.398 * GPAC (3.87)***	+ 0.557 * ACRBASIC (0.09)	+ 3.704 * BCRBASIC (1.72)	+ 1.303 * CCRBASIC (0.50)	121	0.282
GPMICRO =	-0.329 (-0.74)	+ 1.181 * GPAA (7.68)***	+ 1.125 * GPAB (7.26)***	+ 1.109 * GPAC (5.54)***	- 0.154 * ACRBASIC (-0.36)	+ 0.525 * BCRBASIC (3.58)**	+ 0.445 * CCRBASIC (2.51)*	121	0.496
ASSIGNMENTS =	5.160 (5.81)***	+ 1.187 * GPAA (3.85)***	+ 1.300 * GPAB (4.18)***	+ 1.110 * GPAC (2.76)**	+ 0.171 * ACRBASIC (0.20)	+ 0.212 * BCRBASIC (0.72)	+ 0.473 * CCRBASIC (1.33)	121	0.232

[1] $\underline{t}$ ratios are in parentheses beneath estimated coefficients
*$p < 0.05$; **$p < 0.01$; ***$p < 0.001$

Assignment performance. Hypothesis 6 discussed the potentially significant predictors of ASSIGNMENT performance. It can only be partially confirmed because the 'a priori' GPA is indeed a significant predictor ($p < 0.01$), but completion of a mainframe BASIC course is not.

Discussion

This study was designed to assess which of our existing teaching methods would be most effective for students trying to acquire computer literacy. While all methods have merit and should be combined in a complementary manner, one or the other could be expected to emerge as a dominant force. Individual differences in the student population and academic preparation also figure prominently in our ultimate findings.

"Traditional" computer literacy. Different models were tested to see how students would perform in the CAI and CLASS portions of a micro-computer course and what final grade they would receive. As the figures in Table 1 showed, previous completion of a mainframe systems/BASIC programming class was helpful to the student's performance in the CAI section of the course. These results confirm earlier work by Campbell and McCabe (1984) who found that computer science majors with high freshman grades in beginning computing courses performed better and better as they progressed through their academic programs.

On the other hand, credit in a mainframe computer course did not add significantly to the student's GPA for the class section of the micro-computer course. Based on the regression coefficients obtained, it also appears that "traditional" computer literacy helps weaker students (cf. Table 3). Of the students who had previously attended a mainframe computer course, "B" and "C" students did better not only in the CAI section of the micro-computer course but also in their overall class performance. However, this was not true for "A" students. One explanation could be that previous experience with a mainframe computer had made the weaker students less apprehensive, which then helped their performance in tests and traditional class lectures.

Past academic performance. Historic GPA affected the performance of students in all parts of the course, no matter if they were "A" or "C" level students. These findings are not particularly surprising. In earlier research, Butcher and Muth (1985) had found that a student's grade point average achieved in his/her first semester of post-secondary studies predicted his/her performance in subsequent computer science courses. These results support the assumption that student performance is significantly affected by past GPA.

However, an extensive meta-analysis of the literature investigating past academic and current occupational performance was conducted by Samson, Graue, Weinstein & Walberg (1984). They examined about 35 studies done since 1950, relating to the connection between academic and occupational performance in various fields. Predictions of occupational performance based on academic achievement were low in magnitude, making the usefulness of our current results questionable for management. Samson, Graue, Weinstein and Walberg (1984) in fact suggested that other measures need to be developed, allowing for better correlations between academic and occupational performance.

Assignment grade and task performance in a "realistic" setting. It has been suggested that hands-on training, including job-simulating tasks, might be the most effective training method (Menashian, 1985). Unfortunately, there is virtually no data testing and supporting this claim. The results of this study may provide a partial confirmation, though. As Table 3 shows, historic GPA affects individual performance on ASSIGNMENTS in the micro-computer course. However, the coefficients indicate that the differences between "A" and "C" students are relatively small. For instance, when multiplying the coefficients with the GPA, the magnitude of difference between the "A" and "B" students is only about .3 out of a possible ten points.

The effect of a student's past performance in a mainframe BASIC course on current assignment performance is not significant. One interpretation could be that traditional methods of teaching computer literacy may not prove helpful for business applications. Therefore, effective approaches for teaching office workers require methods that are different from those used in traditional computer science courses (see also Lepper, 1985). Computer education needs to reflect its application in the workplace.

Users in office settings usually can draw on commercially available software (Bair & Nelson, 1985). Obviously, knowledge of a programming language and related technical jargon may be required of fewer and fewer employees.

Implications for Management and Future Research

The effective use of computer-based technology assumes a certain level of computer literacy on the part of the operator. He/she thus needs the kind of computer training which allows even the academically less inclined employee to master the required skills. In order to achieve this goal, it is necessary to change not only the current computer education curricula but also the training programs offered in organizations (Cerych, 1985).

This study might be helpful to managers and educators in several ways. For one, it shows that past academic performance might very well determine who should be trained first. Academically gifted individuals can acquire computer skills faster than their peers, making them a resource for their colleagues. The "buddy system" has proven to be effective in encouraging co-workers to increase their computer use in organizational settings (Gattiker & Larwood, 1986).

Furthermore, hands-on work assignments appear to be most helpful for academically less gifted individuals in acquiring computer literacy. Such training also reflects the actual work environment best, where such new skills can then be practically applied (Menashian, 1985). Traditional classroom methods seem better for academically more gifted individuals. However, experience and skills acquired in such settings seldom predict later job performance (e.g., Kulik, Kulik & Bangert-Drowns, 1985; Samson, Graue, Weinstein & Walberg 1984). Based on this study's results, a low-pressure training program for adults using hands-on experience is recommended, including realistic applications in terms of the individual's job as much as possible.

For the researcher, this study raises several issues. It would be of considerable interest to see how much time students spent on doing their assignments. Additional work might show if academic achievers need less time with the computer than their "average" peers. Another area of interest not addressed in this study is gender. Some researchers have found that women are affected differently by computer-based technology, i.e., they have different beliefs about it and show differences in attitudes when compared to their male peers (e.g., Form & McMillen, 1983; Gattiker & Coe, in press). Previous studies have also pointed out that the personality of the instructor may affect student performance (e.g., Brown, Sawyer & Hitchcock, 1984; Eden & Ravid, 1982). This study explains in part which training methods may be more useful when acquiring computer literacy (e.g., Menashian, 1985; Hebenstreit, 1985).

Clearly, the most important result is represented in the fact that academically weaker individuals may profit most from training methods which utilize a hands-on approach. Organizations should design training seminars reflecting actual work situations for the benefit of both the end-user and the company. The overall user effectiveness of new technologies in offices could thus be greatly increased.

References

Amarel, M. (1983). The classroom: An instructional setting for teachers, students, and the computer. In A. Ch. Wilkinson (Ed.), Classroom computers and cognitive science (pp. 15-29). New York: Academic Press.

Bair, J. H., & Nelson, K. J. (1985). User needs as the basis of strategic planning. In K. T. Quinn (Ed.), Advances in office automation (Vol.1) (pp. 3-19). Chichester, UK: John Wiley & Sons.

Bjorn-Anderson, N. (1983). The changing roles of secretaries and clerks. In H. J. Otway and M. Peltu (Eds.), New office technology (pp. 120-137). Norwood, NJ: Ablex Publishing

Bork, A. (1985). Computers and information technology as a learning aid. Education & Computing, 1, 23-35.

Brown, M. E., Sawyer, E., & Hitchcock, M. (1984). Teaching spelling: Microcomputers or small group tuition. Educational Psychology, 4, 269-279.

Butcher, D. F., & Muth, W. A. Predicting performance in an introductory computer science course. Communications of the ACM, 28, 263-268.

Campbell, P. F., & McCabe, G. P. (1984). Predicting the success of freshmen in a computer science major. Communications of the ACM, 27,

1108-1113.
Card, S. K., Moran, T. P., & Newell, A. (1983). The psychology of human-computer interaction. Hillsdale, NJ: Lawrence Erlbaum.
Carter, N. M. (1984). Computerization as a predominate technology: Its influence on the structure of newspaper organizations. Academy of Management Journal, 27, 247-270.
Cerych, L. (1985). Computer education: The problem of policy implementation. Education & Computing, 1, 5-10.
Dierkes, M., & Von Thienen, V. (1984). Kein Ende der Akzeptanzschwierigkeiten moderner Technik? Zum Zusammenhang der informationstechnischen Entwicklung und ihrer Akzeptanz. (Report No. P84-2). Berlin, West Germany: Science Center Berlin.
Doswell, A. (1983). Office Automation. New York: John Wiley & Sons.
Eden, D., & Ravid, G. (1982). Pygmalion versus self-expectancy: Effects of instructor- and self-expectancy on trainee performance. Organizational Behavior and Human Performance, 30, 351-364.
Form, W., & McMillen, D. B. (1983). Women, men, and machines. Work and Occupations, 10, 147-178.
Gattiker, U. E. (1986). Technological change and strategic human resource management. Working paper, The University of Lethbridge.
Gattiker, U. E. (1984). Managing computer-based office information technology: A process model for management. In H.W. Hendrick and O. Brown, Jr. (Eds.), Human factors in organizational design (pp. 395-403). Amsterdam, The Netherlands: Elsevier Science.
Gattiker, U. E., & Coe, L. (in press). Relationship of computer attitudes with perception of career success. Proceedings of the 46th Annual Meeting of the Academy of Management, 37.
Gattiker, U. E., & Larwood, L. (1986). Resistance to change: Reactions to workplace computerization in offices. Paper presented at the annual meeting of TIMS/ORSA, Los Angeles.
Giuliano, V. (Sept. 1982). The mechanization of work. Scientific American, 149-164.
Gutek, B. A. (1983). Women's work in the office of the future. In J. Zimmerman (Ed.), The technological woman. New York: Praeger.
Hebenstreit, J. (1985). Computers in education -- The next step. Education & Computing, 1, 37-44.
Hebenstreit, J. (1983). Training for future office skills. In H. J. Otway and M. Peltu (Eds.), New office technology (pp. 205-220). Norwood NJ: Ablex Publishing.
Kahn, R.L. (1981). Work and health: Some psychosocial effects of advanced technology. In B. Gardell and G. Johansson (Eds.), Working life (pp. 17-37). Chichester, UK: John Wiley & Sons.
Kulik, J. A., Kulik, Ch-L. C., & Bangert-Drowns, R. L. (1985). Effectiveness of computer-based education in elementary school. Computers in Human Behavior, 1, 59-74.
Lieberman, M. A., Selig, G. J., & Walsh, J.J. (1982). Office automation. New York: John Wiley & Sons.
Lepper, M. R. (1985). Microcomputers in education. American Psychologist, 40, 1-18.
Megaw, E. D., & Lloyd, E. J. (Eds.) (1984). Human-computer interaction. A special issue of Ergonomics Abstracts, 16(1).
Menashian, L. S. (1985). Training the office worker. In K. T. Quinn (Ed.), Advances in office automation (Vol. 1) (pp. 161-189). Chichester, Great Britain: John Wiley & Sons.
Morgall, J. (1983). Typing our way to freedom. Is it true that office technology can liberate women? Behavior and Information Technology, 2, 215-226.
Panko, R. R. (1984). Office work. Office: Technology and People, 2,205-238.
Samson, G. E., Graue, M. E., Weinstein, T., & Walberg, H. J. (1984). Academic and occupational performance: A quantitative synthesis. American Educational Research Journal, 21, 311-321.
Siegel, S. (1956). Non parametric statistics for the behavioral sciences. New York: McGraw-Hill.
Taylor, R. P. (Ed.) (1980). The computer in the school: Tutor, tool, tutee. New York: The Teachers College Press.
Wynne, B. (1983). The changing roles of managers. In H. J Otway and M. P. Peltu (Eds.), New office technology (pp. 138-151). Norwood, NJ: Ablex Publishing.

Human Factors in Organizational Design and Management - II
O. Brown, Jr. and H.W. Hendrick (Editors)

THE SITUATIONAL APPROACH TO MANAGEMENT AUDIT

Jack T. Brimer

Systems Management Department
University of Southern California
Los Angeles, California
U.S.A.

INTRODUCTION AND PURPOSE

Management audits should be tied to an underlying body of theory that will facilitate problem identification and provide a scheme for problem solving. This is the purpose of this paper.

The situational approach recognizes that any given management system can be characterized as having structural, conduct, and content dimensions. Structure is the way system elements are tied together. Conduct elements have to do with the manner in which system elements interact. Content deals with the empirical stuff of the particular system studied. In this presentation, these categories are pitted against the traditional management functions forming the framework for a problem solving methodology.

STEPS IN APPLYING THE SITUATIONAL APPROACH

Step I Define the Reference System

To accomplish this it is useful to focus upon the following diagram:

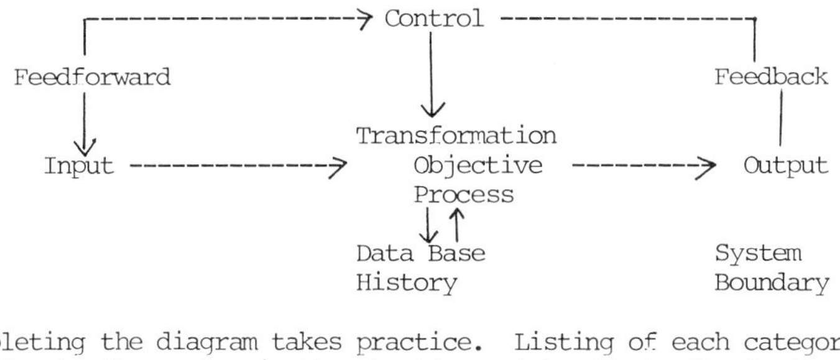

Completing the diagram takes practice. Listing of each category as it applies to the system is the starting point. Next, the feedback and feed-forward aspects are specified. This then leads to an understanding of how the parts function and provides a basis for the problem statement and solution generation steps to follow.

Step II Complete the Management Audit Questionnaire

Figure I presents the Questionnaire. The instrument is designed to assess the degree to which the principles of good management are practiced in the reference system. The questions have been divided into sets relating to the traditional management functions.

Each question should be answered according to the following scale.

Very Very True	Very True	True	Neither True Nor False	False	Very False	Very Very False
(1)	(2)	(3)	(4)	(5)	(6)	(7)

In answering question, place the ranking number to the right of the question answered. Note that the items can be randomized to avoid the obvious functional classification.

Figure 2 presents the Content Matrix showing the cross classification of system dimension against management function. Each question has been pigeon holed to indicate the aspect of the system it addresses. Question 013 applies to the structural dimension of the organizing subsystem.

Step III Score the Questionnaire

The scoring scheme Figure 3 is designed so that each cell in the content matrix can receive a numerical value which reflects the weak and strong areas of the management system. A large number in a leading-conduct block means that there is a potential problem with which the leading function is conducted in the subsystem. This is a potential problem area to be studied further.

Scoring the questionnaire involves: (1) Enter the raw score for the checked block on the questionnaire into the RAW SCORE box to the right of each question. Transfer this score to the square in the Scoring Matrix, Figure 3. The score for Question P5 is entered in the square at the intersection of the Planning-Structure row and the "5" column. (2) Add all the raw scores across the rows to get a row total in the RAW SCORE box down the right hand side of the matrix. (3) Calculate a normalized score by dividing the raw score by the divisor shown in the table. Dividing the raw score for the organizing process row by 54 yields the number to be entered in the normalized score box at the extreme right hand side of the matrix. The larger numbers in this column represent potential problem areas.

Step IV Define Problems and Generate Feasible Solutions

This step involves review of the system in view of the highlighted problem areas to pinpoint the most probable cause of the problem. A problem and cause statement is a convenient tool to use. This can be developed by a detailed review of the items in the questionnaire with reference to the content of the system aspect where the problem occurs to see how the principles of good management are not being applied. Make a list of these problem cause statements.

As an outgrowth of problem definition, solutions to the problem will be suggested. As these occur, write them down and make a list. Problem solutions should take the form of action plans that can solve the problem. These involve solution steps and a suitable time frame, and should be feasible from an operational viewpoint.

Intuition and judgement are involved in this step. There is no mechanical procedure. Knowledge and involvement in the system are the necessary prerequisite in this stage of the audit. The methodology has indicated areas of focus and provided a checklist for analysis. Application depends upon the human participant or consultant to produce a set of problem and solution statements.

Step V Evaluate Solution Alternatives

Evaluation involves the use of the Solution Evaluation Matrix in Figure 4. First, list the problems down the left side and the solutions across the top of the matrix. These can be short, paraphrased statements to facilitate recognition.

Second, assign an evaluation number to each problem-solution cell which reflects the likelihood that the solution at the top of the column will solve the problem to the left of the row. A low number reflects a low probability. The number can be an integer or ranking number or a percentage. There are a number of ways to assign these numbers as described in the literature. They can be assigned by an individual or a committee.

Third, assign weights that reflect the importance of each problem. A convenient weighting scheme is to convert the normalized scores to percentages that sum to one over the problem statement set. This is not necessary, and any weighting scheme can be utilized.

Fourth, compute the weighted average for each column by multiplying each evaluation number by the correct weight and summing the products to obtain a column total for each solution.

Step VI Select the Solution Set

The previous step has provided an ordering of the solution set. It is now up to the evaluator to choose the solutions from the set that have the best collective chance of improving the management system. Once again this is a ognitive process which cannot be reduced to a mechanical procedure.

Step VII Implement the Solution

Space prohibits describing this step in detail. The literature contains discussions of this material.

CONCLUDING REMARKS

This methodology was developed as a pedagogical tool for a masters level introduction to systems management course. It proved to be very useful in this context.

This paper represents a first order development. Further work to computerize the mechanical aspects of the process and develop a second order of more detailed diagnostic questions is being pursued.

FIGURE 1 MANAGEMENT AUDIT QUESTIONNAIRE

PLANNING

P. Planning is consistant with the organizational mission.

P. 2 Planning proceeds action.

P. 3 In making plans, the costs to administer and avoid mistakes are weighted against the benefits of accomplishing organizational objectives.

P. 4 All participants in planning tend to agress on the planning premices.

P. 5 Planning activities are coordinated.

P. 6 Strategies and policies are carefully considered and generally understood by the planners.

P. 7 In formulating plans, the planners focus upon the factors that are critical for goal attainment and ignore irrelavent side issues.

P. 8 Time horizons take into account the forecasting difficulties.

P. 9 The possibility of fulfilling plans is accurately evaluated when plans being determined.

P. 10 A reasonable amount of flexibility is considered during the formulation of plans.

P. 11 The penalities of flexibility are weighted against the risks of not being able to meet future commitments made in the plan statement.

P. 12 Plans are regularly audited and reasonably altered to meet changed conditions.

CONTROLLING

C. 1 Control procedures are focused upon detecting deviations of planned from events and analyzing the reasons for observed deviations.

C. 2 The control process is forward looking in that it utilizes feedforward control instead of relying mainly upon feedback control.

C. 3 Each manager moniters goals attainment of his unit.

C. 4 Overcontrol is not a significant problem in this organization.

C. 5 Controls are not a costly, complex, or burdensome load on workers.

C. 6 Management seems to preceive when a problem exists and takes appropriate action.

C. 7 Control criteria are clear, complete, and concise.

C. 8 Controls adequately reflect organizational goals.

C. 9 Information about deviations of plans from actual results is effectively utilized to make appropriate corrections.

C. 10 Controls incorporate objective, accurate, and suitable standards.

C. 11 Managers tend to focus upon significant positive and negative deviations from the plan.

C. 12 Control standards are established which are independent of the success or failure of the plan.

C. 13 Corrective action and follow up are integral part of the control system.

C. 14 Sufficient flexibility is built into control and they are altered to take account of changing circumstances.

ORGANIZING

O. 1 Individuals are organized so that they work effectively to accomplish the organizational mission.

O. 2 Most participants have a good understanding of all aspects of the organizational mission.

O. 3 People and financial resources are sufficient to "get the job done."

O. 4 Resources are effectively utilized in accomplishing unit objectives.

O. 5 There is a general conscienceness to avoid waste and abuse of resources.

O. 6 Management is generally able to closely supervise work activities and there does not appear to be a tendency to micro-manage activities.

O. 7 Given the job to be done and the ability to do the job, managers have about the right number of people to supervise.

O. 8 Lines of authority are fairly clear and well understood by the workers.

O. 9 Managers have authority and resources to achieve expected results.

O. 10 The lines of authority between superior and subornate are understood and in conducting work related activities.

O. 11 Subordinates generally fulfull their responsibilities.

O. 12 Managers are held accountable and accept responsibility for their subordinate's decision.

O. 13 Generally speaking, each individual worker answers to one boss.

O. 14 Managers and workers have sufficient training and ability to accomplish assigned tasks.

O. 15 Managers tend to avoid referring decisions to superiors when they have the authority and ability to make the decision themselves.

O. 16 Each job description is clear and understood by the incumbant.

O. 17 Task relationships are clearly specified and understood by the worker.

O. 18 The staff does not seem to be overworked nor are they underemployed.

O. 19 Task descriptions are conducive to attaining organizational objectives.

LEADING

L. 1 The present management seems to enjoy a reasonable level of satisfaction from its customers, suppliers, and employees.

L. 2 Personal and job related goals are fairly well integrated for most people in the organization.

L. 3 In general, people feel motivated to do a good job.

L. 4 There is a considerable degree of participation by workers in job decisions regarding work routines and the allocation of duties.

L. 5 Management understands what it takes to motivate workers to do a good job and effectively utilizes this knowledge.

L. 6 Communication between management and workers seems to be clear and free from basis misunderstandings.

L. 7 The communication between workers seems to be open and effectively facilitates task accomplishment.

L. 8 There is an obvious honesty present in communications at all levels.

L. 9 As far as effective communication is concerned, the formal organization does not appear to get in the way.

L. 10 The informal organization between various factions does not impair clear and concise information flow among members of the work force.

L. 11 Misinformation and rumor are not significant problems.

L. 12 The informal organization appears to promote the effective flow of information which cannot be handled by the formal structure.

L. 13 By in large, management communicates fairly well with the employees.

L. 14 Communication is consistant most of the time.

L. 15 Management openly seeks employee comments and suggestions and rewards meaningful suggestions.

STAFFING

S. 1 As a general rule, the right person is hired for the job to be done.

S. 2 People's capacities and potentials are effectively utilized.

S. 3 Job definitions are clearly expressed in job descriptions.

S. 4 Objectives and expected levels of accomplishment are explicitly stated.

S. 5 Employee appraisal is meaningful and fairly conducted.

S. 6 The employee appraisal system adequately measures employee performance relative to specified organizational objectives.

S. 7 The employee appraisal system seems to produce good results.

S. 8 Workers feel they are fairly treated in their performance appraisals.

S. 9 Employees generally feel that their pay and status is commensurate with the work they perform.

S. 10 There seems to be a general feeling purpose and accomplishment among members of the work force.

S. 11 An effective training program exists which improves job skills and promotes the quality of job performance.

S. 12 Employees feel they receive adequate training to do their jobs.

S. 13 There seems to be a reasonable cognizance about the need and purpose of training among the employees and management.

S. 14 There is a stress on continuing education to upgrade job skills and personal development from external sources.

S. 15 Employees actively engage in extramural continuing education programs sponsored by the organization.

Functional Area	Structural Dimension	Content Dimension
Planning	P1, P4 P5, P6, P8	P2, P3 P7, P9 P10, P11, P12
Organizing	O1, O2, O3 O5, O7, O8, O9 O13, O16, O17 O19,	O4, O6, O10, O11, O12, O14, O15, O18 O20
Staffing	S1, S3, S4 S6, S11, S13, S14	S2, S5, S7, S8, S9, S10 S12, S15
Leading	L1, L2, L6, L9 L10	L3, L4, L5, L7, L8 L11, L12, L13, L14, L15
Controlling	C1, C2, C5 C7, C8, C10,C12	C3, C4, C6 C9, C11, C13, C14

Figure 2
Content Matrix and the Correspondence and the Correspondence of Categories and Questions

Functional Area	1	2	3	4	5......	16	17	18	19	20	Raw Score	Normalized Score
Planning Structure		X		X		X	X	X	X	X		
Conduct	X				X	X	X	X	X	X		
Structural Divisors = 24												
Process Divisor = 42												
Organizing Structure				X				X		X		
Conduct	X	X	X		X	X	X		X			
Structure Divisors = 66												
Process Divisor = 54												
Staffing Structure		X	X		X	X	X	X	X	X		
Conduct	X			X		X	X	X	X	X		
Structural Divisors = 30												
Process Divisor = 60												
Leading Structure			X	X	X	X	X	X	X	X		
Conduct	X	X				X	X	X	X	X		
Structural Divisors = 30												
Process Divisor = 60												
Controlling Structure			X	X		X	X	X	X	X		
Process	X	X				X	X	X	X	X		
Structural Divisors = 48												
Process Divisor = 42												

The center columns can be filled in conveniently by the reader.

Figure 3
Management Audit Scoring Matrix

	Alternative Solutions					
Problem Area	Weight	A	B	C	...	Z
Planning Problems						
1						
2						
.						
.						
Organizing Problems						
1						
2						
.						
.						
Staffing Problems						
1						
2						
.						
.						
Leading Problems						
1						
2						
.						
.						
Controlling Problems						
1						
2						
.						
.						
Sum of Weights Times Probability Numbers = Weight Average						

Figure 4
Problem - Solution Evaluation Matrix

Footnotes

1. The management principles utilized are drawn from the text by Harold Koontz and others. Harold Koontz, Cyril O'Donald, and Heintz Weihrich, MANAGEMENT, 8TH EDITION, New York, McGraw Hill, 1984.

2. See for example, J.T. Brimer, The Evaluation of Fuzzy Alternatives, ISSM Report, 1986.

3. See for example, J. T. Brimer, The Situational Approach to Systems Implementation, in H. W. Hendrick and O. Brown, Human Factors in Organizational Design and Management, Amsterdam, North Holland, pp. 553-562.

SAFETY, WORK PRACTICES AND PRODUCTIVITY

Human Factors in Organizational Design and Management - II
O. Brown, Jr. and H.W. Hendrick (Editors)

ARE ACCIDENTS AN ARTIFACT - AN ANALYSIS OF TRANSIENT FACTORS IN DELIBERATE INJURIES

M. J. Sridhar

National Institute of Rural Development
Hyderabad
India

The hypothesis, 'accident behavior is a conscious intention to hurt one-self to achieve the desired goal' was investigated in this study on two groups of-56 High and 60 Low-outpass seekers. The results confirmed that workers of both the groups had a tendency to inflict injuries under conditions of rejection of an outpass. The deliberate injuries were more in the forenoon than in the afternoon. It is suggested that in regard to deliberate injuries, the company should examine the alternate ways of reinforcing their employees rather than inadvertently reinforcing minor injuries.

INTRODUCTION

Accidents are largely the result of momentary physiological and psychological factors like, fatigue, emotionality, irritability or inattention, coinciding with the internal and external factors. Many accidents seem to occur in factories when workers are emotionally upset either on a temporary or more or less chronic basis. Many studies (Hersey, 1936; Himler, 1951; MacIver, 1961) have indicated that emotionally disturbed employees are not only more accident prone but also contribute to 30 per cent of sick absences. Some workers are also beset at times with destructive impulses, some times directed inwardly, some times directed specifically against others and some times directed diffusely against the whole environment. These impulses are, of course, under much better control in some individuals than they are in others. According to MacIver (1961), they are largely unconscious, but, in many instances a degree of conscious awareness of them can be seen. Another hypothesis is that, men have accidents in order to get time off the job (Hill et.al., 1953). The classification of self-destructive acts as deliberate or accidental has been a barrier to the understanding of the dynamics underlying the occurrence of accidents (Rosenberg, 1967).

Studies of this nature are psycho-analytical, where the researchers have tried to define the causes of accidents as due to inadequate or incomplete repression of a hostile id. Apart from these, the unanswered question is whether accidents are due to conscious intentions to achieve the desired goals. These can be readily seen among factory employees, when they feel bored with the routine jobs, take time off the work by inflicting minor injuries on themselves. Unconscious motivation to hurt one-self, in the strictest term, is injurious but conscious intentions to hurt one-self may be because the job is not interesting, or may be, there are more interesting and attractive things outside the job environment. These persons cannot be categorised as psychiatric cases but are normal pleasure seeking individuals. When there is a blockade to their pleasure, these persons become highly emotional and show their aggression towards the management, deliberately inflict injuries on themselves and seek pleasure outside the factory.

The purposes of the present study were: 1) to study whether rejection of an outpass (permission to go out of the factory) makes an employee to get frustrated which subsequently makes him to inflict an injury on himself; 2) to study the important specific causative factors which lead to deliberate injuries.

METHOD

A pretest of accidental injuries in a few departments of Motor Industries Company Limited in Bangalore, India was conducted for a month. During this pretest period, it was observed that workers had a habit of asking for outpasses to go out of the factory during the normal schedule of work activities. The foremen were vested with authority to issue outpasses to their workers for a duration of two to five hours whenever emergencies arose while working. But, whenever an outpass was rejected, the workers were prone to fake minor injuries and request for time off the job. By deliberately inflicting injuries (some times without any apparent injury) workers were seeking accident outpasses to attend to their personal work. The findings of the pretest were used to determine the final design of the observational study of accidents.

The design involved studying two groups of outpass goers and their accident rates: High Group and Low Group. In order to select these two groups, a meeting with twenty foremen from various departments was held. The High outpass seeking group was defined as those who had sought 4 and more outpasses, on an average, per month (both rejected and given outpasses) consistently over a period of 6 months time. The Low group was defined as those who had asked 3 and below outpasses, on an average, per month (both rejected and given) consistently over a period of 6 months time. The sample included 56 High outpass goers and 60 Low outpass goers. Those who participated in the study were males. For valid statistical analysis the following design features were introduced into the overall experimental design:

1) Each worker was observed for 27 days and thus the data base for the study constituted 3132 observational days, of which 1512 observational days (56 workers x 27 days) came from High group and 1620 observational days (60 workers x 27 days) came from the Low group.

2) The High and Low groups observational days were post-facto, further divided into two categories viz., Experimental and Control. The Experimental sub-group included such observation days, during which an injury was reported sequel to an outpass rejection, or an outpass was granted as a result of a 'deliberate' accident. The Control sub-group included such observation days during which either there was no request for an outpass or an outpass was granted only as a consequence of prior 'genuine' accident.

3) Thus, the study had four distinct groups viz., High Group Experimental (HGE); High Group Control (HGC); Low Group Experimental (LGE); and Low Group Control (LGC).

4) The data base had two formats: a) from the foremen who had close supervision on these workers; and b) from unstructured personal interviews with the workers to secure data on immediate and contextual specific factors in accident causation. The following information was gathered for each worker of the High and Low groups:

No. of observation days according to occurrence of accident sequel to rejection of an outpass

Table No	Group	Accident	No Accident	Total	X^2
1	HGE	42* (40.0)	63 (60.0)	105 (100.0)	360.05 (P<.01)
	HGC	20 (1.4)	1387@ (98.6)	1407 (100.0)	
	Total	62	1450	1512	
2.	LGE	16** (35.6)	29 (64.4)	45 (100.0)	329.76 (P < .01)
	LGC	9 (0.6)	1566@@ (99.4)	1575 (100.0)	
	Total	25	1595	1620	
3.	HGE	42 (40.0)	63 (60.0)	105 (100.0)	0.11 (N.S)
	LGE	16 (35.6)	29 (64.4)	45 (100.0)	
	Total	58	92	150	
4.	HGC	20 (1.4)	1387 (98.6)	1407 (100.0)	4.73 (P<.05)
	LGC	9 (0.6)	1566 (99.4)	1575 (100.0)	
	Total	29	2953	2982	

Note: * Includes 21 reported injuries subsequent to rejection of an outpass and also 21 cases who inflicted injuries prior to seeking an accident outpass.

** Includes 10 reported injuries subsequent to rejection of an outpass and also 6 cases who inflicted injuries prior to seeking an accident outpass.

@ Number of outpasses granted just on request 123. On 1264 (1387-123) observation days there were no requests for outpasses.

@@ Number of outpasses granted just on request 157. On 1409 (1566-157) observation days no requests for outpasses were made.

No. of observation days according to time of accident

Table No.	Group	Forenoon	Afternoon	Total	X^2
5.	HGE	32 (76.2)	10 (23.8)	42 (100.0)	4.57 (P < .05)
	HGC	9 (45.0)	11 (55.0)	20 (100.0)	
	Total	41	21	62	
6.	LGE	13 (81.3)	3 (18.7)	16 (100.0)	3.85 (P < .05)
	LGC	3 (33.3)	6 (66.7)	9 (100.0)	
	Total	16	9	25	
7.	HGE	32 (76.2)	10 (23.8)	42 (100.0)	0.004 (N.S)
	LGE	13 (81.3)	3 (18.7)	16 (100.0)	
	Total	45	13	58	
8.	HGC	9 (45.0)	11 (55.0)	20 (100.0)	0.03 (N.S)
	LGC	3 (33.3)	6 (66.7)	9 (100.0)	
	Total	12	17	29	

No. of observation days according to nature of accident

Table No.	Group	Intentional	Unintentional	Total	X^2
9.	HGE	38 (90.5)	4 (9.5)	42	1.21 (N.S)
	LGE	12 (75.0)	4 (25.0)	16 (100.0)	
	Total	50	8	58	

1) Number of outpasses requested; 2) Type of outpass (rejected or granted); 3) The accident statistics (whether occurred subsequent to rejection of an outpass or prior to asking for an outpass); 4) Outpasses given under normal conditions; 5) The time of the outpass (forenoon or afternoon); 6) Time of accident; 7) Reason for an outpass; 8) Reasons for the accident.

The information thus collected were coded first and transferred on to specially designed cards using one card per worker.

RESULTS

For the two groups - High and Low - the number of injuries subsequent to rejection of an outpass (Experimental condition) were found. For comparison purposes, information relating to the number of accidents sustained by the two groups under Control conditions were taken for analysis. Comparisons were also made for other variables like, time of accident subsequent to rejection of an outpass and intentional injuries under rejection conditions. The significance of differences between HGE and HGC; LGE and LGC were found using Chi square analysis. The important specific contextual and situational factors leading to deliberate injuries have been analysed from the data collected through unstructured interviews.

Table 1 relates to the accident rate under Experimental and Control conditions within the High group. The rate of infliction of injury is more among the workers who inflict injuries on themselves as a reaction to outpass rejection. In the Low group also (see Table 2) a similar tendency is perceived between Experimental and Control groups. These two tables together confirm that under rejection conditions, workers prefer to inflict injuries irrespective of the group. Since the rejection of an outpass invites the same kind of reaction i.e., inflicting an injury, the rates in the two groups (Table 3) remain more or less same.

Comparisons of the time of injuries between HGE and HGC (Table 5); LGE and LGC (Table 6) indicated that injuries were relatively more in the forenoon than in the afternoon ($P < .05$). Further, a non-significant Chi square value ($P > .05$) between HGE and LGE with respect to intentional and unintentional injuries (Table 9) indicated that both the groups had the same tendency to intentionally inflict injuries under conditions of outpass rejection, further supporting the view, that outpass rejection brings about intentional injuries.

DISCUSSION

It was observed that accidents were highly rewarding viz., getting an outpass to go out of the factory and also be eligible for accident compensation, which reinforced the workers to inflict many more 'make believe' accidents. Those who had incurred injuries under the Experimental condition were, to a large extent, deliberate or intentional in nature to get time off the job. The primary reasons to go out of the factory were many and varied - as an example, going to the horse racing on the week ends served two purposes: first, getting time off the routine work; secondly, getting the pleasure of betting on horses. There was a manifestation of usually faking the reasons to the foremen in order to get an outpass. Medical outpass did not help workers to seek pleasure, they were sent back after medical assistance. Nobody could question the bonafides of an accident. The foremen had no way of getting at the genuine reasons for "Why the workers wanted time off their jobs?"

Inflicting injuries consciously on one-self depended upon the urgency of the personal work outside the factory and the degree to which the environment stimulated them. By inflicting injuries, they were eligible to get Employees State Insurance (ESI) compensation, if the accident was 'reportable' or 'lost-time' in nature and also were eligible to get factory wages for the day without working. Many old time workers had gone to the extent of influencing other new recruits about this "benefit action game".

However, these destructive impulses were only a transient state and much later (after the outpasses were given) there was no evidence of these emotional upsets. Since there was a desire to go out of the factory, they exhibited the destructive characteristics without inhibitions. At some other time, the same workers who did not have an urgent need to go out, eventhough the outpasses were rejected, withdrew from inflicting an injury on themselves. Workers of both the High and Low groups blamed the organisational policies, in terms of restrictions imposed on leave, lack of better incentives to them and hence thought that it was better to inflict an injury and take an accident outpass to seek pleasure momentarily outside the factory. The Experimental workers of High and Low groups had also learnt to differentiate between the injuries which were rewarding (e.g first aid injuries) and punishments (e.g major injuries). Two workers during their course of inflicting injuries were involved in major accidents. From then onwards, there was no evidence of accidents in them.

This finding is all the more interesting, since variability in other factors like, personality, age, experience have been controlled to a large extent because the same workers have been examined under both the conditions viz., Experimental and Control. The person who became an Experimental subject at one time also became a Control subject at some other time. Thus, there was a 'self control' mechanism operated in each worker, which means the same worker experienced both the Experimental and Control conditions during the 27 days of observation. It was observed, once or twice workers sought outpasses, if they were rejected by the department foremen, they inflicted injuries on themselves and got the accident outpasses. In the subsequent periods, whenever they wanted an outpass, there was no need to ask the foremen, they directly injured themselves with a burr and got time off the job. There was a sort of an achievement and they took pride in telling others about their benefit action game.

It seemed that injuries were relatively more in the forenoon than in the afternoon. According to the Experimental sample, make believe type of accidents in the forenoon were very beneficial. If they had urgent work outside the factory, it was only in the forenoon they thought of faking injuries, as they had the whole day at their disposal to deal with matters, like, legal proceedings, horse racing, side business, house construction, etc., and hence taking outpasses was very useful in the forenoon. Since I shift workers worked from 6 a.m to 2 p.m, taking an accident outpass from 10 a.m onwards, enabled a worker to get 4 hours time off the job. It was very rare that after taking an outpass he came back to resume the work once again. Since, the factory provided free lunch to all the employees, these impulsive or deliberate acts came usually after the lunch.

Whenever a worker wanted time off the job, because of the fear of possible rejection of an outpass and insecurity, gave faked reasons. The workers who intentionally inflicted injuries on themselves reacted very easily to the accident situations. Since the intentions were pre-planned, the injuries were always of first aid in nature. Many workers accepted the factory environment which demanded production and discipline and these workers fell mostly in the Control non-accident group. The conscious intentions to achieve certain immediate goals are not accidents, but just pretentions that an accidental injury has happened. Then, it would follow, "accidents are a means to achieve certain immediate goals".

CONCLUSIONS

These findings do not support the view held by the earlier accident theorists, that, accident behavior is unconscious suicide. The destructive impulses were only transitory. It is suggested that in regard to the deliberate injuries, the safety department in the first instance, should examine the alternate ways of reinforcing the employees rather than inadvertently reinforcing minor injuries. Practical measures to prevent accidents should consider the actual accidents sustained, its real causes, nature and parts of the body injured. The safety department can initiate and devise a few training programmes to change the attitudes of workers. Further, the foremen should have more powers to discipline their department workers in regard to deliberate injuries.

REFERENCES

Hersey, R.W., Emotional Factors in Accidents, Personnel Journal, 1936 15, 59-65.

Hill, J.M.M, and Trist, E.L.A. A consideration of Industrial accidents as a means of withdrawal from the work situation. Human Relations, 6 (1953), 357-380.

Himler, L.E. Human Relations and Accident Prevention. Industrial Medicine, 1951, 20, 121-123.

MacIver, J. Safety and Human Behaviour. Behavioural Approaches to Accident Research, 1961. Association for the Aid of Crippled Children, New York.

Rosenberg, M. On Accidents and Incidents: A Study of Self-destruction. Comprehensive Psychiatry, 1967, 8(2), 108-118.

Shaw, L and Sichel, H.S. Accident Proneness: Research in the Occurrence, Causation, and Prevention of Road Accidents. Pergamom Press Ltd., 1971.

Human Factors in Organizational Design and Management - II
O. Brown, Jr. and H.W. Hendrick (Editors)
© Elsevier Science Publishers B.V. (North-Holland), 1986

DEVELOPMENT OF MAINTENANCE MANAGEMENT SYSTEMS: THE VITAL ROLE OF DEFINING ORGANIZATIONAL INTERFACE REQUIREMENTS*

Marjorie B. Bauman, Jennifer Goodson Stewart, Harold P. Van Cott

Essex Corporation
333 North Fairfax Street
Alexandria, Virginia 22314
U.S.A.

This paper describes case studies of three nuclear power plants that either implemented an automated maintenance management system or held job coordination meetings in an effort to improve communications in the work order process. The effectiveness of the strategies within several communication dimensions was evaluated, and the findings, along with a work-flow analysis and subject matter expert input, were used to identify functional user requirements of an optimal automated maintenance management system.

INTRODUCTION

As the research arm of the electric power industry, the Electric Power Research Institute (EPRI) has sponsored several research projects [1, 2, 3] that demonstrate the critical importance of vertical and horizontal communication in nuclear power plants. This study is an outgrowth of that research. It focuses on communications that affect the efficiency of the maintenance work order process.

The work order process in a plant is complex. Many people, who come from different levels of management and from different departments, are required to implement it. For the work order process to be efficient, there must be good interaction and coordination within and between departments.

This report describes case studies of nuclear plants that implemented intervention strategies designed to improve communications in the work order process. Two intervention strategies were used: (a) an automated work order system (Plants 1 and 4) and (b) job coordination meetings (Plant 2). It was assumed that both strategies would make the maintenance work order process more efficient: the automated system by facilitating and expediting the work order process, and the meetings by improving departmental interfaces.

Before either strategy was adopted, a preliminary survey had shown that five aspects of communication were particularly problematic: timeliness of information; redundancy of information; information withholding or gatekeeping; feedback; and information accuracy and amount.

After the two strategies were implemented, plants were re-surveyed to determine what effect, if any, the strategies had had on improving communications. The survey

*This research is sponsored by the Electric Power Research Institute (EPRI) under RF 2167-3.

results showed that, while there was still room for improvement, both strategies reduced communications problems in several areas.

Although both interventions were found to have strengths, because of the increasing numbers of plants opting for automation of their work order processes and the lack of input into these systems regarding communications needs, this paper focuses on findings and applications for an automated system.

In order to take advantage of the lessons learned from this project, and to transfer the benefits to the industry as a whole, EPRI tasked Essex Corporation to develop a functional requirements analysis for the industry. The analysis outlines the informational and organizational requirements for an optimal automated maintenance management system for general industry use. This paper illustrates how Essex derived the functional requirements for an automated maintenance management system from its knowledge of the work order process and interpretation of the communication survey findings.

METHODS

The project was divided into two phases. The first phase focused on the development, implementation, and evaluation of an organizational communication survey designed to address areas previously found to be problematic at the plants. This survey was used to measure the impact of two different types of intervention strategies (automated work order systems and job coordination meetings) on the communication issues facing the plants.

The second phase of the project built on the survey findings of the first phase. In it, a functional requirements analysis for the design of an optimal maintenance management system was performed. On the basis of the lessons learned from the experimental plants, the functional requirements analysis was designed for use as a guide to other electric power utilities to assist them in the development of a comprehensive automated maintenance management system.

Phase 1: Organizational Communication and Work Order Process Survey

Plants participating in the survey included Plants "1" and "4", which were in the process of instituting automated work order systems. They provided an opportunity to evaluate the influence of automation on communication at these two plants. Plant "2" agreed to hold job coordination meetings for first-and second-line supervisors to examine their effect on organizational communication.

Questionnaires were administered before program implementation and again at 5 to 10 months after implementation. Respondents represented all major departments (Operations, Maintenance, I&C, Health Physics, and Quality Control), as well as all staff levels. Follow-up interviews were used to validate questionnaire findings and to obtain real-world examples of organizational communication issues.

Phase 2: Development of a Functional Requirements Analysis for a Maintenance Management System

After evaluating the respective impacts of the two interventions on organizational communication, the findings were analyzed to determine what changes could be made to the automated system tested to improve on areas still reported as being weak in questionnaire and interview responses. A schematic of a generic work order process, developed by Essex, and depicting departmental roles and interfaces, was used to describe the required sequence of events involved in the performance of corrective maintenance work in the plants under a manual system. It was used as a foundation from which requirements could be defined and upon which improvements could be made.

This portion of the study was supplemented by interviews with system design engineers responsible for the second generation system being developed for use at Plants 1 and 4. These subject matter experts provided input regarding maintenance management "philosophy" that emphasizes principles of system design and organizational factors that should be specifically addressed in the development and implementation of an automated system.

FINDINGS AND APPLICATIONS

More than one-half of the survey baseline respondents indicated that the organizational communication at their plants needed extensive improvement. Five months after interventions were introduced, all plants cited improvements. The automated plants reported a positive impact regarding time to initiate a work order, while the plant that implemented job coordination meetings reported improvements in the receipt of advanced notice of upcoming work. Plants also showed a reduction in the number of duplicate work orders issued with the automated system, although duplication continues to be a problem. QA/QC responsiveness showed notable improvement at all three experimental plants, and Plant 2 (job coordination meetings) respondents reported increased satisfaction with information sharing among departments. Tracking of work orders was significantly improved for the Maintenance and I&C personnel at plants using the automated system. In addition, the accuracy of information on work orders, such as problem descriptions, equipment identification and location, and work assignment, improved at all three experimental plants. Although many gains were made upon implementation of the interventions, the data and interview responses indicate that there is still room for improvement, even in positively impacted areas.

To meet EPRI's goal of effective technology transfer and the electric power industry's move toward increased use of automation, Phase 2 of this study focused on translating the work flow analysis and survey findings into the definition of a user-oriented set of functional requirements to be used as guidelines for future system design.

The survey findings were evaluated from this perspective and were incorporated in an outline of informational and organizational requirements for an optimal automated maintenance management system. An example of existing problems illustrated by the survey and their solutions can be seen in Table 1.

Figure 1 contains an outline of the functional requirements identified, arranged in a format illustrating the type of interactive system necessary to effectively enhance automated maintenance management. From this outline of optimal system capabilities, Essex developed detailed functional flow diagrams of the work order process. These diagrams illustrate the exchange of information between the system and users. For example, the Maintenance History File is queried to define resource and skill requirements and safety requirements, and to determine required tools and equipment. In addition to eliciting information from the system, users input information to keep work order status information current and to provide the most complete records possible of work performed for future reference.

Table 1
Linkages Between Problems With Automated System Tested and Functional Requirements Specified for Optimal System

Dimension	Problem	Solution
Timeliness	Out-of-service tagging less timely because system status reporting gives limited information. Need for OOS is not given.	o Expand "Work Order Monitoring File" to provide more detailed tracking. o Increase interaction with "Planning/Scheduling File" to give operations anticipated start dates for work. (See Figure 1.)
Redundancy	Duplication, attributed to reference index. Trouble Report number assigned initially is changed after review and approval of work order. Difficult to determine if new work order is redundant.	o Retain consistent numbering throughout process. o Institute automated check by equipment I.D. number and problem descripition.
Feedback	Poor feedback regarding problem diagnosis and corrective action because that portion of the work order is torn off prior to close-out.	Expand "Maintenance History File" to make all information relevant to a particular job or job type available. (See Figure 1.)
Information Accuracy/Amount	Insufficient background information and equipment I.D. and location. Existing work order form provides space for problem identification but does not prescribe a standardized format to encourage better detail.	o Format work order form to request specific problem information. o Expand "Equipment Inventory File" to automatically provide detailed equipment information on the work order form. (See Figure 1.)

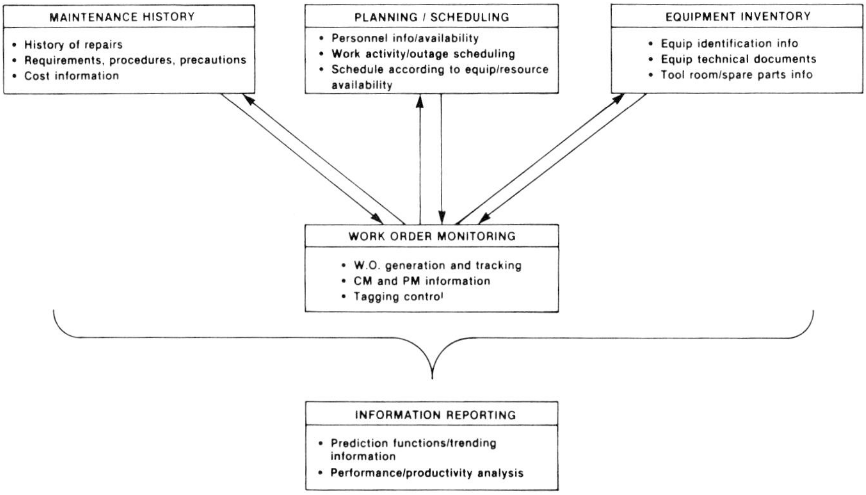

Figure 1. Maintenance Management System — Interactions

The Information Reporting file, shown in Figure 1, has access to all information input into the system that can then be used for trending, analysis, and predictive purposes. This feature has been incorporated into the optimal automated system in response to management reporting requirements which were identified in the functional requirements analysis. For instance, the system can be used to establish a preventive maintenance program and to monitor and evaluate it. Figure 2 illustrates a typical use of the system's information reporting capabilities by maintenance managers. High levels of corporate management would receive the most global data. Each successive lower level would receive more detailed information regarding a particular targeted area of interest. This approach facilitates problem identification for managers while streamlining the technical data they receive.

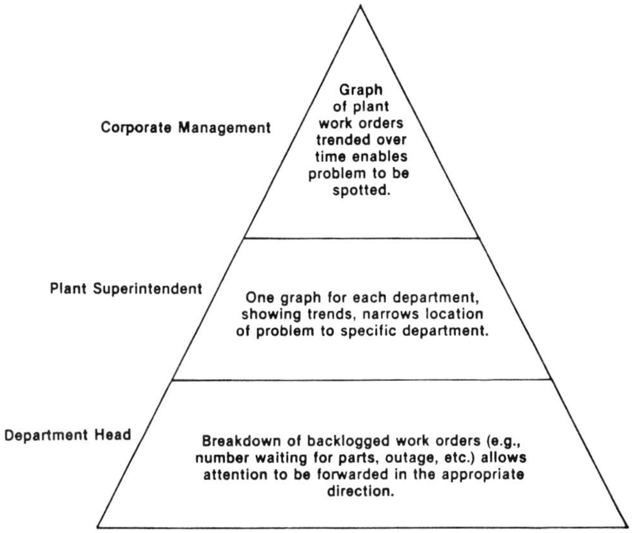

Figure 2. Automated Information Reporting: Workload Assessment in the Plant Management Hierarchy

Other issues defined during the course of this study involve the principles, rather than the content, of the system design. For example, a work order should be formatted so as to be "user friendly" and to elicit specific information (as is referred to in Table 1). It is also important that the system be designed to accommodate the "worst case," as is possible with corrective maintenance work orders, but it should make a point of facilitating more routine maintenance activities. One suggestion in this regard is to provide two work order formats: one "abbreviated," and another, with additional requirements, that would be selected when special procedures and approvals are necessary.

A further recommendation is that automated maintenance management systems be implemented in conjunction with job coordination meetings. The case study results indicate that, while both interventions had a positive impact within the various communication dimensions, each strategy showed some unique strengths in meeting specific communication requirements. Used in combination, the automated system would facilitate and expedite informational aspects of the work order process, while job coordination meetings would complement the system by enhancing face-to-face interactions and those communication areas impacted by them.

CONCLUSION

This paper provides an overview of organizational communication issues impacting nuclear power plant operations and maintenance. The effectiveness of alternative intervention strategies designed to improve organizational communication are discussed.

In addition, the issue of technology transfer of the study findings is addressed in terms of the development of a functional requirements analysis document for a comprehensive automated maintenance management system. The functional requirements analysis will serve as a guide to electric power utilities interested in incorporating the communication needs of a variety of system users into the design of an automated maintenance management system.

REFERENCES

1. Bauman, M. B., Pain, R. F., Van Cott, H. P., and Davidson, M. K. Survey and analysis of work structure in nuclear power plants (NP 3141). Prepared by BioTechnology, Inc., for the Electric Power Research Institute, Palo Alto, California, June 1983.

2. Law, T. M. (Chairman) Workshop proceedings: Power plant maintenance and maintainability (EPRI NP-8-3-LD). Prepared by Pickard, Lowe, and Garrick, Inc. Palo Alto, CA: Electric Power Research Institute, March 1981.

3. Seminara, J. L. Parsons, S. O., Schmidt, W. J., Gonzalez, W. R., & Dove, L. E. Human factors review of power plants maintainability (EPRI NP-1567-SY). Prepared by Lockheed Missiles and Space Co., Inc. Palo Alto, CA: Electric Power Research Institute, October 1980.

Human Factors in Organizational Design and Management - II
O. Brown, Jr. and H.W. Hendrick (Editors)

BENEFITS OF A PARTICIPATORY SAFETY AND HAZARD MANAGEMENT PROGRAM IN A BRITISH COLUMBIA FORESTRY AND LOGGING ORGANIZATION

Bert Painter
B. C. Research
Vancouver, British Columbia, Canada

Thomas J. Smith
School of Kinesiology
Simon Fraser University
Burnaby, British Columbia, Canada

The rationale for involving employees directly in industrial safety and hazard management programs is based upon the fundamental behavioral cybernetic principle that those most directly affected by workplace hazards, the employees themselves, should be given direct opportunity to participate in making decisions regarding how those hazards are managed in relation to work operations. This case study report describes methods and results of a participatory safety and hazard management program, based intensively on employee involvement, which has been implemented in a British Columbia forestry and logging organization. The program was introduced into the six separate company logging camps, covering 50-60 employees per camp, and was structured in terms of six participatory elements: employee involvement, management involvement, follow-up action, trust and communication, planning, and leadership. Employees now directly influence or share responsibility in 16 of 18 distinct decision-making areas affecting logging operation safety. From 1982 to 1985, the company forestry and logging group experienced a 75 % decrease in accident frequency, and a 62 % drop in compensation claim costs, which managers and employees alike attribute directly to the program. We conclude that the human factors approach to organizational design and management, centered on the behavioral cybernetic principle of employee participation, constitutes a compelling model for future development of safety and hazard management programs.

PURPOSE

The concept of industrial safety and accident prevention as an essential management function is a relatively recent idea that has yet to gain complete acceptance across all industrial sectors (Pope, 1981). It can be anticipated that with the current trend in North America towards greater involvement by employees themselves in self-management and self-control of workplace hazards (Bryce, 1981), the present uncertainty regarding the definition, criteria, and terms of reference of the safety management concept will continue and may indeed intensify. One major shortcoming in this area has been the lack of a comprehensive conceptual framework, along with explicit methodology, to guide the organization of hazard management programs based on employee participation. A further lack is the availability of tangible evidence from operating organizations regarding the effectiveness and consequent validity of the participatory approach. These issues were addressed by Smith, Lockhart, and Smith (1983), who delineated the theoretical framework of the cybernetic systems approach to managing job hazards, an approach which relies on a participatory strategy built around employee involvement. This report then went on to detail three different participatory hazard management methodologies which have been developed in different organizations.

The theoretical and methodological review by Smith, Lockhart, and Smith (1983) underscored the lack of concrete evidence, from actual organizations, which could be used to document the effectiveness and validity, or lack thereof, of the participatory approach to safety and hazard management. The purpose of this case study report is to describe methods and results of a participatory safety and hazard management program, based intensively on employee involvement, which has been implemented in the Forestry and Logging Group of Whonnock Industries (Vancouver, B.C.), a major British Columbia forestry company. The program appears to have had dramatic and highly beneficial effects in reducing the frequency and costs of work-related logging accidents over a four-year period, from 1982 to 1985.

THEORY

The theoretical basis of the cybernetic systems approach to safety and hazard management has been developed in a comprehensive series of articles by Smith (1973, 1974, 1975, 1979, 1980) and by Smith and Sauter (1981). The theory has been condensed in the following ten behavioral cybernetic systems principles of hazard management, set forth by Smith and Smith (1983): (1) a safe job is a healthy job; (2) control hazards, not accidents; (3) hazard management is a behavioral cybernetic process; (4) a safe job is a human factored job; (5) base safety standards on operational performance factors; (6) promote self-regulation; (7) involve the employee; (8) integrate safety with production; (9) implement no-fault compliance provisions; and (10) establish equal opportunity for safe work.

As used here, the term safety system refers to the network of individual employees and managers, along with the businesses and industries, unions, governmental agencies, regulatory bodies, insurance groups, and safety specialists and professionals, that collectively define and determine safety and accident prevention policies, programs, and practices. The cybernetic or systems theory of safety and hazard management emphasizes the primary self-regulatory role that those most directly affected by workplace hazards, namely the employees and managers themselves, must play in the hazard management process. The above principles, and most particularly principles two through eight, establish a conceptual framework for the cybernetic approach which we believe can be used to achieve optimal control of workplace hazards.

A starting point for the cybernetic theory is the concept that hazards are defined and subject to control only if they are detected and their potentially adverse effects anticipated. Thus, hazard control is a predictive, feedforward process (Principle 2) which depends on the behavioral and social interactive skills and competency of the entire safety system (Principle 3) in recognizing, evaluating, and controlling hazards (Smith, 1973, 1974, 1975; Smith and Smith, 1981; Tuttle et al., 1973). There is growing scientific evidence to suggest that unsafe or defective working conditions, in addition to unsafe acts, represent primary causal factors in many industrial accidents (Hansen, 1982; Smith, Ginnold, and Brandl, 1982; Sass and Butler, 1978; Hagglund, 1976; Mason, 1973; Swain, 1973a, 1973b). This evidence supports the concept that industrial safety, therefore, is directly and inextricably linked to how job, workstation, and equipment designs are human factored to conform to the behavioral, physiologic, and performance limitations of the employee (Principle 4). Thus, adoption and enforcement of physical safety standards and negligence regulations will not insure safe work unless the job itself is factored to meet human needs (Smith, 1973, 1974, 1975, 1979).

The foregoing concepts pave the way for the two tenets which represent the lynchpin of safety cybernetics: self-regulation and employee involvement (Principles 6 and 7). The essence of the theory is that individual workers and organizations most directly affected by workplace hazards should be primarily responsible for managing and controlling them (Smith, 1973, 1974, 1975, 1980; Cohen, Smith, and Anger, 1979). A major industrial resource for achieving such self-regulation is the employees themselves, whose insight and understanding of their own work operations, conditions, and performance demands can be employed for effective detection and predictive control of the effects and scope of job hazards (Bryce, 1981; Smith, 1974, 1975, 1980; Viaene, 1980).

A variety of objective scientific evidnce now is available to support the theoretic concepts outlined above. For example, a number of studies show that workers are very capable of identifying hazards associated with their work, hazards which often are unrecognized by management or by existing regulations (Coleman and Smith, 1976; Kaplan and Coleman, 1976; Gottlieb, 1976). However, little evidence has yet been reported to clearly document the effectiveness of operational safety and hazard management programs based on employee participation.

WHONNOCK PARTICIPATORY FORESTRY AND LOGGING SAFETY AND HAZARD MANAGEMENT PROGRAM

In 1982, both the accident frequency rate and the workers' compensation (W.C.B.) costs for the Whonnock Industries Forestry and Logging Group exceeded the B.C. industry-wide average, with the former index almost three-fold higher (Figure 1). This pattern, which also had prevailed in other previous years, prompted the company to make a commitment to a revised forestry and logging safety and hazard management program based on two basic principles---accident prevention and employee participation. The program was introduced in 1983 with the cooperation of the union representing company workers---International Woodworkers of America (I.W.A.), Western Region. The following subsections detail results of the program, based on trends in company logging accident statistics from 1982 through 1985, as well as organizational elements of the program identified as having contributed significantly to the results observed.

Program Results

As illustrated in Figure 1, for the years 1982-85, Whonnock Industries experienced a 75 percent decrease in accident frequency, and a 62 percent decrease in total W.C.B. claim costs, for the Forestry and Logging Group. Over sixteen different forestry and logging occupations are covered by these statistics, including faller, hooktender, choker, rigging slinger, landing, chaser, bucker, loader, grapple/spar, truck driver, shop, road, dry sort, boom, and labourer employees, as well as foremen.

The positive trends were particularly evident for the last two years (1984-85), which saw a 62 percent decrease in accident frequency, a 52 percent drop in claim costs, and a reduction in annual serious accidents (30 or more days lost-time) from 17 to 4. With these improvements in logging safety, company records also indicate increases in productivity by at least ten percent, and in total man hours of employment, from 1982 to 1985.

The direct cost saving realized by the company from safer logging has been calculated to be $150,000 per year, which reduces by more than ten percent the company's annual compensation assessment. Company and employee representatives estmate that there have been equally substantial indirect cost savings through such factors as fewer interruptions to production, less equipment damage, better morale, and improved labour-management relations.

Safety trends for the entire Whonnock Forestry and Logging Group (Figure 1) are reflected in individual trends for the six company operations in the Group, each of which employs 50-60 people (data available upon request). The individual operations displayed substantial differences in their rate of progress from 1982 to 1985, and in the severity of the initial conditions which required improvement. This is indicative of other differences among the operations, related to manpower characteristics and logging technologies employed (ranging from helicopter to more conventional high-lead logging). The distinctiveness and relative autonomy of each of the six operations created both a challenge and an opportunity for the development of a participative approach to safety.

Program Development

The process of developing a participative safety program has itself required a participative approach. From the the outset, the Whonnock program had to achieve a balance between the need for flexibility to respond to the individuality of the separate logging opeations, and the need for consistency within the company as a whole and within the terms of one collective agreement with the I.W.A.

Program development began slowly. The first year, 1981, was devoted to periodic discussions and seminars between two levels of company management, the head office managers and the superintendents of the various operations, who together examined the concept of expanded employee participation in company decision-making. These discussions represented a head management initiative, prompted be a general interest in the concept of employee involvement, coupled with specific concerns about the company's safety record. This also was a period of considerable resistance to this organizational cybernetic concept. Nevertheless, leadership in the company supported a second year of limited exploration in 1982, focusing on safety as one area where improvement was necessary and where introduction of employee participation might be particularly useful. A relatively cursory audit of the current safety programs and of the safety performance of the various operations was conducted, and the year culminated with an agreement between Whonnock management and the I.W.A. to initiate an effort to improve safety through an approach emphasizing two principles: (1) prevention of accidents and injuries; and (2) participation of hourly employees.

Within the framework of these two general principles, the third year of the project was devoted to an intensive training and communications effort, to involve elected hourly employee representatives and logging supervisors in determining what specifically would be required in logging operations to improve safety, and to promote the participation of employees in on-the-job safety and hazard management practices and procedures. Employee representatives and supervisors from each logging operation were delegated to establish revitalized joint camp safety committees, that would provide effective organizational structures through which participants could exercise shared responsibility for logging safety. Camp employees, union officers, and camp supervisors also participated with senior management of the company in developing a revised statement of the company's overall safety policy.

A vital part of the program in the next two years, 1984-85, was the continuation of the safety training and communication workshops, bringing together employee representatives and supervisors from the various logging operations. This process generated continuous learning about safety among peers, and supported the emergence of leaders in safety among employee and supervisory ranks. The meetings also reinforced communication up and down the organizational hierarchy, between senior management, union representatives, supervisors, and hourly employees.

Both the employee representatives and the supervisors from the separate logging operations therefore have participated in three levels of program development. They have regularly reviewed with senior management and union representatives the general policy framework. They have concentrated on developing detailed program structures for their own individual operations. Finally, they have supported individual employees and work groups in the development of crew-specific hazard management procedures.

Coordination of the learning process associated with this approach has been the task of an external facilitator, the senior author of this report. Notwithstanding the critical role played by the facilitator in such a program (as set forth by Smith, Lockhart, and Smith, 1983), the key decisions about the direction of the program have always been made by line management and employee representatives. Indeed, employees are now playing an increasing role in the evaluation and redesign of the overall company program.

Program Organization

The organization of the Whonnock safety and hazard management effort has been structured in accordance with three major management objectives, listed below along with the principle subobjectives.

1. Hazard Management by Individuals and Work Groups at the Worksite
   a. Immediate detection and communication of safety problems among crew and foremen.
   b. Constant inspection of worksite and work practices by crew and foremen.
   c. Reporting of, and learning from, near-accidents and close calls.
   d. Effective job safety analysis for purposes of training and education.
   e. Development of engineering guidelines for safer logging.
2. Camp/Operation Safety Program Management
   a. Shared decision-making responsibility by employee and company representatives.
   b. Camp program/performance monitoring and evaluation.
   c. Development of effective safety procedures.
   d. Safety coordination among crews.
   e. Resolution of more difficult safety problems and issues.
3. Company Safety Program Management
   a. Safety policy, commitment, and leadership by management, with union support, and guidance by the Workers' Compensation Board of B.C.
   b. Training, education, information, and other support for individuals and operations to improve safety.
   c. Monitoring of camp safety programs and performance.
   d. Development and evaluation of company safety effort by management and employee representatives.

Woven through this structure are the key elements of 'participation': employee involvement, management involvement, follow-up action, trust and communication, better planning, and leadership development. These factors, explained below, were identified by Whonnock employees, supervisors, managers, and I.W.A. officers, in a survey conducted in 1985, as having contributed most significantly to the improvement in safety in the company forestry and logging operations.

Employee Involvement. Employees on-the-job have been encouraged to identify potential safety problems, and wherever possible, to recommend solutions. Among Whonnock loggers, one of the most effective and concrete ways developed to promote identification of hazards has been a reporting and analysis system for near-accidents. Different operations and different crews have evolved a variety of reporting procedures, with a consistent emphasis that reporting be done anonymously. Over time, the frequency and effectiveness of near-accident reporting has improved significantly. Another useful strategy has been the regular review, and possible re-writing, by loggers themselves of job safety procedures, especially where employees take a major role in job training and inductions.

These employee initiatives depend upon, and are fostered by, good communication between crews and their respective foremen. Moreover, the credibility of the whole process rests upon a balanced and equal participatory role by employees and management in joint safety committees having substantial authority. Despite their lack of official, legal authority, these joint committees exercise de facto authority, because of membership on the committee of the senior management person on-site (usually the camp superintendent).

There have been difficulties in developing this participatory approach. Some management personnel have resisted sharing de facto authority. Some employees have wanted the authority, but not the responsibility that goes with it. However, the sense of direction given to this program by the company and the union has generally prevailed. Especially noteworthy has been the increasing involvement of hourly employees in a broad range of decision-making areas affecting their safety (Figure 2), some of which go beyond the conventional view of 'safety' matters.

Figure 2 lists 18 safety and hazard management decision-making areas, ranging from work planning, to clothing and equipment selection, to education and training, and to safety inspections and investigations, in which Whonnock employees now participate to varying degrees (from no input to shared responsibility). The figure also shows consensus results from a 1985 survey of both hourly employees and foreman, in which respondents were asked to rate both the actual existing, and desired, level of employee involvement in decision-making.

The information in Figure 2 is noteworthy from a number of standpoints. One key observation is that employee participation at Whonnock has been implemented to a degree where employees now directly influence or share responsibility in 16 of 18 major areas of safety program decision-making. Secondly, both employees and foremen appear to share the view that the level of employee participation should be further expanded, to a point that employees are at least consulted in all areas including long term work planning, and are given shared responsibility in 14 of the 18 specified areas. Finally, it appears that sole responsibility for safety decision-making does not represent a desired employee objective, an observation which underscores the point that the participatory approach typically leads to constructive and cooperative sharing, not preemption, of managerial power and authority.

Management Involvement. Safety and hazard management clearly is a two-way process. The effectiveness of employee involvement has depended very critically upon the active involvement in safety matters by supervisors and managers: to listen and respond to employee concerns, to set an example by their own safety practices on-the-job, and to exercise leadership by their attention to safety in the planning of work and the supervision of employees. Excellence in safety has become the standard for performance among Whonnock managers.

Follow-Up Action, Trust, and Communication. These factors represent outcomes of greater employee and management involvement in safety, and they also represent key factors in themselves. The main difference from the past is that more safety problems have been dealt with more quickly and effectively.

What is remarkable in many of the Whonnock logging operations is the openness of communication on safety matters, among employees, and between hourly employees and management. This willingness to communicate has been essential to the identification and solution of hazards on-the-job.

Better Planning. Work planning has been, and will continue to be, a key factor in improving safety. Employees identify it as one area in which they would like to become more informed and involved. An important aspect of safety in relation to planning of operations has been the involvement of company foresters and engineers. In one project they have consulted with employees and supervisors in drafting engineering guidelines for safer logging.

Leadership Development and Safety Education Workshops. Separate workshops have been held each years since 1983 for each of four groups within Whonnock: (1) Safety committee chairmen and senior employees; (2) Union representatives; (3) Foremen; and (4) Superintendents and managers. Employees regard the workshops as especially important in developing awareness of their own duties in safety, and in improving their ability to lead and train fellow employees on-the-job. The workshops have highlighted management's commitment to safety, and have provided a unique opportunity for employees and supervisors from the various Whonnock operations to learn from each other's experience. The workshops for employees have also featured the participation of I.W.A. officers.

Future Directions

At the end of 1985, Whonnock management, after consultation with employees, supervisors, and I.W.A. officers, established an agenda for sustaining the company safety effort. Success in sustaining the program was seen to be based upon: (1) program maintenance (i.e., updated first-aid training); (2) support systems (i.e., safety education and information, performance feedback); (3) program coordination and evaluation; (4) leadership (i.e., renewed objectives, attention to safety); and (5) expansion of employee involvement (into other aspects of the logging operations, such as production planning and quality control).

For a participative approach in safety to be sustained, it needs to comply and be substantially consistent with the larger, continuous context of day-to-day decision-making in company operations, and with the long-term planning process of the company. Otherwise, 'participation' in safety will be discredited, and ultimately rejected, by a more traditional authoritarian or bureaucratic organizational structure, much like a transplanted organ can become alien to its receiving body. Fortunately, the managers and employees within Whonnock, from their experience with the safety program described in this report, have learned a great deal about the nature of employee involvement. They now know more about its limits, its potential, and its requirements. They also have learned more about themselves, their values, their concerns, and their skills. Above all, they have discovered that they must continue to learn about safety and participation, and about how to bring about change in themselves and their work. In our judgement, their ability to sustain what they have achieved will likely depend most upon their ability to change and to deal with change.

CONCLUSIONS

From a behavioral cybernetic perspective, effective organizational safety and hazard management programming represents a behavioral and human factors, not a technical engineering, problem. Thus, Smith (1980) notes that no more than 20 percent of industrial accidents in the U.S. could have been avoided by rigorous industry compliance with standard governmental safety regulations, which are based almost exclusively on technical specification standards. Safety statistics cross North America clearly indicate that occupational accidents and injuries continue to represent a major economic, social, and health burden to society, and that the situation is not improving.

The participatory approach is based on the fundamental behavioral cybernetic principle that those most directly affected by workplace hazards, the employees themselves, should be given direct opportunity to participate in making decisions regarding how those hazards are managed in relation to work operations. We believe that this approach represents a highly promising strategy for bringing about substantive improvements in safety across all organizatrional and industrial sectors. The experience of Whonnock Industries described in this report represents one of the few examples in which beneficial and substantial improvements in industrial safety can be attributed directly to the introduction and implementation of a meaningful participatory safety and hazard management program based on employee involvement. It is noteworthy that the Whonnock program embodies a number of the theoretic systems principles of safety and hazard management outlined in the first part of this report, namely a behavioral cybernetic design, an emphasis on hazard control, attention to ergonomic/human factors aspects of safety (exemplified by a number of the areas specified in Figure 2), promotion of self-regulation and employee participation, and integration of safety with production.

Of course, we cannot prove that if the program had not been introduced, the improvements experienced by the Whonnock Forestry and Industry Group would not have occurred. However, there is no question in the minds of Whonnock employees and managers directly involved that the program itself brought about the beneficial results that were observed. This belief alone represents a critical feature of program acceptance and success. We suggest that the Whonnock record provides strong testimony for the view that the human factors approach to organizational design and management, centered on the behavioral cybernetic principle of employee participation, constitutes a compelling model for future development of safety and hazard management programs.

REFERENCES

Bryce, G.K. (1981). Joint Labour-Management Occupational Health Committees. An Example of Worker Participation in Work Site Health and Safety Programs. M.H.A. Thesis. Ottawa: University of Ottawa.

Cohen, A., Smith, M.J., and Anger, W.K. (1979). Self-protective measures against workplace hazards. J. Safety Research 11: 121-131.

Coleman, P.J., and Smith, K.U. (1976). Hazard Management. Preventive Approaches to Industrial Injuries and Illnesses. Madison, WI: Wisconsin Department of Industry, Labor & Human Relations.

Gottlieb, M.S. (1976). Workers' Awareness of Industrial Hazards: An Analysis of Hazard Survey Results from the Paper Mill Industry. Madison, WI: Wisconsin Department of Industry, Labor & Human Relations.

Hagglund, G. (1976). Causes of Injury in Industry---The "Unsafe Act" Theory. Madison, WI: University of Wisconsin School for Workers.

Hansen, E. (1982). Psychological aspects of accident prevention. Occupational Hazards 44: 115-118 (October).

Kaplan, M.C., and Colemen, P.J. (1976). County Highway Department Hazards: A Comparative Analysis of Inspection and Worker Detected Hazards. Madison, WI: Wisconsin Department of Industry, Labor & Human Relations.

Mason, K. (1973). A Correlation Between Types of Hazard Conditions and Accident Rates. Vancouver, B.C.: Workers' Compensation Board of British Columbia.

Pope, W.C. (1981). The strategy of change and safety management. Occupational Hazards 43: 61-64 (November).

Sass, R., and Butler, R. (1978). The Accident Proneness Theory: A Dead Horse that Won't Lie Down. Regina: Saskatchewan Department of Labour.

Smith, K.U. (1980). The Design and Implementation of a Model Hazard Management Program. Technical Proposal. San Diego, CA: Safety Sciences Inc.

Smith, K.U. (1979). Human-Factors and Systems Principles for Occupational Safety and Health. Cincinnati, OH: National Institute of Occupational Safety and Health.

Smith, K.U. (1975). Behavioral Practices in Risk Management of Industrial Safety and Workers' Compensation. D.I.L.H.R. Interservice Conferences on Systems Approaches to Risk Management of Workers' Compensation, Industrial Safety, and Health Systems. Madison, WI: University of Wisconsin Behavioral Cybernetics Laboratory.

Smith, K.U. (1974). Behavioral Cybernetic Systems Concepts and Practices in Safety, Health, and Risk Management Workers' Compensation. D.I.L.H.R. Conference on Defining and Developing a Model Workers' Compensation Statistics Program. Madison, WI: University of Wisconsin Behavioral Cybernetics Laboratory.

Smith, K.U. (1973). Performance safety codes and standards for industry: The cybernetic basis of the systems approach to accident prevention. In J.T. Widner (Ed.). Selected Readings in Safety. Macon, GA: Academy Press, pp. 356-370.

Smith, K.U., and Smith, T.J. (1981). Cybernetic foundations of preventive behavioral health science. In G.E. Lasker (Ed.). Applied Systems and Cybernetics. Volume IV. New York: Pergamon, pp. 1800-1804.

Smith, T.J., Ginnold, R., and Brandl, W. (1982). Ergonomic basis of disabling injuries in telephone workers. Proceedings of the Human Factors Society 26th Annual Meeting. Seattle, WA: Human Factors Society, pp. 498-502.

Smith, T.J., Lockhart, R.W., and Smith, K.U. (1983). Safety cybernetics: Theory and practice of involving workers in hazard management programs. In International Section of the International Social Security Association for Research and Prevention of Occupational Risks (Ed.). Specialist Day Papers: Analysis of the Risk of Accidents at Work, Methods and Applications. Ottawa-Hull: Xth World Congress on the Prevention of Occupational Accidents and Diseases, pp. 43-60.

Smith, T.J., and Sauter, S.L. (1981). Systems principles of occupational science. In G.E. Lasker (Ed.). Applied Systems and Cybernetics, Volume I. New York: Pergamon.

Smith, T.J., and Smith, K.U. (1983). Behavioral cybernetic systems principles of hazard management. Proceedings of the Xth World Congress on the Prevention of Occupational Accidents and Diseases. Ottawa-Hull: Canadian Centre for Occupational Health and Safety, pp. 218-221.

Swain, A.D. (1973a). A work situation approach to improving job safety. In J.T. Widner (Ed.). Selected Readings in Safety. Macon, GA: Academy Press, pp. 387-411.

Swain, A.D. (1973b). An error-cause removal program for industry. Human Factors 15: 207-221.

Tuttle, T.C., Grether, C.B., Liggett, W.T., Killian, N.K., Margolis, B.L., Kroes, W., and Cohen. A. (1973). Psychological Behavioral Strategy for Accident Control. I. Development of Behavioral Safety Guidelines. Cincinnati, OH: National Institute of Occupational Safety and Health.

Viaene, J. (1980). Organization of prevention within the enterprise. Proceedings of the Ninth World Congress on the Prevention of Occupational Accidents and Diseases. Amsterdam.

Figure 1

Forestry and Logging Safety Statistics, 1982* to 1985.
Whonnock versus Total B.C. Industry

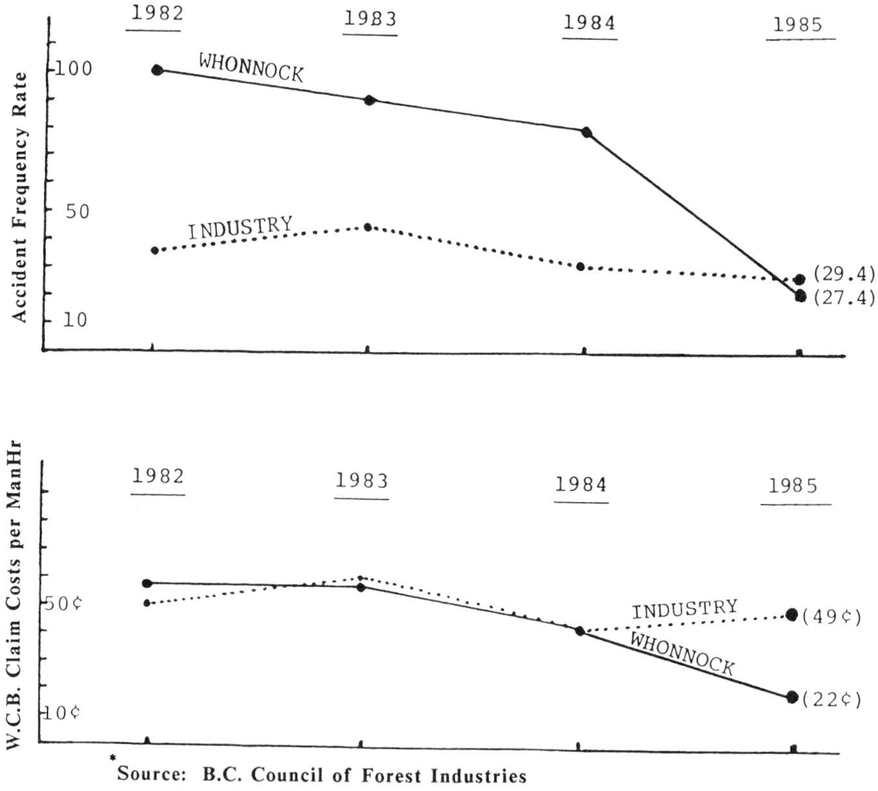

*Source: B.C. Council of Forest Industries

Figure 2

Consensus Opinions of Whonnock Employees and Supervisors About the Actual (X) and Desired (▷) Degree of Employee Involvement in Decision-Making*

	NO INPUT	INPUT	INFLUENCE	SHARED RESPONSI-BILITY	SOLE RESPONSI-BILITY
Short-Term Planning of Work			X - - → ▷		
Long-Term Planning of Work	X - -	- - → ▷			
Housekeeping of Worksite				X - - → ▷	
Protective Clothing - Selection			X - - - - -	→ ▷	
Use				X ▷	
Induction of New Employees			X - - → ▷		
Tools & Equipment - Selection		X -	- - - → ▷		
- Maintenance			X - - -	- - → ▷	
Training			X - - - - - -	→ ▷	
Correction of Unsafe Acts				X - - - → ▷	
Correction of Hazardous Conditions				X ▷	
Work Rules & Standards			X - - - - - - -	→ ▷	
Safety Promotion & Education			X - - -	- → ▷	
Safety Inspections			X - - - - -	- - → ▷	
Accident Investigations			X - - - - -	- - → ▷	
Labour-Management Safety Committee				X ▷	
On-the-job Employee Supervisor Communication			X - - - -	- - → ▷	
On-the-job Communication amongst Crew				X - → ▷	

*Based on responses completed anonymously by a sample of hourly employees and foremen from each operation.

Human Factors in Organizational Design and Management - II
O. Brown, Jr. and H.W. Hendrick (Editors)

THE INTERRELATIONSHIP OF SYSTEM SAFETY AND ORGANIZATIONAL DESIGN AND MANAGEMENT

James O. Pierce, II, Chair

Safety Science Department
Institute of Safety and Systems Management
University of Southern California
Los Angeles, California 90089-0021
U.S.A.

The rapid increase in research, design, development, and implementation of automated technology and robotics is a direct reflection of the worldwide need to increase productivity. It is estimated that 500,000 industrial robots will be operational in the United States by the end of the 1990s. Manufacturers and users of these systems are investing enormous capital in their production environments. For decades this attention has been focused on improving the technology's mechanical, electronic, and communication systems. This attention has left some important areas of research and design poorly attended.

With the increasing trend in application of automation and robotics in industrial and nonindustrial settings, the issue of human factors and safety will be the decisive factor in the further implementation stages of these systems. Contrary to the popular belief, the "total automated factory" cannot be implemented in the foreseeable future. Even the most automated work cells of Japan, Europe and the U.S., maintain a significant group of workers to perform tasks such as supervision, programming, maintenance, repair, trouble shooting, teaching, and back-up. The interaction and safety of these individuals working within their work organizations are essential in successful operation of automated systems. From an organizational point of view, inclusion of these workers within their predetermined roles has been modular with a "seat-of-the-pants" approach. Human Factors Engineering and System Safety are disciplines that provide a systematic and forward-looking approach to identify and evaluate problems at the design stages and processes. System Safety, in particular, ensures an early detection, identification, and prevention of hazards associated with design and interaction of these systems with other system components.

Human Factors in Organizational Design and Management - II
O. Brown, Jr. and H.W. Hendrick (Editors)

STRESS, INFORMATION-FLOW, AND ADAPTABILITY IN INDIVIDUALS AND COLLECTIVE ORGANIZATIONAL SYSTEMS

P.A. Hancock

Department of Safety Science and
Human Factors Department
Institute of Safety and Systems Management
University of Southern California
Los Angeles, CA 90089-0021
U.S.A

ABSTRACT

The central theme of this brief paper is the comparison of the commonalities between the characteristics of individuals and the collective organizational structures within which they operate. Each entity collects, filters, and sequentially transduces information in order to effect optimal adaptive action. Information in this context is distinguished along two axes which represent flow-rate and utility. Each entity seeks to locate itself within this two-dimensional information space at a point which maximizes task-related output at the least energetical cost consistent with successful performance. The transition between normal and failure modes of operation are compared across the human and the organization and can be represented as either gradual degradation or rapid dissolution of adaptability that can be described through the tenets of Catastrophe theory. A compromise between hierarchical, heterarchical, and holarchical structures is posed as one which optimizes response to stress intrinsic to environmental inputs. Manners in which the human and the organization utilize such structures are explored briefly.

I. INTRODUCTION

Successful, goal-directed behavior is perhaps less common than we generally appreciate. There are many occasions in which individual organisms and indeed whole species have failed under the constraints of a specific environment; the result is extinction. Therefore, having happened upon a strategy which results in successful adaptation, either by chance or directed selection, it is reasonable to assume that such strategies will be replicated wherever possible. It is these commonalities among organisms which prove more striking than their immediate morphological or surface differences.

How do these above, somewhat esoteric, observations help the manager and those concerned with organizational design. Simply, it is possible to take successful individual behaviors and to ask how such responses may be utilized in an organization of many individuals. This approach is not new. Among others, DeGreene (1981) has used analogical reasoning to explore the

---

This work was supported by a grant NCC 2-379, from NASA, Ames Research Center. Dr Michael Vidulich was the scientific monitor for the grant.

applicability of theories of physio-chemical activity to behavior in macro systems. The objective of the present paper is to examine three interrelated phenomena; stress, information-flow, and adaptability, and to indicate how reactions manifested by both individual and group behaviors indicate commonalities of strategic response.

II. STRESS, INFORMATION-FLOW, AND ADAPTABILITY

II.1 Stress

In work presented elsewhere, we have attempted to develop a theoretic and predictive view of stress based upon a new descriptive structure oriented toward individual action (Hancock & Chignell, 1985). Such principles that can be derived may be used to distinguish effects of stress under a particular work configuration (Hancock & Rosenberg, 1986). In this approach stress is defined descriptively by the loci of origin and action. Thus an input view of stress is described by a "signature" of the summed physical parameters which compose an environment. A second of the three identified loci features the adaptive response undertaken to compensate the perturbing action of the input experienced. The final locus is reflected as an interruption or cessation of some simultaneous goal-oriented behavior toward which the individual is directed or constrained. Although these loci were identified for an individual response they are paralleled in organizational action. External stress in the form of environmental constraints demand response of the organization. Some specific actions are required to cope with this input and these must be reconciled with the on-going behavior of the organization in order to achieve long-term goals (Hancock & Chignell, 1985).

II.2 Information-Flow

What are the input sources of stress which affect an organization collectively? Basically, these stresses are related to the economic, source-material, and human resources intrinsic to organizational activity. However, the fastest growing form of operational resource which impacts present day organizations is information. Information is the resource raw-material of many contemporary organizations and its prevalence is clearly growing. Unfortunately, the common parlance understanding of information does not adhere, or on many occasions even approximate information as originally envisaged in the landmark work of Shannon (see Shannon & Weaver, 1949).

In classic information theory, information is recognized as that which acts to reduce uncertainty. One characteristic that such strictly defined information metric does not posses is value, i.e., it does not indicate whether a statement is correct or incorrect. Value of information demands judgement on behalf of the receiver as to whether the content of a transmission is relevant to subsequent goal-directed action or not. Therefore, one dimension identified for stress expressed in terms of information is utility which represents a concatenation of uncertainty reduced and perceived value.

It is indispensible to emphasize the dynamic nature of both individual and group response to incoming information in that neither entity is in statis prior to, during, or after the impact of a discrete input. As a result a second and vital dimension of information expressed as a stress is rate of presentation. Both humans and organizations posses inherent limitations in the rate at which they may effectively transduce information. This hypo-hyper stress continuum is integral with the notion of time itself as a resource in organizational action (see Hancock, this volume).

II.3 Adaptability

If information rate and utility are proposed as major forms of both stress and resource or viewed generically as input, then adaptability or the ability to iterate toward optimally transduced this input to effective operational action is the response of interest. Under successful action this adaptability is retained within a range in which negative feedback predominates. Negative feedback is a common method through which to model both individual physiological response (Cannon, 1932), and successful human-machine interaction (Chapanis, 1965). The failure of negative feedback can result in positive feedback in which systems rapidly assymptote toward a non-operational condition. One method through which to model transition across successful and failure modes is through the description given in Catastrophe theory (see Zeeman, 1977).

III. COMPOSITE STRUCTURES MAXIMIZE ADAPTABILITY

There are a number of organizational structures that may be instantiated to cope with the impact of perturbing stress as outlined in the previous section. The hierarchy, with its principle control node represents one structure potentially vulnerable to such adverse effects because of the centralized locus of control. Alternate structures, founded upon an hierarchic template may avoid such weaknesses. Among these are heterarchies and holarchies where both the locus of control may be distributed and information flow may be multi-directional (see Koestler, 1978). It is interesting to note that the advocacy of changing forms of organizational structure almost directly parallels models of human neuropsychological functions directed toward the solution of "multiple degree of freedom" problems posed in complex control (Shaw & Bransford, 1977). It appears that structures with the capability to distribute levels of autonomy and control throughout the organization can ultimately prove most effective in optimal information transduction. Some form of structural multiplex offers the potential to maximize overall system adaptability.

IV. SUMMARY

There are some major observations that might be extracted even from this short synopsis.

1. Successful performance strategies being rare, are replicated in nature wherever possible.

2. Humans are sucessful systems whose strategic capabilities may be replicated at the organizational level.

3. There are a number of similarities between sucess and failure modes of both human and organization.

4. Information is critically important to the human and is

presently a major resource raw material of the organization which often composes an input to which response must be directed.

5. Humans respond to matrices of input information by distribution of autonomy and manipulation of locus of control as initiated by the input.

6. Use of diffuse control forms of management may alleviate an organizations' inability to adapt an thus retain an organizational balance within a region of stability that is marked by close to optimal information transduction.

It is toward the identification of these organizational behaviors through markers such as productivity, lost work time, injury rate, compensation expenditure, and job satisfaction, that the prudent manager is directed. As with many other theory based approaches, it is likely that the astute manager already recognizes such forces, and although not referring to them by specific name, channels such energies to achieve the overall goals of their own specific organization.

V. REFERENCES

Cannon, W.B. (1932) Wisdom of the body. W.W. Norton: New York.

Chapanis. A. (1965) Man-machine engineering, Brooks-Cole: Monterey.

DeGreene, K.B. (1981) Limits to societal systems adaptability. Behavioral Sciences, 26, 103-113.

Hancock, P.A. (1986) On the use of time: The irreplaceable resource. In: H. Hendrick and O. Brown (Eds.). Proceedings of the Second International Conference on Organizational Design and Management, North Holland: Amsterdam.

Hancock, P.A., & Chignell, M.H. (1985) On the principle of maximal adaptability in setting stress tolerance standards. In: R.Eberts and C.E. Eberts (Eds.). Trends in Ergonomics/Human Factors II, North-Holland: Amsterdam.

Hancock, P.A., & Rosenberg, S.M. (1986) The application of the concepts of comfort and adaptability in assessing stress effects of working with display units. In: International Scientific Conference: Work With Display Units, Stockholm, Sweden.

Koestler, A (1978) Janus: A summing up, Vintage: New York.

Shannon, C. & Weaver, W.A. (1949) The mathematical theory of communication. Urbana: University of Illinois Press.

Shaw, R., & Bransford, J. (1977) (Eds.). Perceiving, acting, and knowing. Erlbaum: Hillsdale, N.J.

Zeeman, E.C. (1977) Catastrophe theory: Selected papers 1972-1977. Addison-Wesley: Massachusetts.

Human Factors in Organizational Design and Management - II
O. Brown, Jr. and H.W. Hendrick (Editors)

INTRODUCTION OF ROBOTS INTO HYBRID WORK STATIONS: THE SYSTEM SAFETY CONCERNS

Mansour Rahimi

Safety Science Department
Institute of Safety and Systems Management
University of Southern California
Los Angeles, California 90089-0021
U.S.A.

Increasing number of robots and other automated manufacturing systems are being introduced into industrial work environments. Traditionally, robots have been caged, fenced, enclosed, or interlocked to prevent human injuries. These devices have been proven insufficient in preventing robot related accidents. The very nature of flexible robot task capabilities and motions appear to be contradictory to enclosure devices. In addition, human operators need to access robot danger envelopes for reasons of maintenance, repair, trouble shooting, teaching, and adjustments. Traditional safety measures have not been able to address these "nontraditional" human-machine tasks and work structures.

The need for appropriate safety methodologies has been highlighted by product liability lawsuits following serious robot accidents. Relatively limited amounts of robot accident data have been collected and analyzed to assess the contributing factors for robotic accidents. The discipline of "System Safety" is explained to be an appropriate approach to identifying and analyzing sources of robot accidents. Sources of robot accidents are categorized into Engineering related factors and Behavioral and Organizational related factors. System safety is described as the application of special methodologies for systematic, simultaneous, and forward-looking identification and control of hazards. The major objective is to integrate safety into the system and its components. Qualitative and quantitative analysis procedures are suggested to study safety of automated systems such as robots. A qualitative procedure called Energy Barrier Analysis (EBA) is for robotic application. A generic human-robot (hybrid) system is modeled and used to identify some fundamental safety problems with robotic systems. EBA is a system safety methodology which is predicated on the notion that energy flow is the fundamental cause of accidents. If an energy flow (or transfer) becomes uncontrollable (or unwanted), a potential for damage to other parts of the system exists. Then, safety of a robot system can be evaluated and improved by analyzing:

1. Sources of potential unwanted energy (PUE) in:
   a. robot subsystems and their interrelationships
   b. robot work environment

2. Means of reducing and controlling the levels of PUE from the source.

3. Means of blocking or controlling the transfer of PUE.

4. Means of properly warning operators of the existance of PUE flow.

After an initial analysis, it appears that the most practical barrier solution technique is to block the transfer of PUE to human operators. This may be accomplished by using devices which sense human proximity. The most capable and effective sensing device is a vision sensor. At this point, the costs associated with vision sensors for safety purposes alone do not justify their mass applications into robotic work cells. Future developments and applications will be suggested to improve the available devices.

For the managers responsible for organizing and implementing high technology work cells, the systems approach to safety has been somewhat illusive. To implement an automated work cell, the advantages of higher production flexibility, product quality, and less production life cycle are being weighed against increased capital requirements, increased fixed costs, competition and flexible marketing strategies. Issues of safety and human factors remain at the bottom of the list of important factors in this cost/benefit analysis. Unfortunately, the potential danger for causing injuries to human workers and damage to other expensive machines is not usually considered in this analysis. In particular, traditional retrofitting approaches for safety and human factors prove to be expensive and sometimes impractical. System safety methodologies have been designed to consider these issues at the early stages of design process.

Human Factors in Organizational Design and Management - II
O. Brown, Jr. and H.W. Hendrick (Editors)

SYSTEM SAFETY MANAGEMENT WITHIN THE ORGANIZATIONAL MANAGEMENT STRUCTURE

H.E. Roland

Safety Science Department
Institute of Safety and Systems Management
University of Southern California
Los Angeles, California 90089-0021
U.S.A.

The process of designing and producing high technology systems to ensure that they function with an acceptable state of safety, is neither obvious nor simple. The task is made ever more difficult by increasingly higher acceptable safety thresholds being set by operators and society. System safety is a set of unique management and regulatory structures coupled with specific analysis methods that impose a state of safety on a system during the design process considering the accident potentials of the system over its entire life cycle. The proper incorporation of these processes into the management structure charged with design and development of high technology systems, is critical to their success. The armed forces, NASA, and the Department of Energy, the principal agencies utilizing these methods, each employ a different approach to accomplishing the same safety objective, an acceptable state of safety over the operating life of the system. The separate armed services, operating under the Department of Defense directive, MIL-STD-882B, accomplish the job similarly, but with minor differences. NASA and the Department of Energy function under differing umbrella directives. This panel discussion will briefly outline the unique features of system safety, how management should function to accomplish its purposes, and the differing processes employed by these government agencies to accomplish those purposes.

References:

(1) Roland, H.E., and B. Moriarty. System Safety Engineering and Management (John Wiley & Sons, New York, 1983).

(2) Military Standard System Safety Program Requirements, MIL-STD-882B (Department of Defense, 1984).

PERSONAL AND ORGANIZATIONAL COMMUNICATION

Human Factors in Organizational Design and Management - II
O. Brown, Jr. and H.W. Hendrick (Editors)
© Elsevier Science Publishers B.V. (North-Holland), 1986

Workshop: PERSONAL AND ORGANIZATIONAL COMMUNICATIONS -- WORKING FOR BETTER UNDERSTANDING

Glen R. Gallaway, Battelle Columbus Division;
1637 Forestdale Avenue; Beavercreek, Ohio 45432

Robert F. Bachert
Armstrong Aerospace Medical Research Laboratory
Wright Patterson Air Force Base, Ohio 45433

Larry Ficks, Bell Communications Research Room 2B228
290 West Mount Pleasant Ave.; Livingston, NJ 07039

The transfer of data, information, and knowledge between people, between people and machines, and within organizations is the objective of the communications process. There has been a great deal of technical and conceptual study into the communications process, but little practical, usable, understanding has been developed. The workshop attendees will interact with the presenters to develop a practical conceptual model of the types of personal and organizational communications processes found in the business world.

INTRODUCTION

The processes of communicating between people and of communicating between people and machines often seem to be less than perfect. Sometimes message content is communicated properly; but often the message content is distorted, data is omitted, or extraneous data is added. This results in incorrect content communication, and thus contributes little or nothing to the understanding of the intended subject matter.

In interpersonal communications people are told that improvements are made by speaking more clearly, by listening better, by writing more exactingly, by providing more detail, by knowing the audience, etc. For communications with machines, such as a computer, people are often told to learn to speak the computer's language (instead of designing the computer to learn the person's language). Although, in a practical sense we need to develop these processes/skills, we often don't think about what it is that we are actually expected to accomplish. Each of these processes/ skills is aimed at increasing our knowledge about the receiver of our messages so that we can send messages with content that will be understood. In turn these processes- /skills also increase our ability to understand messages that come back to us. The point is that effective communication requires that both the sender and receiver of a message learn to understand such things as the other's knowledge base about the communications topic, the objective of the communication, and personal knowledge such as a person's process for conceptualization, social background, and speech and body language. In the initial phases of communications these factors are diverse in nature (from one person to another) and likely to cause trouble in communicating. As a dialog continues a common framework for understanding is developed and the process of communication will improve (i.e. people begin to `understand each other'). With a mutual conceptual understanding of the interpersonal communications relationship between people content can effectively flow between the sender and receiver.

BACKGROUND

Many business people have an aversion to scientifically analyzing and/or quantifying work processes such as planning, communicating, and other management functions. To a great extent these limitations are more a "perception" of an impossible task rather than fact. But the "perception" often leads us to unnecessarily and inappropriately govern our activities by emotion, best guess, and/or other subjective ineffective means rather than by the use of tools and techniques that can be applied.

We will begin developing our understanding of communication by expressing the process in everyday examples and by showing how they can be described in a more logical and systematic form. For example, consider a basic conceptual view of a manufacturing system (Figure 1):

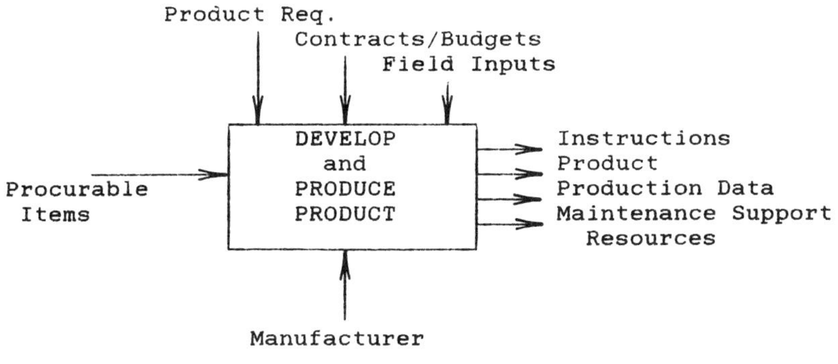

FIGURE 1. Basic Descriptive Model of a Manufacturing System

The arrows represent the interfaces with the business and the outside world (items that are brought to bear on the development process and the output of the process). It is obvious that communications are important to these interfaces. This model is the first step in the analysis of data and information flow between and within systems. Modes of implementation can be designed for the various types of information flow needed for any data arrow. Design considerations may include telecommunications, MIS, public relations, and advertising. Internal communications can be analyzed and designed by decomposing them into more detailed descriptions. Figure 2 illustrates the simple breakdown of the "DEVELOP and PRODUCE PRODUCT" into functional parts.

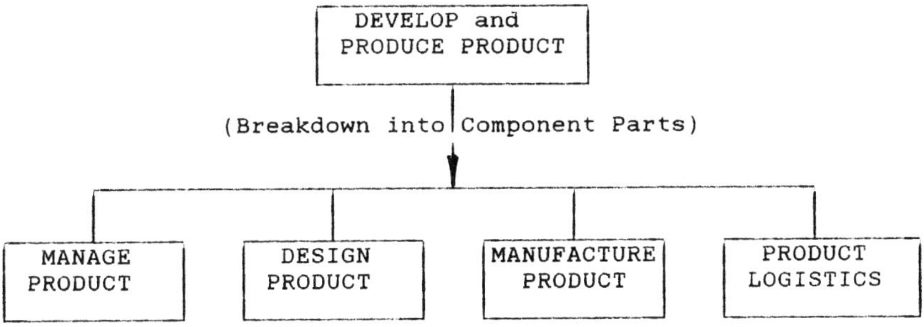

FIGURE 2. Breakdown of Basic Descriptive Model Into Functional Parts

Information flow between the functions shown include such things as directives and reports, requirements, design specifications, scheduling, and maintenance needs.

Organizational and management structure can usually be superimposed on the functional architecture. Organizational communications are usually considered to be vertical, although from a functional viewpoint this is not true. All communications are horizontal or to and from data bases. Management functions generate control arrows (top of boxes) which determine why the activity is accomplished the way it is. Bottom-up communications (suggestions, complaints, etc.) are shown by feed back arrows that usually go to management functions.

There are few, if any, models existing that fully explain and describe the person-to-person or person-to-machine communications processes. These are best approached by examining the what, how, when, and why aspects necessary to construct models.

Data, information, and knowledge comprise the `what' of communication. Data will be defined here to mean a set or collection of stored values. Information will be defined as the term used to describe associations and relationships among data. Knowledge will be defined as the concepts (conceptual models) abstracted from data and information over an indeterminate period of time.

The `why' of communication consists of trying to achieve the goal of transferring either data, information, or knowledge from one person to another, from a person to a machine, or within a group of people.

One way to study communication for a better understanding of the process is to examine the mapping of data, information, and knowledge on to **objects**, **activities**, **problems**, and **systems**. Add to this the `how' (or implementation) and the `when' aspects and further insight will be achieved. The bottom line is improved communications resulting in reduced error, reduced cost, improved effectiveness, and improved relationships. This transfer is related to real world applications depicted sets as shown by the venn diagram in Figure 3.

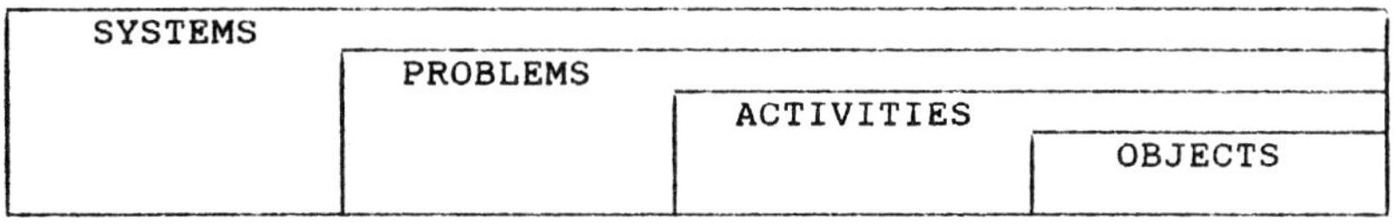

FIGURE 3. Communications Categories

The mapping concepts will be developed by workshop participants. These concepts are the keys (or links) for mapping from one person's brain to another person's brain for transfer of data, information, or knowledge (i.e. communication). For example, consider an object. We can communicate data about the object (attributes, size, shape, etc.). If the object is a chair we can talk of its style, size, and color. Communicating information about the chair may mean where it is used and how the style relates to the size and color. The history of the style, manufacturing process or why that chair is in its present location constitutes knowledge. These mapping concepts are extended to activities, problems, and systems in the workshop and related to participants application areas.

The first point to be discussed is that the mechanics of the communications process is based on data that flows in a network structure. Anything that distorts or impedes the transfer of data between activities or functions often causes bottlenecks and/or breakdowns which impact communication performance and system effectiveness. The second point to be discussed is that problems in the mechanics of communication usually distorts or voids the meaning of the message that is being communicated. Learning to conceptualize the mechanics and what constitutes meaning in a communication will better help us improve our interaction with others.

The discussion so far has focused on the individual to individual or individual to machine communications process. Another concern is communication with and within an organization (group communications). There are many different types and levels of communication existing in the organizational environment which adds to complexity. Discussions of organizational communications by workshop individuals from various backgrounds is expected to help evolve an understanding of the complexities of the overall communications process.

A top down structured approach for describing systems architecture facilitates the examination and analysis of the information, functions and relationships involved in communications. This workshop will develop such an approach and through examples of various systems show how the analysis of communications can be performed. Discussions will include systems analysis and development methodology, systems definitions and structure-/architecture, data and information, functions and goals and data bases which store the systems knowledge resources. Modes and levels of information flow will be discussed. This will include communications between organizations as well as person to person interactions. Participants will examine their own methods of communicating information pertaining to objects, activities, problems, and systems.

WORKSHOP OBJECTIVES AND STRUCTURE

Through the workshop format it is expected that some basic guidelines and premises for communication between people, people and machines, and in the organizational context will be developed. Each attendee will explore ways of defining, analyzing, and working toward communications processes that develop effective solutions. How the communications process effects the structure and performance of management and the organization in the business environment will also be explored. Through the use of simple role playing, lectures, demonstrations of concepts, and discussions, the attendees will gain insight and develop some experience in evaluating their own communication methods and how to improve data, information, and knowledge flow between communication partners.

GENERAL NOTE

To gather information about previous work in the construction of communication models, 5 on-line databases were searched. Based on this search it appears that work in the area of communication models was concentrated in the 1959 to 1970 time frame. Further, this search resulted in only 20 usable references, however these references were rich in descriptions of models. (NOTE: This does not include the wealth of material associated with information theory, which in general treats the communication channel). None of the 26 models reviewed specifically address problems related to organization design and change or to the communication problems surrounding the human interface to office automation.

USER SYSTEMS ANALYSIS

Human Factors in Organizational Design and Management - II
O. Brown, Jr. and H.W. Hendrick (Editors)

USER SYSTEMS ANALYSIS: ORGANIZATIONAL DESIGN FOUNDATIONS OF A MANAGEMENT SUPPORT SYSTEM

Rafik I. Beekun, William H. Glick, and Alan Carsrud
Department of Management, University of Texas
Austin, Texas, U.S.A.

User Systems Analysis (USA) is a comprehensive set of techniques and computer programs for assessing, simulating and graphically representing work and information flows in an organization. User Systems Analysis is guided by previous research in the areas of organizational design and operations management. This paper presents an overview of USA and its origins and describes the contributions of USA to Sociotechnical Systems research.

To produce goods or services, individuals in organizations interact on a recurrent basis, exchanging work-in-process and information. New ways of directing and controlling these complex intra-organizational patterns of interaction are constantly being sought for and developed by researchers in order to improve organizational effectiveness. In modern, complex organizations, however, these efforts frequently encounter problems both in the assessment and graphic representation of these information and work flows.

User Systems Analysis (USA) was designed to address these problems (Brady, 1984; Glick & Beekun, 1984). USA starts with a series of computerized interviews of the user's system leading to the automatic production of work flow diagrams, a summary of organizational design characteristics and simulation statistics about the system performance and resource utilization. These diagrams and organizational characteristics are interpreted using several organizational design theories that form USA's conceptual foundation, and that will be presented in this paper. Subsequent papers (Mukherjee, Saha & Ahmadi, 1986; Saha, Theokisto and Mukherjee, 1986; Ahmadi, Mukherjee & Glick, 1986; Carsrud, Glick, Beekun, Ahmadi, Mukherjee & Saha, 1986) respectively describe the automated questionnaire and data base design, the computerized work flow diagrams, the operations management simulation models, and a case study utilizing USA.

USA: A Relational Network Approach

USA focuses on the systematic patterns of interactions among individuals in organizations. To quantify and measure the exchanges of work and information among organizational members, USA relies heavily on network analysis, a methodology from mathematical sociology (Knoke & Kuklinski, 1982). The primary unit in network analysis is a relation between two people or "actors". The content of the relation can be anything that is exchanged between two people, including, work-in-progress, work-related communication, and goods. A network is a set of relations among a defined set of persons for a defined set of contents (Knoke & Kuklinski, 1982), and is typically represented as a series of square matrices, one matrix for each relational content. The rows and columns in each matrix corresponds respectively to the senders and receivers of the relational exchange. Both organizational participants and environmental actors are included in the networks to represent the organization as an open system.

Theoretical Origins of USA

USA researchers have been applying network analysis techniques to four organizational design theories, task interdependence (Thompson, 1967), differentiation and integration (Lawrence & Lorsch, 1967), loose-tight coupling (Weick, 1979), and Sociotechnical Systems theory (Herbst, 1974). Beekun and Glick (1986) present a complete theoretical development and quantitative operational definitions using network analysis for the key structural, environmental, and technological constructs in the theories of Thompson (1967) and Lawrence and Lorsch (1967). Beekun (1985b) develops a similar set of operational definitions and network analysis equations for the more complex constructs of loose-tight coupling. In a qualitative and quantitative review of empirical Sociotechnical Systems research, Beekun (1985a) discusses many of the strengths and weaknesses of Sociotechnical Systems research, and assesses the generally positive effect of sociotechnical interventions on organizational performance. In the interest of space, readers are referred to these papers. The remainder of this paper is devoted to further contributions of USA to Sociotechnical Systems research.

The power of the relational network approach to organizational design in general and of USA in particular is easily seen by its potential contribution to the Sociotechnical Systems approach. Sociotechnical Systems theory (Rice, 1958; Herbst, 1974) is an open systems approach that focuses upon the relationship between the social and technical systems of an organization. Two key principles are usually emphasized by sociotechnical researchers: joint optimization and organizational choice. Joint optimization implies that neither the technology nor the social arrangements already in place in an organization are to be considered as givens, but may be re-arranged to fit the demands of each other (Herbst, 1974). Organizational choice, the second sociotechnical principle, suggests that for any production process, there is a whole array of suitable technologies available (Pasmore Francis, Haldeman, & Shani, 1982). Similarly, for a given technology, many different social arrangements are possible.

Though these principles are in themselves theoretically sound, the effectiveness of sociotechnical interventions varies considerably because of the gap between sociotechnical theory and implementation. A first issue that has troubled previous reviewers (Pasmore et al, 1982) concerns the nature of both social and technical changes. Indeed, sociotechnical researchers have yet to specify what variables are being manipulated in their change strategies. Second, how is joint optimization measured? No precise and reliable methodology has been suggested. Third, when changes were made in the technical system, sociotechnical researchers relied upon the concept of "variances" (Cummings & Srivastva, 1977) to identify major problems that arose from the nature of the production process. These variances were defined as deviations from relatively arbitrary standards set by managers.

User Systems Analysis offers a workable solution to these three problems in Sociotechnical Systems research. First, when conceptualizing the technical system, Sociotechnical Systems theorists (Rice, 1958; Herbst, 1974) have almost always focused upon one key underlying variable: task interdependence. Except for one forgotten taxonomy of task interdependence patterns (Herbst, 1974), the sociotechnical school has not conceptualized this variable any further. Thompson's (1967) typology of task interdependence which underlies the USA's approach to the technical system of the organization provides a more accurate and reliable operationalization of this variable when translated into relational terms. In contrast to Herbst (1974) and to Thompson (1967), USA does not assume that task interdependence is a categorical variable or that different types of task interdependence cannot combine.

Similarly, the USA's approach to the social system of organizations is more accurate and flexible than the sociotechnical approach. Unlike Sociotechnical Systems, USA does not limit itself to one form of social arrangement, i.e. autonomous work groups. USA uses a variety of dimensions to characterize the patterns for three types of social relations, authority, work-related communication, and friendship.

The second major problem faced by sociotechnical researchers, namely, the assessment of joint optimization, is easily solved by assessing the pattern of similarity between the social (friendship or communication networks) and technical (workflow network) relations. This assessment rests on the measure of organization multiplexity from network analysis (Knoke & Kuklinsky, 1982).

Third, USA improves upon the calculation of variances in 2 ways. The quantitative assessment of joint optimization is a theoretically more appealing approach to sociotechnical variances since multiplexity captures both social and technical variance components. Further, USA includes performance projections and confidence intervals from the simulation subroutines (Ahmadi et al, 1986). These performance projections can be used to provide more accurate estimates of technical variances.

In sum, USA departs from previous approaches to organizational design in that it (1) treats organizations as systematic patterns of exchange among individuals, (2) distinguishes and explicitly measures interactions involving several network contents (i.e., workflows, communication, friendship, and authority), (3) provides an accurate and reliable methodology for tracking and measuring these interactions, and (4) as we shall see in the upcoming presentations, it allows the user to simulate any planned changes ex ante, and to view the impact of these changes both graphically and statistically.

REFERENCES

[1] Ahmadi, R., Mukherjee, S. and Glick, W., Automated simulation and organizational analysis: The transfer of engineering technology, in Hendrick, H. and Brown, O. (eds.) Second International Symposium on Organizational Design and Management (North Holland, Amsterdam, 1986).

[2] Beekun, R., A meta-analysis of Sociotechnical Studies in the USA, paper presented at the Annual Meeting of the Academy of Management (August 1985).

[3] Beekun, R., Towards a contingency framework of loose coupling: A conceptual and quantitative approach, Ph. D proposal, Dept of Management, University of Texas (November 1985).

[4] Beekun, R. and Glick, W., Reinterpreting organizational theories from a network analysis perspective, Paper to be presented at the Annual Meeting of the Academy of Management (August 1986).

[5] Brady, L., Review of the systems methodology of User Systems Analysis, in: Hendrick, H. and Brown, O. (eds.) First international symposium on organizational design and management (North Holland, Amsterdam, 1984).

[6] Carsrud, A., Glick, W., Beekun, R., Ahmadi, R., Mukherjee, S. and Saha, A., Applications of User Systems Analysis technology in the workplace: Case study in work and information flows, in: Hendrick, H. and Brown, O. (eds.) Second International Symposium on Human Factors in Organizational Design and Management (Amsterdam, North Holland, 1986).

[7] Cummings, T. G. and Srivastva, S., Management of Work (San Diego, CA, University Associates, 1977).

[8] Glick, W. and Beekun, R., A theoretical overview of Users' Systems Analysis, in: Hendrick, H. and Brown, O. (eds.), First International Symposium on Human Factors in Organizational Design and Management (Amsterdam, North Holland, 1984).

[9] Herbst, P., Sociotechnical Design (London, Tavistock Publications, 1974).

[10] Knoke, D. and Kuklinski, J., Network Analysis (Beverly Hills, Sage, 1982).

[11] Lawrence, P. and Lorsch, J., Differentiation and integration in complex organizations, Administrative Science Quarterly 12 (1967) 1-47.

[12] Mukherjee, S., Saha, A. and Ahmadi, R., The dialog and database components of User Systems Analysis, in: Hendrick, H. and Brown, O. (eds.) Second International Symposium on Organizational Design and Management (North Holland, Amsterdam, 1986).

[13] Pasmore, W., Francis, C., Haldeman, J. and Shani, A., Sociotechnical Systems: A north american reflection on empirical studies of the Seventies, Human Relations 35 (1982) 1179-1204.

[14] Rice, A., Productivity and Social Organization: The Ahmedabad experiment (London, Tavistock Publications, 1958).

[15] Saha, A., Theokisto, V. and Mukherjee, S., Data based graphics in organizational analysis: An outline, in: Hendrick, H. and Brown, O. (eds.) Second International Symposium on Organizational design and Management (North Holland, Amsterdam, 1986).

[16] Thompson, J., Organizations in Action (New York, McGraw-Hill, 1967).

[17] Weick, K., The Social Psychology of Organizing (Reading, Mass: Addison-Wesley, 1979)

Human Factors in Organizational Design and Management - II
O. Brown, Jr. and H.W. Hendrick (Editors)

THE DIALOG AND DATABASE COMPONENTS OF USER SYSTEMS ANALYSIS

Sambit Mukherjee, Avijit Saha, and Reza Ahmadi
Departments of General Business, Electrical Engineering, and Management
The University of Texas at Austin
Austin, Texas
U.S.A.

The User Systems Analysis (USA) method of organizational assessment requires substantial amounts of data about the work and information flows in organizations. This paper describes the core of the new version of User Systems Analysis, a computerized interview procedure and database component. We describe the system's capabilities, major options, user-friendly interface, general design of the database, and interfacing procedures to other components of User Systems Analysis. We conclude with a discussion of the utility of each component of this system.

The primary purposes of User Systems Analysis are to assess accurately the work and information flows in an organization and to summarize this assessment for managers. This assessment can be used for managing the organization's staffing, coordination, equipment purchases, scheduling, job and organizational design.

The User Systems Analysis organizational assessment begins with a combined interview and questionnaire procedure. Organizational members are asked a series of questions about their involvement in the work and information flows among individuals and across organizational boundaries to suppliers and customers [1, 3]. The questions directed to individuals are fairly simple and straightforward. The final sequencing of questions, however, became very complex as we designed the procedure to ask a minimum number of questions. Given the internal complexity of the interview procedure, a major contribution of the computerized User System Analysis is the user-friendly interface that generates the questions and stores responses in a database.

Figure 1 shows the basic components of this system. First, the Dialog Component provides a user-friendly interface and controls system use. Second, the Database Component stores and organizes the responses in an accessible form. Data extraction routines then process and pass the data to other User Systems Analysis software for generating graphics and simulation statistics. This paper describes each of the components, and the general capabilities of the system.

Dialog Component

The Dialog Component provides a user-friendly interface that presents a complex sequence of questions to interviewers and organizational members in an easy to understand manner. Computerizing this portion of the User Systems Analysis was essential to increasing the reliability and efficiency of

Figure 1

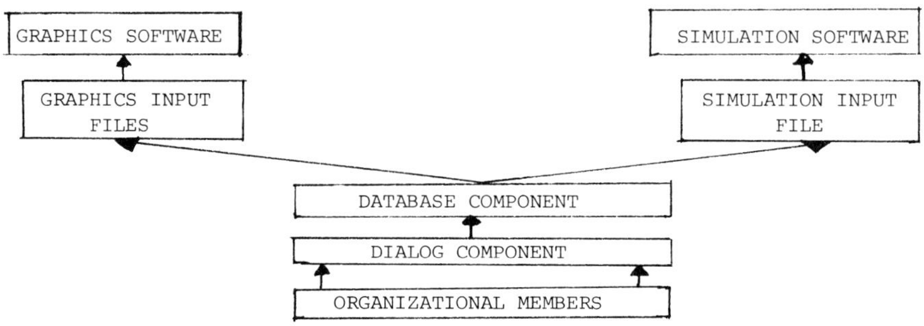

work and information flow analyses. The Dialog Component has two parts, the menu sub-component and the questionnaire sub-component.

Menu Sub-Component

The Menu Sub-Component is implemented as a hierarchy with top-level menu options providing access to submenus. The key idea is to allow flexibility in the use of the system. The major capabilities provided by the menu system include:

1. Running either segments or the complete version of the questionnaire depending on the manager's current goals for the analysis and based on the types of data already entered in the system.
2. Generating selective or comprehensive reports of user responses.
3. Driving data extraction routines that create data files for the Graphics software.
4. Driving data extraction routines that create data files for the Simulation software.
5. Accessing System Documentation and Help files.
6. Updating the database by allowing organizational members to modify or correct previous responses after reviewing preliminary results.

The Menu Sub-Component also provides a protection mechanism in the form of passwords and implicit checks at the submenu level. This ensures respondents' confidentiality on their individual responses and prevents respondents from being biased by prior responses.

Questionnaire Sub-Component

The computerized version of the questionnaire implements the complex hierarchical sequencing of questions required for the interview in a straightforward manner. The underlying software generates a series of panels on the computer terminal based on the user's responses to prior panels. Introductory panels familiarize respondents with the interview protocols. Subsequent panels present the actual questions and allow data entry.

Multiple versions of the questionnaire have been implemented. These allow users the flexibility to utilize a subset of the system's capabilities as dictated by requirements and time constraints. The shortest version takes approximately fifteen minutes per employee and provides a simple diagram of the workflow. The longest version takes approximately an hour and provides a detailed work and information flow diagram, organizational diagnostics, and simulation statistics [1]. All versions provide checks to prevent invalid data entry and user error. Corrective messages are displayed on the screen in case of such occurences and the user is requested to rectify errors.

Database Component

The Database component stores the responses of organizational members for subsequent retrieval and processing. The model used for designing the database was the popular relational model. Relational databases are easy to use and can be supported by commercially available, sophisticated Database Management Systems (DBMS). The current database schema was designed and implemented in two distinct stages, the logical design and the physical implementation.

Logical Design

The modelling method used to design the database is the entity-relationship approach [2,4,6]. There are two stages involved this method [4]. First, the selection of the entities and the relationships between them. Second, the assignment of attributes to the entities and relationships such that a set of tables are obtained that capture all of the required information.
Figure II presents a graphical representation of a subset of the Entity - Relationship scheme for User Systems Analysis. It shows the Product and Activity entities and the relationship between them. The actual relations or tables are shown in Figure III.

Figure II

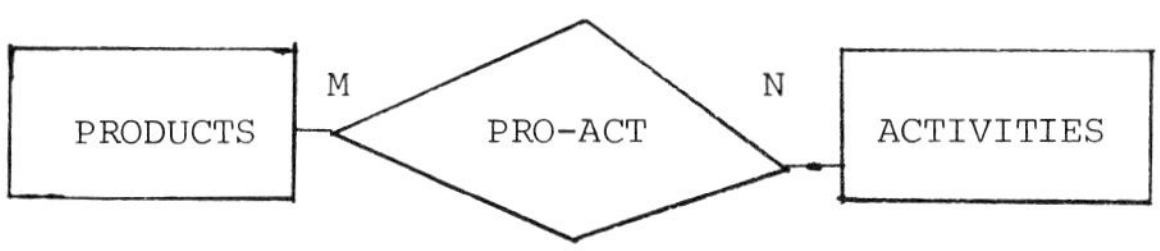

Figure III

PRODUCT

WORKUNIT	PRODUCT ID	PRODUCT NAME	PRODUCT DESCRIPTION

ACTIVITIES

ACTIVITY ID	ACTIVITY NAME	ACTIVITY DESCRIPTION

PRO-ACT

PRODUCT ID	ACTIVITY ID

Physical Implementation

Once the entire logical design of the USA database was completed we mapped the set of logical relations into a physical database using the Data Definition Language of a commercially available Database Management System [5].

We designed the software such that the actual database is completely transparent to the user. Menu driven routines allow users who are unfamiliar with database technology to interact with the system easily.

Interface With Graphics And Simulation Software

One of the major capabilities of USA is its ability to provide a graphical representation of work and information flows in an organization. This is done by means of a computer-aided diagramming technique. Required data for driving this program is extracted from the database by a menu driven routine which creates four formatted ascii files. These files are then used as input by the graphics software.

The simulation component of USA calculates average resource utilization, average idle time and other statistics. A separate data extraction routine is used to create the input file for this program. The file gives the resources utilized and the time required to complete activities for each person performing the activity.

Conclusion

We have presented the conceptual outline and implementation issues involved in the design of a management support system. The Dialog Component is the software that provides the user interface of USA. It essentially supports the use of of the system by organizational members. The Database component provides data storage and data access capabilities. Data extraction routines allow interface to other system software.

We have adopted an iterative design approach allowing us to refine the system based on user comments. There is definitely scope for further improvements, both in the Dialog Component and the Database Component. We feel that user criticisms and reactions will provide invaluable guidelines for improving the system.

References

[1] Beekun, R. I., Glick, W.H., Carsrud, A. L., User Systems Analysis : Organizational Design Foundations of a Management Support System. Brown, O. and Hendrick, H. W. [Eds]. Second International Symposium on Human Factors in Organizational Design and Management. (North - Holland, Amsterdam, 1986).

[2] Chen, P.P., The entity-relationship model : Toward a unified view of data. ACM Transactions on Database Systems , 1(1) : 9 - 36, 1976.

[3] Glick, W. H., Beekun, R. I., A theoretical overview of User Systems Analysis. Brown, O. and Hendrick, H. W. [Eds], First International Symposium on Human factors in Organizational design and Management. [North - Holland, Amsterdam, 1984].

[4] Howe, D.R., Data Analysis for Database design. [Arnold, Baltimore, MD, 1983].

[5] Microrim, Rbase 5000 User's manual. [Microrim, Bellevue, WA, 1985].

[6] Ullman, J.D., Principles of database systems. [Computer Science Press, Rockville, MD, 1982].

Human Factors in Organizational Design and Management - II
O. Brown, Jr. and H.W. Hendrick (Editors)

DATABASED GRAPHICS IN ORGANIZATIONAL ANALYSIS: AN OUTLINE

Avijit Saha, Victor Theokisto, and Sambit Mukherjee

Departments of Electrical Engineering
Mechanical Engineering and General Business
The University of Texas at Austin
Austin, Texas
U.S.A.

One of the major strengths of User Systems Analysis is the methodology for consistently and efficiently diagramming patterns of work and information flows in organizations. The graphics component provides a flexible routine that automatically draws diagrams of workflows from data collected with the USA automated interview program. The graphics component also is capable of diagramming any arbitrary relations given a properly formatted input file describing the relations among the nodes. This package is more flexible than the commercially available tools.

The graphical component of User Systems Analysis reliably and efficiently diagrams patterns of work flows in organizations. This capability is a major selling point of USA for managers and organizational members. The work flow diagrams provide an intuitively appealing picture of potentially complex patterns of workflows that can be interpreted by any manager or organizational member.

In the design of the graphics portion of USA , numerous decisions were made. This paper describes several different forms of representing work and information flows and our rationale for using directed graphs. In the latter sections of the paper some basic design issues and the general methodology for obtaining the relationship graph are discussed. Issues regarding user friendliness and user transparency are also addressed.

Different Forms of Representing Work Flow Information

There are three major forms of graphically representing work flow information, petri nets, information control nets and directed graphs. A **petri net** description of any system consists of four elements viz., places, tokens, transitions, and directed arcs. As illustrated in Figure 1, a place, $\mathbf{P_i}$, corresponds to the conditions that may hold in the system. When the condition is true, the place holds a token as indicated by a solid circle in the place. Transitions, $\mathbf{T_i}$, correspond to events that may occur. Directed arcs connect places to transitions and transitions to places. A transition may fire when all the input places linked to the transition by a directed arc are full, i.e. they all have a token. After firing, a token is removed from every input place and added to every output place. In Figure 1, a token is located in $\mathbf{P_3}$, the only input place linked to transition $\mathbf{T_2}$. The other transition, $\mathbf{T_1}$, cannot fire since all the input places connected to this transition does not have a token. After $\mathbf{T_2}$ has fired, places $\mathbf{P_1}$ and $\mathbf{P_2}$ will now have tokens and therefore $\mathbf{T_1}$ is ready to fire.

Figure 1: A Simple Petri-Net

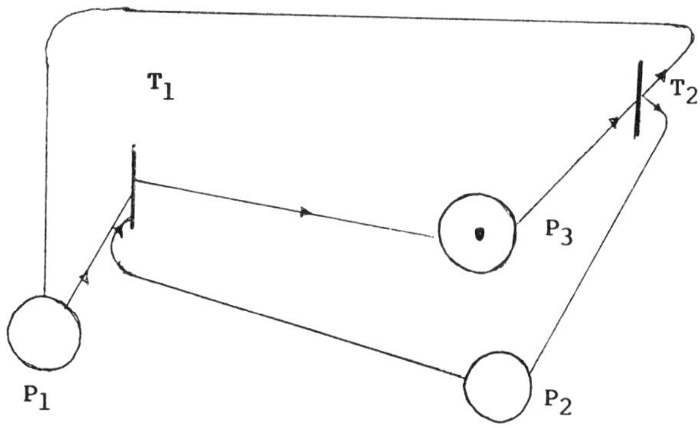

The second form for representing work flow information is the **Information Control Nets [ICN].** As shown in Figure 2, this form of representation essentially uses the directed graph approach . However, there are two types of nodes- one representing activities and the other information repositories. Also, there are two types of arcs for representing information as well as control flow. Tokens are placed on arcs or nodes of the graph to represent the current location of information or work in the system.

In the generalized ICN model proposed by Ellis [3], the basic elements for the representation are OBJECTS depicting goals, activities, or resources; TOKENS to represent flow of data or flow of control; and MAPPINGS between objects to relate activities, resources and goals. Figure 2 shows a simple ICN.

Figure 2: Information Control Net

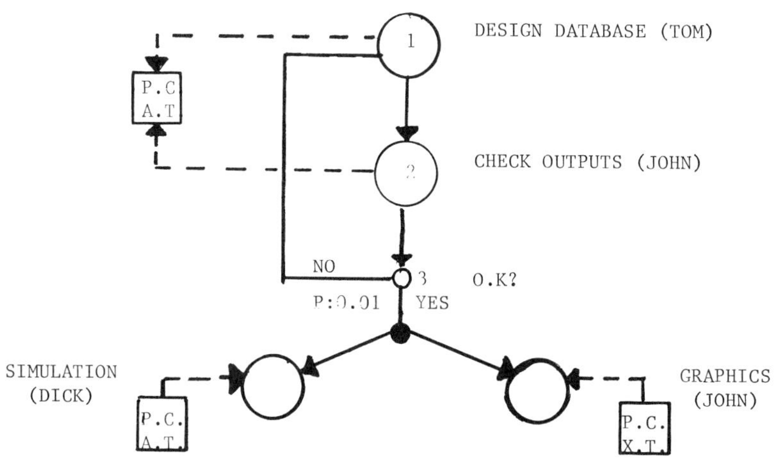

The final form for representing work flow information is the generic Directed Graph. In our representation of work flow we use a simple directed graph (see Figure 3). A job is viewed as a result of atomic elements which are the activities. An activity is associated with one or more persons to constitute a distinct person-activity entity, which are the nodes in our model. The nodes represent the execution of the activity. The arcs indicate the transfer of control over the work and the binding relationship that links these person-activity entities together. This model can then be looked upon as a petri net model with tokens available at all the initial places. Also it is assumed that firing from one node (traversal of an arc) produces the appropriate intermediate results and no intermediate checks are necessary. Petri nets are more useful for dynamic modelling of an information flow system where it is nescessary to observe the actual change of control from one activity node to another. Since our model is a static one, the directed graph model as described above was found to be appropriate. Moreover, this choice simplified the implementation details in the graphic representation of our system.

Figure 3: Generic Directed Graph

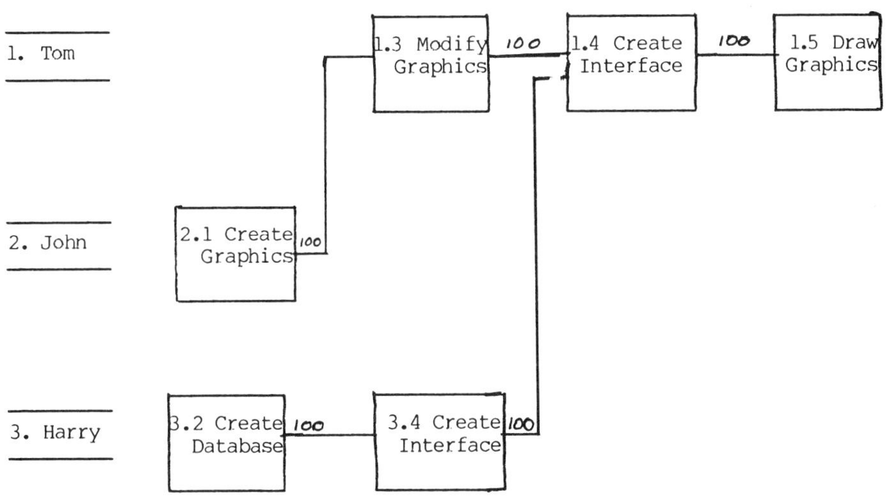

DESIGN ISSUES

The graphics package has been developed as an independent program with minimum interface with the database. The database provides a list of the number of activities, the names of people, and a list of relations among all person-activity entities. All information for the graphics programs are retrieved from the database under program control with minimum user intervention [2]. This information is received by the graphics program in a formatted file. A parser program then computes the nodes (person-activity boxes) that are necessary and the arcs that link these boxes .

A cluster analysis approach is used to create a directed graph with a minimum number of lines crossing and overlapping each other. This results in a much cleaner and more interpretable diagram than is possible in manual methods of drawing work and information flow diagrams.

For easier analysis, a set of rows are allocated to a single person only and a set of columns to a single activity. Function primitives are called upon to calculate the co-ordinates for the appropriate location of the person-activity boxes and to draw the links between the appropriate boxes. An adjacency matrix is used for this purpose. All the person activity boxes are labeled with the activity name and the person activity number. The person activity number appearing as a label in each person activity box is the concatenation of the person number and the activity number. The database program calculates the probability of execution for each arc connecting to a next person-activity box. This probability appears next to the arc in the person-activity box on the final diagram. Provisions have been made to provide additional information in the form of labels.

CONCLUSIONS

Given a properly formatted data input file from the USA database, and one command from the user, THe graphics program automatically generates a complete workflow diagram for an organization. The minimal data requirements are the person and activity lists and the set of relations among the person-activity entities. Because the database is completely screened form the praphics program, it is possible to use the graphics component of USA independent of the database. Thus , it is possible to draw a flow diagram of any arbitrary set of relations among two dimensional entities(eg., person-activities or department-product entities). To use the graphics component seperately, users must specify the set of relations, "persons", and "activities" in the proper format and execute the graphics command. This will produce a flow diagram without the actual existence of any database.

Future modifications will make a page of the flow diagram appear on the screen before being printed and will allow an user the added flexibility of being able to provide additional labels before the final printout.

REFERENCES

[1] Ellis, Clarence A. and Marc Bernal, " Officetalk-D : An experimental Office Information System", Proceedings of the SIGOA conference on Office Information Systems, pages 131-140. University of Pennsylvania, Philadelphia, PA, June 1982.

[2] Mukherjee, S., Saha, A. , and Ahmadi , R.," The Dialogue and Database Components in User Systems Analysis", In Brown, O., and Hendrick, H. W. (eds.). Second International Symposium on Human Factors In Organizational Design and Management, (North Holland, Amsterdam, 1986).

[3] Peterson, J. L., Petri Nets, Computing Surveys, 9, (September 1977), pages 223 - 252.

Human Factors in Organizational Design and Management - II
O. Brown, Jr. and H.W. Hendrick (Editors)

AUTOMATED SIMULATION AND ORGANIZATIONAL ANALYSIS: THE TRANSFER OF ENGINEERING TECHNOLOGY

Reza Ahmadi, Sambit Mukherjee, and William H. Glick

Departments of Management and General Business
The University of Texas
Austin, Texas
U.S.A.

The complexity of many production and operation systems restricts the applicability of closed form analysis techniques. In many of these instances simulation analysis provides a good approximation of a system's performance. The technical difficulties associated with developing a simulation model for the real life system has prevented many users to employ this modeling and analysis tool. This paper discuses the development of automatic simulation code for systems which can be represented as a directed graph. This user friendly simulation modeling technique provides an important tool for analysis of the performance of complex real life systems.

Managers seeking to improve productivity have found that many, if not most, manufacturing and service oriented systems are too complex to be analyzed mathematically. Mathematical and closed form results developed by operations researchers (Suri & Hildebrant, 1984) can only be applied to a limited range of systems. Results on queueing networks have been successfully applied to some systems, but the results only hold for a limited set of cases. Further, the mathematical formulations for queueing networks require simplistic assumptions, such as markovian arrivals and service time distributions, identical service rate for systems servers, etc.

An alternative to these closed form analysis techniques is provided by simulation modeling and analysis. These techniques can be applied to most manufacturing and operations systems, especially systems that are event driven. Simulation results provide many insights regarding performance measures of the different aspects of the system, particularly the mean and variance of servers' utilizations, queue length, waiting time, and number of throughputs. Simulation analysis also can be used in a "what if" type of analysis. System designers can use a "what if" analysis to evaluate the performance of the system under different parameter specifications with different components and resources without actually purchasing or implementing the different system configurations in real life.

Wide spread usage of simulation modeling and analysis in the past has been limited due to its computer and technical requirements. Developing a proper simulation code manually requires a great degree of familiarity with simulation modelng and system analysis. As a part of User System Analysis we have developed a pascal based program which automatically generates the simulation code for a high level simulation language, SLAM II (Pritsker, 1984). This automation relieves the user from the burden of developing the simulation code manually. The user by answering a series of questions in an automated interview session defines the characteristics of the systems under study. The questions are designed for users who are not familiar with the systems modeling.

Directed Network Representation

The USA simulation package models a broad variety of physical systems that can be represented as directed networks or graphs. Each network has two components, the set of edges or arcs, and the nodes or connectors. For many stochastic systems, the arcs could represent the duration of an activity based on some probability distribution. The nodes may specify the relationships among a set of activities that are required to complete a particular process. Figure 1 shows a simple example of a system consisting of seven arcs and seven nodes.

Figure 1: System Directed Network

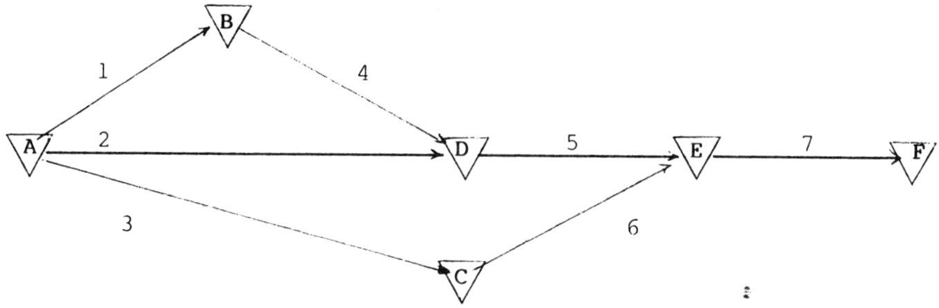

The particular environment for a given arc is shown in Figure 2. In this environment, the three main branches represent the labor, equipment, or other resource requirements of the system. In this example, both A and B are required for task completion in conjunction with either C or D.

Figure 2: Sample Arc Specification

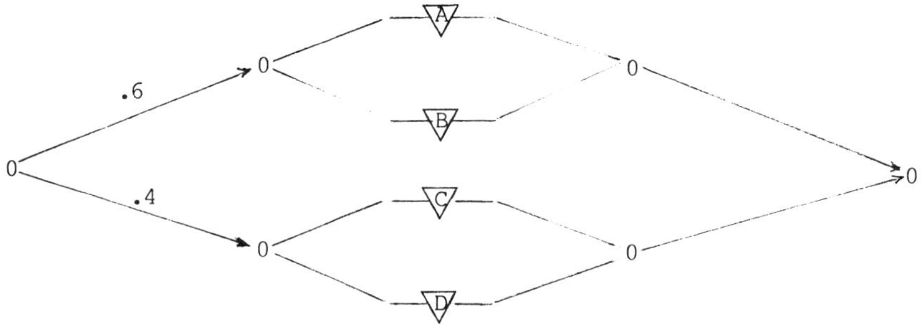

Decomposition Approach

We adopted a decomposition approach to generate the required SLAM II simulation code for these systems. The procedure decomposies the network to its component arcs, models each arcs with the appropriate procedures, and, finally, synthesizes the component arcs into a simulation model. This modularization procedure requires developing different routines for creating SLAM II code for a variety of specific arrangements of resources, equipment, and operators for any given activity. Specific models have been developed for the most commonly occuring arrangements. We are able to model a wide variety of complex systems because we divided the system at the first step and treated the unique arcs in isolation. This decomposition approach reduces the parallelism and repetition involved in manually developing simulation code.

Module Specifications for the Arcs

The first step in developing the exact simulation code for each arc is to determine the specific arrangement of the resources for the arc. Parameter specification for an arc involves questions such as "Are the resources (labor and equipment) arranged in parallel, in series, or in series-parallel?" "What are the alternative routings of the work-in-process?" "What are the possible branches?" ... Figure 2 shows an example of this arrangement of parameters of an arc. An array of records maintains the set of parameters of each arc in the network. A algorithm is used to detect the relevant modules based on the special characteristics of the parameter set such as the number of non-zero entries in each data field, the non-zero position of some key parameters, etc. When an exact representation of a parameter set is not available, the algorithm automatically selects the closest available approximation. The effect of this approximation is negligible for modeling most systems.

Synthesizing the Arcs

The system is initially modeled by an adjacency list that defines the precedence relationships among the activities in the system. After developing the individual arcs, the moduls are synthesized to reconstruct the entire network. Developing this system requires detailed accounting for the precedence relations among the arcs and the parallelism in the performance of activities. Our algorithm resembles a CPM-PERT type algorithm.

Several difficulties were encountered in constructing the entire code, such as the merging or matching of the activities incident to a node and the treatment of arcs with the same precedence node. To overcome these problems, a dynamic labeling system has been developed that controls the execution of the parallel arcs, and keeps track of the merging requirement of the activities. To complete the simulation code, special procedures also were developed to construct the conditions under which the system is to be simulated. The simulation code is executed on the PC version of SLAM II, and the summary statistics are recorded.

Conclusion

The USA simulation package automates the generation of simulation code and analyzes the labor, equipment, and resource utilization of a wide range of systems in a detailed analysis of the system's performance. This model can be used for many different systems that can be modeled as a directed network.

References

[1] Pritsker, A. B. Introduction to Simulation and SLAM II. (2nd. ed.). West Lafayette, Ind.: Halstead, 1984.

[2] Suri, R., and Hildebrant, R. R. Modeling Flexible Manufacturing Systems Using Mean-Value Analysis. SME Journal of Manufacturing Systems. 1984, Vol. 3., pp. 27-38.

Human Factors in Organizational Design and Management - II
O. Brown, Jr. and H.W. Hendrick (Editors)

APPLICATIONS OF USER SYSTEMS ANALYSIS TECHNOLOGY IN THE WORKPLACE: A CASE STUDY IN WORK AND INFORMATION FLOWS

Alan Carsrud, William Glick, Rafik Beekun,
Reza Ahmadi, Sambit Mukherjee, and Avijit Saha

The University of Texas at Austin

THE USA PACKAGE

User Systems Analysis (USA) is the marriage of state-of-the-art technology from the engineering and computer sciences with the theories of the behavioral sciences. The goal was to design USA to be a highly sophisticated management support system and as an applied research methodology for assessing work and information flows in organizations. The system was designed to provide researchers and managers with accurate and timely assessments of the activities within a broad range of work units. USA is especially useful in assessing the impact of various managerial decisions or equipment additions on the flow of work and/or information within the organization. The reports and diagrams automatically generated by the USA package of programs are designed to be understood easily by practicing managers and interpretable in terms of existing, previously validated theories or organizational design.

The previous papers in this symposium have described the recent developments of the organizational theories underlying USA (1), the nature of the data collection system (2), the graphic display technology (3), and the simulation modeling techniques (4). These theories, data-base questionnaires, graphic displays, and simulation techniques should provide managers with useful, accurate recommendations for improving organizational effectiveness.

The practical design of USA is to provide a management support system that does not require extensive education or training to be implemented. In order to provide the manager-consumer with readily understandable information USA relies on the use of "pictures" or diagrams that "draw" the flow of work and information. What makes USA unique from other manually based systems is that it is systematic, reliable, and a data-based computer system. The program takes the information from the data-base questionnaire and then actually draws out a pictorial representation of the organization, or "information map", as perceived by each of its members clearly showing their interrelationships. The true test of any management support system or applied research methodology, however, is not in the intentions of the designers, but in the field (5).

THE ALPHA TEST CASE

The analysis of an on-going new business venture involved in typesetting services was the initial test for USA's computer assisted organizational analysis methodology. This test site was chosen because of:

a) the relatively small size of the firm (under 25 employees),
b) its assumed sequential nature of work production,
c) the level of cooperation provided by its owners, and
d) a belief by the owners and the employees that the firm could operate more efficiently and effectively.

The applied goal of this analysis for the business was to provide the company with information concerning possible areas in which the firm would benefit from organizational, personnel, and equipment changes. That is, the

USA system was serving as a management support and decision assistance system to this new venture. This privately held firm had undergone several major organizational and equipment changes in its eighteen months of existence. It was assumed that these events had an effect on the flow of work and information within the organization and that the USA system would be able to identify these effects. The events that occurred included:

1) resignation of two partners who were active in the typesetting process.
2) rapid growth in the number of employees.
3) acquisition of new computer equipment to assist in typesetting operations, and
4) the expansion of the firm's physical space by nearly one hundred percent.

DATA COLLECTION

The primary owner of the firm served as the initial key respondent for the questionnaire and initially defined the nature of the tasks and the equipments being employed. Then all remaining employees of the firm completed the data-base questionnaire. Their responses updated the data matrices and more clearly defined the relationships between individuals, activities, and equipment. The resulting person-activity data matrix was then transferred to the draw program to diagram the person-activity relationships. The resulting "picture" was used for analysis to determine potential solutions to work and information flow problems identified by the draw program.

THE DIAGRAM

Below is a portion of the "drawing", or map, produced by the USA package for the typesetting firm. Results indicated that the firm was less a sequential process than initially thought. The cross-utilization of employees, typical in a small business, was obvious in the diagram. In addition, staff overloads were easily identified through the questionnaire and the graphic presentation of the data.

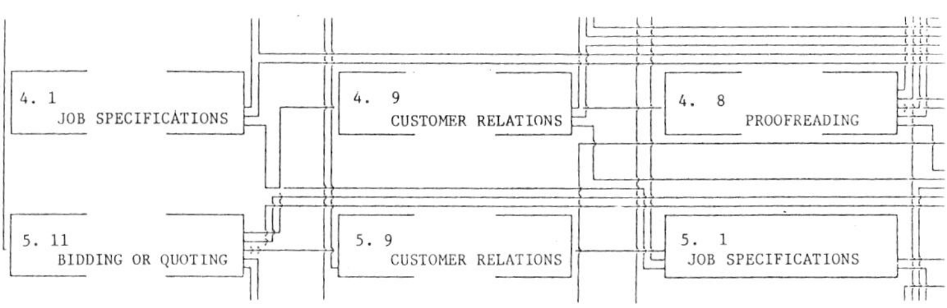

RECOMMENDATIONS

Various recommendations were made concerning job redesign and organizational structure changes to reduce problems noted in the findings produced by USA. These included:

a) individuals taking orders and estimating billing may be the sources of errors because they are "spread-too-thin."
b) functional work groups should be set up within each general employee category (typesetter, proofing, etc.) to help coordinate activities and insure uniformity in the performance of job related tasks.
c) work flow buffers, such as in/out boxes, need to be broken down into different categories until the advent of a computerized tracking system. This should assist in maintaining a timely flow of priority work through the typesetting process.

CONCLUSIONS

Although this experimental prototype of User Systems Analysis has been applied initially to one test site, its use as a management support system was well received. USA did identify problem areas and provided the management team with easily understandable data in the form of a "picture" with which to take corrective action in an effective and efficient manner. Staff were re-assigned and jobs re-designed in order to improve the flow of work and information within the firm. Further refinements of the components of USA will enhance its applicability to other work units with much different patterns of interaction and organizational problems. USA has successfully demonstrated the integration of technology from a variety of disciplines to the systematic and empirically based solution of complex organizational problems.

REFERENCES

[1] Beekun, R., Glick, W. & Carsrud, A., User Systems Analysis: Organizational Design Foundation of a Management Support System, (this volume).

[2] Mukerje, S., Saha, A. & Ahmadi, R., The Dialog and Database Component of User Systems Analysis, (this volume).

[3] Saha, A., Thektisto, V. & Mukherje, S., Data-based Graphics in Organizational Analysis: A Picture is Worth a Thousand Words, (this volume).

[4] Ahmadi, R., Mukherje, S. & Glick, W., Automated Simulation and Organizational Analysis: The Transfer of Engineering Technology, (this volume).

[5] Carsrud, A., Organizational Analysis: Idiosyncratic Methodologies Searching for Holistic Solutions, In Hendrick, H. W. & Brown, Jr., O. (eds.) Human Factors in Organizational Design and Management (North-Holland, Amsterdam, 1984).

ORGANIZATION AND MANAGEMENT OF HUMAN FACTORS

Human Factors in Organizational Design and Management - II
O. Brown, Jr. and H.W. Hendrick (Editors)

ORGANIZATION AND MANAGEMENT OF HUMAN FACTORS IN HIGH TECHNOLOGY ORGANIZATIONS

Ogden Brown, Jr., Chair

Institute of Safety and Systems Management
University of Southern California
Los Angeles, California, USA

This panel discussion will examine the organizational characteristics of successful firms which have human factors departments or groups within their formal structure. A similar discussion was held during a panel at the First International Symposium on Human Factors in Organizational Design and Management. Again of particular interest is the organizational design utilized by the companies represented on the present panel. Also of interest is how the human factors function is incorporated into that design. Of further interest as well is the degree of differentiation and the extent to which integrating functions are present. With the emergence of macroergonomics, an approach which focuses on the impacts of technology upon organizational systems, the human factors function within organizations must be able to adapt to the turbulent environments characteristic of high technology organizations. In order to learn about organizational design, alternative management systems, and problems encountered in such organizations, the panel is constituted by those who have responsibility for the management of human factors programs in large, high technology organizations which have incorporated the human factors function into their formal organizational structures.

Each panelist will discuss the organization and management of human factors within their respective organizations.

ERGONOMIC CONSIDERATIONS IN TECHNOLOGY TRANSFER AND ORGANIZATIONAL DEVELOPMENT:

CROSS CULTURAL IMPLICATIONS

Human Factors in Organizational Design and Management - II
O. Brown, Jr. and H.W. Hendrick (Editors)

ASSESSING CROSS-CULTURAL SUPERVISORY SKILLS TRAINING: PRELIMINARY RESULTS COMPARING AMERICAN AND CHINESE SAMPLES

Andrew S. Imada

Institute of Safety and Systems Management
University of Southern California
Los Angeles, California
USA

This paper reports the preliminary findings of a study comparing the results of a supervisory skills training program conducted in the United States and China. Both samples were operating in the same international hotel organization. Managerial assessments of supervisor's behaviors before and after training were statistically different. These results will be discussed in greater detail when all analyses are completed. For the moment, the preliminary success of this training offers hope for developing organizational and ergonomic intervention strategies that can be developed in one culture and used in another.

INTRODUCTION

The maturation of ergonomics as a practical discipline has taught us that we cannot hope to make a significant impact on organizational systems without considering the entire system. This systems approach was used in an organizational intervention to improve organizational functioning in a large international hotel chain. These interventions included executive assessment, supervisory skills training, management reinforcement, and worker involvement in quality circle activities. (For more details see Brown, 1985; Hendrick; 1985; Imada, 1985; and Van Slyke, 1985).

The effectiveness of the supervisory skills training is of particular interest in the present paper. Imada (1985) reported that managerial ratings indicated that they perceived a definite improvement on the five skill areas trained: (1) Handling work performance problems; (2) Handling work habits problems; (3) Following-up on previous discussions; (4) recognizing and reinforcing employee actions; and (5) Taking disciplinary action. The differences between before and after training ratings were all statistically significant. All supervisors in the American sample were working in properties in Hawaii and California.

In the past few years a new market has opened in the hospitality industry in the People's Republic of China. Sheraton Hotels has been one of the pioneers in the hospitality industry in China. To maintain the consistent quality of service the organization and its customers have come to expect, the same supervisory skills training was implemented in China. There are certainly varying viewpoints on the generalizability of concepts or theories that are developed in one culture and introduced into another (e.g., Hofstede, 1980). Cross cultural interventions are difficult to assess and make generalizations about. However, it is abundantly clear that China provides a stringent test for the generalizability of any managerial or organizational intervention because of its isolation from these ideas over the past few decades. It is also clear that China is entering a period of change--both economically and educationally (See Limerick, Davis, & Fitzroy, 1985).

SUPERVISORY SKILLS TRAINING IN CHINA

The same supervisory skills training utilized in the American sample was also used at the Great Wall Sheraton Hotel Beijing. Forty eight supervisors participated in the training. The training was conducted by two American instructors who were also involved in the American sample training. These supervisors at the Great Wall Hotel were exposed to the same supervisory skills training developed by Development Dimensions International. The same basic skills were covered in a somewhat modified format; These variations were made to accomodate language difficulties and scheduling constraints. Because of these scheduling constraints, the supervisors completed the training in a compressed schedule. The supervisors were exposed to the same behavior modeling program. The supervisor is first exposed to a videotaped model of the desired behavior. Next, the behaviors are summarized into specific learning points that are discussed and reviewed. Third, the supervisors practice these behaviors in structured role playing situations. Finally, they are given positive reinforcement by their peers and the trainers. Together, these elements have been found effective in changing a wide range of supervisory behaviors. For more details on this procedure see Goldstein and Sorcher (1974).

Six months after the supervisors completed the training, their immediate superiors (e.g., department heads, Resident Manager, Executive Assistant Manager, supervisor) rated their behaviors on the skill areas covered by the training. The same questionnaire that was used in the American sample was also used in this sample. Managers were first given a definition of the skill, some example behaviors that illustrate the skill, and then asked to assess the supervisor's behaviors before and after the training. All ratings were made on a six-point scale. The results from both American and Chinese samples are summarized in Table 1.

As a preliminary analysis, before and after training means were compared using t tests for related samples. All before and after training differences are statistically significant at the .001 level in the Chinese sample. The patterns of means in both samples are illustrated in Figures 1 and 2. While both samples show statistically significant differences before and after training, two observations can be made: First, the American sample shows a more consistent pattern among means--both before and after training. Second, the American sample shows slightly higher levels both before and after training. These preliminary analyses cannot determine the causes for these differences. At the time of this writing these results only indicate that there were statistically significant differences for each of the skill areas as assessed by the managerial ratings. More detailed analyses on these and supervisors' self ratings will be presented at the Second International Symposium.

Figure 1. Before and After Training Means as Rated by Managers for American Sample

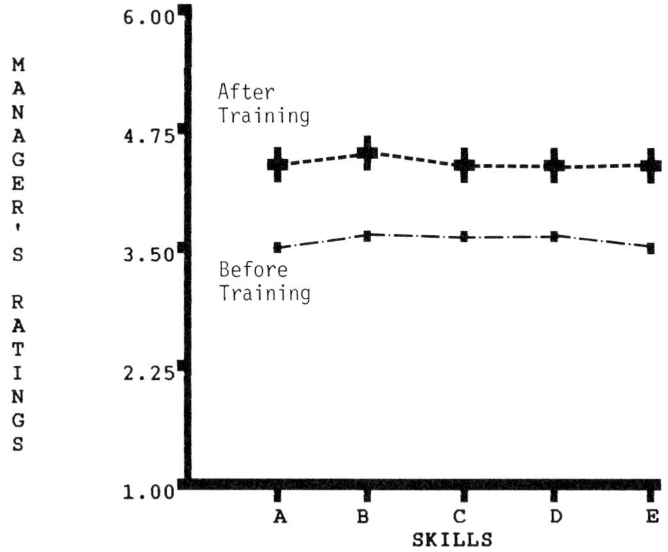

Figure 2. Before and After Training Means as Rated by Managers for Chinese Sample

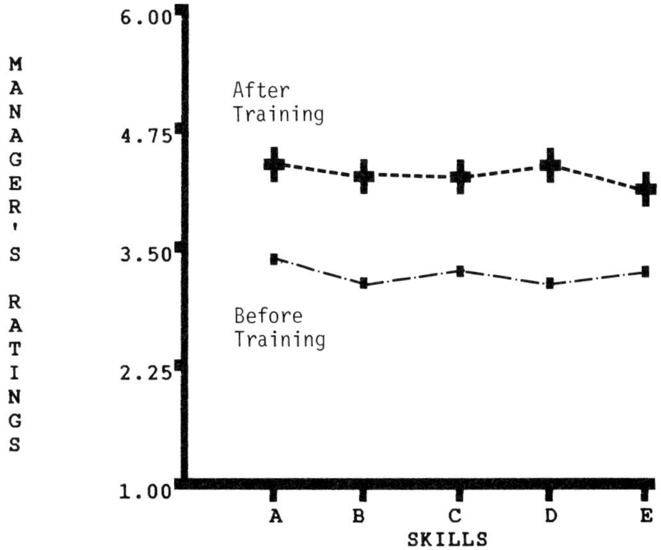

Table 1. Means for Before and After Training on Five Skill Areas as Rated by Managers in Chinese and American Samples.

SKILL	CHINESE SAMPLE		AMERICAN SAMPLE	
	Before	After	Before	After
A. Handling Performance Problems	3.3	4.5	3.6	4.5
B. Handling Work Habits Problems	3.2	4.3	3.7	4.5
C. Following-up on Previous Discussions	3.3	4.3	3.6	4.4
D. Reinforcing Employee Behavior	3.2	4.5	3.7	4.5
E. Taking Disciplinary Action	3.3	4.1	3.6	4.4

While these results are still preliminary, they offer hope in the generalizability of training developed in one culture and applied to another. This generalizability may be attributable to many possible reasons; the universal appeal of this particular training, the training method, the organizational climate, and the skill and motivation of the supervisors themselves are all contributing factors. Whatever the reasons, we need to explore the reasons for these behavioral changes to duplicate these efforts in a broad organizational context.

REFERENCES

(1) Brown, O., Jr. Increasing participatory ergonomics and organizational functioning through supervisory skills training. In I. D. Brown, R. Goldsmith, K. Coombes, & M. A. Sinclair (Eds.) Ergonomics International 85 (Taylor and Francis, 1985), 887-889.

(2) Imada, A. S. Results of supervisory skills training: Preliminary analysis of a system intervention. In. I. D. Brown, R. Goldsmith, K. Coombes, & M. A. Sinclair (Eds.) Ergonomics International 85 (Taylor & Francis, 1985), 886-888.

(3) Van Slyke, M. D. Effects of management commitment and reinforcement on a supervisory skills training program. In I. D. Brown, R. Goldsmith, K. Coombes, & M. A. Sinclair (Eds.) Ergonomics International 85, (Taylor & Francis, 1985), 880-882.

(4) Hofstede, G. Motivation, leadership, and organization: Do American theories apply abroad? Organizational Dynamics, Summer (1980) 42-63.

(5) Limerick, D., Davis, J., & Fitzroy, P. Management education in China, Journal of Management Development 4 (1985) 3-14.

(6) Goldstein, A. P., & Sorcher, M. R. Changing supervisor behavior. (Peragamon, New York, 1974).

Note: The author wishes to acknowledge the work and time that Dr. Mark D. Van Slyke contributed to a successful training and follow-up; to Mr. Andy Lum for his coordination on this project; to the staff and management at the Great Wall Sheraton Hotel Beijing; to Tom Litant for his help with the graphics; and especially to Dixie Imada for her understanding, help, and encouragement throughout the project.

Human Factors in Organizational Design and Management - II
O. Brown, Jr. and H.W. Hendrick (Editors)

THE EFFECTS OF HUMAN FACTORS ON THE SUCCESS OF TECHNOLOGY TRANSFER PROJECTS TO INDUSTRIALLY DEVELOPING COUNTRIES: A REVIEW OF REPRESENTATIVE CASE STUDIES

Najmedin Meshkati
Michelle M. Robertson

Human Factors Department
Institute of Safety and Systems Management
University of Southern California
U.S.A.

Human Factors considerations play a key role in the survival, effectiveness, and success of technology transfer projects to the Industrially Developing Countries. Four representative case studies of such projects are eclectively chosen and analyzed. Major relevant ergonomical issues are identified. It is concluded that ergonomics of the local infrastructural factors, technology adjustment, cultural variables, and managerial and organizational methods transfer are the primary determinants for the human factors analysis.

INTRODUCTION

According to a National Research Council study for the U.N. Conference on Science and Technology for Development (1979); "Development is concerned with two important goals: (1) eradication of extreme poverty by satisfying the most basic needs for food, shelter, health, employment and education; and (2) modernization and growth of national output both for domestic consumption and to earn income through export."

Infrastructure and the technological capabilities of the Industrially Developing Countries (IDC) is essential to achieving both of the above goals. However, no developing economy can wait until its people create the required knowledge base for the advanced technologies needed for modernization, since much of this already exists in and is available from developed countries. Thus, the optimal solution to development problems is to acquire the technology from existing sources, and not try to re–invent the wheel. This process leads one to raise the unavoidable question of which technologies to acquire, how to master and control them, how to adapt them to specific resources and environment of a given country, and how to build upon them (Denver Research Institute, 1983).

Industrially developing countries, in general, have special inherent sociocultural and environmental conditions which significantly affect the productivity and success of every transferred technology project. For instance, many of the IDC countries have the extremely hot and dry climate. Problems of the working environment in hot climate are even more numerous and of higher magnitude than those in cold climate. Unless correct measures are taken during the conceptual phase of designing the work environment, the creation of a physically and mentally satisfying workplace will not occur. It is generally too late and costly to incorporate any type of design modifications at the later stages of technology transfer (Sen, 1982).

Wisner (1985) studied the negative aspects of imperfect technology transfer and its effects on the health and safety of the human element in the production process. These included; a high rate of work accidents, a greater frequency and variety of occupational illnesses, and an increase in parasitic diseases. The frequencies of these unwanted outcomes in IDC are significantly higher than of the originating country of the technology. Furthermore, Krishna (1980) considered the negative by–products of inappropriate technology transfer as; low volume of production due to machines not being used to capacity, poor product quality, and equipment deterioration due to unskilled workers.

In this review, representative cases of technology transfer are eclectively chosen and systematically analyzed to demonstrate the significant effects of the presence (or absence) of human factors considerations on the success (or failure) of such projects. The attempt has been made to present at least one related case to each one of the major areas discussed in the following section.

MAJOR HUMAN FACTORS CONSIDERATIONS IN TECHNOLOGY TRANSFER

There are four primary areas of human factors considerations in technology transfer (Meshkati, 1986). (1) Ergonomics of the existing and general industrial–related infrastructural parameters. This includes special physical and environmental variables, anthropological and anthropometric considerations, labor force characteristics, and idiosyncratic human–machine relationships. (2) Technology adjustment; products and production techniques utilized in developed countries need to be adjusted to accommodate different needs and conditions encountered in developing countries. The nature and extent of adjustment effort should be determined by the complexity of the production and process, the IDC's trade and industrialization policies, production resources availability, and the technological capabilities of the IDC. (3) The effects of cultural variables on technology utilization. Culture in its broad sense, as the collective mental programming of the people who live in the IDC, determine not only the extent of the success, but also the survival of the technology transfer project. A representative list of culturally–related variables may include work habits, general set of attitudes towards the nature of machinery and its operation, and individualism versus collectivism. (4) Managerial and organizational (methods) transfer. Without suitable and careful analysis of the management factor in any technology transfer transaction, the prospects for success diminish rapidly. Moreover, without a matching of transferor and technology acquirer management styles and methods any prospects for successful technology utilization may evaporate entirely (Denver Research Institute, 1983). Many of the Western organizational control methods such as; production planning and control, and quality control are considered to be too rigid and inappropriate to be used "as is" in the IDC (Baranson, 1969).

REPRESENTATIVE CASE STUDIES

Union–Carbide Pesticide Plant in Bhopal, India

On December, 1984 a toxic cloud escaping from a Union Carbide pesticide plant in Bhopal, India brought death to at least 2500 residents, injured perhaps 200,000 and brought terror and suffering to hundreds of thousands more. In reviewing what went wrong in Bhopal, industrial experts have placed human error caused by carelessness, ignorance, confusion, and fright high on the list. In this regard the Technical Director of the Chemical Manufacturers Association commented; "You can design the best system, but when you deal with people you can create problem." Moreover, an official of the United Nations Environment Program, referring to the issue of traditional agriculture–oriented Indian workers and their reaction to the imported technology stated, "They have not internalized the technological culture" (Engler, 1985). This phenomenon of the "internalization" of Western technology by the workers from IDC has also been studied by Pierce (1963). According to his work, technological sophistication of Western culture produces a general set of attitudes concerning the nature of machinery and its utilization. This results in facilitation of the acquisition of skills needed to work efficiently with modern machinery. Ethnic groups not being exposed to mechanization and not having an historical background of industrialization, may not readily understand and accept the concepts of; compatibility, display/control relationship, and machine operations. Weltman (1963), considered such an acceptance to be a prerequisite for the efficient integration of human operator into the human–machine system.

Bhopal plant was an extremely complex system with many tightly coupled units and feedback loop systems. In contrast to simple and linear production systems wherein it is relatively easy to pinpoint the difficulty and bypass the trouble spot. According to Perrow (1984), problems with complex system are not immediately comprehensible and predictable. Multiples of small failures, making unexpected connections and moving along unplanned or hidden paths may set off a large and catastrophic accident leaving the managers uncertain about how or where to intervene. This led Engler (1985) to conclude that; the conventional explanations presented after Bhopal and other disasters, blaming design or operator failure may be inadequate and deceptive. In his judgement, the trouble certainly lies not in Indian genes but in Western technology and industrial systems.

Finally, having not included the qualified ergonomists in the plant design, plus the lack of a collaborative effort between the Indian design engineers and their Union–Carbide counterparts in the U.S., resulted in the decision of storing of the highly volatile chemical in large bulk tanks. This decision has been cited as being partially responsible for the accident. "The engineering department of Union–Carbide of America imposed that decision over the objection of the engineers of the Union–Carbide India Ltd" (Engler, 1985).

Sulfur Extraction Plant; Mosul, Iraq

In 1978, a multidisciplinary team of experts from the International Labor Office (ILO), and World Health Organization (WHO) visited a sulfur extraction plant located in the hot and reltively humid city of Mosul, Iraq. According to their report, the technology along with the equipment of this plant had been acquired from a northern European country. It was originally designed for the northern European climate and it was installed with "very little or no modification to the European design". Due to this fact, the level of sulfurous gases around the plant were very high and indeed damaging to workers health. Moreover, the original design did not have any provisions to reduce the radiant and convectinal heat which is required for the thermal comfort of the workers in that particular environment. Furthermore, another climatic feature which was not considered was the air movement affected by down–draft winds at the plant site. This caused the sulfurous gases to be blown back down on to the plant area. This could have been avoided by installing taller chimneys and incorporating other design provisions.

In addition, there was high level of sulfurous gas leaks which resulted from poor plant maintenance. The insufficient technical training of the local industrial and maintenance operators, plus the unmodified original maintenance guidelines, were cited as the cause of this problem.

Personnel protection equipment to be used by the workers in a plant of this nature is an essential item to preserve the employees health and safety. However, it was reported that the equipment was far from being suitable for the climatic condition of Iraq.

This is a classic manifestation of the aftermaths of transferring an unadjusted and unmodified technology from a developed country to the IDC. All of the above compounded with many other problems led a senior staff member of expert team to informally comment that; "the whole plant should be closed down."

European Chemical and Pharmaceutical Projects, Asian Region

Lasserre (1982) analyzed thirty–three cases of technology transfer in the chemical and pharmaceutical industries from European countries to the Association of Southeast Asian Nations (ASEAN) companies. In his analysis,

Lasserre presented the convergence of both points of view of the European and local managers and their attitudes toward operational problems. The European technology transferor officials and the local expatriate managers metioned the social and cultural environments of the technology acquirer country as the major cause of most of the problems which significantly affected the operational effectiveness of their firms. This was primarily due to the unfamiliarity of the European side with the cultural variables of the ASEAN. This fact led the author to recommend training and orientation toward the development of "cultural skills" for the non–local managers. The training programs should include; "language training and the initiators to sociocultural aspects of the local country." The local partners' perception of the primary source of the problem which hindered efficiency was; "the behavior of the European managers". They cited the poor communication between the European managers and their employees as one of the indications of the aforementioned problem. They also attributed this to the foreign managers' lack of knowledge and understanding of the subtle cultural variables within the local countries

Organizational and Managerial Systems of a Multinational Corporation, 40 Subsidiaries around the World

Hofstede (1980) surveyed more than 116,000 employees concerning their attitudes, perceptions, and reactions to the organizational policies and managerial styles of subsidiaries of a large U.S.–based multinational corporation located in 40 countries around the world. He concluded that the national cultures and the affected managerial practice in those subsidiaries differ on four basic dimensions: power distance, uncertainty avoidance, individualism–collectivism, and masculinity–femininity. Power distance indicates the extent to which a society accepts the fact that power in institutions and organizations is distributed unequally. Uncertainty avoidance indicates the extent to which a society feels threatened by uncertain and ambiguous situations and tries to avoid these situations by providing greater career stability, establishing more formal rules, not tolerating deviant ideas and behaviors, and believing in absolute truths and the attainment of expertise. Individualism implies a "loosely knit" social framework in which people are supposed to take care of themselves and their immediate families only, while collectivism is characterized by a tight social framework in which people distinguish between in–group and out–group; they expect their in–group (relatives, clan, organizations) to look after them, and in exchange for that they owe absolute loyalty to it. Masculinity dimension expresses the extent to which the dominant values in society are "masculine' – that is – assertiveness, the acquisition of money and things, and not caring for others, the quality of life, or people. Femininity is the opposite pole of the masculinity–femininity continuum.

Furthermore, Hofstede has argued many of the American motivation, leadership, and organizational theories do not function as expected in foreign countries. For instance, his evaluation of Maslow's hierarchy of human needs indicated that it is not the description of an universal human motivation process. He considered it to be a description of a value system, "the value system of the U.S. middle class to which Maslow belonged." In addition, he studied the concept of achievement and achievement motivation set forth by McClelland (1961). He reported that this word (achievement) is hardly translatable into any language other than English, for this reason, the concept could not be used in the organizational and managerial contexts of multinational corporations Also, Spence (1985) argued that the conceptions of achievement are culturally bonded to American society and can not be over generalized to other cultures.

CONCLUSION

As it appears from the above case studies, the collective human factors considerations, in its broadest sense, plays a key role in the survival and successful operations of technology transfer projects to the IDC. Considering and incorporating ergonomical related factors influencing local workers performance represents an essential factor in determining the appropriateness and choice of technology. The importance of this fact is emphasized and reiterated by Kogi (1985) in a recent survey to identify the priorities for improvement in conditions of work, welfare, safety and occupational health of eight developing countries in Asia. He found that there was a critical lack of appropriate technological design to incorporate the local workers' physical dimensions and capabilities.

Technology adjustment and modification according to the ergonomic study of the abilities and limitations of the infrastructural, environmental, and anthropological of the IDC, presents a challenge to both the technology tranferors and acquires. It certainly affects the productivity as well as health and safety of the local population.

Cultural and traditional variables are major determinants of applicability and effectiveness of technological endeavors in the IDC. Kogi (1985) reported that certain aspects of local traditional culture or habits represent obstacles to the promotion of technological advances. Furthermore, managerial and organizational theories and methods are also subject to cultural influence. As noted philosopher, Peter Ducker (1974) has stated; "Management is a social function and is embedded in a culture –a society – a tradition of values, customs and beliefs and in governmental and political systems. Management is– and should be– culture conditioned."

REFERENCES

Baranson, J., Industrial Technologies for Developing Economies (Praeger Publishers, New York, 1969).

Denver Research Institute, International transfer and acquisition of technology, final report, Prepared for: Agency for International Development (AID), Washington D.C., 1983.

Drucker, P., Management: Tasks, Responsibilities, Practices, (Harper and Row, New York, 1974).

Engler, R., Technology out of control, The Nation, (1985) 488–500.

Hofstede, G., Motivation, leadership, and organization: do American theories apply abroad?, Organizational Dynamics, (1980) 42–63.

Kogi, K., Improving working conditions in small enterprises in developing Asia (International Labor Office, Geneva, 1985).

Krishna, The economic development of India, Scientific American, 243 (1980) 118–193.

Lasserre, P., Training: key to technological transfer, Long Range Planning, 15 (1982) 51–60.

McClelland, D.C., The Achieving Society (D. Van Nostrand Company, Inc., Princeton, N.J., 1961).

Meshkati, N., Major human factors considerations in technology transfer to industrially developing countries: An analysis and proposed model, in: Hendrick, H. W. and Brown, O. (eds.), Human Factors in Organizational Design and Management II (North Holland, Amsterdam, 1986).

National Research Council, U.S. Science and technology for development: A contribution to the 1979 U.N. Conference; a background study on suggested U.S. initiative for the U.N. Conference on Science and Technology for Development, Vienna, (1979).

Perrow, C., Normal accidents, living with high–risk technologies, (Basic Books, Inc., New York, 1984).

Pierce, B.F., The ethnic factor in the man–machine relationship, General Dynamics Corporation, (San Diego, California, 1963).

Sen, R.N., Certain ergonomic principles in the design of factories in hot climates, Proceedings of the International Symposium on Occupational

Safety, Health and Working Conditions, Specifications in Relation to Transfer of Technology to the Developing Countries, International Labor Office, Geneva (1982) 123–147.

Spence, J.T., Achievement American style: the rewards and costs of individualism, American Psychologist, 40 (1985) 1285–1295.

World Health Organization and International Labor Office, Report to the government of Iraq on the work of the ILO/WHO multidisciplinary mission, International Labor Office, Geneva (1978).

Wisner, A., Ergonomics in industrially developing countries, Ergonomics, 28 (1985) 1213–1224.

Human Factors in Organizational Design and Management - II
O. Brown, Jr. and H.W. Hendrick (Editors)

MAJOR HUMAN FACTORS CONSIDERATIONS IN TECHNOLOGY TRANSFER TO INDUSTRIALLY DEVELOPING COUNTRIES: AN ANALYSIS AND PROPOSED MODEL

Najmedin Meshkati

Human Factors Department
Institute of Safety and Systems Management
University of Southern California
U.S.A.

Appropriate technology is considered to be the optimal choice for the Industrially Developing Countries (IDC) in quest of economic freedom and development. However, it is found that there has not been a systematic approach to analyze and ensure the appropriateness of technology. Human factors as a unique multidisciplinary subject is potentially capable of tackling this problem through incorporating major infrastructural as well as technology–related variables in the analysis. An integrated model based on these premises is developed an proposed.

INTRODUCTION

Technology is a key factor in economic development. Furthermore, no developing country can wait until its people create the required knowledge base for the advanced technologies needed for modernization, while much of this already exists in and is available from developed countries. Thus, the optimal solution to development problem is to acquire the technology from existing sources, and not to try to re–invent the wheel. This process leads one to raise the unavoidable question of which technology to acquire, how to master and control it, how to adapt it to specific sources and environment of a given country, and how to build upon it. According to a study by Denver Research Institute (1983), these issues comprise the major concern of any developing country; acquirer of technology.

Transferring appropriate technology seems to be the optimal solution to the above problem. However, according to Evans and Nogg Adler (1979), "Appropriate technology does not offer an absolute prescription but rather a process of choosing from among a broad range of options. The appropriate technology approach seeks to optimize solutions, Wherever possible, through reliance on problem–solving capabilities of local people as well as a sensitivity to environmental and cultural impacts."

Human Factors, due to its multidisciplinary nature, is capable of systematically identifying and addressing most of the influential factors determining the appropriateness of a technology for a developing country. Works of Sen (1982 and 1984), Wisner (1985), Shahnavaz (1985) are only a representative sample. Moreover, in a recent study, the author (Meshkati and Robertson, 1986) examined the vital role of the human factors considerations in the success (or failure) of four technology transfer projects. One of the cases studied was the Union Carbide pesticide Plant in Bhopal, India. The catastrophic accident of December 1984 at this plant resulted in the death of 2500 individuals and the injury of approximately 200,000 nearby residents. In reviewing what went wrong in Bhopal, industrial experts have placed human error caused by carelessness, ignorance, confusion, and fright high on the list. In this regard the Technical Director of the Chemical Manufacturers Association contented: "You can design the best system, but when you deal with people you can create problem." Moreover, an official of the United Nations Environment Program, referring to the issue of traditional agriculture–oriented Indian workers and their reaction to the imported technology stated, "They have not internalized the technological culture." (Engler,1984).

The above example reiterates the observed aftermaths of the Three Mile Island accident. It, once more, echoes the ironic concept: "Human Factors is like a vacant seat in a class, most noted when it is absent."

Nowadays, some of the IDC governmental officials have become cognizance and started to realize and acknowledge the vital role of human factors in ensuring the appropriateness, safety, and effective utilization of the transferred technology. Indonesian officials took a lead in this respect. In a recent speech, the Vice–President of the Republic of Indonesia referred to the above factors as: "We all know from experience that taking over modern technology as such without considering social, cultural, and anthropological factors prevailing in the receiving country can lead to economic as well as social sufferings. Therefore, it is the obligation of the developing countries themselves to research and develop ergonomics to suit their own peculiar situations" (Wirahadikusumah, 1985). The Minister of Manpower of Indonesia has also stated that; " Developing countries do need ergonomics which emphasize the human aspects of work as well as to improve welfare, productivity, and the quality of life." (Sudomo, 1985)

The purpose of this study is twofold. First, identify major human factors considerations influencing appropriateness of a technology for IDC. Second, introduce a conceptual model to ensure the systematic integration of human factors analysis in technology transfer process.

APPROPRIATE TECHNOLOGY AND HUMAN FACTORS

There are numerous studies on the concept of "Appropriate Technology", its required characteristics, and benefits. Stewart (1979), in a study for the World Bank, argued that, appropriate technology consists of technology more in line with developing countries needs and resources. In a subsequent study, she characterized a technology to be appropriate only if it reflects the environment in which it is to be operated (Stewart, 1981). However, in developing policies to secure more appropriate technology, she only repeated her original assertion by stating; "More appropriate technology may be roughly defined as technology whose resources use is more in keeping with less developed country resource availability and whose products are more suited to low-income consumers."

Evans and Nogg Adler (1979), presented a thorough case history of appropriate technology for development. They referred to the analysis process of the technologies that may be available for transfer to a developing country as an "essential" step in ensuring the appropriateness of the technology. They concluded: "Analysis must consist educational, social, cultural, economical, infrastructural, and political aspects to the maximum extent possible." They also suggested that the technology from the developed world requires modification, adaptation, or redesign when utilized in developing countries.

United Nations Conference on Trade and Development (UNCTAD), has published a guideline for the study of the transfer of technology to developing countries (1972). According to this document, the selection of the appropriate technology is a "difficult task" for developing countries. In another study by the UNCTAD Secretariat, Transfer of Technology (1978), the role of human factors is emphasized more explicitly: "Only if people are considered in the transfer processes, the technology is likely to make a genuine contribution to the well-being of the community and not fall into disuse after the departure of the experts."

A report by the Development Center of the Organization for Economic Cooperation and Development (OECD, 1973), regarded technology diffusion centers in developing countries as the essential element in technology transfer. It also suggested that technology diffusion is made more efficient by linking the technology suppliers' capabilities and with the receivers' human resources characteristics and limitations.

Appropriate technology is also considered to be one of the basic four technological objectives of the World Bank operations. An exhaustive report entitled; Technology and Science in World Bank Operations (1982), refers to this goal as: "to ensure that the most appropriate technologies are used in the projects it finances". The report continues by arguing that: "the Bank's techniques for preparation and appraising projects are intended to ensure that the technology used is tailored to the social, cultural, and environmental needs of specific situations."

MAJOR HUMAN FACTORS CONSIDERATIONS OF TECHNOLOGY TRANSFER

In analysis of the major human factors considerations in technology transfer to the IDC, four preliminary areas have been identified. These would draw upon the intertwined principles of macro and micro ergonomics. Moreover, due to the dynamic and integrated nature of the ergonomic analysis, some of the following areas may overlap and could be interrelated.

1– Ergonomics of the Existing and General Industrial–Related Infrastructural Parameters

This includes special physical and environmental variables, anthropological and anthropometric considerations, labor force characteristics, and idiosyncratic Human–Machine relationship. Sen (1982), evaluated the problems of the working conditions in hot cilmates as "more numerous and of higher magnitude" than those in cold climates. He recommended thorough study of the developing countries, prior to any industrial (e.g. factory) technology transfer. He identified primary ergonomic factors to be studied as: (1) The site in relation to human habitation, landscape, vegetations, altitude, etc. (2) The orientation of the building, including roofs, walls, windows, etc. (3) The insulation and thermal capabilities of the building materials. (4) The use of sound material and form of construction. (5) Correct design of workplace, windows, doors, stairs, corridors, etc.

Regarding the importance of the anthropological and anthropometric considerations in technology transfer, Sen (1982) referred to the effects of body shape or the surface to volume ratio on thermal preferences. According to his studies, a thin person, generally found in hot climate, has a much greater body surface than a short, fat person of the same body weight, and he or she can dissipate more heat and will tolerate and prefer a higher temperature. Moreover, the color of the skin of the people in hot climates containing the pigment melanin prevents the penetration of damaging ultraviolet rays and increases the heat emission from the body in the same proportions as it affects absorption (Ibid).

The differences in the amount of oxygen consumption of the pilots from different ethnic backgrounds were studied by Pierce (1963). According to his work, it has been traditionally a requirement for American pilots to take supplemental oxygen when flying in an unpressurized plane above 10,000 feet. Peruvian pilots, on the other hand, who are natives to high lands, did not need the additional oxygen due to special development of the lungs and cardiovascular system. They could fly in comfort at 20,000 feet without the use of the oxygen equipment. Physiological differences among various people of the earth play a major role in determining (transferred) technology performance and success. Wyndham (1975), observed significant physiological differences between the work capacities of indigenous Bantu people (a major South African tribe) and European workers. These differences were so great that the Bantus' work standards have to be adjusted downward as compared with those of Europeans (Chapanis, 1974).

Another major important factor in technology transfer is anthropometric variables and its effects on technology design, transfer, and utilization. According to Chapanis, (1974), problem of fitting equipment to people due to anthropometric differences is more complex than it appears at the first glance. He asserted: "The problem is not solved by simply taking a piece of equipment built for one population and scaling it up or down for another population. The difficulty is body proportions also differ among the people of earth." For instance, Germans compared to other people of the world, have much longer legs in proportion to their heights whereas Japanese have much shorter legs in proportion to their statures. Furthermore, when applying American anthropometric design standards to other people of the world, it fits nearly 90% of Germans, but only 80% of Frenchmen, 65% of Indians, 45% of Japanese, 25% of Thais, and 10% of the largest Vietnamese.

An important area of application of anthropometry data is in the aircraft cockpit. According to Pierce (1963), because of Japanese pilots shorter stature, they were unable to fly the American made aircraft. Since layouts of cockpits and flight decks in American aircrafts were designed for American pilots and even when both the controls and seats were adjusted to the limits, still many Japanese pilots were incapable of reaching some controls. Another anthropometrical problem faced by Japanese pilots obliged to use American equipment was that of acquiring a workable full–pressure suit with a "good fit" which is a critical factor in its usefulness. Even the smallest–sized suites were too long for many of the Japanese pilots and it triggered an anthropometric survey of Japanese pilots involving similar measurements as those taken for the United States Air Force Survey (Ibid).

Other characteristic affecting productivity of the labor force and consequently technology utilization in the IDC is manual dexterity. It has been argued that most Japanese have a high degree of manual dexterity imparted by early training in writing with pen and the art of calligraphy (Baranson, 1963). This is cited as one of the contributing factors to the success of the postwar electronic industry in Japan which required labor to assemble miniature parts. According to a study by Calder (1961), Eskimos are highly proficient as watch–makers and radio engineers. Their high degree of mechanical aptitude is attributed to the development of special mental faculties in order to survive in a hostile environments: acute eyesight, photographic memory, nimbleness of hands, and special abilities to improvise. Differences in the characteristic of labor force working in different regions of the world led Kaminaka (1985) to also recommend special consideration of ethnic variables in the design of U.S. agricultural equipment. These differences are primarily limited to acquired skills and those skills that could be improved later by training and practice. It has been proven that there is no difference in basic cognitive capacity between men belonging to different societies and cultures (Wisner, 1985). Although it should be noted that there exist some differences in performance levels of human operators attending to various tasks. However, it is not caused by differences in cognitive capabilities of the individuals, the primary factors responsible for such disparities are individuals cognitive complexity, information processing behavior, and psychomotor skill (cf., Meshkati, 1983; Meshkati and Driver, 1984; and Meshkati and Robertson, 1985). Moreover, in a recent comprehensive study, Gould (1981) rejected the concept of "biological determinism" which holds that shared behavioral norms, cognitive abilities, and the social and economic differences between human groups, primarily races, classes, and sexes arise from inherited, inborn distinctions. A comparative analysis of operators' behavior working on the same equipment in two different countries clearly demonstrates the universal character of human cognitive qualities (Wisner, 1985). For example, the Boeing company reported that Vietnamese pilots and maintenance operators trained in the U.S. perform as well as their counterparts working for U.S. airlines (Chapanis, 1974).

Idiosyncratic Human–Machine relationship in the IDC results from socio–cultural variables and also is partially affected by the evolution and diffusion processes of technology in their societies. Technological sophistication of Western societies results in a general set of attitudes concerning the nature of machinery and its operation which facilitate the acquisition of skills required to work efficiently with modern equipment. For ethnic groups not having a historical background of mechanization, ready acceptance of the concept of compatibility in design, naturalness of the display and control relationships, and machine's response may not come easily. Weltman (1963), considered such acceptance as the prerequisite for the integration of the human operator into the man–machine system.

Perceptual differences of different people also affect their responses toward display–control relationship. Perceptual response is a representation of a particular combination of features, determined by levels of stimulus dimensions, triggering a categorical response (Wickens, 1984). It is also affected by needs, values, group opinion, prior experience, and personal utility system (Hochberg, 1975). Winter (1963), reported the unintelligibility and misunderstanding of the European safety posters by the Bantus. Chapanis (1974) has referred to this fact and recommended special precautions in the design of visual display for different populations.

One of the areas usually studied in Human–Machine relationship is the working posture of the human operators interacting with machine. Workers in the IDC have different working postures which should be heeded, studied, and integrated into design of technology (e.g. machinery) transferred to these countries. For instance, in India, workers normally interact with machines while in crouching position whereas operational controls on Western designed machines are at the standing level (Baranson, 1969).

2– Technology Adjustment

According to a recent OECD (1982) report, 80% of the transferred technology to developing countries is carried out by Multinational Corporations (MNC). A good majority of these MNC are American owned, based, and promote

American technology. Furthermore, it should be noted that this technology is interwoven with American history and also with many other local ingredients; personal, political, and institutional (Ferguson, 1979). American technology strives always for more things, for more people to plan always in terms of an expanding market (Ibid). A European has observed that "Quantity in America is not a fact, as it is with us, it is a value" (Lyell, 1855). To go for the numbers in American resulted in mechanization of manufacturing processes and when processes were mechanized, the ideal of efficiency demanded that machines to run continuously. This precocious outpouring of automatic high–volume production machines led directly to the methods of mass production, which are America's chief contribution to world technology (Ferguson, 1979). American Technology draws upon American workman's traits of initiative, drive, and originality, an eager resort to machinery, a "go–ahead" spirit , and striving for efficiency (Ibid).

According to an Austrian Chief Executive (The Conference Board, 1975), acquisition of technology and import it "as is" when there is an essential lack in basic infrastructure of an IDC is not a transfer, but only a "transplant with questionable results". Shahnavaz (1985) also labeled not ergonomically adjusted technology as "transplanted technology" and due to negative effects, doom to failure.

Technology adjustment is a brand concept which includes activities at different levels of project design, implementation, and utilization. It covers technology hardware and software adjustment, development of specialized training manuals and procedures, etc.

A joint team of experts from World Health Organization and International Labor Office (1978) visiting Iraq, reported that disastrous outcomes resulted when a sulphur extraction technology designed and built in a northern European country (with cold climate) transferred "as is" to a developing country; Iraq (with hot climate), without modification on the basis of anthropometric, climate, socioeconomic, and workplaçe variables.

Production techniques adjustment is an essential component in a successful technology transfer project. A vast array of technical knowledge and manufacturing know how are often required to duplicate a single industrial process. This requires a highly trained engineering and technical personnel familiar with IDC local needs and capabilities at both ends of technology transfer, i.e, transferor and acquirer. It requires, according to Baranson (1969), "conversion" engineers to adapt product designs and production techniques to local environments and deficiencies.

Transferred technology maintenance requirements also should be considered in the context of technology adjustment. For a variety of reasons, the level of achievable maintenance (both preventive and regular) in IDC is different from the expected level in a developing country. Moreover, improper translation of maintenance manuals and paucity of technical terms in a particular language aggravate the problem. A poor quality translation of a maintenance manual, according to Sinaiko (1975), resulted in performance that was about seven times worse than the performance demonstrated by the operators who used the higher quality translation.

3– The Effects of Cultural Variables on Technology Transfer

To anthropologists, culture is the way of life of people; the sum of their learned behavior patterns, attitudes, customs, and material things. Kluckhohn (1951) offered a consensus of anthropological definitions on culture as: "Culture consists in patterned way of thinking, feeling and reacting, acquired and transmitted mainly by symbols constituting the distinctive achievements of human groups, including their embodiments in artifacts; the essential core of culture consists of traditional (i.e. historically derived and selected) ideas and especially their attached values." Culture could operationally be defined as the collective mental programming of people's minds (Hofstede, 1980a). Cultural milieu of IDC represents a double–edge sword to technology transfer project. If provisions for cultural compatibility is considered in advance, then the result of the transfer would be positive, otherwise the failure is inevitable. Cultural variables reactions to technology transfer projects are complex, interrelated, and hard to single out. However, generally it is reflected through such processes as; attitudes toward work, technology, and organization, working habits, working group dynamics, religious beliefs and customs.

Attitudes toward work strongly affect industrial productivity in IDC. In countries like India, a job is generally regarded as a family responsibility to such extent that its performance is considered to be collective and is shared by all the family members (Chapanis, 1974). Moreover, in India individuals and group are accustomed to craft–oriented industry. They had considerable difficulty in adjusting to the techniques and organizations associated with modern mechanized factories. Working on automatic machines degraded them from "craftsman" to "laborer" and thus they had very low efficiency (Baranson, 1969). As reported by Pierce (1963), Somoans are skilled craftsmen and energetic workers, but their dislike for factory–type jobs is based on their negative attitude toward having to put in time even when there is no immediate task to be performed. In Somoa, work is done (and enjoyed) to complete a job, after which relaxation during any time of the day is normal.

In some IDC some jobs are culturally regarded as; "degrading". Engineers and technicians of some IDC consider tasks or occupations associated with "dirty hands" degrading. This may be one of the reasons why many of whom, according to Baranson (1969), lack practical experience. Similarly, effectiveness of many professions is hindered by their surrounding cultural values. For many years attempts to teach industrial safety in Latin America foundered on cultural reefs when it was discovered that safety engineers had to wear overalls and "demonstrate" safety measures on machines in the plants (Hall, 1959).

Religion which is an intertwined variable with culture of any society, represents another influential factor in technology utilization paradigm. The Protestant work ethic has been cited frequently as a major contributor to the U.S. advance industrial and technological status (Spence, 1985). Religious customs and rituals of the IDC are not exceptions and the attesting observations are mentioned below.

In Moslem countries, during the holy month of Ramadan, people fast from dawn to dusk. This drastically affects workers performance and consequently production's quantity and quality. Moreover, Shiite Moslem workers' mood and temper are subject to many other chronological and historical events. For instance, prior to and after the anniversary of the Imam Housien's martyrdom the religious fever is high and its effects on workforce are reflected through their sadness, restlessness, and unwillingness to work. At the onset of its operations, General Motors of Iran had not originally predicted and did not have a systematic plan to face this issue. As a result it experienced sever animosity and extreme inefficiency of production in the first year of operation (Meshkati, 1976). A relatively smoother and more positive religion and transferred technology has been recorded by Sediono (1979). His study of the cement plant located in Gresik, Indonesia, indicated that; the presence of the plant has gradually changed some of the old habits and local customs. The traditionally rigid and orthodox Moslem beliefs were soften and a "new wave of religious revival took place." Also, he reported that conservative Moslems, especially younger generation, as a result of the introduction of the new technology to their society; were becoming more tolerant to the Christian community.

4– Managerial and Organizational (Methods) Transfer

The important role of the managerial and organizational factors in the success of any technology transfer project is an undisputable fact. A recent study by the Denver Research Institute (1983) for the Agency for International Development (AID) concludes; "Without suitable and careful analysis of the management factor in any technology transfer transaction, the prospects for success diminish rapidly."

As it has been demonstrated by Haire, Ghiselli, and Porter (1966); Baranson (1969); Kaurt (1975); Negandhi (1973); and Hofstede (1980a and 1980b), Western managerial and organizational methods and practices encounter serious operational problems when applied "as is" to the IDC. This fact led developing countries, international agencies, and many concerned social scientists to recommend that; no managerial and organizational method should be transferred without prior adjustment and adaptation to the country's local conditions (Kubr and Wallace, 1983). It was also suggested that the decision on what management (system or) method to promote, should be based on specific human, cultural, social, and deeply rooted local behavioral patterns.

Consequently, just as matching foreign technologies to the IDC's industrial factor endowment is critical, so is matching management methods and techniques (Denver Research Institute, 1983).

Those aspects of managerial methods which are used in managing the human element of the organization relate directly to cultural and social values. Many of these methods are difficult to transfer, or their transfer may cause adverse effects. For instance in this regard, Kubr and Wallace (1983) argued that; remuneration packages and financial incentive plans which reward individual performance rather than collective performance are not well received in collectivist societies; problem solving approaches employing democratic values of one society, do not properly function in autocratic cultures; and matrix organizational structure does not work well in cultures in which people prefer unity of command, hierarchy, and well defined formal relationships.

According to a hearing on Human Factor in Innovation and Productivity (1982), before the Subcommittee on Science, Research and Technology of the U.S. House of Representatives; Western (and American) managerial and organizational methods are; " traditionally hierarchical and decisions are made at the top with little, if any, consideration of the value of on–the–job knowledge acquired by employees." Furthermore, Western managerial methods is often criticized by the IDC practitioners for containing "value Orientations"; being "overly complex" and "design oriented", and requiring "excessive planning" (Ingle, 1981). For example, universality of the Maslow's theory of hierarchy of human needs has been studied by Hofstede (1980b). He concluded that, this theory is not the description of a universal human motivation process, it is the description of value system; "the value system of the U.S. middle class to which Maslow belongs to." Similar arguments on the lack of universality have been presented regarding the concepts of achievement and achievement motivation (cf.; Spence, 1985).

INTEGRATED TECHNOLOGY TRANSFER MODEL

Background

As mentioned before, 80 percent of the transferred technology to nonsocialist countries is carried out by Multinational Corporations (MNC), (OECD, 1982). However, the MNC can not be counted on to create the types of technology that are most useful for the developing countries (Magee, 1981). Therefore, the ultimate responsibility for development–related technology transfer inevitably rests with the policy maker officials involved in the IDC's development planning.

In order to ensure the optimality of the choice and the appropriateness of the transferred technology to the IDC, a systematic and integrated approach seems to be instrumental. Kogi (1985), has argued that it should consist of joint consideration of technological, administrative, and economical factors. A previously cited UNCTAD (1972) document suggest; "it is necessary for developing countries to have efficient institutional machinery" to formulate and implement appropriate national policies concerning transfer of technology. Black (1972), has even proposed the establishment of a center staffed with experts to assist the IDC in their quest of appropriate technology. The aforementioned study of the Denver Research Institute (1983) recommended the development of a strong local Research and Development (R&D) institution in the IDC to advise local industry, entrepreneurs, and governmental agencies in acquiring foreign technology. Wigglesworth (1985), considered the host countries' systematic and independent evaluation (audit) of the transferred technology projects as a key to their success. These facts call for creation of centralized and integrated teams of responsible officials from technology transferor entities, associated international donor agencies, and technology acquirer countries. The primary mission of such teams is to conduct thorough analysis and evaluate the economic as will as ergonomic aspects of the proposed technology, adjust it to the IDC's local variables determined by their socio–technical infrastructure, oversee all the operational activities, and finally be prepared for on–line incorporation of further needed modifications.

The Model

In order to address the above needs for a cohesive approach and also alleviate most of the aforementioned problems associated with sub–optimal technology transfer, a generic and integrated model, herewith is proposed. It systematically identifies the five inherent phases comprising the life cycle of a technology transfer project as; Initiation, Analysis, Implementation, and Utilization. Moreover, this model is in accordance with the System Development Cycle concept introduced by Meister and Rabideau (1965). The System Development Cycle is a time–oriented, sequential process in which factors involved in the design, production, and testing of a man–machine system interact within the framework of an industrial setting to accomplish system requirements established at the start of the cycle. Figure 1 depicts a system analysis version of the model plus a representative list of the required activities associated with each phase.

The objective of this model is twofold. First, to demonstrate the need, feasibility, synergy, and potential for a systematic integration of economic (i.e., ECON in Figure 1.) and ergonomic (i.e., ERGO in Figure 1.) analyses which are needed to be performed in any technology transfer transaction. Perpetual collaboration of ergonomists with economists and other officials involved in the economic planning and development of the IDC is a key factor in the future success of those transactions. Ideally, it should start at the design conceptualization stage and continue throughout the life cycle of the project. Second, the proposed model attempts to initiate and establish a preliminary algorithm for collaborative and certainly mutually beneficial efforts of the two involve parties in the transfer process, i.e.; Technology Acquirer and Technology Transferor. Furthermore, due to the dynamic nature of technology transfer

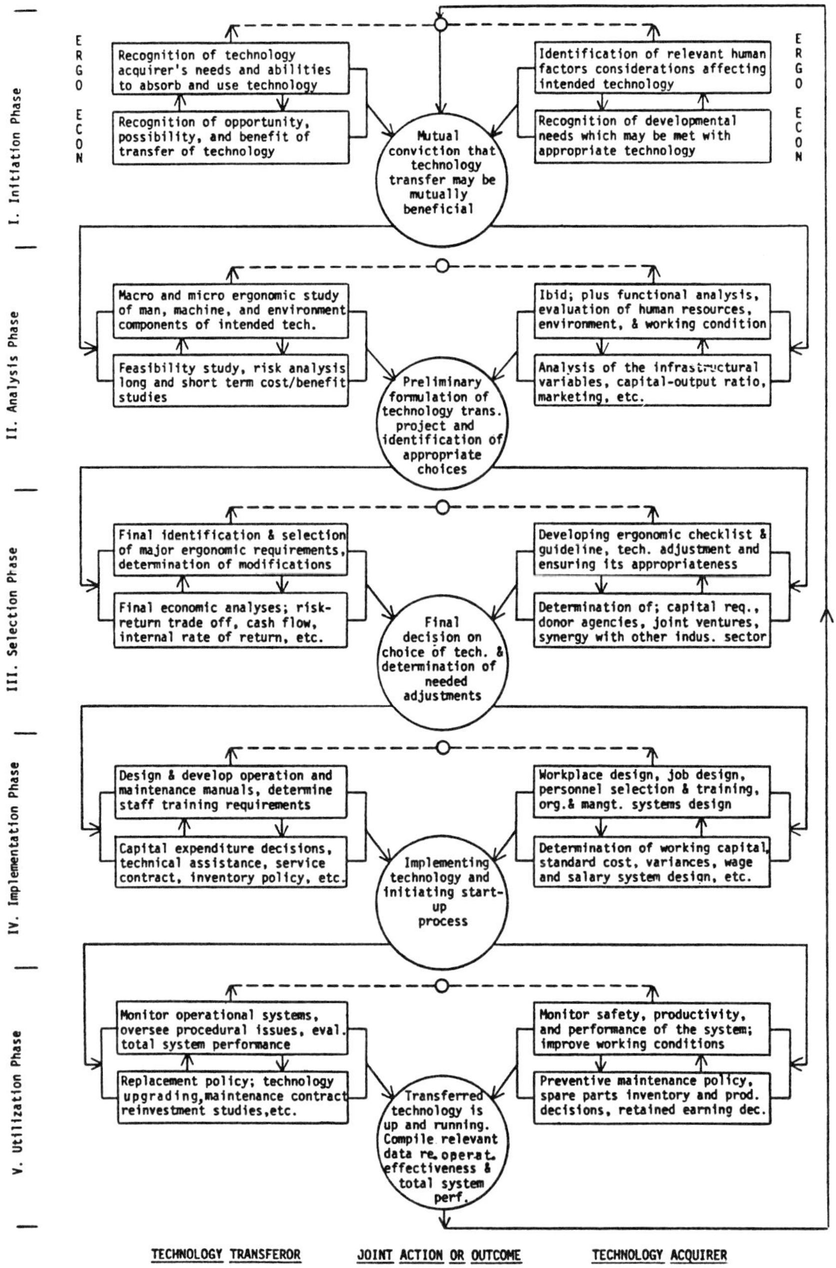

Integrated Technology Transfer Model
Figure 1

process, it is proposed that; since the acqiurer's technology utilization is a continuous process and the further interactions with transferor; indeed at different levels (e.g., technology upgrading, service, spare parts, etc.), are required to keep the project running, then the model should have provisions to also monitor these activities. Thus, a feedback process to facilitate this perpetual effort is an essential and integral element in the model.

This model presents a viable alternative to verify and reify the thoroughness of initial study, subsequent selection, and final utilization of the appropriate technology. Also, it enhances the potential for proactive and on–line diagnosis and further removal of possible anomalies which may be caused by IDC's unpredictable environmental, cultural, and social factors.

REFERENCES

Baranson, J., Economic and social considerations in adapting technologies for developing countries, Technology and Culture, IV (Winter, 1963) 22–29.

Baranson, J., Industrial Technologies for Developing Economies (Praeger Publishers, New York, 1969).

Black, R.P., Mechanism for the Efficient Transfer of Technology to Developing Countries (Stanford Research Institute, Menlo Park, 1972).

Calder, R., After the Seventh Day: The World Man Created (Mentor Books, New York, 1961).

Chapanis, A., National and cultural variables in ergonomics, Ergonomics, 17 (1974) 153–175.

Denver Research Institute, International transfer and acquisition of technology, final report, Prepared for: Agency for International Development (AID), Washington D.C., 1983.

Drucker, P., Management: Tasks, Responsibilities, practices, (Harper and Row, New York, 1974).

Engler, R., Technology out of control, The Nation, (1985) 488–500.

Ferguson, E.S., The American–ness of American technology, Technology and Culture, 20 (January 1979) 3–24.

Evans, D.D. and Nogg Adler, L. (eds.), Appropriate technology for development: A discussion and case histories (Westview, Bolder, Colorado, 1979).

Gould, S.J., The Mismeasure of Man (Norton, New York, 1981).

Haire, M., Ghiselli, E., and Porter, L.W., Managerial Thinking (Wiley, New York, 1966).

Hall, E.T., The Silent Language (Fawcett Pub., Greenwich, Conn., 1959).

Hochberg, J.E., Perception (Prentice–Hall, Englewwod Cliffs, N.J., 1965)

Hofstede, G., Culture's Consequences, International Differences in Work–Related Values (Sage Publications, Beverly Hills, 1980a).

Hofstede, G., Motivation, leadership, and organization: do American theories apply abroad?, Organizational Dynamics, (1980b) 42–63.

Human Factor in Innovation and Productivity, Report by the Subcommittee on Science, Research and Productivity, U.S. House of Representatives (U.S. Gov. Printing Office, 1982).

Ingle, M.D., Appropriate Management Technology: A Development Administration Perspective. Prepared for Agency for International Development, (Practical Concepts Inc., Washington, D.C., 1981).

Kaminaka, M.S., Research needs in American agricultural equipment industry, Applied Ergonomics, 16 (1985) 217–220.

Kluckhohn, C., The study of culture, in: Lerner, D. and Lasswell, H.D. (eds.), The Policy Sciences (Stanford University Press, California, 1951).

Kogi, K., Improving working conditions in small enterprises in developing Asia (International Labor Office, Geneva, 1985).

Kraut, A.I., Some recent advances in cross–cutural management research, Academy of Management Journal, 18 (1975) 538–549.

Krishna, The economic development of India, Scientific American, 243 (1980) 118–193.

Kubr, M. and Wallace, J., Success and Failures in Meeting the Management Challenge, Strategies and Their Implementation. World Bank Staff Working Papers, No. 585 (The World Bank, Washington, D.C., 1983).

Lyell, 1855, C., A Second Visit to North America, 3d ed. (London,

1855).

Magee, S.P., The appropriability theory of the multinational corporation, in: Heston, A.W. and Pack, H. (eds), Technology Transfer: New Issues, New Analysis (The Annals of The American Academy of Political and Social Science, Vol. 458, 1981).

Meister, D. and Rabideau, G.F., Human Factors Evaluation in System Development (Wiley, New York, 1965).

Meshkati, N., Notes and observations on Iranian industries, Personal Archives (1976).

Meshkati, N., A conceptual model for the assessment of mental workload based upon individual decision styles, Unpublished Ph.D. Dissertation, University of Southern California (1983).

Meshkati, N. and Driver, M.J., Individual information processing behavior in perceived job difficulties: A decision style and job design approach to coping with human mental workload, in: Hendrick, H.W. and Brown, O. (eds.), Human Factors in Management and Organizational Design I (North– Holland, Amsterdam, 1984).

Meshkati, N. and Robertson, M.M., Individual differences in experiencing mental workload: A guide for cockpit design evaluators. Proceedings of the Eight Annual Meeting of the Los Angeles Chapter of the Human Factors Society, (Human Factors Society, Los Angeles, June 1985).

Meshkati, N. and Robertson, M.M., The effects of human factors on the success of technology transferred projects to industrially developing countries: A review of representative case studies, in: Hendrick, H. W. and Brown, O. (eds.), Human Factors in Organizational Design and Management II (North Holland, Amsterdam, 1986).

National Research Council, U.S. Science and technology for development: A contribution to the 1979 U.N. Conference; a background study on suggested U.S. initiative for the U.N. Conference on Science and Technology for Development, Vienna, (1979).

Negandhi, A.R., Modern Organizational Behavior (The Kent University Press, Ohio, 1973).

Organization for Economic Cooperation and Development (OECD), Transfer of technology for small industries. Proceedings of Seminar

Held at Schloss Hernstien, (Austria, 5–8 July, 1973).

Perrow, C., Normal accidents, living with high–risk technologies, (Basic Books, Inc., New York, 1984).

Pierce, B.F., The ethnic factor in the man–machine relationship, General Dynamics Corporation, (San Diego, California, 1963).

Sediono, S, Transfer of technology in Indonesia, in: Technology to Improve Working Conditions in Asia (International Labor Office, Geneva, 1979)

Sen, R.N., Certain ergonomic principles in the design of factories in hot climates, Proceedings of the International Symposium on Occupational Safety, Health and Working Conditions, Specifications in Relation to Transfer of Technology to the Developing Countries, International Labor Office, Geneva (1982) 123–147.

Sen, R.N., Applications of ergonomics to industrially developing countries, Ergonomics, 27 (1984) 1021–1032.

Shahnavaz, H., Ergonomics and industrially developing countries, Center for Ergonomics of Developing Countries (CEDC), Institute of Human Work Sciences, Lulea University, Sweden (1985).

Sinaiko, W.H., Verbal factors in human engineering: Some cultural and psychological data, in: Chapanis, A. (ed.), Ethnic Variables in Human Factors Engineering (John Hopkins University Press, Baltimore, 1975).

Spence, J.T., Achievement American style: the rewards and costs of individualism, American Psychologist, 40 (1985) 1285–1295.

Stewart, F., International technology transfer: Issues and policy options, World Bank Staff Working Paper, No. 344, (The International Bank of Reconstruction and Development, Washington, D.C., 1979).

Stewart, F. Arguments for the generation of technology by less–developed countries, in: Heston, A. and Pack, H. (eds.), Technology transfer: New issues, New Analysis (The Annals of the American Academy of Political and Social Science, Los Angeles, Vol. 458, 1981).

Sudomo, Keynote address at the International Symposium on Ergonomics in Developing Countries, Jakarta, Indonesia, (November, 1985).

The Conference Board, International Transfer of Technology, A Worldwide Survey of Chief Executives (The Conference Board, Inc., New York, 1975).

United Nations Conference on Trade And Development (UNCTAD), Guidelines for the Study of the Transfer of Technology to Developing Countries, A study by the UNCTAD Secretariat, (United Nations, New York, 1972).

United Nations Conference on Trade And Development (UNCTAD), Transfer of Technology: Its Implications for Development and Environment, Study by the UNCTAD Secretariat, (United Nations, New York, 1978).

Weltman, G., Cultural biotechnology: A letter from Israel, Human Factors Society Bulletin, 6, (May 1963) 1–3.

Wickens, C.D., Engineering Psychology and Human Performance (Charles E. Merrill Pub. Co., Columbus, 1984).

Wigglesworth, D.C., Auditing technology transfer, Training and Development Journal, (October, 1985) 42.

Winter, W., The perception of safety posters of Bantu industrial workers, Psychologica Africana, 10 (1963) 127–135.

Wirahadikusumah, U., Keynote address at the International Symposium on Ergonomics in Developing Countries, Jakarta, Indonesia (November, 1985).

Wisner, A., Ergonomics in industrially developing countries, Ergonomics, 28 (1985) 1213–1224.

World Bank, Technology and science in World Bank operations, (The International Bank for Reconstruction and development, Washington, D.C., 1982).

World Health Organization and International Labor Office, Report to the government of Iraq on the work of the ILO/WHO multidisciplinary mission, International Labor Office, Geneva (1978).

Wyndham, C. H., Ergonomics problems in the transition from peasant to industrial life in South Africa, in: Chapanis, A. (ed.), Ethnic Variables in Human Factors Engineering (The Johns Hopkins University Press, Baltimore, 1975).

Human Factors in Organizational Design and Management - II
O. Brown, Jr. and H.W. Hendrick (Editors)

ATMOSPHERE, DOGMATISM AND LOCUS OF CONTROL: COMPARISONS OF PSYCHIATRIC FACILITIES IN THE U.S.A. AND INDIA

Devendra Bhagat
Timpanogos Behavior Control

Bruce L. Brown and Warner P. Woodworth
Brigham Young University

Provo, Utah
U.S.A.

This study investigates pre and post test attitudes of patients and staff in a U.S. psychiatric ward which went through a process of reorganization. Researching this as a natural experiment, data were collected on the chaos of change and the reestablishment of order and control one year later. Cross cultural comparisons were also obtained with staff and patients of a similar facility in India using scales which measured several variables--dogmatism, ward atmosphere, and internal/external locus of control. Significant similarities were found as well as marked differences between the two cultures.

INTRODUCTION

An understanding of human behavior, whether normal or maladaptive, involves not only individual dynamics, but also the milieu in which health or illness is developed. The role that the treatment environment plays in the adjustment of mental patients has come into sharp focus with the advent of "milieu therapy" and "therapeutic community." Both are concerned with enriching the treatment environment to facilitate patient growth. In the field of mental health and also in business, there have been important developments with respect to optimizing the environment.

The "therapeutic community" (Jones, 1953) has been one very important step in this direction in the context of psychiatric units. So too has been the concept of "job redesign" (Hackman and Oldham, 1980) in the context of business organizations. Concepts such as unconditional regard, genuine understanding, communication, support, autonomy, program clarity, appropriate confrontation, and soft control have become key phrases in describing the therapeutic environment in the wards of state hospitals or other therapeutic community settings. Concepts such as skill variety, task identity, task significance, autonomy, feedback, democracy, job enlargement and redesign, recognition, and achievement have become key phrases for job enrichment in business.

Lehman and Ritzler (1976) have referred to this approach to mental health as the "community model", in contrast to the "medical model". Various studies (Moos & Houts, 1970; Sidman & Moos, 1973; Moos, Shelton & Petty, 1973; Lehman & Ritzler, 1976) have been conducted to assess the relative effectiveness of these two models. As a general summary, it can be said that the community model seems to be clearly superior when one measures variables related to patient satisfaction and personal growth, but a more rigid orientation produces higher ratings on such things as orderliness and clarity of ward expectations. Also, readmission rates are higher for the more open wards, which is not really too surprising when one considers that patients view such wards as more pleasant places to be.

THE STUDY

Undoubtedly, the personal attitudes and beliefs of the staff of a psychiatric facility would have a great deal to do with their preference for a community model as compared to a more traditional medical model. Conversely, one might expect that the social climate that develops in such a facility would have some impact upon the attitudes and beliefs of the staff as well as those of the patients. For example, scores on a measure such as Rokeach's (1960) dogmatism scale would be expected to be quite related to ward atmosphere. That is, all else being equal, one would expect that a psychiatric facility would tend to be more democratic if it is conducted by staff with low dogmatism scores. On the other hand it would be more autocratic and restrictive if conducted by staff who are high in dogmatism.

To a much lesser extent one might also expect that a democratic climate in such a facility would have a tempering effect upon the dogmatism of both staff and patients. It might also be argued that a more democratic ward climate could help patients (and perhaps even staff to some extent) to become more inner directed, and thus receive lower scores on Rotter's (1966) I-E scale which measures internal/external locus of control.

One purpose of this study was to take advantage of a major reorganization of a psychiatric facility in the U.S.A. as a kind of "natural experiment" to measure dogmatism, internal vs. external locus of control, and ward atmosphere (Moos & Houts, 1968) both before and after the reorganization took place. This psychiatric facility was measured on a number of dimensions (including these three) both in 1979, before the reorganization, and also in 1980, thirteen months after reorganization was complete. This natural experiment would have been ideal if the intent of the reorganization had been to move from a traditional model to a community model. But this particular reorganization was prompted more by local administrative needs rather than a major change in philosophy.

However, one would expect any major organizational realignment to have measurable effects upon perceived ward atmosphere and perhaps even upon the attitude-belief measures of dogmatism and locus of control. It was hoped that something could be learned from a comparison of the relative magnitude of effects of the reorganization upon each of these variables. In order to assess the impact of dogmatism upon ward atmosphere, we decided to compare this facility in the U.S.A. with a similar psychiatric facility in India. Our rationale was that the Indian facility would be considerably more authoritarian in structure, because of an hypothesized lag of development in psychiatric attitudes in Third World countries.

Rokeach's D Scale, Rotter's I-E Scale and Moos and Houts' Ward Atmosphere Scale were administered to staff and patients at the psychiatric facility in the United States both before and after the reorganization. The same scales were administered to staff and patients at the psychiatric institution in India. Data were gathered form 103 respondents at the U.S. facility before reorganization, of which 52 were patients, 35 were nonprofessional staff and 15 were professional staff. Thirteen months after reorganization of the wards, the scales were administered to 123 respondents, 56 of which were patients, 46 nonprofessional staff and 21 professional staff. In the Indian psychiatric facility, the scales were administered to 32 patients, 40 nonprofessional staff, 26 female nurses and 4 professional staff.

FINDINGS

A summary of results is provided in Table 1. The means were calculated for each of the subgroups in the Indian facility and the U.S. facility, before and after reorganization, on each of the ten subscales of the Ward Atmosphere Scale. The most obvious finding was that all three groups (patients, non-professional staff and professional staff) of respondents in India scored much higher on the dogmatism scale than their counterparts in the U.S.A. The mean Rokeach D scale scores for the patients, nonprofessional staff and professional staff from India were 221.9, 186.4, and 162.0 respectively, whereas the corresponding D scale scores for these three categories in the U.S. facility were 161.5, 146.6 and 127.3. The differences between the United States mean and the India mean for each of these corresponding groups were all significant beyond the .01 level, with F-values of 19.53, 9.34 and 11.88 for patients, nonprofessionals and professionals respectively (with 222 degrees of freedom in the pooled error term).

Similarly, the mean values on the Rotter I/E scale were higher for all three Indian groups than for the corresponding U.S.A. groups (with a high score indicating a more external locus of control). Patients, nonprofessionals and professionals in India had means of 10.28, 10.25 and 10.03 respectively while the corresponding U.S. groups had means of 9.06, 7.15 and 7.31 respectively. However, these differences were only significant for two of the three groups, nonprofessional (F = 4.94) and professional staff (F = 6.25). The difference for patients was not statistically significant.

Respondents in the U.S. facility came from three different wards which were quite different in their composition. The first ward consisted of those who were chronic and needed long term care, the second those who were believed to be capable of responding to a behavior modification program, and the third those who needed only short term care. In order to assess the interactive effects of the reorganization, the ward of the respondent, and the status of the respondent (patient, nonprofessional staff or professional staff), a 2 X 3 X 3 three-way analysis of variance was computed for the dogmatism scale scores and for the locus of control scale scores. In the three-way analysis of variance for dogmatism, none of the two-way or three-way interaction terms were statistically significant. As shown by the means referred to above, the effect was for patients to be the highest in dogmatism, professional staff the lowest, and nonprofessional staff in the middle. None of the simple effects and none of the interactions were significant for the three-way analysis of variance of locus of control scores. The only factor that was close to being significant was status of the respondent, with patients having a higher mean I-E score (indicating an external locus of control) than either the professional or the nonprofessional staff.

Table 1

MEANS FOR THE RESPONSES OF DIFFERENT SEGMENTS OF THE STATE HOSPITAL POPULATION IN U.S.A. AND INDIA ON EACH OF THE TEN SUBSCALES OF WARD ATMOSPHERE SCALE

	Involvement	Support	Spontaneity	Autonomy	Practical Orientation	Personal Problem Orientation	Anger and Aggression	Order and Organization	Program Clarity	Staff Control
1979 USH Patients:										
ward 1	5.65	5.82	4.65	5.59	6.35	4.18	4.71	6.12	5.47	6.29
ward 2	7.50	6.08	4.50	5.58	5.75	3.75	4.92	6.83	5.92	6.33
ward 3	4.82	5.00	3.45	5.09	6.36	4.82	6.18	4.82	5.09	4.73
1980 USH Patients:										
ward 1	6.59	6.24	4.35	5.18	6.41	5.65	5.41	6.71	5.94	5.47
ward 2	6.07	5.86	4.21	5.14	6.50	4.86	5.43	6.71	5.50	5.07
ward 3	7.62	6.31	4.15	5.85	7.23	5.85	6.08	6.92	5.46	6.08
1979 USH Nonprofessional:										
ward 1	6.10	6.00	4.70	5.70	5.30	6.30	7.00	4.50	5.80	3.80
ward 2	6.75	6.75	4.00	5.50	7.00	6.00	7.25	5.25	4.50	3.75
ward 3	6.93	6.29	5.07	6.21	7.21	5.57	6.14	5.21	5.57	4.07
overall	7.00	4.00	5.00	5.00	6.00	8.00	7.00	6.00	7.00	5.00
1980 USH Nonprofessional:										
ward 1	6.09	6.64	5.36	5.27	6.18	6.00	5.64	6.00	6.73	3.82
ward 2	5.50	6.57	5.79	5.36	7.71	4.36	5.79	5.57	5.79	3.64
ward 3	7.93	6.93	4.79	6.64	7.57	6.64	6.50	6.86	11.71	3.21
overall	6.75	5.25	4.75	6.00	7.25	4.75	4.50	6.75	6.50	4.00
1979 USH Professional:										
ward 1	4.63	5.25	3.63	5.50	6.38	5.50	6.25	3.00	5.13	5.38
ward 2	5.00	6.50	4.50	6.50	6.50	6.00	7.00	6.50	5.00	3.00
ward 3	8.25	8.50	6.00	6.25	7.75	8.25	6.75	6.25	6.25	3.00
overall	6.00	5.75	4.25	7.25	7.50	6.25	6.50	5.50	6.25	4.00
1980 USH Professional:										
ward 1	7.00	6.00	4.67	6.17	7.00	5.50	7.33	5.00	5.50	4.50
ward 2	5.40	6.20	5.60	7.20	8.20	5.00	6.60	5.40	5.60	3.20
ward 3	8.40	8.60	6.00	6.40	8.00	7.20	5.40	7.60	7.20	6.00
overall	6.00	7.17	4.50	6.17	8.67	6.83	6.17	8.00	7.50	4.17
1983 RMA (India):										
male patients	4.71	5.93	4.57	3.71	5.43	4.86	4.79	5.93	5.21	6.36
female pat.	4.72	5.56	4.78	5.39	5.39	5.11	5.06	5.17	5.44	5.39
male atten.	6.11	5.30	4.00	3.60	4.50	5.00	4.90	5.60	4.50	5.50
female atten.	4.17	5.33	4.75	7.42	4.58	4.17	5.42	5.75	4.50	5.25
nurses	6.00	7.00	4.20	4.50	6.10	5.70	5.20	6.20	6.10	4.70
profess. staff	5.25	6.00	4.75	3.25	6.50	6.25	7.00	5.50	5.50	4.25

Although the before vs. after reorganization did not have a statistically significant effect upon either the dogmatism scale scores or the locus of control scale scores, it did have a marked and significant effect (on both the multivariate and the univariate significance tests) upon most of the subscales of the Ward Atmosphere Scale.

Figure 1 is a principal components biplot (Everitt, 1978; Brown, Williams & Barlow, 1984) that displays the total multivariate pattern of these results graphically. The factor scores for the mean ratings of ward atmosphere by _patients_ and the mean ratings by _professional staff_, both before and after reorganization, are displayed in Figure 1 for psychiatric facilities in India and the United States. The semantic space showing mean ratings by _nonprofessional_ staff (before and after reorganization in the U.S.) and the nonprofessional and _nursing staff_ at the Indian facility are not displayed here due to space limitations, but the results will be drawn upon for discussion purposes. This factor solution had eigenvalues of 4.18 and 1.73 for Factors I and II respectively, indicating that the biplot contains about 60 percent of the total information in the mean scores on the ten subscales. The biplots are useful to display the total gestalt of the results on the ten subscales, but this overall pattern can also be verified one subscale at a time from the means in Table 1.

The factor structure shown here was graphically rotated from the original factor solution so that the horizontal factor (Factor I) would reflect trust and cohesion, with high trust and cohesion to the left. In this orientation the vertical factor (Factor II) represents a dimension of orderliness and control with "well organized" at the top of the figure and "chaotic" at the bottom.

Wards 1 and 2 would be expected to have more serious problems with maintaining order, since these patients are more chronic and more seriously disturbed. It is interesting to note that the mean ratings by the patients in those two wards are near the top of the figure, indicating that patients see the wards as orderly, controlled and well-organized. The professional staff assigned to these two wards, on the other hand, see them as chaotic suggested by mean ratings toward the bottom of the figure. From the other biplot it appears the nonprofessional staff for these two wards portray their units as chaotic before the reorganization but much more orderly after it. In contrast, the professional staff see the change from before to after as primarily an increase in the direction of more trust and cohesion. The patients in these two wards rate their respective wards about the same before and after the reorganization, but if anything, a little less orderly after.

DISCUSSION

The data of this study are consistent with a number of other researchers (White and Lippett, 1968) who suggest that authoritarian, dogmatic administrative styles may lead toward conformity, submission, and external dependency. A more democratic approach, in contrast, nourishes a milieu or organizational culture which is more open, self-determining, and psychosocially integrated.

The dogmatism scores of respondents in the U.S.A. reflect the fact that the psychiatric facility had experienced a prolonged authoritarian

Figure 1

MEAN WARD ATMOSPHERE SCALE RATINGS
FOR PROFESSIONAL STAFF AND PATIENTS

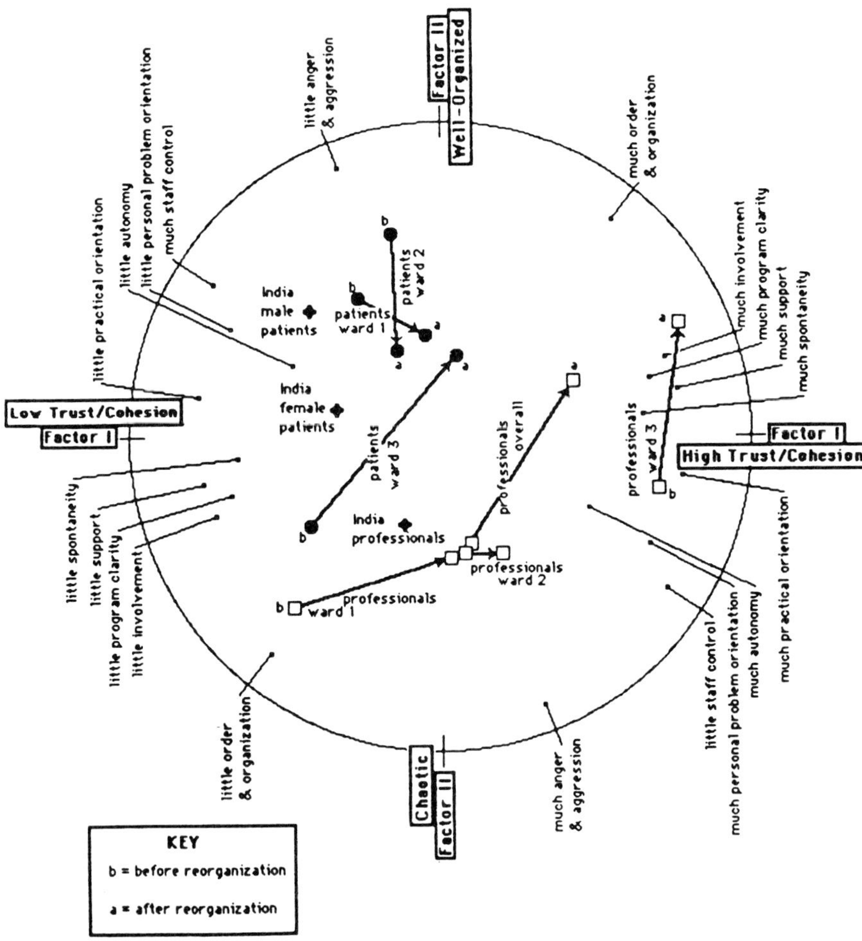

administration for several decades. Despite diluted programs which mirrored the form but not the substance of a therapeutic community, the reality tended to reflect an authoritarian mandate from top management, eliciting organizational incongruence. Democracy, to the extent it existed, was imposed on patients through dogma. The emphasis, especially by non-professional staff, was on maintaining order and control, attending to patient needs only secondarily. This tended to impede the emergence of a therapeutic milieu, creating instead a schism between staff and patients.

With respect to locus of control, the patients among the three groups felt the least internal direction. This externality could be attributed to their life experience, rigid routine, and confinement. Their infrequent short-term freedoms attained through passes and privileges were controlled by the staff, causing them to perceive themselves to be mere pawns, shuffled about by outside authority figures.

On site observations of the origin and outcome of reorganization suggest additional insight into the problem of control. The new administrator of the facility met with considerable staff resistance when he first took over. Consistent with the argument of Diamond and Allcorn (1985), this administrator may have been attempting to alleviate his own anxiety by reorganizing and reshuffling the hospital wards, thereby exerting control over the staff. To some degree, he seemed to act out of his own frustration, overcoming staff resistance by placing conforming subordinates in the right places. As can be seen from the biplot, the patients bore the brunt of this strategy. Apparently the staff resorted to alleviating their own anxiety by expanding their degree of control over the patients.

CROSS CULTURAL IMPLICATIONS

Marked differences between Indian respondents and U.S. counterparts are not surprising. The American staff reflects an overall sophistation in terms of knowledge of mental heatlh, various treatment modalities, and cultural tendencies toward status quo and stability. With respect to dogma, for instance, all three groups from Indian are not as open-minded as those from the U.S. Their experience is limited to a very conservative medical model of therapy. The psychiatric faculty has only a limited provision for individual psychotherapy, behavior modification, or group therapy, to say nothing of a more progressive therapeutic community which attempts to achieve psychosocial congruence and effectiveness. Indian culture generally is more authoritarian than is American society, India having only a few decades of experience with democracy.

However, locus of control values did not reveal such consistent differences between the two cultures. Based on apparent disparity in therapeutic environment, one might have expected significant differences between patient groups, but this is not the case. The Indian facility could be characterized as a psychonoxious milieu. To a large extent patients are forgotten people, simply dumped there, shunned by relatives, maintained by staff-administered drug treatment. Many suffer from learned helplessness (Maier and Seligman, 1976), unable to extricate themselves from an antitherapeutic situation because they have no other options. Lacking a concerned and/or resourceful family and in the absence of halfway houses or other support systems, they eventually tend to stop trying any further. In short, they have no sense of inner control. The American experience is largely different, yet the patient locus of control is not significantly different. Perhaps this intimates that while the psychonoxious milieu in India is quite overt, the U.S. case is equally brutal in outcome. But it is camoflaged by outward trappings which make it seem more humane. This hypothesis is additionally supported by comparing ward atmosphere across cultures. The biplot clearly suggests bright contrasts of both staffs and patients between the U.S.A. and India.

CONCLUSION

This study tracks changes after psychiatric facility reorganization. As a general observation, all staff see improvement one year after, either in terms of more orderliness or more trust/cohesion. Of the patients, only those in ward 3 agree that there has been such improvement. As a general cross-cultural observation, respondents from India, whether patients or staff rate their own facility as lower in trust and cohesion, but higher in orderliness and staff control than those from the U.S. facility rate theirs. This fits with the finding of much higher dogmatism scores across the board in the Indian respondents. It is important to note as an additional observation that the reorganization had a marked effect upon perceptions of ward atmosphere, but not on the more enduring predispositions of staff and patients as measured by the dogmatism scale and the locus of control scale.

The study delineates therapeutic variables both in the treatment environment and in the sphere of management. It suggests several human factors which may block effective treatment of patients in psychiatric facilities, and outlines elements to be used in redesigning the sociopsychological environment. A genuine therapeutic community model enriches the qualtiy of life for mental patients. It not only brings them greater personal satisfaction, but expedites their improvement and rehabilitiation as well.

REFERENCES:

Brown, B. L., Williams, R. M. and Barlow, C.D., PRIFAC: A Pascal factor analysis program, Journal of Pascal, Ada and Modula 2. 3 (1984) 18-24.

Diamond, M.A. and Allcorn, S. Psychological dimensions of role use in bureaucratic organizations, Organizational Dynamics, 14 (1985) 35-59.

Everitt, B. S., Graphical Techniques for Multivariate Data. (North Holland, New York, 1978).

Hackman, R. and Oldham, G. R., Work Redesign. (Addison-Wesley, Menlo Park, CA., 1980).

Jones, M., Therapeutic Community. (Basic Books, New York, 1953).

Lehman, A. and Ritzler, B., The therapeutic community inpatient ward: Does it really work? Comprehensive Psychiatry, 17 (1976) 755-761.

Maier, S. F. and Seligman, M.E.P., Learned helplessness: Theory and evidence, Journal of Experimental Psychology, 105 (1976) 3-46.

Moos, R. H. and Houts, P. S., Assessment of the social atmospheres of psychiatric wards, Journal of Abnormal Psychology, 73 (1968) 595-604.

Moos, R. H. and Houts, P. S., Differential effects of the social atmosphere of psychiatric wards, Human Relations, 23 (1970) 46-60.

Moos, R. H. Shelton, R. and Petty, C., Perceived ward climate and treatment outcome, Journal of Abnormal Psychology, 82 (1973) 291-298.

Rokeach, M., The Open and Closed Mind. (Basic Books, New York, 1960).

Rotter, J. B., Generalized expectancies for internal versus external control of reinforcements, Psychological Monographs, 80 (1966) 609.

Siman, J. and Moos, R., On the relation between psychiatric ward atmosphere and helping behavior, Journal of Clinical Psychology, 29 (1973) 74-78.

White, R. and Lippitt, R., Leader behavior and member reaction in three 'social climates,' in Group Dynamics: Research and Theory by D. Cartwright and A. Zander (eds.) (Harper and Row, New York, 1968).

TRANSLATING RESEARCH ON NEW TECHNOLOGY INTO PUBLIC POLICY

Human Factors in Organizational Design and Management - II
O. Brown, Jr. and H.W. Hendrick (Editors)

TRANSLATING RESEARCH ON NEW TECHNOLOGY INTO PUBLIC POLICY TRENDS & ISSUES IN JAPAN, SWEDEN AND THE USA

Panel Convenor: Benjamin C. Amick, III
Office of Technology Assessment
Congress of the United States
Washington, DC 20510

While taught in many courses that research should contribute to the development of a rational public policy, little effort is given to the types of research that should or that are conducted to inform public policy issues. Alternatively, many people do not consider the organizational issues involved in implementing policies concerning new technology. Two consecutive panel sessions have been developed to address various apsects of the the relationships between research and policy. To more fully elaborate the relationship each panel session will allow for the cross-cultural evaluation of the trends.

The first panel will address the more general issue of how we can design research to guide a humane approach to public policy. This assumes that at the level of the national government there is variability in how this issue is addressed and the types of research that are utilized. The three countries represented are commonly held to have different approaches; Japan, Sweden and the USA.

Panel I: TOWARDS A MORE HUMANE PUBLIC POLICY

- Sadao Sugiyama, Professor, Kansei Gakuin University, Japan
- Elizabeth Lagerlof, Work Environment Attachee, Swedish Embassy, Washington, DC
- Tapas Sen, AT&T, USA

A second panel will be more focused and address the issue of "VDT Regulations". At this point in time this is one of the major areas of policy activity. Again, there are differences between the three countries stated above. In this session the government official responsible for the development of the VDT regulations in each country will speak about the role of research and the researcher in the development of the regulations.

Panel Session II: VDT REGULATIONS AS PUBLIC POLICY

- Kiyoshi Kajikawa, The Bureau of Labor Standards, The Ministry of Labor, Japan
- Ricardo Edstrom, National Board of Occupational Safety & Health, Sweden
- Roger Stevens, Occupational Safety and Health Administration

PARTICIPATORY ERGONOMICS, QUALITY OF WORK LIFE AND ORGANIZATIONAL EFFECTIVENESS

Human Factors in Organizational Design and Management - II
O. Brown, Jr. and H.W. Hendrick (Editors)
Elsevier Science Publishers B.V. (North-Holland), 1986

JAPANESE QUALITY CONTROL: IMPLICATIONS FOR JOB AND ORGANIZATION DESIGN

Steven L. Dockstader

Navy Personnel Research and Development Center
San Diego, California 92152-6800

The Japanese approach to total quality control (TQC) is examined regarding the impact it would have upon job and organization in American companies. Sources of resistance to control systems (Lawler, 1976), and requirements for systems change to meet the requirements of TQC (Metz, 1984) are discussed. Job characteristics theory is used as a conceptual framework to determine the motivating potential of jobs created by adopting the TQC approach.

INTRODUCTION

The purpose of this paper is to examine the characteristics of the Japanese approach to total quality control (TQC), and to determine whether they can effectively be adopted by U.S. organizations. This question was prompted by questions arising in the Department of Defense as to whether the methods used by the Japanese would be applicable to maintenance and service organizations in the public sector. The paper will examine the following questions or issues:

1. What are the characteristics of TQC?
2. What are the principal sources of resistance to TQC?
3. What are the implications of TQC for job and organization design?
4. Can jobs be enriched, and organizations designed, in ways to overcome these sources of resistance?

THE BASIC CHARACTERISTICS OF TOTAL QUALITY CONTROL

The concept of total quality control (TQC) represents an integration of approaches from marketing analysis, process control, and statistical quality control into a comprehensive system of management. Armand Feigenbaum (in Ishikawa, 1985) provided an early and still currently used definition:

> "an effective system for integrating the quality development, quality maintenance, and quality improvement efforts of the various groups in an organization so as to enable production and service at the most economical levels which allow for full customer satisfaction"

Ishikawa goes on to say that quality of product is too narrow a concept and has limited the full elaboration of the total quality management concept in the West. He points out that:

> "Broadly interpreted, quality means quality of work, quality of service, quality of information, quality of process, quality of division, quality of people, including workers, engineers, managers, and executives, quality of systems ... and objectives. To control quality in its every manifestation is our basic approach"

One can begin to understand the meaning of "total" from Ishikawa's definition. Unfortunately, however, these concepts have either been too subtle or their virtues have not been in evidence to the degree required for U.S. organizations to have adopted this management approach. At the heart of total quality control are the methods of statistical quality control. The fundamental objective of these techniques is to achieve control over the process which produces the product or service, such that quality defects or unsatisfactory services never occur. This quality control process can be illustrated in Figures 1 and 2.

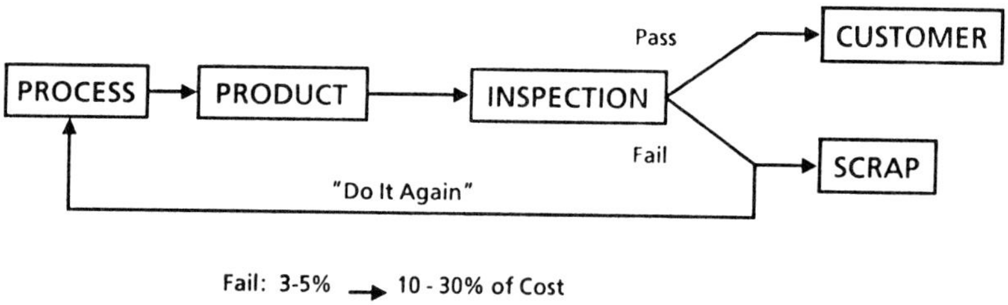

Figure 1
A Flow Diagram of the Inspection Approach to Product Quality Control

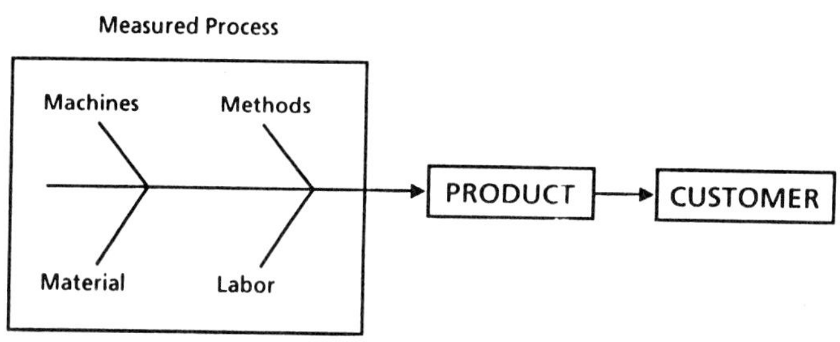

Figure 2
A Flow Diagram of the Process Control Approach to Product Quality

Figure 1 depicts the typical case in the manufacturing environment. Here, products are produced, some of which are defective and, in order to meet demand, reprocessing is necessary or, in the case of major defects, the product is lost to scrap. In Figure 2, the scenario does not typically include inspection because the components of the process which can produce defects are monitored and controlled. These process components are identified here as four, but these are just the major categories of variables that are measured and controlled using this approach.

Getting back to the definition of "total", the idea is that process quality controls are directed at all of the significant processes in an organization, not just those principally identified with manufacturing. In addition, the philosophy is extended to management processes as well--so that the general strategy is one that can involve all employees at all levels in the organization.

The primary reason for attempting such a comprehensive approach can be found at the bottom of Figure 1: the cost of maintaining quality using the inspection approach can be as high as 30% of the product cost. This was (and is) a particularly critical factor for the Japanese because they have virtually no natural resources and could not afford to scrap any material that would result from faulty processes. When they found that process control methods also lowered labor costs (no reprocessing, less process problems to deal with), this convinced them to apply the approaches outside the manufacturing environment. The rest is history.

SOURCES OF RESISTANCE TO THE TQC APPROACH

Lawler (1976) has analyzed the nature of control systems in organizations and the kinds of resistances to their development. In his analysis, he points out that resistance to control systems is most likely when:

1. The control system measures performance in a new area.
2. The control system replaces a system that people have a high investment in maintaining.
3. The standards for control are set without participation.
4. Control system feedback does not go to those who are measured.
5. Control system feedback goes to higher levels in the organization and is used in the reward system.
6. People affected by the existing system are relatively satisfied and see themselves committed to the organization.
7. The people affected by the system are low in self-esteem and authoritarianism.

Using Lawler's analysis, we can examine the resistance to change from the product inspection approach to the process control approach. Following that, an argument will be made on the basis of contemporary work motivation theory as to why the process control approach should be very successful, once implemented.

In the following discussion, Lawler's seven sources of resistance will be considered in turn for the operations of a large aircraft overhaul facility in the Navy. Although the observations made here are not based upon empirical data, they have been corroborated by senior managers in that organization and by on-site research personnel.

1. The process control approach does measure performance in new areas. In fact, the essence of the approach involves measuring several significant features about the production process prior to completion of a product or service.

2. Ultimately many of the personnel who are currently used in the quality control department will be deployed to other parts of the organization, or be conducting quality control activities not currently being performed (e.g., incoming supplies, customer services, etc.). This displacement and/or retraining of personnel is viewed as a threat by those in the current quality control function.

3. Quality control standards are usually established by engineers or quality technicians. In the case of process control, however, a fixed standard has no meaning. Control is defined by taking actions to keep the process within variability limits which are determined by the process itself. Because the limits change as a function of improvements in the system, no fixed standard can be applied.

4. Using the process control approach, the basic data is collected by the performer. In this sense, feedback is immediate. Furthermore, because the information gathered is typically a historical record with relational information on the record (e.g., a control chart), the worker can evaluate the data and determine what actions, if any, need be taken.

5. Whether the data is fed to higher levels and used within the reward system depends upon a number of factors. The most significant is the degree to which the worker has discretion to make decisions concerning corrections to the system. This, in turn, is usually based upon the extent of the system changes and their costs, but could also be a reflection of the management philosophy of the organization. This will be considered in greater detail in a subsequent discussion.

6. This is the "status quo" factor, and it can be said that a change in the inertial state of the organization will be determined by whether or not a "critical mass" (Deming, 1985) can be developed to overcome the status quo. The state of inertia in most bureaucracies, such as those in most large bureaucratic organizations, is at steady state and resistant to change under normal workload conditions.

7. It is difficult to assess this factor. The people most affected by the quality control system are those in the "production" area. As a group, they are the largest in number and exert the greatest influence on achieving the mission of the organization. However, under the current product inspection approach, they receive the most censure when product quality does not meet specifications/test. Managers have been of the opinion that this has led the workers to lose identity with the quality of their products because someone else has been responsible for detecting it.

Lawler has indicated that, to the extent that these factors hold for workers, they will engage in non-productive or even counter productive behaviors. Using his analysis and the previous discussion, it appears that the process control approach should meet less resistance in terms of factors 3, 4, 6, and 7. That is to say that it (a) does not deal with standards per se, (b) provides feedback information to the performer, (c) is of greater benefit to most of the work force than the existing system and (d) can enhance the self esteem of the worker as he begins to take charge of the quality of his work.

Of the other factors, only the second appears to be of significant concern in terms of resistance to change. In the organization under study, quality control is vested in a functional department. While performing inspections or audits of the work conducted in the production area is not their only function, it does define their central raison d'etre. In addition, this organization is one of several which reports to a headquarters. Both the headquarters and the sister organizations contain quality control functions based upon inspection and audit. Resistance here would have to be overcome.

Factors 1 and 5 are potentially areas of resistance because of the new measures and added work required (1) and because the information could be used to evaluate performance of the workers (5). Neither of these is necessarily negative, but the work force is often wary concerning the use of performance measures. If the management philosophy and the culture of the organization is one that rewards improvement then there will be little resistance.

PROCESS CONTROL AND WORKER MOTIVATION

Our discussion thus far has focused on the desirability of changing from product inspection to process control and the nature of resistances in making such a change. While it appears obvious that such a change is both desirable and feasible from a management standpoint, what is in it for the worker? After all, with the exception of some of the existing quality control personnel, the major job changes will be that of the worker and perhaps his immediate supervisor. If this change is not seen by the worker as having incentive value, then it will very likely be resisted.

Job Characteristics Theory (Hackman & Lawler, 1971; Hackman & Oldham, 1976) provides a conceptual framework to evaluate the design of a workers job to include the process control approach to quality control. The theory is based upon a plethora of research which has revealed that there are three psychological states which contribute to worker motivation. These are feelings of meaningfulness, responsibility, and knowledge of results. The theory goes on to describe what job characteristics will result in these feelings. The theoretical relationships can be schematicized as follows:

Table 1
A Model of the Effects of Job Characteristics
(After Hackman & Oldham, 1975)

Five Core Characteristics	Psychological States	Outcomes
Skill Variety Task Identity Task Significance	Feeling of meaningfulness	High intrinsic motivation High quality work High satisfaction Low absenteeism and turnover
Autonomy	Feeling of responsibility	
Feedback	Knowledge of results	

Considering each of these characteristics in turn, we can determine the motivating potential, or incentive, of the workers job when process control becomes a part of the job. Skill variety is obviously increased because the job will now involve collection of data, charting of data, and reporting process aberrations. Task identity should increase because attention will be focused on aspects of the process which were previously receiving less formal, e.g., measurement, attention. The perceived significance of the task may also be enhanced because taking process control actions should occasion interaction with supervisors, staff, and managers which would not ordinarily occur. Autonomy will be increased because quality control actions and responsibilities will now be formally placed in the hands of the worker. Finally, feedback will be immediate in terms of the things being measured. Feedback as a result of process changes will, in most cases, be immediate as well.

From this logical analysis and the model displayed in Table 1, we can predict the outcomes displayed there. During the course of the forthcoming year, these hypothetical relationships will be tested in the Navy maintenance environment. The use of process control as a method to enrich jobs has not received attention in the empirical literature, but the aforementioned analysis suggests that it should be an effective way to motivate workers as well as increase the quality of their efforts.

SOURCES OF RESISTANCE IN ORGANIZATION AND MANAGEMENT

Job characteristics theory paints an optimistic picture with regard to the success potential of the TQC approach--at least at the level of basic process control. However, unlocking the motivating potential of TQC means more than forming quality circles at the worker level--it means taking a systems view of processes and how they are managed. The remainder of this paper will be directed at exploring four arguments for why, despite the proven benefits of the TQC approach, it will be greatly resisteu in most organizations. These are:

1. Seven of the most fundamental assumptions of TQC are counter to the existing quality assurance culture.

2. For several decades, management development in the U.S. has been based upon financial control systems and those managers currently in power have no stake in change.

3. Most organizations fail to recognize that productivity and quality improvements of the order implied by TQC require an integrated and strategic change process.

4. Except in organizations where an economic threat is perceived, there is no felt need for change.

In his article, Metz (1984) quite correctly argued that TQC requires a holistic organizational view that may involve a cultural change. In work conducted by myself and my colleagues (Dockstader, 1984; Houston, Shettel-Neuber & Sheposh, 1986) the elements of cultural change for a sample of major American corporations and for large Navy organizations were identified as:

1. Changing the production focus to an emphasis on the quality dimension from the current quantity focus.
2. Adopting process control and eliminating inspection.
3. A need to understand the systemic causes of quality problems and to change the attitude that most are the direct result of poor workmanship.
4. Process control decisions should be data based rather than quick reactions to unknown causes.
5. Continuous improvement of processes does not lead to increased cost--the opposite is true.
6. The benefits of quality improvement accelerate with time, so a long term commitment is necessary.
7. Quality is a fundamental responsibility of all employees, not just production workers and/or quality assurance personnel.

In the culture of most U.S. organizations, these "facts" of TQC are resisted because they are counter to existing management styles and beliefs. Deming (1985) explains American reluctance to accept these facts in terms of (a) attitudes formed by management which were largely superstitious in nature and (b) a style of management that has evolved resulting from selecting and promoting managers with financial management training.

Deming's argument concerning superstitious behavior is based upon his observation that the U.S. controlled most of the world markets following World War II and, in this environment, even the most wasteful and non-controlled of production systems were rewarded due to market demand. This was further exacerbated by the fact that the U.S. had tremendous natural resources and a large and willing work force so that any concern with scrap, waste, or even human resources was very low. Any management practice, he says, could succeed in this environment.

His second argument is that during the same period, banking and commerce became increasingly important in our society and, with the boom going on in production, corporate attention moved from production systems to those of financial management. Emphases in education then shifted from engineering and production management to the business schools. With more and more corporate managers being drawn from these ranks, there was less and less talent involved in production management. This situation resulted in an elaboration of financial management systems by people who had little process and product knowledge. This view has been supported by the findings of Tuttle, Sink & DeVries (1984) who have indicated that the model of organizational effectiveness used by most U.S. businesses is one which is driven (backwards) by financial outcomes, rather than by operational objectives.

If Deming's assessment is correct, the preponderance of managers are not knowledgeable concerning the processes upon which their enterprise is based. If this premise is true, then they are probably driven by demand for short term profits and evaluated in terms of short term gains which is at odds with a TQC philosophy that views quality improvement as a long term and continuous activity. In addition, the audit orientation of financial managers would probably find valid the idea of inspection the only "real" way to assure quality. Finally, the annual system of appraisal which has rewarded this kind of thinking and behavior for the last several decades will continue to promote it--or encourage job hopping when things appear to be on the down side.

In view of these observations, and the pressures that schedule places on the production managers, it is difficult to view the managers job as one that would be enhanced--in the near term--by process control methods. Even though an argument can be made that controlled processes dramatically reduce the myriad of daily problems that face managers--most managers are unaware of the need for or the methods by which to consciously redesign an organizational system or culture. This latter point was made by Metz (1984) in his analysis of the characteristics of approaches for implementing quality improvements. They range from short term "programs" with little management involvement to long term socio-technical interventions with continuing top-level involvement. In his analysis, TQC (identified as statistical quality control) is one of the latter. I have paraphrased, within the context of TQC, the set of prescriptions he makes as the basis for undertaking a redesign effort:

1. Development of a quality management philosophy.
2. Definition of goals and values to introduce clarity, consensus, and commitment to quality.
3. Organization of a transition steering committee which will define and plan for the future state.
4. Development of a strategic change plan which, when fully implemented, would place TQC as the paramount management system.

While these prescriptions make a great deal of sense regarding a "roadmap" for cultural and organizational change, they do not deal with a basic assumption of all change theory--the perceived and felt need for change. Clearly, organizations studied by Houston et al. (1986) all felt the loss of market share and the threat of even greater disaster. But, for organizations not currently threatened by the successes of Japan and those relatively immune to the free market economy (e.g. most public sector organizations) the only hope of overcoming the inertia of history is continuous leadership with great vision.

SUMMARY AND CONCLUSIONS

The feasibility of extending statistical process controls to management of all organizational processes was examined in the context of (a) resistance to control systems and (b) social and historical factors influencing management and organizations. Job characteristics theory was used to explain the motivating potential of such an approach at the level of basic operations, where it was revealed that great potential exists for the enrichment of the job of the great number of workers making up the base of the organizational pyramid. For most managers, however, it is much less obvious how the approach could have the immediate positive impact upon their work, requiring them to re-examine their existing assumptions and approach. Finally, although a roadmap for socio-technical change to the total quality approach can be developed from existing organizational development knowledge, it is not apparent to most managers that the change is required.

REFERENCES

(1) Deming, W.E. Transformation of western style of management, Interfaces, 15 (1985) 6-11.

(2) Dockstader, S.L. What to do when there are more than five deadly diseases. Paper presented at the MIT conference on Managing Systems and People for Quality Improvement, San Diego, 29 June 1984.

(3) Hackman, J.R. and Lawler, E.E. Employee reactions to job characteristics, Journal of Applied Psychology, 55 (1971) 259-286.

(4) Hackman, J.R. and Oldham, G.R. Development of the job diagnostic survey, Journal of Applied Psychology, 60 (1975) 159-170.

(5) Houston, A., Shettel-Neuber, J., and Sheposh, J. Management methods for quality improvement based on process analysis and control. Technical Report, Navy Personnel Research and Development Center, San Diego, CA (1986).

(6) Ishikawa, K. What is total quality control the Japanese way? (Prentice-Hall, Englewood Cliffs, NJ, USA, 1985).

(7) Lawler, E.E., Control systems in organizations, in: Dunnette, M.D. (Ed.) Handbook of Industrial and Organizational Psychology (Rand McNally, Chicago, IL, 1976).

(8) Metz, E.J. Managing change: implementing productivity and quality improvements, National Productivity Review, Summer (1984) 303-314.

(9) Tuttle, T. Sink, S. and DeVries, P. Productivity measurement and evaluation: what is available, National Productivity Review, Summer (1984) 265-287.

Human Factors in Organizational Design and Management - II
O. Brown, Jr. and H.W. Hendrick (Editors)

THE HUMAN FACTORS/APPLIED PSYCHOLOGISTS ROLE IN THE CORPORATE CULTURE IMPROVEMENT PROCESS

Henry E. Bender*
Member Technical Staff
Bell Communications Research
Piscataway, New Jersey 08854

Vincent Conte*
Senior Human Resources Consultant
Insurance Services Office
New York City, New York 10038

Human Factors/Applied Psychologists have enjoyed expanded recognition and acceptance in industry. Their skills have proven useful to Human Resources organizations in conducting employee attitude/opinion surveys. This article reviews key elements and decisions professionals encounter in performing these surveys.

Human Factors/Applied Psychology is an evolving profession. At its inception, practitioners were primarily concerned with systems analysis and the man-machine interface. Work efforts were associated with military systems, through funding derived from the Department of Defense. As the usefulness of such efforts were recognized, government funding began to support non-defense applications, and Fortune 500 corporations recognized the benefit of employing these specialists. They were employed in research and development groups participating in projects requiring systems analysis, man/machine interface, research, training, and applied psychology-psychometric knowledge.

The role of the HF/Applied Psychologist has continued to expand! ODAM's existance attests to the increased involvement of the Human Factors professional in organizational design and management/Human Resource related efforts. In our labor intensive, service oriented work force, the same analytical, research design, and problem solving skills which were primarily exercised in R & D and manufacturing have been recognized as useful in the white collar cultures of organizations across all functional areas.

The acceptance of Human Factor/Applied Psychology's participation in industrial life, especially in the corporate personnel department, has in part, been spurred on by troubling trends in American business. The early 70's witnessed a sharp decline in the productivity of U.S. workers. As Business Week noted: "the decline in the U.S. position in the 1970's alone amounted to some $125 billion in lost production and a loss of at least two million industrial jobs."(Business Week, 1980) "Futurists", Toffler (1980) and Naisbitt (1984) have recognized, as have others, (Starr, 1986) that the key management issue for the 80's remains the reversal of lost productivity and the establishment of "new" corporate systems and processes to ensure productivity gains.

---

*The concepts presented represent the views of the authors, and not their representative organizations.

As America's business leaders, politicians, and educators have become absorbed in this productivity crisis, the unique tools and training of the HF specialist have come to the fore as useful in the search for better systems and work methods. HF practitioners have been enlisted in diagnosing the problems associated with human behavior in the workplace and prescribing changes in the sociotechnical structure underlying the work effort. As a result, special emphasis has been directed to effective selection and motivation of workers. Studies using the skills of the HF/Applied Psychologist have cemented the inroads made by these specialists and seemingly assured their presence in personnel management.

In their role as personnel specialists, applied psychologists have functioned as action researchers. The action research data is often generated by measuring employee affective reactions to the organization or some aspect of the sociotechnical system within the organization. This activity is closely tied to attempts to stem the slide in worker productivity.

In the past few years, employee audits in particular, have received increasing attention in publications. By 1983, Locke (1983) estimated that a minimum of 3,350 articles have been written on the subject of "job satisfaction" alone. Recognition of the importance of the attitude of the worker to his or her place of work has accounted for much of this attention. Lawler and Porter (1967) note that the concept of high job satisfaction being related to high performance is consistent with the value system found in those personing the human relations movement. Further, it was believed that the previous studies in Western Electric supported such an assumption.

Presently, findings are inconclusive. Data exists which support both sides of the issue. For example, Iaffaldano and Muchinsky (1985) state that while the belief that "a happy worker is a productive worker" has been dispelled in research concerning the phenomenon, "there still exists residual support for its veracity." They conclude that satisfaction and performance are, at least, slightly related. Payne and Pugh (1983), in summarizing the research of others notes that: "different climates have had predictable effects on satisfaction." Also, organizational structure may have been related to organizational performance. From these studies, there is the indication that a relationship appears to exist between job satisfaction, organizational structure, and performance; although it is not a strong one.

Presently in an era of rapid organizational realignments due to mergers and acquisitions, the concern with merging corporate cultures has fostered a renewed interest in the area. The rapid increase of mergers and acquisition activity has resulted in major shifts of corporate cultures, with resulting decreases in job satisfaction and employee productivity levels.

Data documenting a worker's tasks/activities, frequently associated with industrial engineering, can be collected using a variety of methods. The most commonly used techniques of in-depth interviews, the Delphi, Nominal and Focus group techniques, work/task observations and questionnaire instruments are described below.

Interviews are useful in obtaining relatively "free form" data. This approach is particularly helpful to researchers who are unfamiliar with an organization and may not have a hypothesized set of organizational problems.

Interviews are ususally based on a small percentage of the corporate work force, due to the length of time required to conduct them. As such, they are prone to sampling bias. Of course, the validity of interview data depends on the "honesty" of individuals interviewed, and the skill level of the interviewer. Interviews can provide rich data, since they allow for two-way communication and selective probing for detail refinement. Often, the resulting interview transcripts are difficult to code for analysis, however (Locke, 1983).

The interview method can be made more cost effective by using a group application of this technique. Group interviews are susceptible to the influence of peer pressure, status differentials and unequal levels of participation among group members. The interviewer must, therefore, apply skill to retain control and direction in group situations.

The Delphi Technique was developed in an attempt to structure a method that had "freedom from group pressure and effects (via anonymity), and consensus (attained via feedback and iteration).(Nelms, 1985) This method requires "expert" participants to be isolated from one another. They are presented with a questionnaire soliciting their opinion and the underlying reason(s) for this opinion. A consolidation of the findings are then recirculated in an attempt to reach some form of consensus (Mac Crimmon, 1983). The success of the Delphi method depends on the initial set of questionnaire items. This requires an expert questionnaire designer and some preliminary time to "research" appropriate items. Also, the isolation inhibits a more subtle level of understanding that might come about through non-verbal or paralinguistic communications channels available in a face-to-face situation.

The Nominal Group Technique requires participants to generate ideas in private; then engage in a "round-robin" meeting in which each offers one idea at a time; discuss key points associated with the problem; and finally; vote to determine the group's answer. This variant of the Delphi method allows for face-to-face interaction and further in-depth understanding of the results obtained. However, decisions might tend to rest more on majority rule than concensus.

Focus Groups are group interview based. A number of "experts" are brought together and under the direction of a practiced leader to discuss the issue at hand. The format is under the control of the group leader who is responsible for maintaining an even discourse with all group members. As a group process, it allows more individuals to participate and is therefore more cost effective than individual interviews. The skill of the group leader is critical to the flow of discussion. Focus groups tend to generate a synergy that operates below the conscious level of thought and, therefore, tap into participants belief structures and values.

Work/Task Observations can be performed by a familiar set of methods described by McCormick (1980). For the present discussion, the observation method seems most appropriate for inclusion here. The objective of this technique is to obtain a sample of behaviors which is representative or descriptive of the individual's general work related behavior. The observer records observations of behavior as they occur. In this case, however, the observer is recording actual behaviors of the individual, and not the affective component of the individual in the work place. As with interviews, the observations are labor intensive and prone to bias.

Questionnaire Instruments require a skilled practitioner to design, develop and implement the technique. Questionnaire studies have the ability to reach a large number of potential participants at a low unit price. This resulted in a vast literature about the method. Bouchard (1983) has listed 21 questions that must be considered in the development and implementation of a questionnaire study, while Lockhart (14) has just completed the editing of a text oriented to the mailed questionnaire. While such documentation supports the use of the questionnaire, it is recognized to have some limitation. One such limitation is the typically low response rate, which can result in sampling bias. It is usually worth the time and effort to compare respondents to the pool of people solicited, but who did not participate, with the aim of identifying possible sources of bias.

A summary of attitude/opinion survey methods to determine corporate cultures can be seen in Table 1 below. The authors believe that no one method is perfect for all situations and research needs. In general, they favor the questionnaire method as the most cost-effective means of gathering data. Furthermore, we concur with Mischkind (1980) that "the way in which data is collected has a mjaor influence on the level of morale. Face to face interviews, accompanied with opinion survey items, yield the most favorable response distributions." In our work, we have used face to face interviews for an initial understanding of the organizational "climate" and potential sources of organizational ineffectiveness. These exploratory efforts are followed by data collection using the questionnaire method.

Table 1
Comparison of Data Gathering Methods

Technique	Advantages	Shortcomings
Interview	Rich data. Two-way commun.	Limited Sample. Time consuming. Requires skilled interviewer.
Observation	"Objective" viewpoint. Respondents not required to have strong verbal skill.	Limited to repetetive tasks. Time consuming. Requires expert observer.
Questionnaire	Broad participation. Fast turn-around. Low unit cost. Ease of data analysis.	Expert up-front design. Results no better than questions asked. Response bias can flaw generalizability.
Nominal Group Analysis	Minimal up-front development. Rich data. Group synergy.	Not anonymous. Difficult data. Time consuming in large groups. Expert group leader.
Delphi Analysis	Anonymous. Rich data. Group synergy.	Difficult data. Time consuming, even in smaller groups.
Focus Group	Minimal up-front development.	Expert leader. Small groups only.

An essential process in the development of any questionnaire is selecting the categories of items or "domains" to be included in the data collection form (Nunnally, 1980). Well designed questionnaires are built around theoretical models which attempt to explain employee behavior on the job. Payne and Pugh (1983) developed the categories of: organizational context (charter, size, resources, technology, dependence, and ownership); organizational structure (role activities, authority systems, status systems, configuration of roles within the structure;)organizational climate (progressiveness and development, risk taking, warmth, support and control); the individual (personality, needs, abilities, satisfaction and goals); and the individual's immediate environment (boss, colleagues, reward system and tasks) to define the major influences on organizational structure and climate. It appeared to the authors that sub-components of these categories are somewhat independent.

Locke (1983) identifies the typical job dimensions that have been studied as: work (intrinsic interest, variety, opportunity for learning, difficulty, amount, chances for success, control over pace and methods, etc.); pay (fairness or equity, method of payment, etc.); promotions (opportunities for, fairness of, basis for, etc.); recognition (praise for accomplishment, credit for work done, criticism, etc.); benefits (pension, medical, annual leave, paid vacations, etc.); working conditions (hours, rest pauses, equipment, temperature, ventilation, humidity, location, physical layout, etc.); supervision (leadership style, influence, technical ability, human relations ability, and administrative skill, etc.); co-workers (competence, helpfulness, friendliness, etc.); and company management (concern for the employee as well as pay and benefit policies). This lengthy listing represents a mixing of events or conditions with agent factors. While the primary categories might be "pure", they appear to mix or use different sub-component elements. Conte (1984) further notes that "factor analytic studies of job dimensions have further reduced these five: satisfaction with work, satisfaction with pay, satisfaction with promotions, satisfaction with supervision, and satisfaction with coworkers", which form the scales of the familiar Job Description Index.

The authors have re-evaluated the most commonly used categories and developed an item pool based on the category listing presented in Table 2. Recognizing the possibility that some categories were not independent, it was hoped that further data analysis would clarify the factor questions raised.

Finally, a single job satisfaction item based on the one developed by Habbe (1947) was included. It asked the respondents "how they feel about their job." This general item on job satisfaction has appeared in various forms on most job satisfaction questionnaires since the early 50's.

This listing is relative comprehensive and whole categories are often excluded during specific implementations. Management teams are frequently reluctant to include items which are not "fixable" or might in some way cause embarrassment to a particular individual or group within the organization. There is wisdom in using discretion when considering categories for inclusion in a questionnaire. Raising highly sensitive, unaddressable issues might, in fact, lower morale. However, avoiding discussion of visible and important problems will undermine the credibility of the survey process and possibly rob the organization of the two-way communication which might work to solve the problem. There is no easy resolution to this type of problem. Management must be sensitized to the trade-offs involved.

And, care must be taken in assessing the impact of the data gathering and feedback process on employee morale.

Table 2
Domains of Items for Employee Surveys

- Advancement Opportunities
- Employee Benefits
- Corporate Communications
- Corporate Image
- Employee Suggestion Program
- Future with Company
- Goal Setting Practices
- Job Security
- Level of Productivity
- Management
- Performance Review Process
- Problem Solving Practices
- Quality of Work
- Reward Recognition
- Salary Level
- State-of-the-art Technology
- Supervisor
- Work Itself

The most popular scale points used in surveys appear to be 4, 5, 7, and 11. There is a vast literature on the pros and cons of each set of scale points. One author has used a 10 point scoring scheme where the participants are requested to estimate their "extent or percentage of agreement" with the items. Our decimal based mathematical system, as well as our tendency to incorporate percentiles in our day to day judgements,lends intuitive appeal to a 10 point scale. This technique is not commonly used in existing opinion questionnaire forms, however. Experience to date with the 10 point scale indicates that many of the participants relate well to the concept and that it facilitates responses that are less ambiguous.

In an earlier review, it was noted that the questionnaire technique often suffers from low response rates. Several factors can be critical to improving this rate. They include the scaling factors mentioned above, the physical layout of the instrument, securing top management commitment, communication of the purpose of the survey to employees, and protecting confidentiality and/or anonymity of responses. Touching briefly on these, it is recognized that:

- the physical layout of the instrument can increase response rates when it is easy to read, with scoring categories easily determined. Further, the more "professional" the presentation (visual appeal), the higher the response rate.
- no attitude/opinion data gathering can be meaningfully implemented without the support of top management. It is important that potential participants recognize that the effort has the support of top management, and that use will be made of the findings.
- the respondents must understand that the survey is one critical method of communicating their attitudes and feeling concerning their jobs and the workplace, to upper levels of management, with the potential for change and improvement.
- If the potential respondent feels that there is the possibility of being identified, with resulting punitive actions for "negative"

responses, response rates will decrease. Respondents can, however, be given the opportunity to voluntarily identify themselves on the questionnaire. This, coupled with self-addressed return evelopes for delivery to a non-company address, and with written assurances of confidentiality have tended to assure respondents of the privacy of their responses. Resulting return rates appear to be higher than those normally encountered in many attitude-opinion surveys.

Data reduction strategies can be categorized as user oriented vs. research related (see Table 3 below). The former includes the traditional descriptive statistics (mean, median, s.d., etc.) which are easy to generate and easy to interpret. Frequency/percentile scores are eminently understandable but tend toward oversimplification of the results and may lack interpretability. Measures of central tendency also share the same strengths and weaknesses. They are especially vulnerable to the influence of "outlier" responses which create skewed response distributions.

Table 3
Comparison of Data Reduction Strategies

Methods	Advantages	Shortcomings
Frequency/%tile	Ease of understanding Easy to generate	Oversimplification Limited interpretability
Means/S.D.	Easy to interpret Easy to generate	Prone to distortion Masks distribution
Correlation	Theory building	Not useful to managers
Multivariate Analysis	Powerful data reduction capabilities	Difficult to understand and apply.

The mean and standard deviation are, indeed, the staple of survey reports. However, these cannot provide the theoretically interesting information which can assist researchers in fine-tuning the questionnaire itself, or in establishing salient relationships which might guide management to a deeper understanding of their organization's character.

Correlational and multivariate analyses are commonly used by researchers and theoreticians. They can enhance theoretical insight when appropriately applied without violation of the key assumptions of each technique. However, these tools are misleading to laymen who may, at best, achieve a superficial grasp of what the statistical tools can and cannot do. And, there is the danger of "turning off" the employees who might take offense at being talked down to.

The choice of data analysis tools, we believe, should be presented in terms that are most meaningful and most faithfully convey information to the user population. Applied researchers should always take the time, prior to preparing a presentation of the findings, to examine the characteristics of the data, thus avoiding misinterpretations and oversimplication. Where possible, the researcher should use varied analysis tools on the same data to prevent statistical bias in interpreting the data.

Most employee surveys elicit information regarding the individual's likes or dislikes about a work related issue. However, it can be of equal importance to measure the value of each issue to the respondent. Unless we understand an employees value system, we cannot rationally understand the effects of these issues on the employee's behavior. For example, an employee may rate satisfaction with the company's benefit program very high and satisfaction with salary very low. Until we know how important benefits are relative to pay for this individual, we cannot tell what the net effect on the employee is. Scoring the value of each item can be accomplished in a number of ways. Perhaps the most straightforward method is to multiply an item's satisfaction score by its value score. Care should be taken to ensure that polarity or direction from low to high on each of these items is correct.

After the data has been collected, analyzed, and interpreted, it is important to supply feedback to the respondents. The authors have seen the following methods used: standardized report to all participants; corporate news release; small group feedback sessions; and major feedback sessions by a corporate officer.

Releasing standardized reports to all personnel or the use of the corporate in-house media are efficient, though impersonal. Only one report need be produced. Unfortunately, there is no opportunity to use the feedback process for a dialogue.

Good, small group feedback sessions overcome the last shortcoming. They allow in-depth probing concerning results and potential uses of the data. On the negative side, it suffers from the potential for the group leader to gloss over or ignore certain findings, while emphasizing those responses which support the biases of management.

Regardless of the method used, it is important to provide participants with survey results. This encourages a better response rate, for initial survey and in future follow-up administrations. It also gives management an opportunity to discuss action plans for correcting problems, as well as an opportunity to explain why some policies and procedures exist. Management should anticipate the "hard questions" prior to each session. And, presenters should bone up on the essential facts and figures which might clarify a company's business strategy or gameplan. Most importantly, feedback group leaders should be willing and able to apply good listening skills, demonstrating empathy for the employee and focussing in on his/her concerns. The objectives of the feedback process should include: the creation of action plans to remedy fixable problems; the sharing of information to help employees understand how decisions are made and why; to gain valuable information from lower levels of employment which are not traditionally consulted.

In conclusion, the survey feedback method provides a useful tool for measuring employee morale, for two-way communication, for monitoring an organization's processes, and for diagnosing problems. The HF/Applied psychologist is well equipped to provide industry with the necessary research tools to successfully achieve these ends.

REFERENCES

(1) Bouchard, T.J. Field research methods. In M. Dunnette, Ed., Handbook of Industrial and Organizational Psychology 2nd Ed. New York: Wiley, 1983.

(2) "The Decline of U.S. Industry," Business Week, June 30, 1980, p. 59.

(3) Conte, V.A., A multivariate model of the turnover decision of data processing professionals. Unpublished dissertation, Hofstra U., 1984.

(4) Habbe, S. Job attitudes of life insurance agents. Journal of Applied Psychol., 1947, 31, 111-128.

(5) Iaffaldano, M.T., and Muchinsky, P.M., Job satisfaction and performance. Journal of Applied Psychol., May 1985, pp 251-273.

(6) Lawler III, E.E., Porter, L.W. The effect of performance on job satisfaction. Industrial Relations, October, 1967.

(7) Locke, E.A., The nature and causes of Job Satisfaction. In, M. Dunnette, Ed. The Handbook of Industrial and Organizational Psychology. New York: Wiley, 1983.

(8) Lockhart, D.C., Making Effective Use of Mailed Questionnaires. Evaluation Research Society, no. 21, 1984.

(9) Mac Crimmon, K.R., and Taylor, R.N., Decision making and problem solving. In, M. Dunnette, Ed. Handbook of Industrial and Organizational Psychology 2nd Ed. New York: Wiley, 1983.

(10) McCormick, E.J., Human Factors Engineering. 3rd Edition, McGraw-Hill, 1970.

(11) McCormick, E.M., Job Analysis: Methods and Applications. AMACOM, New York City, 1979.

(12) Mischkind, L.A., Is employee morale hidden behind statistics? Personnel Journal, February 1986.

(13) Naisbitt, J., New York Times, Section 3, p. 1. December 23, 1984.

(14) Nelms, K.R., and Porter A.L., EFTE: An Interactive Delphi Method. Technological Forcasting and Social Change 28, 43-61, 1985.

(15) Nunnally, J.C., Psychometric Theory. New York: McGraw-Hill, 1980.

(16) Payne, R., and Pugh, D.S., Organizational structure and climate. In M. Dunnette, Ed. Handbook of Industrial and Organizational Psychology 2nd Ed. New York: Wiley, 1983.

(17) Starr, S., '86 Survey Results. Industrial Engineering, January, 1986.

(18) Toffler, A., The Third Wave. New York: Bantam, 1980.

Human Factors in Organizational Design and Management - II
O. Brown, Jr. and H.W. Hendrick (Editors)
© Elsevier Science Publishers B.V. (North-Holland), 1986

PARTICIPATORY ERGONOMICS: METHODS FOR IMPROVING INDIVIDUAL AND ORGANIZATIONAL EFFECTIVENESS

Andrew S. Imada
University of Southern California
Los Angeles, California USA

Kageyu Noro
Waseda University
Tokyo, JAPAN

Mitsuo Nagamachi
Hiroshima University
Hiroshima, JAPAN

This paper summarizes three areas of application to be covered in a workshop on participatory ergonomics. The psychological and managerial theories that underlie the participatory method and the philosophy of this approach are covered. The primary emphasis is placed on demonstrating specific methods and techniques that workers and managers can use to improve productivity, safety , and efficiency. These methodologies have been found effective in diverse organizational and cross-national applications.

INTRODUCTION

Over the past few years there has been increasing interest in involving different levels of an organization when introducing ergonomic concepts. Noro and Kogi coined this intervention strategy "Participatory Ergonomics". The first formal presentation of this approach was made at a workshop by the same name at the First International Symposium on Human Factors in Organizational Design and Management (Noro and Imada, 1984). Since then, there has been a growing interest in this approach and its methods. As an example of this growing interest, the Ninth Congress of the International Ergonomics Association had four sessions devoted to Participatory Ergonomics or had "participatory ergonomics" in its title. Moreover, participants in these sessions represented a broad cross section of interests and applications. Researchers presented work conducted in the United States, Japan, Germany, Australia, Thailand, the United Kingdom, Sweeden, France, the Philippines, India, and the Soviet Union.

The conceptual underpinnings of Participatory Ergonomics can be traced to psychological and motivational theories. Industrial/organizational psychologists, sociologists, management theorists, and now, ergonomists, are contending that organizations can improve productivity, safety, occupational health, satisfaction, and quality of work life by allowing various levels of the organization to participate in the introduction and implementation of ergonomic principles. Until recently, introduction of ergonomics into the workplace was a unilateral process. That is, management, or human factors specialists, analyzed the way work was being performed, recommended solutions, and then told workers to implement these ideas. The worker was only involved in the implementation phase; never the problem identification or solution analysis phases. This approach has two distinct disadvantages: First, the worker has little or no ownership in the solution and, consequently, feels little commitment to seeing the ideas succeed. Second, unaware of the reasons or ergonomic rationale, the worker is unable to generalize the solutions to other problem areas. Thus, for each unique problem, a specialist must be brought in to analyze the problem and recommend solutions. In strong contrast, participatory ergonomics involves the worker in all phases of the "technology introduction". Workers are made aware of the ergonomic principles, analyses, and rationale for implementing solutions. This involvement can be expected to generate greater commitment, interest, and real skills that workers can use to solve immediate problems as well as prevent future problems from occuring.

The difficulty with this approach, as with any emerging concept, is the lack of formal structure, rules, and formalization. In one sense this is to be expected--unique problems require unique solutions. Moreover, each organization has different needs and workers with different skill levels. Therefore, the absence of any rigid superstructure is not problematic. What is somewhat troubling is the absence of a wide array of methodologies that can help workers understand and solve problems. At present there are very basic techniques that allow people to outline the generalities of the problem. There are also highly sophisticated analytic techniques that are the domain of academicians and researchers. What we need are a set of intermediate "tools" that can help people in several levels of an organization solve problems. These tools should meet at least three criteria: 1) they need to be adaptable to different kinds of problems--they should not be domain specific; 2) they should be easy to use by many levels in the organization; 3) they should be instrumental in helping people solve problems. In so doing, the technique motivates one to use it again in other situations. Tools that meet these criteria will not only help people solve problems, they will also provide a common language that many levels of an organization can use to describe, communicate, and solve ergonomic problems.

The workshop on this topic area describes some successful applications of participatory ergonomics to on-going organizations. Three different areas of emphasis are presented: Improving productivity through ergonomics, Quality Control Circles used in solving ergonomic problems, and cross-cultural applications of participation in implementing ergonomic changes. Together, these case studies demonstrate specific skills and methods that ergonomists can use in making changes in work and organizational design.

NEW METHODS FOR ENHANCING PRODUCTIVITY

Ergonomic techniques for enhancing productivity are introduced. A unique feature to these methods is that the primary purpose for introducing these ergonomic concepts is to enhance productivity. Three cases are presented. At Yokohama Rubber Company, for example, small groups in each division were involved in the introduction of these methods. Details of these group activities are presented. Two other examples, from Toyota Motors and Nippon Steel demonstrate the practical utility of participatory methods in enhancing productivity.

Together, the specific methods from these three cases demonstrate how Japanese companies introduce ergonomics to various levels in a participatory manner and its effects on productivity. These methods are sufficiently robust to be used by other practitioners interested in implementing these new techniques to enhance productivity.

QUALITY CONTROL CIRCLES AND PARTICIPATORY ERGONOMICS

Quality Control Circles (QCCs) are based on psychological and motivational theories that are highly conguent with participatory ergonomics. Two QCC cases demonstrate the diversity and utility of involving workers in solving ergonomic problems.

The first case involves a refrigerator manufacturer that needed a systemm-wide method for addressing issues of aging workers. An assembly line for medium-sized refrigerators was selected as a model for job redesign. A project team was made up of staff from Personnel, Production Engineering, the labor union, managers, and workers from the line. This team was instructed on ergonomic principles; with special emphasis on aging factors, working posture, and job redesign for older workers. Members of a QC circle from the assembly line and the project team proposed a number of solutions for improving the assembly line. For example, the workers developed a lift platform that allows them to adjust the height according to their own needs and allows them to maintain good working posture. The improvement in the assembly line had a dramatic effect -- a 143% improvement. Additionally, it allowed older workers to be involved in all phases of the work process.

The second QCC case involves an engine assembly plant that had problems with lumbago on the assembly line due to bad working posture. All managers in this plant were trained in ergonomics, industrial medicine, and how to redesign jobs to prevent lumbago. They discussed methods for preventing lumbago and wrote a manual for preventing lumbago in the workplace. In turn, each manager trained the members of QC circles on ergonomics, industrial medicine, and industrial engineering principles using this manual. The QC circles then used these methods to evaluate working posture and, by doing so, improved the jobs. Cases of job improvements were presented in a meeting of managers and a project team. This is also a form of QC Circles that can be very effective in solving job redesign problems. About one hundered cases were redesigned to prevent lumbago. The results include improved performance and not a single case of lumbago since the redesign.

CROSS-CULTURAL APPLICATIONS

Several cases will demonstrate how different analytic methods can be used as tools to help workers identify problems and implement solutions. In one case, the combination of cause-and-effect diagrams and a link analysis of the work layout were used in concert to help identify service problems in an international hotel operating in China. This case demonstrates how ergonomic and QCC tools can be used in a different cultural context. Cases from other industrially developing countries are used to illustrate specific methods and the generalizability of participatory ergonomics. The work by Kogi (1985) is particularly instructive. Kogi and his colleagues have been working in India, the Philippines, and Thailand and have developed checklists and work methods that have practical use in these different cultures.

Together these three sections demonstrate how we can use ergonomics to improve productivity, quality, safety, and health. This session integrates the knowledge, methods, and management techniques necessary to implement participatory ergonomics as a systematic organizational change effort.

REFERENCES

[1] Noro, K., & Imada, A. S., Participatory Ergonomics. A workshop presented at the First International Symposium on Human Factors in Organizational Design and Management, Honolulu, Hawaii (August, 1984).

[2] Kogi, K. Participatory Approach in Applying Ergonomics for Workplace Improvement in Developing Countries. In I. D. Brown, R. Goldsmith, K. Coombes, & M. A. Sinclair (Eds.), Ergonomics 85. Proceedings of the Ninth Congress of the International Ergonomics Association, 2-6 September 1985, Bournemouth, England.

Human Factors in Organizational Design and Management - II
O. Brown, Jr. and H.W. Hendrick (Editors)

A HEALTH PROMOTION MODEL FOR ORGANIZATIONS: THE EFFECT OF HEALTH PROMOTION ON ORGANIZATIONAL EFFECTIVENESS AND ABSENTEEISM

William H. Hendrix and Terry L. Leap

Clemson University
Clemson, South Carolina
U.S.A.

Robert P. Steel

Air Force Institute of Technology
Dayton, Ohio
U.S.A.

Recent research indicates that health promotion programs can assist management in improving organizational effectiveness and reducing absenteeism, however, no empirically derived model exists to guide health promotion programs toward increasing these goals. This research developed a preliminary model to guide future research in evaluating the effectiveness of health promotion interventions. The model links a variety of organizational and personal characteristics to absenteeism and productivity.

INTRODUCTION

There has been an increased interest by the business community as well as various academic groups in health promotion and wellness programs. The effect of these programs on the organization and its employees is thought to be desirable although most have no systematic means of defining and measuring these benefits. Organizations are not presently provided with adequate information for evaluating the utility of various health promotion interventions on organizational outcomes (e.g., absenteeism, turnover, and productivity) because no empirically derived evaluative model has been established.

The purpose of this research is to develop a preliminary empirically derived health promotion model to guide future research in evaluating the effectiveness of health promotion interventions. In addition, the research has to have implications for managerial behavior, and organizational and job design.

A number of theoretical health-related models have been proposed which have implications for developing a comprehensive health promotion model (Brief, Schuler, & Van Sell, 1981; Cooper, & Marshall, 1976; Hendrix, 1985; Ivancevich, & Matteson, 1980). The major components suggested by these models have included: stress and health related determinants (i.e., intraorganizational, extraorganizational, and individual factors), job and life stress, psychological and physical outcomes, and organizational outcome factors. Therefore, it would appear reasonable to include these components initially when developing and testing a more comprehensive health promotion model.

The present research adopts a health promotion model proposed by Hendrix (1985). This model depicted in Figure 1 suggests that there are several categories of determinants of job and life stress. These categories

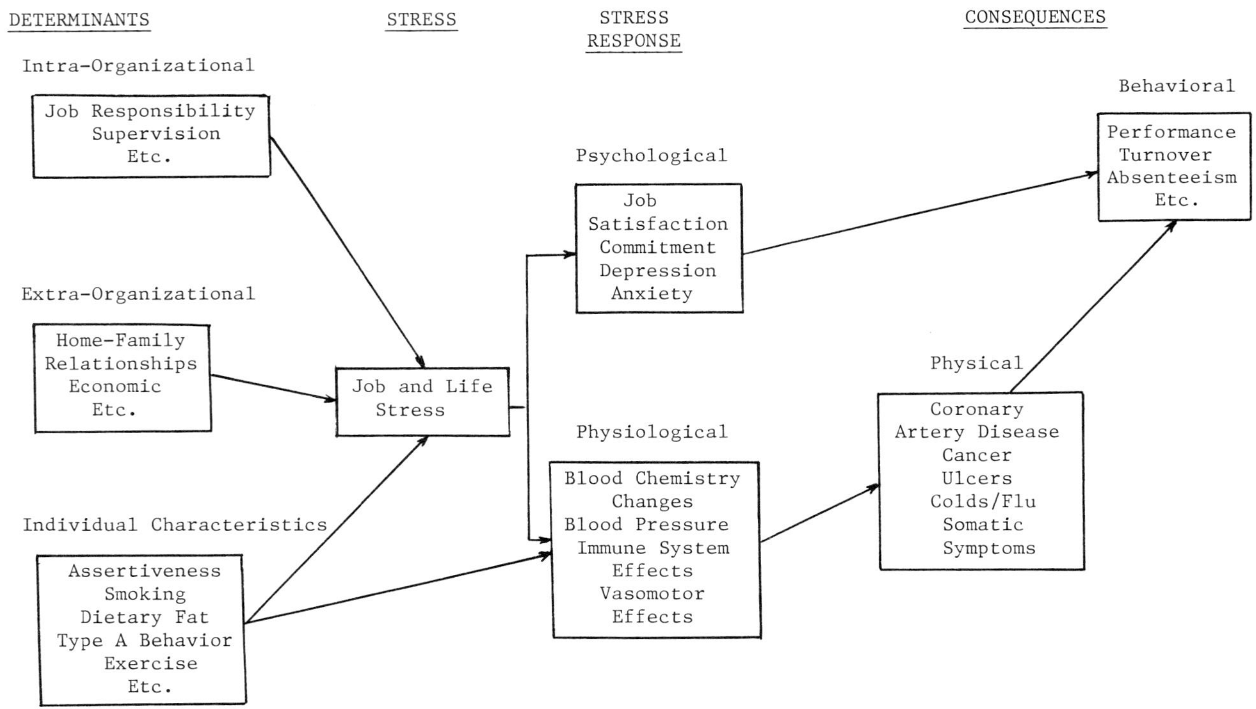

Figure 1. Research Model

Reprinted with permission from Aviation, Space, and Environmental Medicine, 1985, Vol. 56, p. 655.

include intraorganizational and extraorganizational factors, and individual characteristics. In turn, these factors affect both job and life stress. Some of the individual characteristics also affect stress as well as physiological factors such as blood chemistry and blood pressure. Job and life stress, on the other hand, lead to effects that are considered to be stress responses. These include psychological factors (e.g., job satisfaction, organizational commitment, anxiety, and depression) and physiological factors (e.g., blood chemistry factors such as cholesterol and cortisol). Psychological factors also have an effect on individual behaviors such as absenteeism, performance, and turnover rates. In addition, the physiological factors affect individual behaviors, although the effects are not direct; physiological factors cause physical changes (e.g., diseases such as coronary artery disease, colds and flu, and somatic symptoms) which in turn affect individual behaviors.

METHOD

Subjects and Procedure. A sample of 463 individuals completed a health promotion survey and had a blood sample drawn. Subjects consisted of U.S. Department of Defense civilian employees located in Colorado, Florida, and Texas. Approximately 55 percent were female, 3 percent black, 87 percent white, 5 percent hispanic, and 5 percent were either listed as asian, indian or failed to provide their race. The mean age was approximately 43 years with a range of 18 to 70 years.

Measures. The single physiological variable used in this study was the cholesterol ratio which is a ratio of total serum cholesterol divided by high density lipoprotein (HDL) cholesterol. This measure has been found to be one of the best single laboratory predictors of coronary heart disease (Malaspina et al., 1981; Uhl, et al., 1981). The behavioral outcome variables incorporated in this research included employee absenteeism rates and individual performance measures. Absenteeism was operationalized as the number of sick days taken in the last 6 months while individual performance was operationalized as the employee's last merit performance report. The single physical factor included is the number of cold and flu episodes for each subject during the past year.

The remaining variables incorporated in the hypothesized health promotion model included intraorganizational and extraorganizational factors, individual characteristics, and stress related factors.

Analysis. Exploratory path analysis was conducted to determine which paths in the hypothesized causal chain (Based on Figure 1) were statistically signigicant and to develop a revised model for future testing. The first phase of this analysis involved the regression of each variable in the model (other than the hypothesized determinants) against all preceding variables. The resulting beta weights served as the path coefficients and also indicated the relative strength of each variable in the model.

The second phase involved eliminating all variables not found to be significant ($p < .05$), and rerunning the regressions. This resulted in the revised model provided in Figure 2.

RESULTS

Investigation of the revised model (see Figure 2) reveals that many of the

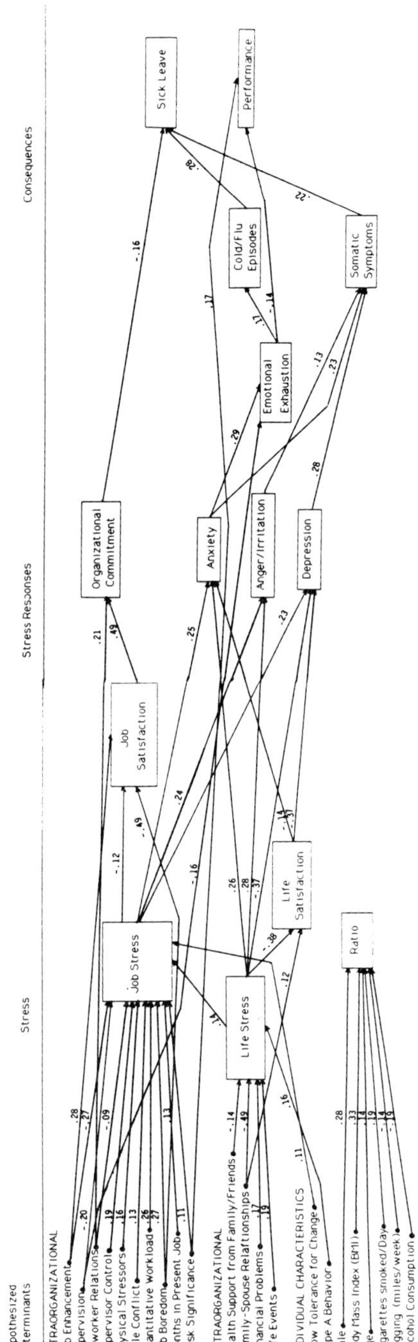

Figure 2. Path Model.

hypothesized relationships were not supported. The revised model indicates that some of the intraorganizational variables not only affected job stress but they also had a direct effect on job satisfaction and organizational commitment. In addition, one variable, supervision, affected emotional exhaustion; the better the supervision the less the emotional exhaustion a subordinate experienced. Extraorganizational factors primarily affected life stress with the major contributing factor being family-spouse relationships.

The individual characteristics variables were the major factors contributing to changes in the cholesterol ratio. Individuals tend to have higher cholesterol if they were males, were overweight (i.e., had a high BMI index), smoked cigarettes, and were older. The cholesterol levels tended to be lower for individuals who jogged and consumed alcohol in moderation. One individual characteristic, Type A Behavior, however was related to life stress; with a Type A person reporting higher life stress than others.

The results of the upper route of Figure 1, the psychological route, suggests that job stress leads to job satisfaction and job satisfaction in turn affects the level of one's organizational commitment. Organizational commitment, in turn, inversely affects absenteeism (i.e., sick leave days taken); the higher the commitment the lower absenteeism. Job stress tends to result in higher levels of anxiety, anger/irritation, depression, and emotional exhaustion while life stress is associated with lower life satisfaction, increased anxiety, and increased depression.

The results associated with the lower route of Figure 1 (the physiological route) suggests that cold and flu episodes were influenced by emotional exhaustion and anxiety. That is, cold and flu episodes increased with increase in reported levels of emotional exhaustion and anxiety. Somatic symptoms (e.g., headaches, and fatigue) were positively related to emotional exhaustion, depression, anger/irritation, and anxiety. As cold and flu episodes and somatic symptoms increased, absenteeism rates also increased.

Performance, as measured by the individuals last merit performance rating, was affected primarily by two variables, emotional exhaustion and task significance. The more significant the task, the higher the performance rating. However, with increase in emotional exhaustion there was a decrease in performance levels.

Discussion

The results provide an empirically derived model for future research. The data provide links between the hypothesized determinants of stress and job and life stress. Intraorganizational variables were indicated as the primary causal factors associated with job stress. Life stress, however, was a result of extraorganizational variables, with family-spouse relationships being the dominant factor contributing to life stress. The two major variables that are probably of most concern to organizations are absenteeism and productivity. The revised model suggests that absenteeism is affected primarily by one's commitment to the organization and the extent that individuals experienced illness (i.e., cold/flu eipsodes, and somatic symptoms). Performance, on the other hand, was higher when the task was considered significant and performance was lowered with increases in emotional exhaustion.

Organizations which create a hospitable work environment that has low physical stressors, low role conflict, enriched jobs, and good supervision will tend to have employees who have less stress, are more commited to the organization, and who have lower absenteeism rates. The specific causal paths established here have implications for health and wellness programs, management development programs, as well as for organizational and job design. In addition, the health of individual employees might be improved by screening for high cholesterol ratios and then providing action based on the results of this research to reduce the level of those with high ratios.

The model derived from this research suggests that by modification of the variables contained in the model both individual health and organizational productivity should be increased.

REFERENCES

[1] Brief, A.P., Schuler, R.S., & Van Sell, Managing job stress (Little, Brown, Boston, MA, 1981).

[2] Cooper, C.L., & Marshall, J., Occupational sources of stress: A review of literature relating to coronary heart disease and mental ill health, Journal of Occupational Psychology, 49, (1976) 11-28.

[3] Hendrix, W.H., Factors predictive of stress, organizational effectiveness, and coronary heart disease potential, Aviation, Space, and Environmental Medicine, 56, (1985), 654-659.

[4] Ivancevich, J.M., & Matteson, M.T., Stress and work: A managerial perspective, (Scott-Foresman, Glenview, IL, 1980).

[5] Malaspina, J.P., Bussiere, H., & LeCalve, G., The total cholesterol/HDL cholesterol ratio: A suitable atherogenesis index, Antherosclerosis, 40, (1981), 373-375.

[6] Margolis, B.T., Kroes, W.H., & Quinn, R.P., Job stress: an unlisted occupational hazard, Journal of Occupational Medicine, 16, (1974), 659-661.

[7] Quick, J.C., Quick, J.D., Organizational stress and preventive management, (McGraw-Hill, New York, 1984).

[8] Russek, H.I., Zohman, B.L., Relative significance of heredity, diet, and occupational stress in coronary heart disease of young adults, American Journal of Medical Science, 235, (1958), 266-275.

[9] Uhl, G.S., Troxler, R.G., Hickman, J.R., & Clark, D., Relation between high density lipoprotein cholesterol and coronary artery disease in asymptomatic men, American Journal of Cardiology, 48, (1981), 903-910.

Human Factors in Organizational Design and Management - II
O. Brown, Jr. and H.W. Hendrick (Editors)

HOW TO SUCCESSFULLY INTRODUCE PREVENTIVE HEALTH MEASURES IN THE CONSTRUCTION INDUSTRY

Ernst A.P. Koningsveld
Foundation for Occupational Health Care
in the Construction Industrty
P.O. Box 8114
1005 AC Amsterdam
The Netherlands

In the Netherlands the occupational health care for the construction workers is carried out by one organization, BGBouw. In the past decade BGBouw was primarily concerned with individual health care, distribution of information and instruction. The information, brochures and leaflets were in great demand. However, results in the sense of less accidents, less absenteeism or improved working conditions could hardly be perceived.
Discontented with these results we have developed a new approach which proves to be successful.

INTRODUCTION

The construction industry is quite different from other branches of industry: the situation at the work place changes continuously and workers come and go day by day, hour by hour. Therefore health and safety programs require a specific approach. In this paper several suggestions are given, which could also be applied in other situations.
I will first give a brief description of our organisation and will discuss the way it has been working until about a year ago. Lastly the new approach is explained.

BGBouw

"BGBouw" stands in Dutch for the Organisation for Occupational Health Care in the Construction Industry. It is a nation-wide organization, whose main objectives are to protect and promote the health and safety of the 250,000 construction workers in the Netherlands and to improve their working conditions. It is boarded by both employers organizations and unions.
For a correct understanding of this paper the following characteristics are important:

- BGBouw is a nation-wide organization, that does not have direct connections with individual companies;
- BGBouw deals only with the construction industry;
- Employers nor employees can be obliged by BGBouw in whatever sense, so the objectives have to be achieved by motivation;
- The only organised contacts with target groups are the medical examinations of individuals, which are on a voluntary basis.

Next to these characteristics, those of the construction industry have to be considered. This combination restricts the ways in which the objectives could be achieved.

The contents of our health and safety messages are based on existing knowledge of the work related health and safety aspects and research and development.
Up to now the chanals of information could be divided into:
1. information to the workers;
2. information to the manufacturers of materials and tools;
3. information to the management of construction companies and sites.

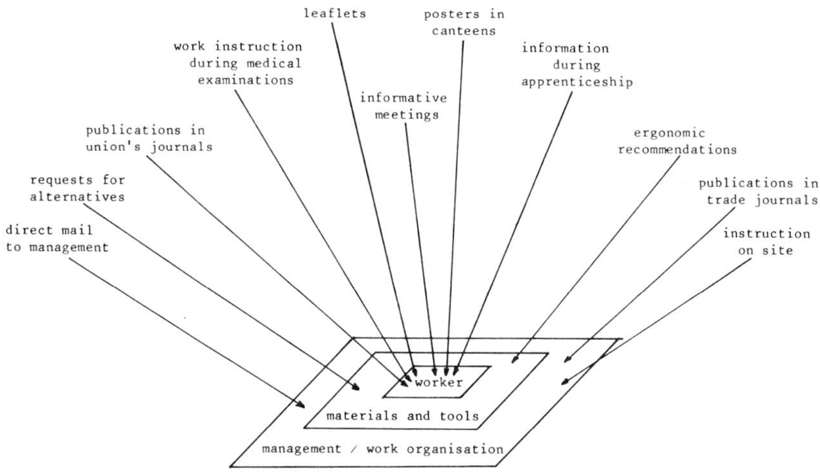

Fig. 1: Chanals of information to target groups.

1. information to the workers

Figure 1 shows the different actions that have been undertaken in the past decade. Most actions have been directed to the workers. The recurrent theme was: 'take care of yourself, work better, healthier and safer'. However, in many cases the worker is not in the position to change aspects of his work. For example:
- the advise to lift correctly is correct from a theoretical point of view, but do the working conditons allow this? Is the load easy to handle?
- "do not grind asbestos-cement materials, nibble them". Also a theoretically good advise. But, are there adequate tools at hand, or is the worker allowed to go and buy some?

In other cases the worker is advised to protect himself e.g. by wearing good working clothes. Yet, it is still very difficult to convince the workers to do so. Younger workers for instance hardly recognise health hazards when they receive instruction during their apprenticeship.

2. information to the manufacturers of materials and tools

In the field of building materials and tools the realisation of our objectives also meets with obstructions. Examples are:
- the 50 kg cement bag, that should be replaced by 25 kg bags. A nation-wide action failed because of the European Community's regulations, which protect producers in member countries against unfair competition;
- the use of a hand-saw versus the circular electric saw for the shortening of gipsum blocks. This measure was suggested in order to prevent serious dust exposure. The use of the hand saw resulted in a price rise of a house by 0.5 ‰ ($20 !!), which turned out to be insuperable;
- ergonomic recommendations for tower-crane cabins. This study was widely appreciated, but after seven years results in the form of better cabins are hardly perceptible;
- at our request mechanical equipment has been developed and introduced for the laying of concrete blocks with a weight of 23 kg and more. However, many managers keep this development at bay on account of so called "organizational problems".

Therefore the improvement of materials and tools is hampered by considerations of economy, organization or competition. Besides, it is very difficult to convince manufacturers of the necessity to design their products according to health and safety demands.

3. information to the management of construction companies and sites

When trying to attain progress by organizational interference you can expect the following comments:
- regarding alternative specifications for materials, etc.: only in a few cases suggestions like this will lead to success; the argumentation is: "materials have already been ordered; alternatives are too expensive; etc.". Besides suggestions will be rejected on the basis of improper argumentation that is the result of resistance against renewals by the middle management.
- it is a fact that a better working organization and work place arrangements can reduce the physical and mental workload. However, thoughtlessness dominates and work is arranged in a hurry. The worker does not permit himself to address his common sense. Calculations are only fixed on efficiency, and not on an optimal input of manpower from a human factor's view.

Above I have indicated several factors that have hampered a successfull prevention of health and safety hazards.
Unsatisfied with the poor results of our efforts, we have been discussing alternative ways to achieve our goals and objectives. The first step was to draw up so-called action programs on the basis of new research results. All conclusions and recommendations of these studies are translated into specific actions towards specific target groups (fig. 2).

TARGET GROUPS:	workers	employers	manufacturers	legal autorities	principals	occ. health centres	etc.
Recommendations:							
working postures	instruction	improve machines	improve new machines	-	-	instruct workers	-
noise reduction	-	purchase policy	development new machines	-	-	-	-
toxic paints	hygiene	instruction to workers	less solvents in paints	-	order less toxic paints	assess act. exposures	-
safety	instruction	-	-	legal measures	close raod if possible	-	police control

Fig. 2. Part of an action program based on a survey of the working conditions of road markers

Moreover four lines have been developed in order to point out what the responsibilities are for each target group:

1. workers

Our message has to be changed: information to workers should only consist of matters they can deal with. Workers rightly object when preventive measures are only suggested to them, whereas source solutions are not discussed. Therefore it should be made clear that other target groups have been approached in order to improve working conditions.

2. manufacturers of materials and tools

Do not only put your finger on the spot of hazardous materials or non-ergonomical tools, but co-operate with the manufacturers in research and development of good alternatives, e.g. by supplying risk-bearing capital.

3. management

In 1985 unions and employers organizations inserted several articles on health and safety in the collective agreement for the construction industry. Apart from an official regulation of all aspects of the occupational health care program, each company employing 35 workers or more is obliged to draw up a policy plan on working conditions. This plan must consist of nine measures, such as actions to protect health and prevent unsafe circumstances; purchase policies; delegation of responsibilities; systematic safety controls; education and instruction.
For relatively large construction works an action program has to be developed, consisting of specific measures for each seperate stage.

Thus companies that already took care of good working conditions will no longer be in an unfavourable position when competing with other companies.

4. principals

It turned out to be possible to convince great investors, that better working conditions are worth while. As a first step we have agreed with the Pension Fund for the constuction industry that:

- the management on site will be trained in safety;
- on their sites some materials, like asbestos, will not be applied;
- cement will only be used when packed in bags of maximum 25 kg;
- heavy materials have to be handled mechanically;
- where-ever possible noise reduction will be applied;
- etc.

CONCLUSION

Although we yet have little experience with the new approach, the first results show good prospects. For similar organizations in the field of health, safety and human factors, this approach may contain elements that can be considered as examples for a model that is designed to their specific situation.

Human Factors in Organizational Design and Management - II
O. Brown, Jr. and H.W. Hendrick (Editors)

EVALUATING OFFICE ENVIRONMENTS: THE CASE FOR A MACROERGONOMIC SYSTEMS APPROACH

Alan Hedge
Applied Psychology Division,
Aston University, Birmingham
B4 7ET, UK

Elia M. Sterling,
TDS Ltd.,
Vancouver, B.C.
V6J 2E2

Theodor D. Sterling,
Faculty of Interdisciplinary Studies,
Simon Fraser University,
Burnaby, B.C. V5A 1S6

A comparative field study of the work-related health problems of office workers in a public sector organization revealed significant differences between those in a naturally ventilated building and those in an adjacent air conditioned building. Measures of ambient conditions including indoor air quality alone could not account for these differences and it seems that individual and organizational variables may play a major role. The results highlight the importance of adopting a macroergonomic systems approach to evaluating office environments.

Introduction

White-collar workers now comprise over one-third of the UK workforce and for the foreseeable future their numbers will continue to increase. Likewise, in the USA it has been estimated that by 1990 over 50% of all new jobs will be white-collar work (Wineman, 1986). These trends are clearly indicative of the transition from an industrial to an information society dominated by the information and 'service industries', such as banking, insurance, etc. (Christie, 1985). Given these changes research on issues of health and productivity among white-collar workers is of particular importance.

One problem here is that research on office workers usually has been conducted at one of three different levels of analysis: either the immediate workstation, the office work area, or the office building, and consequently it is often difficult to integrate the different findings. There is, for example, a considerable literature on the possible health hazards of VDTs at work (Pineault & Berthelette, 1984). There is also controversy on the impact of the ergonomic redesign of VDT workspaces on productivity and health. Some studies e.g. Ong (1984), report significant improvements in productivity and health among data entry VDT workers following the redesign of their workspace, while others e.g. Tijerina and Rockwell (1984), have failed to show productivity changes following workstation redesign even though comfort ratings may improve. It seems that at best, simply providing workers with ergonomically designed workstations may be a necessary though not sufficient condition to improve workers' health and organizational productivity.

Other research has looked more broadly at the impact of the general office work environment. These studies invarioably have shown that, irrespective of workstation ergonomics, when compared with cellular offices, open-plan/landscaped office layouts can create serious problems for certain groups of workers such as managers who often report reduced productivity because of greater problems coping with the frequent disturbances and with loss of privacy (Hedge, 1982, 1986). Such mismatches between what the individual worker requires to do their job effectively and what the office environment actually provides may also prove to be stressful.

More recently studies of the health problems of office workers have shown that a variety of chronic symptoms of work-related ill health are consistently reported by workers in certain types of office buildings. These symptoms have been variously referred to as the 'sick building syndrome' (WHO, 1983; Finnegan et al., 1984; Molhave, 1985; Robertson et al., 1985), the 'tight building syndrome' or 'building illness' (Sterling et al., 1984; Hedge et al., 1986). The range of health problems reported typically includes headache, lethargy/fatigue, eye, nose and throat irritation, and cold/flu-like symptoms. These symptoms have been shown to be more prevalent in air conditioned office buildings than in naturally ventilated buildings (Turiel et al., 1983; Hedge, 1984a,b; Sterling et al., 1985). Because the incidence of complaints appears to systematically vary with the type of ventilation system in the office building researchers have looked for contaminants in indoor air, though to date, specific indoor air pollutants have not been found which alone might be responsible for these problems. Indeed, one of the paradoxes of 'sick buildings' is that the indoor climate of such buildings often appears consistent with the ergonomic recommendations for a healthy and comfortable workplace.

A worker who feels unwell at work is unlikely to be at his or her most productive, and given that the prevalence figures for complaints of such ill-health typically exceed 40% of the workforce and may even be as high as 80% of workers, the costs of a 'sick building' to any organization potentially are very high.

Many of the studies of ill health among office workers appear to have presumed a simple medical model whereby the presence of a symptom is assumed to reflect the presence of some adverse condition or noxious agent in the environment (e.g. Turiel et al., 1983; Finnegan et al., 1984; Robertson et al., 1985) even though none has yet been identified. Preliminary indications that such a model may be unsatisfactory have been reported by Hedge (1984 a,b) who found that work-related illness varies according to individual factors such as gender (proportionally more women report symptoms) and job type (proportionally more clerical workers report symptoms). However, as yet no single study has adequately explored the possible interrelationships between the broad range of individual, organizational, and environmental factors and workers' health.

The Research Study

In an attempt to broaden our research approach we chose to investigate six sets of factors identified as possibly contributing to ill-health in offices (see figure 1). Because this work was being conducted within a single organization we did not anticipate the need to build a more comprehensive array of organizational variables into the design.

Two adjoining office buildings occupied by a public sector organization were studied. One of these buildings is a turn-of-the-century office development comprising conventional cellular office accommodation, and occupied by a large department of the organization. Workers in this building have local control of heating (hot water radiators), lighting, and ventilation (openable windows). The other building is a mid-1970's energy efficient air conditioned (variable air volume system) building in which some workers have modest control of temperature (thermostatically controlled zones) but no control over other ambient conditions.
A detailed survey of the ambient environmental conditions in each of these buildings over a two-week period was undertaken and data were collected on the range of environmental factors and atmospheric pollutants shown in figure 1.

Following the environmental survey a self-administered questionnaire was internally distributed to and collected from employees working in two offices. This questionnaire, the 'Office Work Environment Survey' comprises 151 questions, including asking for information on environmental factors, job stress, office equipment, personal/lifestyle and workplace variables (see figure 1). A total of 489 workers in the air conditioned building and 97 workers in the naturally ventilated offices completed the questionnaire.

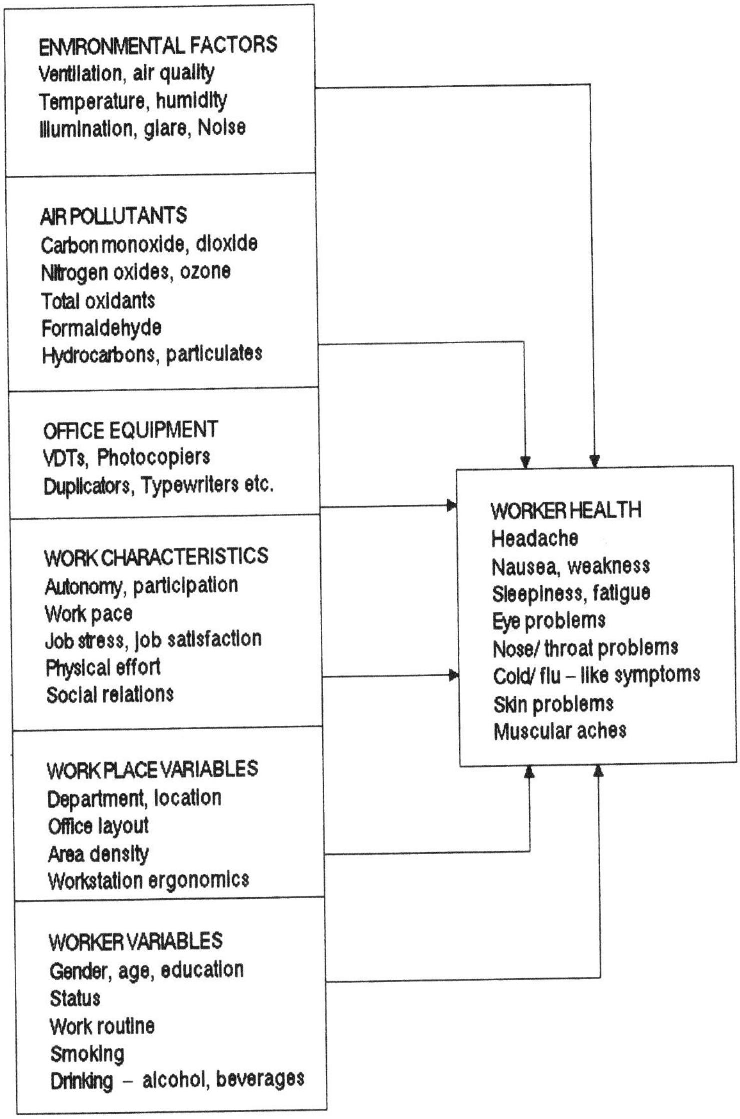

Figure 1 Factors possibly contributing to office worker ill - health

Results

The survey of ambient conditions and air pollutants failed to uncover any major differences in the indoor climate in the two buildings at the time of survey. None of the pollutants measured were at concentrations normally deemed hazardous. The results from the questionnaire survey did, however, show significant differences between the responses of workers in these two buildings. More workers in the air conditioned offices reported problems with environmental factors (ventilation, temperature, humidity, and illumination) while more workers in the naturally ventilated offices reported poor job control and low job satisfaction (see table 1). More workers in the air conditioned offices also reported symptoms of ill-health (see table 2). These differences could not be accounted for by the office equipment workers were using, nor were they related to factors such as the amount of time spent in the building, smoking or other activities such as drinking beverages. At the level of the office building these results appear consistent with the assumption that the mode of ventilation is a prime determinant of ill-health among office workers. However, when the data were analysed by department a somewhat different picture emerged. To illustrate this point we have selected data from two departments (A & B) served by the same air delivery plant but on different floors of the air conditioned offices. The results for these departments are also given in tables 1 and 2. What emerges is that one of these departments presents a response profile that in fact is significantly better than that shown by those working in the naturally ventilated offices. Inspection of the sample structure in these two deparments revealed that one (A), which presented the most favourable response profile, comprised predominantly male, professional workers who expressed greater job satisfaction, while the other (B),which presented a markedly less favourable profile, comprised predominantly female clerical workers who reported lower levels of job satisfaction, and in these respects was more comparable with the workforce in the naturally ventilated offices.

		Naturally ventilated offices	**Air conditioned offices**		
			Total	**Dept.A**	**Dept.B**
		(n=97)	(n=489)	(n=28)	(n=66)
Ventilation	Good	22%	9%	36%	6%
	Poor	31%	42%	14%	44%
Temperature	Good	19%	9%	36%	6%
	Poor	10%	8%	0%	11%
Humidity	Good	23%	12%	46%	9%
	Poor	27%	45%	14%	42%
Lighting	Good	36%	22%	50%	15%
	Poor	13%	41%	18%	50%
Job control	Good	1%	4%	7%	4%
	Poor	57%	41%	43%	38%
Job satisfaction	Good	42%	42%	61%	26%
	Poor	26%	15%	4%	17%

Table 1 Significant differences between the two office buildings in the environmental and job indeces

	Naturally ventilated offices	Air conditioned offices Total	Dept.A	Dept.B
	(n=97)	(n=489)	(n=28)	(n=66)
Sleepiness	50%	69%	39%	70%
Difficulty concentrating	38%	55%	36%	50%
Cold/Flu symptoms	37%	52%	32%	51%
Eye focussing problems	31%	44%	32%	42%
Nose irritation	30%	42%	46%	38%
Skin dryness, itching etc.	13%	24%	18%	23%

Table 2. Significant differences between reports of ill-health in the two office buildings.

Discussion

Although we have confirmed the findings of previous researchers that the prevalence of work-related ill-health is greater in air conditioned than naturally ventilated offices we failed to find any evidence to support the view that this is simply a direct consequence of exposure of workers to noxious substances in the indoor air. Neither did we succeed in identifying any systematic differences in ambient conditions that alone could account for the differences in reported ill-health. Rather, what we found was that by treating the self-report data at the level of the building we were masking important individual and probably organizational differences that may also influence prevelance rates. In fact by analysing the data by department and comparing results within the building we found marked changes in responses that were not reflected by concomitant changes in ambient conditions.

It seems that workers' perceptions of the adequacy of the ambient conditions at work and the prevalence of reports of ill-health are not a direct correlate of the prevailing physical conditions, but these seem to be mediated by various individual and organizational variables not normally included for study in investigations of this type. It would, however, be wrong to interpret these data as showing that individual and organizational factors alone can account for differences in prevalence rates, rather we suggest that what these emphasise is the need for more sophisticated models of user responses to their workplace than the traditional dose -response model adopted by many health researchers..

We believe that to better understand the complex array of factors that seem to be mediating the impact of the environment on the health and productivity of office workers requires the development of an integrative framework to guide research and the development of theory. Such a framework must systematically encompass the role played by environmental, organizational, and individual variables in influencing workers' perceptions of and reactions to their workplaces. Here, the macro-ergonomic approach to analyzing and solving organizational problems proposed by Robertson and Hendrick (1984) seems particularly attractive. Based on systems analysis, this approach emphasises the need to analyse the organization-technology interface in a fairly formal and systematic way starting with a detailed definition of the problem by, for example, graphically representing this as a problem factor tree showing the possible interrelationships between causal factors, subproblems, and the major problem, and then progressing through a further five stages of setting objectives, developing, modeling, evaluating, and selecting alternatives, through to planning for implementation. We suggest that applying such an approach in future research will be particularly helpful in clarifying the salient variables influencing office workers' health, and will also greatly assist the development of a suitable ameliorative strategy for any organization wishing to safeguard the health of its workers and improve its productivity.

References

[1] Christie, B., (ed.) Human Factors of Information Technology in the Office (John Wiley, Chichester, 1985).

[2] Finnegan, M.J., Pickering, C.A.C. and Burge, P.S., The sick building syndrome: prevalence studies, Brit. Med. Jrnl. 289 (1984) 1573-1575.

[3] Hedge, A., Suggestive evidence for a relationship between office design and self-reports of ill-health amongst office workers in the United Kingdom, Jrnl. Arch. & Plan. Res., 1 (1984) 163-174.

[4] Hedge, A., Ill health among office workers: an examination of the relationship between office design and employee well-being, in: Grandjean, E. (ed.), Ergonomics and Health in Modern Offices (Taylor & Francis, London, 1984).

[5] Hedge, A., Open versus enclosed workspaces: the impact of design on employee reactions to their offices, in: Wineman, J.D. (ed.), Behavioral Issues in Office Design (Van Nostrand Reinhold, New York, 1986)

[6] Hedge, A., Sterling E. M. and Sterling T.D., Building illness indeces based on questionnaire responses, Paper presented at ASHRAE Conference, Indoor Air Quality '86.

[7] Molhave, L., Volatile organic compounds as indoor air pollutants, in: Gammage, R.B. and Kaye, S.V. (eds.) Indoor Air and Human Health (Lewis, Chelsea, 1985).

[8] Ong, C.N., VDT work place design and physical fatigue: a case study in Singapore, in: Grandjean, e. (ed.), Ergonomics and Health in Modern Offices (Taylor & Francis, London, 1984).

[9] Pineault, R. and Berthelette, D., Health hazards of VDTs, in: Grandjean, E. (ed.), Ergonomics and Health in Modern Offices (Taylor & Francis, London, 1984).

[10] Robertson, A.S., Burge, P.S., Hedge, A., Sims, J., Gill, F.S., Finnegan, M.J., Pickering, C.A.C. and Dalton, G., Comparison of health problems related to work and environmental measurements in two offices with different ventilation systems, Brit. Med. Jrnl. 291 (1985) 373-376.

[11] Robertson, M.M. and Hendrick, H.W., A systems analysis approach to integrating OD and ergonomic solutions to organizational problems, in: Hendrick, H.W. and Brown Jr., O. (eds.) Human Factors in Organizational Design and Management (Elsevier, New York, 1984).

[12] Sterling, E.M., McIntyre, E.D. and Sterling, T.D., The effects of sealed office buildings on the ambient environment of office workers, in: Grandjean, E. (ed.) Ergonomics and Health in Modern Offices (Taylor & Francis, London, 1984).

[13] Sterling, E.M., McIntyre, E.D., Collett, C.C., Meredith, J. and Sterling, T.D., Sick buildings: case studies of tight building syndrome and indoor air quality investigations in modern office buildings, Env. Health Rev. 29 (1985) 11-19.

[14] Tijerina, L. and Rockwell, T.H., The effects of VDT workstation design on operational productivity, work behavior, and comfort. Hawthorne revisited, in: Hendrick, H.W. and Brown Jr., O. (eds.), Human Factors in Organizational Design and Management (Elsevier, New York, 1984)

[15] Turiel, I., Hollowell, C.D., Miksch, R.R., Rudy, J.V. and Young, R.A., The effects of reduced ventilation on indoor air quality in an office building, Atmos. Env. 17 (1983) 51-64.

[16] WHO, Indoor air pollutants: exposure and health effects, EURO Reports and Studies 78 (1983).

[17] Wineman, J.D. (ed.), Behavioral Issues in Office Design (Van Nostrand Reinhold, New York, 1986).

Human Factors in Organizational Design and Management - II
O. Brown, Jr. and H.W. Hendrick (Editors)

BUT I CAN NEVER REACH YOU... ASSURING CUSTOMER SERVICE THROUGH WORKFORCE PLANNING AND MANAGEMENT

James W. Suzansky, Ph.D.

President
Work Systems Associates, Inc.
65 West Valley Brook Road
Long Valley, New Jersey 07853

High quality customer service requires promptness and reliability from an organization's employees. Whether the service is sales, account inquiry, installation, or repair, scheduling the proper number of employees at the times they are needed to meet customer demands is essential to the organization's continued success. This paper reviews the processes of "**Workforce Management**" which enable both long- and short-term employee scheduling to assure that the proper number of employees are available to efficiently meet customer needs. Originally developed only for large organizations, these processes are increasingly cost-effective in small- and medium-sized organizations through personal computer applications programs.

The notion of "optimum scheduling" of employees is not a new one. Managers and Industrial Engineers have long attempted to predict when they need to add or remove employees, and how to arrange their schedules to meet work requirements. Statisticians and Operations Research specialists have provided assistance with queuing, modeling, and routing solutions.

The processes of "**Workforce Management**" share these origins, but they are most closely associated with the telecommunications industry where they are known as "Force Management" and "Traffic Engineering." This industry has worked for decades at arranging efficient and effective scheduling to provide twenty four hours-a-day and seven days-a-week customer access to critical services such as telephone operators and repair groups. Having realized the financial and customer satisfaction benefits of these processes, they have been extended to many other work groups in this industry.

The purpose of this paper is to overview the basic components and concepts of Workforce Management processes. The overall premise is that application of these techniques will lead to improved customer service with reduced labor and opportunity costs.

Workforce Management Phases

The components of a Workforce Management system can be divided into three phases based on their temporal relation to the actual performance being scheduled.

Phase 1: Planning

The planning activities are all completed **before** the actual work to be performed. They result in the completion of a "game plan" as preparation for management of the workforce when the planned time actually occurs. Four related activities are conducted as part of the planning phase.

1. **Data Base Management**

   Most customer service work activities can be related to some "driver" event which initiates the need for work to be performed. Usually, these drivers are a customer contact by mail, telephone, or in person. An initial step in implementing a Workforce Management system for a work group should be the identification, through task analysis, of the drivers of its work. Then, records should be established which summarize the times and days these drivers occur and the volumes of work they cause to be performed.

   As an example, consider a repair service "hotline" for an appliance manufacturer. The primary drivers of work are telephone calls to an "800" toll-free number staffed for twenty four hours every day. The work to be performed is to document the nature of the customer's problem and refer the customer to a convenient repair facility. A data base should be established and maintained to summarize the number of calls received each hour of each day of the week and the average time to complete a call. Computer-based aids are readily available which will prevent such record-keeping from becoming a tedious manual task.

2. **Long-Range Planning**

A long-range plan should be prepared for as little as a few months to as long as a year in the future. This plan must prepare the work group for overall staffing requirements, and should account for paid times when employees will not actually be present. These "overhead" times should include realistic allowances for vacations, holidays, absence, training, and any other requirements for time not directly performing the group's primary work activities. Computer-based aids are available which enable the creation of a variety of scenarios to evaluate alternate staffing plans.

In addition to specifying staffing requirements, this long-range planning activity will enable distribution of the controllable "overhead" time to prevent negative impacts on future schedules. If in the appliance "hotline" example it is known that December is a peak month for customer calls, training and vacations can be scheduled in other months.

3. **Short-Range Forecasting**

This activity is usually performed one to two weeks in advance of the time being planned. It consists of reviewing the trends shown in the data base to produce a forecast of the quantity of "drivers" which will occur during specific time intervals of a future day, and an estimate of the volume of work they will cause to be performed. Both manual and computer-based procedures exist which aid the creation of work volume forecasts that have proven to be extremely accurate (within 5% of actual) in a variety of customer service work groups.

A portion of the forecast for Monday, September 15, 1986, for the "hotline" group could appear as follows:

HOUR	NUMBER OF CALLS	WORK VOLUME MINUTES
6:00	250	750
7:00	300	900
8:00	275	1100
9:00	300	1200

4. Employee Scheduling

The final activity in the planning phase is to define and distribute a schedule for each employee. For mutual convenience, this is usually done near the end of a work week and describes the schedule for all days of the following week. The objective is to match employee availability with the forecast of work volume, while still meeting the requirements of labor laws or agreements. This match is rarely "perfect," but can be improved by establishing flexibility in the way work hours are assigned.

For example, even though all "daytime" employees may work for eight paid hours, they could be scheduled to have varied starting times, break times, and lunch times in order to provide a balance between when they actually work and when customers are forecast to make telephone calls to the group. In addition, part-time employees and overtime work can be scheduled to match the forecast. Alternate work schedules, such as three twelve-hour days or four nine-hour days each week for some employees, can also be successfully included. As with the other workforce planning activities, both manual and computer-based procedures exist to reduce the actual labor required to produce an efficient and effective schedule.

Phase 2: Execution

The execution activities are performed **during** the actual work day. They result in compliance with the plan, and realistic adjustments when unexpected conditions occur as a day progresses. Two intraday Workforce Management activities should be conducted.

1. Schedule Compliance

The initial part of managing a workforce should be activities to assure the schedules are understood by the employees - not only what they mean but also why adherence to schedule is important for customer satisfaction. Depending on the organization's climate and culture, activities to encourage compliance could range from management oversight to voluntary cooperation maintained by employee groups.

2. **Intraday Adjustment**

It is unlikely that a forecast will ever be a perfect match for a day's experienced work volume. External events such as advertising or new product introductions can change the trend unexpectedly, and weather can change the pattern within a day or week. The intraday adjustment process enables changing from the original plan to react to the work volumes and patterns actually experienced.

When work volumes exceed the original forecast, work schedules should be adjusted to provide more employees performing the primary group functions. Part-time or temporary employees can be added; other activities such as training or special projects can be deferred; and break, lunch, and overtime plans can be adjusted. Alternately, times with too many employees in the primary function can be adjusted by granting requests for time off or changing employees to other support functions temporarily. As with the original planning activities, the objective remains to efficiently match employee availability with customer requirements.

Phase 3: Analysis

Analysis should be performed **after** a work day has been completed in order to gain from the experience of each day and improve the accuracy of future forecasts and adjustments. While many indicators can be employed to track the quality of the Workforce Management process, two key indicators provide an overall summary of the process.

1. **Customer Access**

Since the primary objective of these processes is customer service, a measure of the speed with which customers receive service should be devised. For telephone or walk-in service centers, this measure could be the average delay between being present for service and actually receiving it. The turnaround time between receipt and response provides a viable access measure for mail driven work groups. Regardless of the method selected, it should be designed to capture delay **from the customer's perspective.** It should not be contaminated by excuses or obscure mathemathical manipulations.

2. Forecast Accuracy

The daily ratio of actual work volumes experienced to the original forecast of work volume provides a convenient summary of the quality of the forecasting process. Experienced work groups have repeatedly demonstrated the ability to maintain this ratio in the 95-105% range.

Organization for Workforce Management

Introducing and maintaining a Workforce Management process requires providing dedicated staff resources who become the experts and managers of the process. These employees will generally have other responsibilities, but should be the ones to be relied on to complete the Workforce Planning activities each week. The activities could be assigned to one employee or divided among several, and could be performed by either management or occupational personnel. These processes could also become part of the "self-management" tools of a participative, autonomous work group.

Conclusion

The Workforce Management processes described have been introduced and tested in a variety of work groups. They have repeatedly demonstrated the ability to simultaneously improve both customer service and operating efficiency. Recent innovations in personal computer capabilities have resulted in software which supports cost-effective use of Workforce Management in most organizations. As excellence in customer service becomes a competitive prerequisite, the application of these processes becomes an increasingly essential component of sound management practices.

Human Factors in Organizational Design and Management - II
O. Brown, Jr. and H.W. Hendrick (Editors)

FROM FUNCTIONAL PRODUCTION TO THE PRODUCT SHOP: A FOLLOW-UP STUDY OF WORK CONTENT, WORK LOAD, AND PHYSICAL ENVIRONMENT

Soili Klemola, Pentti Seppälä, and Ilkka Kuorinka

Institute of Occupational Health,
Department of Physiology
Helsinki, Finland

A follow-up of the work load and job content was undertaken in mechanical workshop where extensive changes of the production technology were made. The results showed that individual characteristics of a workstation and the degree of automation correlated better to work load and job content than the switch from functional production to product shop itself.

INTRODUCTION

A medium-sized mechanical workshop manufacturing household appliances underwent a major change in production technology in 1983. The old serial production, which had previously taken place in three different, functionally-organized shops, was redesigned mainly due to high costs and lack of production flexibility. In the new production design a product shop with flow production lines were constructed in the same premises. All products were to be made on two lines, thereby allowing greater flexibility and better loading efficiency on the production line.

The study concentrated on the following aspects: job content; autonomy; amount of work and time pressure; social contacts; the physical and chemical environment; and the skill and physical condition required by the work.

METHODS

Two mutually supplementary methods of job description were used along with a questionnaire study. The first job description method was based on description and registration of the work and production technology. In the other method, known as the AET method, an observer analyzed the work, using rating scales that yielded a work profile. The questionnaire study was conducted in order to reveal the workers´ subjective assessments of job characteristics and physical and mental symptoms of their work load.

RESULTS

Group work had been implemented along the new flow production line where practicable. The conveyor system had buffer stocks between the work stations. Some tasks had been automatized so as to avoid monotonous work and diminish exposure to physical and chemical agents. In addition, various types of ergonomic improvements had been implemented at the workplace.

The analysis of job content showed that most previously repetitive tasks had become less repetitive. These supervisory tasks had increased amounts of control and inspection activities.

Most tasks underwent no change in autonomy. In principle the new production technology allowed flexibility in the scheduling of individual tasks, but due to the increased production rate this flexibility had been used to advantage. The total amount of work per worker had increased significantly. Time pressure had also increased somewhat. Opportunities for social contacts had been good in the old factory, and they were still better in the new one. The physical environment was definitely better in the new factory.

The psychological requirements of the work had increased in most tasks. In all tasks the new factory involved increased skill requirements, more sensory load, and a greater need for fine motor - sensomotor skills. The need for memorization and other cognitive demands had not changed much.

Physical load was fairly low in the new factory, and it had become more varied. Back strain had increased in some tasks but decreased in others. The same was true for the relation between static and dynamic work. Inspection tasks were a clear exception with respect to physical work load: both the number of bad postures and the amount of static work had increased.

This case study analyzing changes in the production technology indicates that there is no clear-cut correlation between change in production technology and variables of work content or work load. In principle, the new production line should offer more varied and more favorable job content, increased skill requirements and, consequently, more interesting work. The final results, however, suggest that the design of individual tasks and workplaces has a more profound effect on job content and work load than production technology as such. In this case study it became evident that the basic characteristics of the work tasks had not changed much. From the points of view of job content and work load, some tasks had become better whereas others had become worse. Thus the mean changes were small and did not match the changes in production technology, work organization, and the physical environment.

REFERENCES

(1) Klemola, S., Kuorinka, I., Seppälä, P., Koskinen. P., and Siikaranta, R., Job stress before and after the redesign of semipaced production lines, in: Brown, I.D., Goldsmith, R., Coombes, K., and Sinclair, M.A. (eds.), Ergonomics International 85, Proceedings of the Ninth Congress of the International Ergonomics Association 2-6 September 1985, Bournemouth, England.

(2) Klemola, S. and Koskinen, P., Redesign of production systems; a case study (in Finnish), Institute of Occupational Health, Helsinki, Finland.

Human Factors in Organizational Design and Management - II
O. Brown, Jr. and H.W. Hendrick (Editors)

PARTICIPATORY ERGONOMICS: HISTORICAL PERSPECTIVES, TRENDS AND EFFECTIVENESS OF QWL PROGRAMS

Ogden Brown, Jr.

Institute of Safety and Systems Management
University of Southern California
Los Angeles, California, USA

In 1970, Quality of Work Life (QWL) programs were practically unknown in the United States. Today they represent a major element in the management strategies of such major corporations as AT&T, Digital Equipment, Ford, General Motors, Hewlett Packard and Honeywell. QWL programs are no longer limited to change and development in a single plant. QWL is becoming an industry wide strategy in steel, automobiles and communications. This paper will define and examine QWL programs as a general management strategy. Forces accounting for the trend toward greater use of QWL programs will be discussed as will the effectiveness of QWL programs.

HISTORICAL PERSPECTIVES

Nadler and Lawler (1982) have postulated six definitions which have evolved over the years. The first definition emerged during the early seventies: QWL as an outcome. Many people saw quality of work life as one's reaction to work, or personal consequences associated with work experience. The concern was how to improve QWL for the individual. What was unique then about QWL was that it focused upon individual outcomes like job satisfaction and emphasized the impact of work on the person. A common view was that organizations should be evaluated on the quality of work life they afford. Again during the early seventies a number of projects were begun, principally to get management and labor to work together to improve the quality of work life. These included United Automobile Workers (UAW) projects with Harmon Industries in Tennessee and with General Motors (GM) at the Tarrytown, NY plant. As a result of such projects, quality of work life became associated with certain approaches. A second definition then emerged which defined QWL as an approach (Nadler and Lawler, 1982). There was still a focus on the individual person rather than on organizational outcomes, but QWL also tended to be regarded as cooperative management-labor ventures aimed at improving both individual and organizational outcomes.

Within the last fifteen years, the number of cooperative management-labor programs has dramatically increased. These programs have been instituted in all types of organizations. In fact, management-labor QWL programs have been established not only within entire companies but also throughout certain industries such as steel, automobile and communications. Ford and General Motors have instituted company wide plans with the UAW. Joint

GM/UAW programs began in two plants in the early 70s, and they have expanded to well over one hundred plants. Ford and the UAW began a joint program called "Employee Involvement" later, but expansion was more rapid than at GM. Joint projects were created in most Ford plants by the early 80s.

Some non-union experiments were also conducted during this same period of time which employed different techniques. Job design and restructuring of work methods at the General Foods plant in Topeka (Walton, 1972) and other similar programs received a great deal of publicity and focused attention on specific ways to change the workplace and its impact on people. These efforts led to a third definition, QWL as methods (Nadler and Lawler, 1982). Here, QWL was regarded as a group of methods or approaches to enhance the work environment, making it more satisfying and productive. QWL was then associated with things like job redesign, job enrichment, and the design of new plants as integrated sociotechnical systems.

In the late seventies QWL activity decreased somewhat. Many felt that interest in the subject had declined with the energy crisis and economic problems. At this time some people became concerned with creating a group of interests which would support the QWL movement. Organizations were constituted to further the concept of QWL, and out of these activities came the fourth definition, QWL as a movement (Nadler and Lawler, 1982). QWL was perceived as more of a statement concerning the nature of work and the employee's relationship to the organization. "Participative management" was often conceptualized as an ideal of the QWL concept. The development of QWL as a movement might have caused some problems between those who use the term QWL and those who use other terms to describe their work. As Nadler and Lawler (1982) observed: "With movements, all too often one is either part of them or 'against' them, there are no other possibilities".

A renewed interest in quality of work life programs came about in the late 70s and early 80s. During this time a fifth definition of QWL appeared. It is best described by the following situation:

> "The chairman of the board of this large company is an enlightened and thoughtful individual who is concerned about the future of his/her organization. He attends some seminars or talks with some people at a cocktail party. He has a discussion with a consultant, and he becomes aware of this thing called QWL. He may read about experiments at a particular company, or read about some things done overseas. Coincident with this, internal surveys in his own organization indicate that people are feeling less satisfied with their work, more uncertain about the future, and more alienated from management and the ideals of the company. So the chairman then decides that there is a need to improve QWL; he sends out memos or makes speeches discussing his concern. While everyone agrees, no one is sure exactly what it means, and typically there is some passive resistance from senior management, who are uncertain about this new development. Finally, the chairman decides that his firm will have a QWL Program, and sends out instructions or puts QWL into the objectives of senior management. Senior managers turn around in the next level down and say "we need some QWL." In turn, those folks turn to their people on the next level down

> and convey the same message, until a human resources manager is told by his operating manager, "we need QWL. They're very interested in it upstairs. Get me some." In this case, having been faced with this situation, the human resources manager has several choices. One choice is to go out and buy some. Indeed, there have been an array of vendors willing to sell "QWL packages." Another approach is for the human resource manager to survey the various activities that are going on, including organization development work, internal consulting, organizational effectiveness, etc. He then goes back to his supervisor and says, "Boss, we already got some. We just didn't know we had it." Out of this emerges the fifth definition then, which is QWL equals everything." (Nadler and Lawler, 1982, pp. 6-7).

In this kind of culture organizational effectiveness efforts become synonymous with QWL. It is perceived as a global concept and often is viewed as a "cure-all" for coping with everything, such as quality control, low productivity, foreign competition and grievances. Obviously the problem here is that no one approach can successfully address all of these issues. This definition also tends to make managers wonder about what it is. Due to expectations created by the definition that QWL equals everything, a possible sixth definition of QWL could appear: QWL equals nothing (Nadler and Lawler, 1982). As the inevitable failures of some QWL projects take place, and if an inability to fulfill some of the promises occurs, even the skeptics who reluctantly accepted QWL will back down.

As a working definition, I view quality of working life as basically a way to think about people, work, and the organization. It is concerned with the impact of work on the people in the organization, with the effectiveness of the organization, and with employee participation in organizational decision making. Based upon this view, there are two things to remember: (1) the thrust of QWL efforts is not just on how people can better do their work, but also on how the work may make the people better (different from other organizational change and development programs because it is concerned with outcomes for the person); (2) an important aspect of QWL is participation in making major organizational decisions. This is not the same as complete "participative management". All organizational decisions are not made in a participative fashion, but individuals are involved in making certain decisions which affect them.

TRENDS IN QWL PROGRAMS

People from three or four different organizations would probably describe their QWL programs using three or four different conceptions of QWL. It might even be hard to understand why such different programs might bear the same name. Historical factors and the very complexity of QWL serve to contribute to its diversity. QWL represents the blending of several historical trends but nothing much new. QWL programs have employed organizational development technologies like participative management, job enrichment and job redesign, and survey feedback techniques. They have also adopted the sociotechnical systems model, the concept of autonomous work groups, and industrial democracy approaches which have their origins in European research (Trist, 1981). The term QWL has also been applied to management strategies for the design of innovative new plants, to the cooperative management-labor programs mentioned earlier, and to alternative management strategies such as quality control circles and guaranteed employment (Pascale and Athos, 1981).

Because of the differing histories and several definitions of QWL, it is no surprise that the term has contradictory and varied meanings. As the practice of quality of work life programs has matured in this country, however, a relatively stable and standard set of objectives, principles, and strategies have evolved.

There are three important trends which currently appear in the literature and in various business publications. QWL programs do not seem to be utilizing any new or radically different practices than those of the 70s. Second, there seem to be many more QWL programs than in the past. In fact, over 1,500 companies are using one single type of quality of work life program, the quality control circle (Business Week, January 11, 1982). And third, many QWL programs appear more comprehensive than in the past, both in the number of interventions and in the degree of change within a given organization (Cammann and Ledford, 1983). It would appear, then, that quality of work life programs are seemingly more comprehensive than in the recent past.

EFFECTIVENESS OF QWL PROGRAMS

The organizations and even entire industries which incur the greatest challenges from the environment are the most prone to adopt some type of QWL program. The recent and major decline of the steel and automobile industries has led management to call into question many basic assumptions, has encouraged labor to cooperate with management to save jobs, and has even resulted in the formation of several joint ventures between management, labor, and Japanese automobile manufacturers. Deregulation of the telecommunications industry and the "divestiture" at AT&T have contributed to the development of major management-labor QWL programs (Cammann and Ledford, 1983).

Organizations in turbulent environments must be able to adapt quickly to changing conditions. Without the cooperation and help of the labor force, this can be difficult at best. QWL programs are one way to develop employee commitment to required organizational changes. An increasing number of reported successes serves to encourage the adoption of QWL programs. A variety of external environmental forces are encouraging organizations in the United States to adopt QWL programs. Market conditions, foreign competition (and success), and high growth markets are all environmental considerations for the adoption of QWL programs. It is interesting to note that the experience of successful QWL programs within a given organization may have les to do with the adoption and expansion of QWL efforts than with these external environmental factors just described.

The quality of work life movement has entered an important and interesting phase. There is really no basically new QWL approach to organizational change since most of the approaches and techniques have been used for years. What is new is the commitment on the part of the organization to employ QWL methods in an integrated fashion to realize fundamental changes in the way the organization functions (Cammann and Ledford, 1983). The crucial question now is whether QWL idelogy will become the basis for development of more productive, effective and satisfying organizational designs and structures. When such programs involve organization-wide change and are conducted by those who understand and accept the values and strategies of QWL programs, their chances of success are excellent. Typically, an aura of trust and mutual respect between labor and management, and a commitment on the part of the organization to effectively employ QWL programs will lead to improved productivity, product quality, and quality of work life (Brown, 1985).

REFERENCES

Brown, O. Jr. (1985). Some perspectives on participatory ergonomics in high technology organizations. Proceedings of the Human Factors Society 29th Annual Meeting, 880-883.

Business Week (1982). Will the slide kill quality circles? January 11, 1982, 108-109.

Cammann, C. and Ledford, G.E. Jr. (1983). Quality of work life programs in the 1980's. Los Angeles: University of Southern California.

Cummings, T.G. and Malloy, E.S. (1977). Improving productivity and the quality of work life. New York: Praeger.

Fuller, S.H. (1981). Employee participation for productivity: a management view. In J.M. Rosow (Ed.), Productivity: prospects for growth. New York: Van Nostrand Reinhold.

Hackman, J.R. and Suttle, L. (Eds.)(1977). Improving life at work. Santa Monica: Goodyear.

Lawler, E.E. III (1982). Strategies for improving the quality of work life. American Psychologist, 37(5), 486-493.

Lawler, E.E. III and Ledford, G.E. Jr. (1982). Productivity and the quality of work life. National Productivity Review, Winter, 1, 23-36.

Mohrman, S. (1982). Employee participation programs: implications for productivity improvement. Los Angeles: University of Southern California.

Nadler, D.A. and Lawler, E.E. III (1982). Quality of work life: perspectives and directions. Los Angeles: University of Southern California.

Pascale, R.J. and Athos, A.G. (1981). The art of Japanese management. New York: Simon and Schuster.

Trist, E.L. (1981). The evolution of socio-technical systems as a conceptual framework and as an action research program. In A.H. Van de Ven and W.F. Joyce (Eds.), Perspectives on organization design and behavior. New York: Wiley.

Walton, R.E. (1972). How to counter alienation in the plant. Harvard Business Review, Nov-Dec, 70-81.

Human Factors in Organizational Design and Management - II
O. Brown, Jr. and H.W. Hendrick (Editors)

HOW TO MAKE A GOOD WORK PLACE-WORK "GOOD" (WELL)

Henry M. Taylor

R.E.C., Incorporated
Chester, Virginia
U.S.A.

Chemical yields were not sufficiently uniform to detect small spirals of improvement for market position (± one standard deviation) - until process design and man-machine interfaces were more uniformly controlled. Statistical control for such uniformity alone could not guarantee the managerial teamwork needed. Consequently, the direction of the organization was reoriented towards quality, pride, planning, a customer concept, and jobs for people. Now, managers are better situated to work for a competitive market position for their manufacturing site.

INTRODUCTION

As a nation of would-be managers, we are told authoritatively by a Japanese manager "to stop whining and to get to work" (Karatsu (1985)). In a like manner, we have all heard slogans such as: "We need to work smarter not harder" and that the road to competitive position is "to do it right the first time". For many years, I've wanted to exclaim: show me the road map! According to A Passion For Excellence the road map to business wellness has a number of lanes (Peters (1985)).

The purpose of this paper is to share the experience of a small pharmaceutical manufacturing plant in Petersburg, Virginia, as their managers find the most effective lanes on the road to competitive position. At least one of the lanes along the road to competitive position is behavioral, abstract, and subjective (Taylor (1983)). As this single lane was traveled, by Infracorp, Ltd. and Lee Laboratories, Inc. they demonstrated that this way helped to make a good workplace (Taylor (1984)). However, it became evident that a good workplace, in and of itself, did not provide sufficient impetus when traveling the road to become cost effective.

It has been shown by Juran (1964) and Deming (1982), that another one of these lanes is more scientific, chartable, and objective. With a wider concept, Peter F. Drucker (1986) says "managerial effectiveness boils down to being reasonably competent and highly conscientious". Isn't this both scientific and behavioral? Tom Peters (1985) infers that competitive companies travel both lanes effectively. On the other hand, Blanchard and Lorber (1984), in Putting The One Minute Manager To Work, suggest that the behavioral lane is more beneficial to making the good work place - work "good" (well). In agreement, William Ouchi (1984) summarizes in Theory Z the characteristics for the best organizations as "their entire focus is on human qualities - ". This paper is about the application of these concepts to make a good work place work "good". What may happen at "good work places" needs further description.

Experience helps us realize that although a good work place may have all of the program elements in place to position employees for site effectiveness - its product may not be cost effective. Such a good place, somehow, may not be able to achieve a consistent potential for competitive position. Such a good place is often driven on a daily basis by a benevolent system, short on effective mid-management leadership experience (Deming (1982)).

PROBLEM ANALYSIS

Employee injury and turnover, machinery breakdowns, explosion, and widely irregular process yields are all events that reflect system quality failure. Management's challenge today is to identify and to remedy the vital quality points for system success. Managers must scratch where the <u>real</u> itch is! During the past thirty years, in the United States it has been a common mistake to always blame the worker for "human" errors. More recently supervisors have been overwhelmed with training programs about "how to supervise". All too often managers require immediate supervisory behavior without providing effective system authority and system training, or even a reasonable system goal. In short, it seems many American managers have not been consistent, nor have they scratched at the proper location. The revelation here is that - the conduct of the managerial system is the <u>real</u> itch point.

Many managers dutifully do what they preceive their boss wants. In a real way, the boss often becomes their main customer. All too often, knowingly and unknowingly, they lose independent behavior and play a survival game of "yes sir".

In addition to its existence as a good work place, a site that works "good", has also nurtured responsible and accountable managers who volunteer for system quality (Taylor (1986)). These employees are oriented within their system to identify needs based upon their own statistical analysis, and to then make just decisions. This system is often driven by a team type of managerial leadership (Yoshihazu (1986)).

To summarize problem analysis: most managers - while at work - almost always - please the boss - by any means - for personal survival. This creates the need to bring both good lanes - the scientific and the behavioral - together in a system to help market position.

SITE EVENTS AND EVALUATIONS - CASE STUDY

Infracorp, Ltd. and Lee Laboratories, Inc., in Petersburg, Virginia, manufacture bulk pharmaceutical products to be distributed primarily under the A.H. Robins label. Sixty-three employees are involved in making the active ingredients for Robitussin and other fine drugs. The plant is jointly owned by A.H. Robins Company and Boehringer Ingelheim of West Germany. The site is now eight years old. The organizational design during their first six years was presented as a case study to the Human Factors in Organizational Design and Management symposium two years ago in Hawaii (Taylor (1984)).

At that time, a unique kind of statistical decision analysis audit to measure the effectiveness of the operations department for loss prevention was shown. The audit had been conducted during April of 1983. The audit report suggested the alarming possibility that the system which surrounded the shift evaluated had about a forty percent risk of not being able to help technicans avoid unforeseen catastrophy.

Almost immediately, the site management undertook intensive supervisory training. A weekly decision making course began. Job decriptions were revised. Participatory projects were initiated. Juran's (1964) Managerial Breakthrough was studied. A unique oral temperature evaluation was conducted to estimate shifter's stress from 12-12 shift rotation times. Deming's (1982) fourteen top management obligations for improvement of quality, productivity, and competitive position were introduced, as was the idea for "managing by walking around" (Blanchard (1984)).

However, a split in top level site management hampered much of this work. There were then two presidents with widely differing managerial experience, methods, and conduct at this small site. Emotional conflict gradually increased among all employees. In June of 1984, the Board reacted to retain just one of the presidents.

A catastrophic explosion occured at the site during August 1984. Repairs and restart effort demanded the attention of everyone within the managerial system. Their exemplary organizational schedule for repair was fullfilled. Restart was underway in September 1984. A period of penetrating analysis of the explosion followed and ended in the conclusion that ethylamine had come in contact with a chloroalcohol - within a closed drum. The resultant exotherm blew off the drum lid. The flying drum lid cut into a protected overhead light conduit causing ignition - and a boiling liquid expanding vapor explosion (BLEVE) ensued.

Managerial will was to revitalize the site managerial system to continue its farsighted growth practices. The concept of random variability and use of control charts, to record quality graphically, was started for one chemical process. Environmental toxicology to inform employees about risk control continued. An innovative program, PRIDE, was initiated to help stir managerial awareness for this participatory need. Quality was emphasized, as was retraining, education, and operational process guides. In October 1984 more managerial time in operations was justified. Meetings and workshops were held to discuss quality. Stress-less sensor pads were provided to managers for a biofeedback method to help manage stress. A high level position to find new product business was also filled. Near the end of 1984, the basis for an automated blood pressure machine for employee use was begun. A site wide comprehensive medical examination was also conducted. At the same time, another industrial hygiene engineering evaluation of employee exposures to chemicals and noise was completed. The only significant medical finding was a reduced forced expiratory volume (FEV-1) which, as in the past, correlated with employee smoking habits. Nevertheless, a supervisor's educational program to control chemicals even better was initiated.

The year 1985 began with a quality workshop. In February, an innovative muscle memory program to maintain sound employee back strength was renewed. Although none had occured, the risk from back stress continued to be recognized. An employee attitude survey was conducted in March that found the level of frustration for many employees at the site conflicted with quality craftsmanship. In early April, the Manager of Operations attended The Deming Course. Also in April, the shift rotation was changed to 8-8 and oral temperature evaluations confirmed reduced stress. Additional managerial help was provided in the operations group. Managerial work continued to remove words like "slowly" from hands-on work instructions. Renewed effort was made to write all instructions for the same conduct using the same wording. A workshop dealing with employee frustration about job attitudes and the managerial system was also conducted during April. By May, managerial recognition was to swing from supervisory causes for frustration toward managerial system conduct. Management addressed peer management to emphasize the need for Deming's fourteen top managerial obligations. By June it was evident that managerial cooperation was drowning in "gotcha" games. All site managers were reminded of the site customer concept and the importance of minute to minute attention to technician coaching. It was also suggested to demand managerial constancy of purpose. In September the hazard determination program, which had been site practice for seven years, was formally written.

In October 1985, showing began of the Deming Videotapes. Their purpose was to point out what managers must learn to do in order to achieve cost competitiveness. All site managers, including the president, agreed to attend each quality workshop. As part of this effort, employee insecurity and fear was discussed. An outline was made to curb fear by modification of managerial action and conduct.

As the Deming managerial tapes continued into 1986: Risk-Rework and Loss Prevention for the site's past, present, and future was highlighted. This emphasis was upon recognition of system insecurities - the long term - as well as flow charts to explain the system and control charts to show a graphical record of quality (ANSI-1975). At the end of January, the president conducted a workshop about basic statistical concepts. This was the system turning point - from reaction to crisis towards a managerial system to become the best in the world. There was also a renewed emphasis upon the high cost of cigarette smoking to the site and its employees because of the associated reduction in employee breathing capacity. The ANSI (1975) guide for quality control and control chart method of analyzing data was used for this problem. In addition, the management team developed a helpful flow chart to introduce new products into production. This helped to turn management by conflict toward managerial cooperation. By February, measurable results from managerial teamwork were evident. An urgent customer delivery date for a new product was fullfilled ahead of schedule, without a single unpleasant incident and without managerial conflict! In addition, the maintenance group started to develop control charts, and texts for unnatural levels of variance were consulted (AT&T (1956)). Technicians were involved with statistical tools! As a result it was found, as Juran (1964) reported, that eighty percent of the problems were system related and only twenty percent of the problems were people related. The safety system for new products was also made more explicit. In addition, there was renewed attention given to the system loss prevention checklist from March 1983 (before the explosion) with additional focus upon the control of rework, managerial purpose, and competitive market position. There was also new painful awareness of the need in operations for even more technician involvement in the decision process. By April, the managerial quality workshops with the Deming videotapes were complete.

To conclude this case study: most managers - while at work - almost always -

are thorough - if shown the system method by their president - for survival.

SUMMARY - SOLUTION CONTROL

The logic tree on the following page summarizes the events perceived to be vital for the site's business success. This paper has dealt with the decision logic A from the tree and has reviewed the evolution of a managerial system which moved from narrow departmental views toward a team view. The managerial team now recognizes a new customer concept - its on-site customer is the manufacturing operations group, not the president(s).

The site has evolved toward acceptance of a concept that an unwanted incident of any kind is a result of premeditated system carelessness. New incidents are very rare events. Product yields are increasing. Pride in craftsmanship is much improved and has real base in quality. Absenteeism is almost nil. The number of jobs has increased nine percent. Manufacturing capability is increased forty percent. New orders increase. Profits are increasing.

The site is learning that there are no easy solutions. It must make its own "do it right" road map to competitive success. The controlling forces vital to making their map include constancy of managerial purpose, customer concepts, teamwork, flow and control charts, knowledge, experience, training, and quality. How to position and orient these vital forces rightly is up to the innovation of the managers at the site.

To summarize solution control: most managers - while at work - almost always - use capacity thinking - by orienting themselves toward the system method for team quality - for system survial.

ADDENDUM

It is everyones responsibility not to allow insidious time pressure to reduce the quality of their work. Today, the associated risk for competitive position, safety, health, and the environment, as a consequence of rework, is both alarming and unaffordable. The apt use of control charts for micro-ergonomics (process design and man-machine interfaces) provides a method for hands-on workers to regulate vital problem criteria exactly. Thus, they forecast their success by doing it right the first time. Achieving this control stimulates PRIDE in hands-on workers for their craftsmanship. The manager is then free from constant auditing of worker's craftsmanship. He is free from crisis and can then relearn how to adapt the system to the following method - the fourteen management system obligations (Deming (1984)).

CONDENSATION OF THE FOURTEEN (DUTY) POINTS FOR MANAGEMENT*

The Fourteen Points apply anywhere, to small organizations as well as large ones, to the service industry as well as manufacturing. They apply to divisions within a company.

1. Create constancy of purpose toward improvement of product and service,

* W. Edwards Deming (1984)

PRIDE

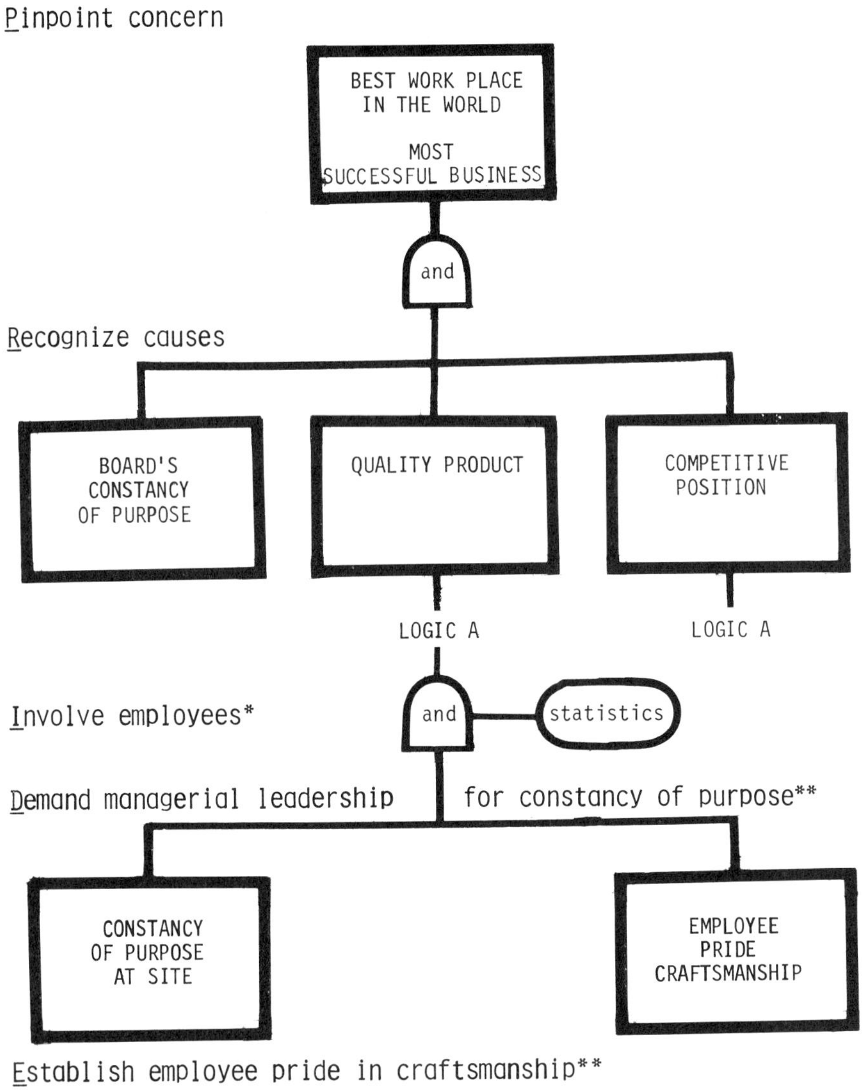

*Encourage employees to use statistical control charts to develop freely and with natural variation up to each person's own best potential.

**Demand constant managerial behavior to help hands-on employees be proud of their craftsmanship.

R.E.C., Inc. 4/86

with the aim to become competitive, to stay in business, and to provide jobs.

2. Adopt the new philosophy. We are in a new economic age, created by Japan. Transformation of Western style of management is necessary to halt the continued decline of industry.
3. Cease dependence on inspection to achieve quality. Eliminate the need for inspection on a mass basis by building quality into the product in the first place.
4. Stop awarding business on the basis of price tag. Purchasing, design of product, manufacturing, and sales must combine to work with the supplier chosen. The aim being to minimize total cost, not merely initial cost.
5. Improve, constantly and forever, every activity in the company, to improve quality and productivity, and thus decrease costs.
6. Institute training and education on the job for all employees, including management.
7. Institute supervision. Supervision should be aimed at helping people and machines and gadgets to do a better job.
8. Drive out fear, freeing everyone to work effectively for the company.
9. Break down barriers between departments. People in research, design, sales, and production must work as a team, to foresee problems of product or service.
10. Eliminate slogans, exhortations, and targets for the work force asking for zero defects and new levels of productivity. Such quality and low productivity are caused by system ineffectiveness, and thus lie beyond the power of the work force.
11. Eliminate work standards that prescribe numerical quotas for the day. Substitue aids and helpful supervision, using statistical methods.
12. a. Remove the barriers that rob the hourly worker of his right to pride of workmanship. The responsibility of supervisors must be changed from sheer numbers to quality.
    b. Remove the barriers that rob people in management and in engineering of their right to pride of workmanship. This means (among other things) abolishing the annual or merit rating and management by objective.
13. Institute a vigorous program of education and retraining. New skills are required for changes in techniques, materials, and service.
14. Put everyone in the company to work in teams to accomplish the transformation.

The criteria which are selected to score a frustrating problem are the key elements to positive results. Their successful completion forcasts a successful outcome for the problem. It is important to learn how to work smarter and harder to identify vital "itch" points and to then word the chosen criteria in a way that they can always be improved. Thus, to improve a job, we need only to narrow the job's criterion range for success.

As for macroergonomics: (defined here as the organizational system focus necessary to implement microergonomics). The managers at Infracorp, Ltd. and Lee Laboratories, Inc. are becoming aware that, while not intentional, most of the stress yet in the workplace is the consequence of past oversight by the managerial system. However, their belief in the worth and dignity of human potential provides clarification and confirmation for how they wish their system to work. They also recognize that their constant application of Deming's fourteen management obligations (managerial operational definitions) will work (scratch) to dispell disruptive stress: to make this good work place work well and thus to become the best site system in the world for pharmaceutical manufacture.

REFERENCES

American National Standard Institute: guide for quality control and control chart method of analyzing data, (American National Standard Institute, Inc., NY, 1975).

AT&T, Statistical Quality Control Handbook, (Delmal Printing Co., Charlotte, NC, 1958) 25-27.

Blanchard, K. and Lorber, R., Putting the One Minute Manager to Work, (William Morrow and Company, Inc., NY, 1984).

Deming, W.E., Quality, Productivity, and Competitive Position, (Massachusetts Institute of Technology, Cambridge, MA, 1982) 37-133.

Deming, W.E., What America's managers must do to improve quality, productivity, and competitive position, (Massachusetts Institute of Technology, Videotapes, Cambridge, MA, 1982).

Deming, W.E., Condensation of the fourteen points for management, personal correspondence, (1984).

Drucker, P.F., Timeless truths about performing at your best: Working Smart (Learning International, Petersborough, NY, April 1986) 3.

Juran, J.M., Managerial Breakthrough, (McGraw-Hill, NY, 1964).

Karatsu, Hajime, U.S. Managers: stop whining and work!, Quality Circle Digest (Red Bluff, CA, December 1985) 24-28.

Ouchi, W.G., Theory Z, (Avon Books, NY, 1982) 156.

Peters, Tom, A Passion for Excellence (Random House, 1985).

Taylor, H.M., How to make things work, Hazard Prevention, (March-April 1983) 7-9.

Taylor, H.M., How to really make things work: Human Factors in Organizational Design and Management Proceedings (North-Holland, Amsterdam, 1984).

Taylor, H.M., Loss prevention (R.E.C., Inc., Chester, VA, 1986) 8.

Yoshihazu, T., Tribus, M., and Newcomb, J.E., Changing the corporate culture, (Massachusetts Institute of Technology, Boston, MA, 1986).

BIOGRAPHICAL SUMMARY

Henry M. Taylor is president of R.E.C., Inc., a management consulting firm located in Chester, Virginia. R.E.C., Inc. uses a system approach to evaluate risk and control safety, health, and environmental hazards chiefly through prevention of rework. Mr. Taylor has been associated with innovative tasks at Uncle Ben's, Inc., Allied Chemical Corp., Stanford Research Institute, the Ethyl Corporation, and TVA. He received a B.A. in chemistry and zoology at North Central College and an M.P.H. in environmental science from the University of North Carolina. He has also served as an adjunct college/university faculty member to instruct in safety, industrial hygiene, and human factors.

Human Factors in Organizational Design and Management - II
O. Brown, Jr. and H.W. Hendrick (Editors)

THE SELF-MANAGEMENT CONDITIONS AND IMPROVEMENT OF QUALITY OF WORKING LIFE ON ASSEMBLY-LINE

Vuksan P.Bulat
Department of Industrial Engiaeering
Faculty of Mechanical Engineering
University of Belgrade
Yugoslavia

The analysis of working conditions and experiences of workers on assembly-lines regarding technology and organization under the self-managements relations,shows the features of man's alienation. Establishing the model of the Autonomous Self-Management group working on assembly-line and possessing some crisis parameters,it is shown that the improvement of Quality of Working life could be acquired by substantial forms improvement under self-management conditions.

PROBLEM

An implement of the assembly-line system was exercised simultaneonsly with the development of the self-management relations, resulting from the Yugoslav modern technology.The researches performed on the range of the assembly-line system,proved that 41,8% of industrial workers are employed on assembly-lines /1/. The structure of labour has the same characteristics as in those milieus which technological transfer was performed from. Namely,the female labour is that which prevails-61,5%,68,6%, consisting of the young-under the age of thirty,as well as the semi-skilled ones - of 66,3%.Taking all these facts into accounts,there is a good reason for already stated adversenesses due to the assembly-line system,to be expected.Owing to the presence of the self-management system,though,already being introduced in Yugoslavia,a hypothesis that workers position in the self-management social system can compensate such a negative effects,is put up.

RESEARCH METHOD

A research-approach was developed on the ground of a survey of the assembly-line system ranging different industrial branches as well as the checking of certain relevant specific manifestations,charaeterised by the following traits:

1.The technological and organizational solution of the assembly-line system,jointly with other conditions,determines the totality of requirements for worker as an individual and member of a working group while performing his working task.

2.Worker as human being experiences these requirements and conditions,reacting individually,wich is followed by the appearance of multidimensional interaction with group members and the others as well.

There was concieved a set of instruments for the research accomplishment,comprising the following:

- general view of assembly-line performances,working conditions and other elements about assebmbly-line functioning as a whole.

- working task details,all relevant to subject matters,both quantitatively and qualitatively,and also conditions and material elements which worker as a performer is exposed to.

- experiences and reactings of worker related to his working task and the conditions under which it is performed,seen from various aspects/psycho-physiological,sociological and others/.

All elaborated set of instrument was applied to the thirty five selected assembly-lines,involving the complex of metal,chemical pharmacological,foot-wear,ready-made garments,food and tobacco industries.

The results of the eighteen representative assembly-lines out in four groups belonging to analogous enterprises,will be partly displayed in this work:

(a)industry of metal:assembly of vehicles,assembly of motors,assebly of trucks,assembly of tractors, assebly of washing-machines,assembly of electrical counters,assembly of etectro-motors,assembly watch-mechanismes.

(b)industry of shoes:production of boots,production of shoes,operation of soles.

(c)indutry of garments:production of jackets,production of pants,production of training-dress.

(d)food industry:production of soft drinks,production of crackers,filling of bonbonnieres I and II.

On the basis of the gained results and corresponding comparative analysis,the project of the Model of the Autonomous Self-Management working groups /ASMWG/ was undertaken as a sort of a specific overcoming of man's alienation concerning the work on assembly-lines under the self-mahagement conditions.

The application of the ASMWG on assembly-lines was done for:

- assembly of watch-mechanismes

- filling bottle with fluids

the relevant results of which will be consequently shown, in order to clarify man's role in the improvement of Quality of Working Life under the self-management conditions.

RESULTS AND DISCUSSION

The relevant results deriving from the research project and the set hypothesis,will be shown further on.

The conditions of performing working tasks are mainly the same everywhere,they reccur showing slight differences in subjects, depending on production model,which relate to three assembly-lines.

The cycle time as a kind of time determination of the possibility of diversificated content of working task,is ranged into groups as follows:

-industry of metals:15 up to 150 sec,
-industry of shoes: 67 up to 120 sec,
-industry of garments: 84 up to 150 sec,
-food industry: 1,2 up to 14,6 sec.

Futhermore it is characteristic that the rate of working hours usage is higher /cca 80%/,as the cycle time is shorter (assembly of washing-machines - 36 sec. and assembly of watch - mechanismes - 19 sec).The similar thing happens with assembly-lines in the group of food and ready-made clothing industries. A dictated working rhythm is remarkably expressed with the assembly-lines in the foot-wear and food industries.Extra breaks are exceptional in 25% elaborated assembly-lines.

As for the ecological factors of working conditions,the microclimate state is as follows:

-the temperature is above 22% found with thirteen assembly-lines,

-the humidity above the comfort zone appeared in two cases /production of soft-drinks and filling of bonbonnieres/,

-there is no pollutioners on the seven assembly-lines,in other cases various vapours,dustes and wastes deriving from production process are present,though.

The noise is on the level of 80 dB and even more on ten assembly-lines.

A comparative survey on ten elaborated assembly-lines shows that if the requests concerning the cycle time repetitive dictate and the like are unsuitable,even more difficult are working conditions,as for instance:

-production of soft-drinks /cycle time of 1,2 sec,

dictated rhythm,no extra breaks,temperature-26°C, humidity being of 71%,noise-94 dB/,

-production of crakers/cycle time of 5 sec.,dictated rhythm,no extra breaks,temperature-25°C,humidity -46%,noise of 84 dB/,

-assembly of match-mechanismes /cycle time that of 19 sec.dictated rhythm,no extra breaks,temperature-24°C,humidity ranging from 40 to 60%,noise not equal even up to 94 dB/.

In the frame of the study on worker's experience of his working task and his role within the self-management context and from the point of view of possible Quality of Working Life improvement,the further explication will be concetrated on questions and answers,selected from a questionary,obtained basicly from corresponding data,for eighteen elaborated polled out assembly-lines.

To the question:"How do you feel about your personal experiencing of your own working post",out of the offered multiple-choiee answers such as:badly,fine,very well,excellent,the results mark out some pecularities such as:

-the majority of people working on assembly of vehicles-71%,contrasting with others,decided on the answer "badly",although considering their working conditions in total,they are not in the worst position,even on the contrary.The similar happens,not expressed to that extent though,with those working on other assembly-lines,belonging to the metal industries,which can be justified by a particular developing level achieved by metal workers and by their greater autonomy in the working tasks accomplishement,and concerning this matter,the possibility of expressing their discontent,

-quite the contrary,the majority of those working on assembly-lines in the food-industry feel well,and that on the production of soft-drinks-100% respectively,production of crakers-69%,filling of bonbonniers I -56%,including "very well" too:filling of bonbonnieres II -60%,which is explicable by already commented very aggravating conditions,and,first of all,by the fact that merely women are employed there meaning for them the only chance of getting a job, on the very low degree of qualification,though.

But,to the questions:"Do you feel tired after work?" "If so can you define more precisely what kind of a fatigue could it be?",within the reference of offered answers such as:"Normal fatigue,great fatigue,and slight fatigue",meaning "physical, mental,both mental and physical",relying on the obtianed results,there appears to be a similarity comfortably to answers about general feelings,with the remark that,concerning the assembly-lines for:

-assembly of watch-mechanismes, the percentage of great fatigue is positively high-72,72%, and as for physical, mental and physical and mental fatigues, they are almost equally disposed-27,4%, 36,3%, 36,3% respectively, due to existing differences between working posts and to various individual inclinations and capabilities for corresponding working experience as well.

-filling of bonbonnieres II with unsufficiently expressed fatigue, but the mental and physical one is found with all workers-as a results of intensive work on very limited working subject matter, and the cycle time of 2,4 sec.

The answers of people working on assembly-lines to the question "Would you gladly change your working post?"testify of their general disposition. On the six assembly-lines, workers made positive statements-60% and even more, and on the four assembly-lines-6o and more per cent workers answered negatively.

The given answers on the state and development of the self-management -"How would you judge the development of the self-management relations in your work organization?", according to the offered answers such as: poor, good, very good, excellent", these results follow:

-the majority of respondents decided on the answer "poor" and "good" thus making the score of higher than 50%, on the eight assembly-lines.

The answers to the question "Has the self-management reached certain positive results of improving working conditions on assembly-lines?" were polarized in the way that in nine cases they were positive-of 50 and more per cent, and in the same time there appeared an equal number of negative ones as well, indicating that, according to workers concern, something is really happening there.

Workers engagement in self-management authotorities is quite usufficient, the prof of which is the fact that on the eight of assembly-lines, more than 8o% are not members of self-government authorities. In certain connection with this are the answers to the question: "Do you consider yourself of having a possible influence on decision-making in your work organization?" offered to state in this way: "I do; partially; little; no influence at all", the summed up answers to these two last ones, for thirteen cases, were more than 50%, and in only three cases of assembly-lines, the answers of "I do" happened to be found of 2o%.

On the ground of the exposed matter dealing with work experiecing by workers employed on assembly-lines, in confrontation with requirements and working conditions with the regard to the judgement of self-management possibilities of contributing, as a system of social relations, to work humanisation, the concept of ASMWG was developed and applied to assembly-lines for watch-mecanismes. /3/

An active cooperation of assembly-lines workers in working out of this concept,then in self-management authorities making corresponind decisions as well,gave another direction from the former sceptic humour into enthusiasm to the introduced changes. This has made manifest through an insisting those changes to become definite and to be spread to all other departments where work on assembly-lines is being done.The like happens in a factory where this experiment was led out on assembly-lines for filling-up bottles of fluids.

CONCLUSION

Performed research and obtained results make possible drawing of the following conclusions:

- -the actual state of self-management forms development does not meet the need of a sufficient effective confrontation with negativenesses affecting worker on the assembly-line system,
- -the model of ASMWG gives the chance of activating workers in positive level rising of Quality of Working Life.

REFERENCES:

/1/ Bulat V. and coautors,Istraživanje humanizacije rada na proizvodnim trakama u samoupravnim uslovima,Centar za ergonomska istraživanja Instituta tehničkih nauka SANU, Beograd,1982.

/2/ Bulat V. and coauthors,Istraživanje ljudskog faktora na proizvodnim trakama u samoupravnim uslovima,Centar za ergonomska istraživanja Instituta tehničkih nauka SANU, Beograd,1985.

/3/ Bulat V.and coauthors,Autonomous Self Management Work Group as a Solution for Humanization of Paced Work,Proceedings IX Congress IEA,Bournemouth,England,1985.

Human Factors in Organizational Design and Management - II
O. Brown, Jr. and H.W. Hendrick (Editors)

A PHASED TECHNIQUE FOR DIAGNOSING AIR QUALITY AND RELATED VENTILATION PROBLEMS

E.M. Sterling[1], E.D. McIntyre[1], C.W. Collett[1], J. Meredith[2]
T.D. Sterling[3]

We report on a phased system of integrated measurements of air quality, ventilation and occupant complaints that is relatively inexpensive yet leads to effective diagnosis of a building's IAQ and related ventilation problems. A case study is presented to illustrate the technique through case studies.

INTRODUCTION

Often a buildup of airborne contaminants due to the malfunctioning of the building mechanical ventilation system is suspected to be the cause of occupant comfort and health problems. The detailed monitoring of ventilation and air quality required to document possible system malfunctioning can be a time consuming and expensive undertaking, often with limited practical results. On the other hand, a phased program of gathering information from occupants combined with determining concentrations over time in various locations of easily measured key substances (CO and CO_2) plus inspection of other easily observed ventilation parameters can lead to more or less detailed judgments about the adequacy of the performance of a building's ventilation system. (1)

Phase 1 of the screening process includes:

a) General overview of the performance of the existing mechanical systems including site inspections and review of plans.

b) Administration of a Work Environment Survey Questionnaire to all building occupants as a means of documenting environmental problems and symptoms reported in the building and to locate clusters of complaint areas for detailed monitoring.

Phase 2 of the screening process includes:

a) Measurement of carbon dioxide as an indicator of a build-up of contaminants generated indoors from occupants and equipment.

b) Measurement of carbon monoxide as an indicator of combustion byproducts infiltrating from the outside, especially from parking garages or from other sources of indoor combustion.

Phase 3 of the screening process includes:

a) Measurement of air exchange rates with tracer gas to determine the total ventilation rate within the building as well as the pattern of air movement through the building including air leakage from potential indoor pollutant sources such as parking garages.

One complete building evaluation is used here to illustrate the methods used.

The Building

In addition to the building under study, a control building was also included in the investigation for comparison purposes. Each building is occupied by approximately 130 office employees.

The study building is a 4 story, 10 year old, sealed, mechanically ventilated and air conditioned office building. This building is similar to much of the low-rise office space existing throughout North America.

The control building is also a 4 story office building. However, it is approximately 50 years old and is equipped with operable windows for ventilation and radiators for heat.

PHASE 1: SITE INSPECTION AND QUESTIONNAIRE SURVEY

Description of Mechanical Systems

The study building has two mechanical ventilation systems, one services the core, the other the perimeter office space. The core area offices are ventilated by a constant volume ducted system providing approximately 2,000 litres per second (l/s) of outdoor air distributed equally between the four office floors. Outdoor air is delivered directly from the roof top mechanical penthouse to the centre of the ceiling plenum space on each floor. Ceiling mounted heat pumps then draw a mixture of outdoor and return air from the ceiling plenum space, temper it and diffuse the mixed air into the office area through grills in the suspended ceiling.

A second system provides a mixture of outdoor and return air to the perimeter office space through perimeter console or fan coil heat pump units. The perimeter console units are located below the windows on three sides of the building. All console units draw air directly from the outdoors, except six located along the east facing side of the first floor. These six units draw outdoor air from a duct that passes through the basement parking garage.

The Survey of Occupants' Perceptions

The survey questionnaire was administered to all of the office occupants of both buildings. A predominance of environmental complaints such as inadequate ventilation, dry and stuffy air were identified among building occupants of the study building compared to the control building. Symptoms reported most prevalently among occupants of the study building were of mucous membrane irritation such as the eyes, nose and throat. Symptoms also included headaches and fatigue. Of the four floors of the study building the third floor was found to have the highest incidence of both environmental complaints and symptoms.

PHASE 2: ENVIRONMENTAL MEASUREMENTS

Carbon monoxide and carbon dioxide were measured in both buildings two times a day on work days during a continuous three week period. Measurements were taken at selected sites in the occupied office space on each floor in the study building and on two floors of the control building. Outdoor measurements were taken on the roof of the study building.

Carbon Monoxide

Table 1 shows concentrations measured inside the study building (including the basement parking garage) outdoors on the roof and in the control building. The mean concentrations of CO do not vary greatly by floor or by morning or afternoon. The highest CO concentrations were found in the basement parking garage, with morning and afternoon means of 11.4 ppm and 15.3 ppm. The CO concentration on all floors of the study building were similar and did not vary significantly by time of day. Concentrations were, however, slightly lower in the control building.

The CO concentrations encountered in both buildings are high in comparison with those measured in similar buildings. Figure 1, extracted from the Building Performance Database[4](2), shows the distribution of CO in similar buildings. (3)

Results of statistical tests of CO measurements taken in the parking garage, outdoors and at sites on each occupied office floor of the study building, showed that concentrations of CO in the office space of the study building are affected by both outdoor sources and fumes from the parking garage in the basement. The outdoor contribution to indoor CO is consistent throughout the building. CO from the parking garage, although permeating the whole building, is strongest on the first floor and progressively weakens by floor level. Insofar as CO is an indicator of vehicle exhaust fumes, the conclusion is compelling that part of the indoor problem is created by the many irritants in exhaust gases and aerosols.

Carbon Dioxide

Carbon dioxide measurements taken in the study building, outdoors on the roof and in the control buildings are shown in Table 2. Two trends are apparent in the study building:

1. CO_2 tends to be higher on the upper two floors.
2. CO_2 concentrations increase in the afternoon.

The patterns of daily variation and variation between floors for CO_2 are shown in graphs of Figure 2. There are sharp contrasts between morning and afternoon as well as differences in CO_2 between floors. Morning concentrations of CO_2 show a strong positive relationship between indoor and outdoor levels. However, this relationship is not evident for afternoon CO_2 concentrations.

In the mornings, indoor CO_2 concentrations are strongly influenced by the outdoor levels. However, this changes over the day, with outdoor CO_2 no longer a main determinant of indoor CO_2 in the afternoon. Results show that outdoor CO_2 has a strong effect on the increase of indoor concentrations in the morning. However, that effect is no longer present in the afternoon after the building has been occupied. It is therefore likely that the ventilation system is unable to keep pace with occupant generated CO_2 and, by extension, that the ventilation system is unable to keep pace with accumulation of other contaminants generated by sources in the building.

PHASE 3: VENTILATION SYSTEM PERFORMANCE

Phase 3 consisted of a number of tests and evaluations performed to:

1. Measure the amount of air being provided to the building occupants by tracer gas and smoke pencil evaluations.

2. Determine the path by which polluted air from the basement parking garage enters the occupied office floors, by tracer gas and smoke pencil evaluations.

Measurement of Evaluation of Outdoor Air Supply

Tracer gas was used to measure the amount of outdoor air supplied to the entire building, the amount of outdoor air provided to occupants of each floor and to the first floor conference room. In addition, the percentage of outdoor air provided by two of the perimeter console units was determined.

Movement of air currents in the building were visually determined by following the movement of "smoke" produced by smoke pencils.

Table 3 shows the total air supply on each floor provided by both perimeter and core mechanical ventilation systems. Ventilation rates for each floor comply with the current ASHRAE standard of 10 l/s/person.

Table 4 presents results of testing the amount of outdoor provided by two of the 88 console units that comprise the perimeter mechanical system. The total air delivered by each console unit is approximately 100 l/s on maximum setting and the fresh air component is 18-20 l/s. By extrapolation, if all 88 console units were operating at maximum capacity, then they would supply approximately 1,760 l/s/person of outdoor air to the building.

Table 5 shows that while the supply of outdoor air to the first floor as a whole is adequate, the supply of outdoor air to the conference room on the first floor (presuming a population of 20 people) is inadequate. While the ASHRAE standard suggests a supply of 17.5 l/s/person of outdoor air for conference rooms, the existing system supplies only 3.5 l/s/person. From an analysis of the building plans, it was also determined that due to problems with ventilation design and efficiency other densely populated areas of the building may also be poorly supplied with fresh air.

Qualitative tests of air circulating and mixing patterns made using smoke pencils showed good mixing of air in the occupied office space.

Determination of Pollutant Pathways from Garage

Carbon monoxide measurements collected as part of Phase 2 of the three-part screening process showed a strong correlation between the concentrations of CO in the garage and in the first floor adjacent to six perimeter console units. A sheet metal duct providing outdoor air to the six perimeter console units located along the east side of the first floor was suspected to be the pathway for the passage of CO and other pollutants from the garage into the building. This duct passes through the garage and was found by smoke pencil tests to entrain air from the garage when the consoles were operating. A tracer gas test was conducted to verify the penetration of air from the parking garage into the first floor office space through this duct.

Table 6 shows the results of releasing tracer gas into the garage and then measuring the concentrations of tracer gas penetrating the first and second floors. Tracer gas was detected throughout the first floor, with higher concentrations on the east side near the six consoles supplied with outdoor air by the duct passing through the garage. No tracer gas, however, had penetrated into the second floor. These results show that the duct passing through the parking garage is the main pathway of penetration of contaminated air from the garage into the first floor.

DISCUSSION

The phased screening approach to field monitoring of indoor air quality and ventilation system performance now tested on a wide variety of buildings in the U.S., Canada and England, provides the components for development of a standard method for undertaking building evaluations. Adoption of this standard methodology for evaluating indoor air quality and ventilation system performance, will benefit the investigator as well as the building owner/tenant by allowing the comparing of measurement collected in any building to data from other buildings. As more studies are undertaken, a baseline of data will also be available for comparison purposes. In addition, use of standardized approach will provide the investigator with a check against overlooking pertinent information that may lead to practical and effective solutions.

REFERENCES

(1) Sterling, E.M., McIntyre, E.D., Collett, C.W., Meredith, J., Sterling, T.D., Sick buildings: Case studies of tight building syndrome and indoor air quality investigations in modern office buildings, Environmental Health Review. (1985) Vol 29(3):11-19.

(2) Theodor D. Sterling Limited (TDS), The Building Performance Database User's Guide, (1984). (Available on request from TDS Limited.)

(3) Sterling, E.M., Sterling, T.D., Pollution concentration in buildings, Proceedings. AIA Research and Design '85, Los Angeles, CA., March 14-18, 1985. 279-184.

FOOTNOTES

1 Theodor D. Sterling Limited, 70-1507 W. 12th Avenue, Vancouver, B.C. V6J 2E2

2 British Columbia Buildings Corporation, 3350 Douglas Street, Victoria, B.C. V8W 2T4

3 Faculty of Interdisciplinary Studies, Simon Fraser University, Burnaby, B.C. V5A 1S6

[4] Information about use of and access to the Building Performance Database can be obtained by contacting Theodor D. Sterling Ltd., #70-1507 W. 12th Avenue, Vancouver, B.C. Canada V6J 2E2, (604) 733-2701.

TABLE 1

MEAN CARBON MONOXIDE CONCENTRATIONS AND (NUMBERS OF OBSERVATIONS) IN PARTS PER MILLION, BY FLOOR FOR STUDY BUILDING, INCLUDING BASEMENT, CONTROL BUILDING AND OUTDOORS

	Study Building						Control Building	
	Basement Parking Garage	Floor 1	Floor 2	Floor 3	Floor 4	Roof	Floor 3	Floor 4
AM	11.4 (14)	8.9 (30)	7.8 (15)	8.2 (15)	7.8 (15)	8.7 (15)	6.8 (15)	7.5 (15)
PM	15.3 (13)	7.9 (30)	7.5 (15)	7.8 (15)	7.4 (15)	5.9 (15)	7.3 (15)	7.6 (15)

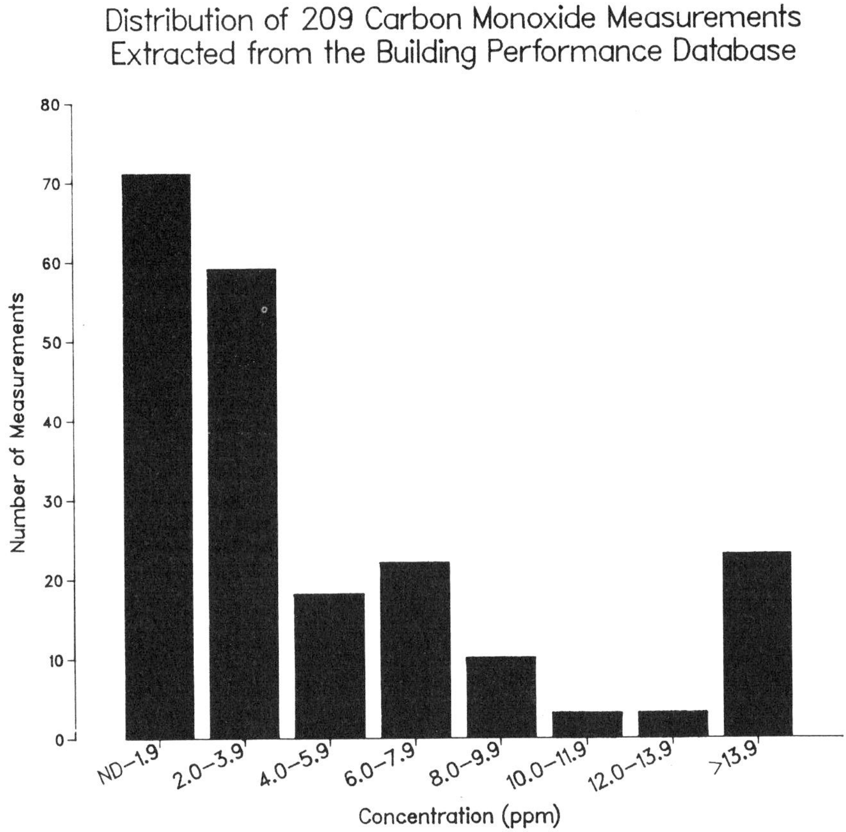
Distribution of 209 Carbon Monoxide Measurements
Extracted from the Building Performance Database
Number of Measurements
0
10
20
30
40
50
60
70
80
ND–1.9
2.0–3.9
4.0–5.9
6.0–7.9
8.0–9.9
10.0–11.9
12.0–13.9
>13.9
Concentration (ppm)

FIGURE 1

TABLE 2

MEAN CARBON DIOXIDE CONCENTRATIONS AND (NUMBERS OF OBSERVATIONS) IN PARTS PER MILLION BY FLOOR FOR STUDY BUILDING, CONTROL BUILDING AND OUTDOORS

	Study Building					Control Building	
	Floor 1	Floor 2	Floor 3	Floor 4	Roof	Floor 3	Floor 4
AM	434.9 (106)	468.4 (103)	516.2 (84)	509.6 (104)	255.6 (15)	533.3 (15)	483.3 (15)
PM	471.2 (104)	491.6 (105)	545.5 (86)	509.6 (15)	288.9 (15)	408.9 (15)	433.3 (15)

FIGURE 2 Mean Morning and Afternoon Carbon Dioxide Concentrations Measured in the Study Building

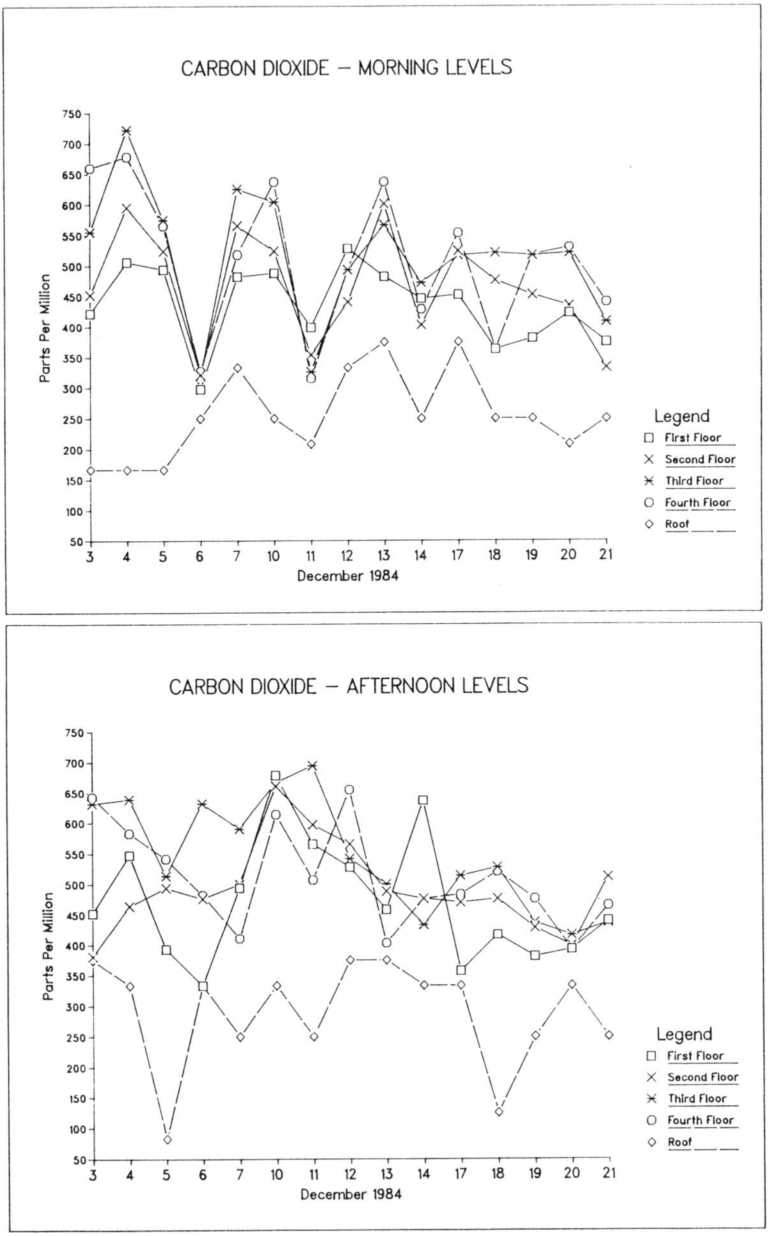

TABLE 3

TRACER GAS TESTS OF OUTDOOR AIR SUPPLIED TO EACH FLOOR OF THE STUDY BUILDING*

	Total Outdoor Air Flow	Actual Population	Outdoor Air Supply Per Person
First Floor	630	45	14
Second Floor	737	33	22
Third Floor	800	43	18
Fourth Floor	676	28	24

*In litres per second, l/s.

TABLE 4

TRACER GAS TESTS OF OUTDOOR AIR SUPPLY FROM TWO CONSOLE UNITS

	Unit #1	Unit #2
% Outdoor Air Supplied By Console Units	18%	22%

TABLE 5

OUTDOOR AIR SUPPLY

	Total Outdoor Air Flow	Outdoor Air Supply/Person	Air Change/Hour
Conference Room	69	3.5**	1.5
First Floor	630	14.	0.6

*In litres per second, l/s.

**Based on 20 persons (ASHRAE Standard for a conference room is 17.5 l/s/p)

TABLE 6

TRACER GAS TESTING OF LEAKAGE PATTERNS FROM THE PARKING GARAGE

	Nitrous Oxide Levels	
	First Readings*	Second Readings*
Parking Garage	3.52	2.40
First Floor East	0.68	0.65
First Floor West	0.32	--
Second Floor	--	ND**

*Readings are in absorption units and are only relative

**Non-detectable

--No data collected

MACROERGONOMICS

Human Factors in Organizational Design and Management - II
O. Brown, Jr. and H.W. Hendrick (Editors)

MACROERGONOMICS: A CONCEPTUAL MODEL FOR INTEGRATING HUMAN FACTORS WITH ORGANIZATIONAL DESIGN

Hal W. Hendrick

Human Factors Department
University of Southern California
Los Angeles, CA 90089-0021
U.S.A.

During this fifth decade of our profession, a new focus of ergonomics is developing which is concerned with organizational- machine interfaces, and thus is concerned with ergonomics at the _macroergonomic_ level. Macroergonomics represents a top-down approach to system design based on a sociotechnical system perspective. The organizational factors important to macroergonomics are discussed. The sociotechnical research on technological, personnel subsystem, and human factors as they affect macroergonomic design are reviewed.

The field of ergonomics, or human factors as it is known in North America, is now in its fifth decade as an identifiable profession. Our profession began in the United States during World War II with the design of controls, displays, and workspace arrangements to improve aviation safety. In Europe and Japan, the origins of ergonomics centered around the rebuilding of industry following World War II. For the next three decades, ergonomics was concerned primarily with human physical and perceptual characteristics, and translating these data into recommendations for the design of human-machine interfaces. The development and application of _human-machine interface technology_ has had tremendous impact on the safety, efficiency and comfort of many systems throughout the world. Of particular note has been the application of this interface technology to the design of transportation systems and to individual workstations in factories and offices. As a relatively small profession in numbers, I believe that we can be justly proud of the contributions that were made by ergonomics to the betterment of society during those first three decades.

Late in the third decade of our profession's history, a second focus began to emerge in ergonomics. This second focus was a shift to the cognitive nature of work as reflected in system design. Much of this new shift in emphasis was driven by the technology developments of the late 1960s and 70s. In particular, the development of the silicon chip and the resulting work on computer systems may be regarded as the prime mover of this new generation of emphasis. As we began work on computers and VDU systems, the way people think and process information became increasingly more important to system design. Software structures, as well as knowledge about how people process and use information has made an impact on the "user-friendliness" or functional utility of computer-based systems. In spite of its very short period of existence, this second generation of emphasis already has had some dramatic influences on job design through the development and application of user-interface technology.

The application of first generation human-machine technology and second generation user-interface technology will continue to be the major part of our profession and to contribute to the betterment of human-machine system design for generations to come. It should be noted, however, that these first two generations of ergonomics focused on the individual workstation or microergonomic aspects of systems. With increasing developments of new technology for automating factory and office systems, there also has been a concurrent increase in our awareness of the impacts of technology on organizational systems. Conceptually, we have begun to realize that it is entirely possible to do an outstanding job of microergonomically designing a system's components modules and subsystems, yet fail to reach relevant systems effectiveness goals because of inattention to the macro-ergonomic design of the system. Perhaps no where else is that better understood than by this audience with your concern for human factors in organizational design and management.

It is my belief that there now is an emerging third generation of ergonomics that is focused at the macroergonomic or overall sociotechnical system level. It is concerned with ergonomics research, development and application of organization-machine interface technology. I have labeled this third generation of focus descriptively as macroergonomics.

ORGANIZATIONAL DIMENSIONS OF MACROERGONOMICS

An organization may be defined as "the planned coordination of two or more people who, functioning on a relatively continuous basis and through division of labor and a hierarchy of authority, seek to achieve a common goal or set of goals"1(p.5). This concept of organization, with its division of labor and hierarchy of authority, implies structure. The structure of an organization, including how it is managed, may be thought of as having three major components: complexity, formalization, and centralization.5

Complexity

Complexity refers to the degree of differentiation and integration that exist within an organization. Three major kinds of differentiation are inherent in an organizational structure. These are (a) _horizontal differentiation_ or the degree of horizontal separation between units, (b) _vertical differentiation_ or the depth of an organization's hierarchy, and (c) _spatial dispersion_ or the degree to which the organization's facilities and personnel are dispersed geographically. Increasing any one of these dimensions will increase the organization's complexity. Integration refers to the extent to which structural mechanisms are included in an organization's design for facilitating communication, coordination and control across the differentiated elements of the system.

In general, there is a direct relationship between the degree of differentiation that exists within an organization and the extent to which integrating mechanisms are required for effective system functioning. As the degree of differentiation increases, the need for integrating devices also increases. These mechanisms are needed because with greater differentiation of an organization's activities, the difficulty of communication, coordination and control increases. Some of the more common integrating mechanisms that can ergonomically be designed into an organization's structure are formal rules and procedures, liaison positions, committees, system integration offices and computerized information and decision support systems. Vertical differentiation, in itself, also serves as a key integrating mechanism for horizontally and spatially differentiated units.

a. _Horizontal differentiation_. Horizontal differentiation refers to the degree of departmentalization and job specialization that is designed into the organization. Although it has the inherent disadvantage of increasing organizational complexity, job specialization is common to high technology systems because of the inherent efficiencies in the division of labor. This was demonstrated over 100 years ago by Adam Smith2 when he noted that ten men, each doing particular tasks (job specialization) could produce about 48,000 pins per day. On the other hand, if each man worked separately, performing all of the production tasks, these ten men would be lucky to make 200. Division of labor creates groups of specialists. The way in which ergonomists group these specialists in system design is known as departmentalization. Departments can be designed into the organization's structure on the basis of (1) function, (2) simple numbers, (3) product or service, (4) client, and (6) process. Most large organizations use all six1.

b. _Vertical differentiation_. Vertical differentiation refers to the number of hierarchical levels separating the chief executive position from the jobs directly involved with the system's output. In general, as the size of an organization increases, the need for vertical differentiation also increases3. The primary factor underlying this size-vertical differentiation relationship appears to be _span of control_. Put simply, the number of subordinates that can be directly controlled effectively by any one supervisor is limited. Thus, as the number of subordinate positions required to meet an organization's goals increases, the number of first level supervisory positions also must increase. This increase, in turn, increases the number of second, third and higher order positions required. Thus, the number of higher levels needed also ultimately increases. Factors which effect the optimal span of control in an organization will be discussed later.

c. Spatial dispersion. Spatial dispersion may be operationally defined as the degree to which a location of an organization's facilities and personnel are dispersed geographically from the main headquarters. The three major measures of spatial dispersion are (1) the number of geographical locations within the organization, (2) the average distance to the separated places from the organization's headquarters, and (3) the number of employees in the separated locations in relation to the number at the headquarters location4.

Formalization

From an ergonomics perspective, formalization can be defined as the degree to which jobs within organizations are standardized. In highly formalized organizations, jobs are designed so as to allow for little employee discretion over what is to be done, when or in what sequence tasks will be performed, and how they will be accomplished1.

The management system includes explicit job descriptions, extensive rules, and clearly defined procedures covering work processes. Often, the design of a system's hardware and human-machine interfaces, in themselves, restrict employee discretion.

Organizations having low formalization allow employees more freedom to exercise discretion; jobs are ergonomically designed to permit considerable autonomy and self management; instead of extensive rules and procedures, employee behavior is relatively unprogrammed; and design of the human-machine interfaces usually allows for considerably greater use of one's mental capacities than in more formalized systems.

Centralization

Centralization refers to the degree that formal decision-making is concentrated in an individual, unit or level (usually high in the organization) thus permitting employees (usually low in the organization) only minimal input into decisions affecting their jobs1.

In general, centralization is desirable (a) when a comprehensive perspective is required, such as in strategic decision-making, (b) when operating in a highly stable and predictive environment, (c) for financial, legal and other decisions where they clearly can be done more efficiently when centralized, and (d) when significant economies clearly can be realized.

Decentralization is to be preferred (a) when operating in a highly unstable or unpredictable environment, (b) when the design of a given manager's job will result in taxing or exceeding human information processing and decision-making capacity, (c) when more detailed "grass roots" input to decisions are wanted, (d) for providing greater motivation to employees by allowing them to participate in decisions that affect their jobs, more fully utilize their mental capacities, and increase their sense of personal control and psychological significance to the organization, (e) for gaining greater employee commitment to, and support for decisions by involving them in the process, and (f) for providing greater training opportunities for lower-level managers.

SOCIOTECHNICAL SYSTEM CONSIDERATIONS IN MACROERGONOMICS

To more adequately convey the nature of complex human-machine systems, Emory and Trist5 coined the term" socio-technical system." The sociotechnical system concept use organizations as open systems engaged in transforming inputs into desired outcomes. Organizations are viewed as open because they have permeable boundaries exposed to the environments in which they exist. Organizations bring two major sociotechnical system components to bare on the transformation process: Technology in the form of a _technological subsystem_ and people in the form of what, in ergonomics parlance, is known as a _personnel subsystem_. The design of the technical subsystem defines the _tasks_ to be performed. The design of the personnel subsystem precribes the _ways_ in which the tasks are performed. These two subsystems interact with one another at every human-machine interface. They, thus, are interdependent and operate under _joint causation_, meaning that both subsystems are affected by causal events in the environment. The technical subsystem, once designed, is relatively stable and fixed. It therefore falls to the personnel subsystem to further adapt to environmental change. Joint causation gives rise to the related sociotechnical system concept of _joint optimization_. Joint optimization means that, since the technological and personnel subsystems respond jointly to causal events, optimizing one subsystem and fitting the second to it will result in sub-optimization of the joint _system_. Joint optimization thus requires _joint design_ of the technical and personnel subsystems, given the objectives and requirements of each, and the nature of the relative external environments.

As inferred above, the design of an organization's structure and related management system involves consideration of three major sociotechnical system components that interact and affect system functioning: (a) the _technological subsystem_, (b) the occupational roles and relationships, _personnel subsystem_, and (c) characteristics of the _external environment_ that permeate the organization. Each of these major sociotechnical system components has been studied in relation to its effect on the elements of the fourth major component, organizational structure, and empirical models have emerged.

Technology

Technology, as a determinant of organizational and management system design, has been operationally defined in several distinctly different ways such as by the mode of production6 6 and by the Strategies for Reducing Technological Uncertainty7. To date, however, the most thoroughly validated and generalizable model of the technology-organizational design relationship is that of Perrow8 which utilizes a knowledge-based concept of technology. In this classification scheme, Perrow begins by defining technology as the action that one performs upon an object in order to change that object. This action requires some form of technological knowledge; hence, technology can be categorized by the required knowledge base. Using this approach, he has identified two underlying dimensions of knowledge-based technology. The first of these is task variability or the number of exceptions encountered in one's work. For a given technology, these can range from routine tasks with few exceptions to highly variable tasks with many exceptions.

The second dimension of knowledge-based technology has to do with the type of search procedures one has available for responding to task exceptions, or task analyzability. For a given technology, the search procedures can range from tasks being well-defined and solvable by using logical and analytical reasoning to being ill-defined with no readily available formal search procedures for dealing with task exceptions. In this latter case, problem-solving must be based on experience, judgement and intuition. The combination of these two dimensions, when dichotomized, yields a 2-by-2 matrix having four cells. Each cell represents a different knowledge-based technology as is shown in Exhibit 1.

Exhibit 1. Perrow's Knowledge-Based Technology Classes

		Task Variability	
		Routine with few exceptions	High variety with many exceptions
Problem Analyzability	Well defined and Analyzable	Routine	Exceptions
	Ill defined and Unanalyzable	Craft	Nonroutine

a). Routine technologies have few exceptions and well-defined problems. Mass production units most frequently fall into this category. Routine technologies are best accomplished through standardized coordination and control procedures, and are associated with high formalization and centralization.

b). Non-Routine technologies have many exceptions and difficult to analyze problems. Aerospace operations often fall into this category. Most critical to these technologies is flexibility. They, therefore, lend themselves to decentralization and low formalization.

c). Engineering technologies have many exceptions, but they can be handled using well-defined rational-logical processes. They thus lend themselves to centralization, but require the flexibility that is achievable through low formalization.

d). Craft technologies typically involve relatively routine tasks, but problems rely heavily on experience, judgement and intuition for decision. Problem-solving, therefore, needs to be done by those with the particular expertise. Thus, decentralization and low formalization are required for effective functioning.

Personnel Subsystem

At least two major aspects of the personnel subsystem are critical to an organization's design. These are (a) the degree of professionalism or the skills and training requirements of the organization in its constituent sub-units, and (b) the psycho-social characteristics of the workforce.

Degree of professionalism. It should be noted that formalization can take place either on the job or off. When done on the job, formalization is external to the employee. Rules, procedures and human-machine interfaces are designed to limit employee discretion. This tends to characterize unskilled and semi-skilled positions, and is what is meant by the term "formalization". In contrast, professionalism creates internalized formalization of behavior through a socialization process that is an intrical part of formal training and education. In this case, values, norms and expected behavior patterns are learned before the employee enters the organization1.

From an ergonomics standpoint, in system design there is a trade-off between formalizing the organizational design and professionalizing the jobs and related human-machine interfaces. As positions in the organization, or in selected sub-units are designed to require persons with considerable training or education (i.e., professionalism), they can, and should be designed and integrated to allow for considerable employee discretion and a management system based on low formalization. To the extent that positions and related human-machine interfaces are designed to be performed with little skills training or education, formalization also should be built into the system's design.

Psycho-social factors. Perhaps the most useful integrating model of psycho-social influences on organizational design is that of cognitive complexity. Harvey, Hunt and Schroder9. have identified the higher order structural personality dimension of concreteness-abstractness of thinking or *cognitive complexity* as underlying different conceptual systems for perceiving reality. Regardless of our culture, we all start out in life relatively concrete in our functioning. As we gain experience, we become more complex in our conceptualizing, and this changes our perceptions and interpretations of our world. What is important to us here is the fact that different cultures or subcultures tend to foster reaching relatively different levels of cognitive complexity and related conceptualizing. In general, the degree to which a given culture or subculture (a) provides, through affluence, education, communications and transportation systems, an opportunity for *exposure* or diversity, and (b) encourages, by its child-rearing and educational practices, an *active* exposure to this diversity, the more cognitively complex the persons of that cultural group will become. An active exposure to diversity increases one's opportunity to develop new conceptual categories, and shades of grey within categories in which to store experiential data, or *differentiation*. With an active exposure to diversity, one also learns new rules and combinations of rules for *integrating* information and deriving more insightful conceptions of problems and solutions. A lack of exposure to diversity precludes this opportunity. Similarly, the closed-minded approach to new experiences also blocks conceptual development, even though the opportunity for diverse experiences may be present. Abstract or cognitively complex functioning, then, is characterized by a relatively high degree of differentiation and integration in one's conceptualizing.

Relatively concrete adult functioning consistently has been found to be characterized by a high need for structure and order and for stability and consistency, closedness of beliefs, authoritarianism, absolutism, paternalism, and ethnocentrism. Concrete persons tend to see their views, values, norms and institutional structures as relatively static and unchanging. In contrast, abstract adults tend to have a low need for structure and order, openness of beliefs, relativistic thinking, a high degree of empathy and a strong people orientation. They tend to be less authoritarian and to have a dymanic conception of their world -- they *expect* their views, values, norms and institutional structures to change.9, 10.

I have found evidence to suggest that relatively concrete workgroups and managers function best under high centralization, vertical differentiation, and formalization.

In contrast, abstract or cognitively complex workgroups and managers function best with low centralization, low vertical differentiation and little formalization[11,12].

ENVIRONMENT

The survival of an organization depends on its ability to adopt to its external environment. Stated in open systems terms, organizations require monitoring and feedback mechanisms to follow and sense changes in their relevant task environments, and the capacity to make responsive adjustments. Of particular importance to us is the fact that all specific task environments vary along two dimensions that influence the effectiveness of an organization's design. These are the degree of (a) change and (b) complexity[13]. The degree of change is the extent to which a given task environment is dynamic, or remains stable over time. The degree of complexity refers to whether the components of an organization's environment are few or many in number. These two dimensions in combination determine the environmental uncertainty of an organization.

Two major models have been derived empirically for assessing environmental dimensions as determinants of organizational design. The first model, developed by Burns and Stalker[14] focuses solely on environmental uncertainty. The second model, developed by Lawrence and Lorsh[15] treats environmental uncertainty as the most important of several key environmental dimensions affecting structure.

Burns and Stalker: environmental uncertainty. Based on studies of twenty English and Scottish industrial firms, Burns and Stalker found that the type of organizational structure that worked effectively in a relatively stable and simple organizational environment was very different from that required for a more dynamic and complex environment. For stable and simple environments (low uncertainty) mechanistic structures, charcterized by high complexity, formalization and centralization were the most efficient. Mechanistic organizations typically add routine tasks, program behaviors, and were slow in their compacity to respond to change. In relatively unstable and complex environments (high uncertainty) organic structures, characterized by flexibilty and adaptability were the most successful. Organic firms placed emphasis on lateral rather than vertical communication, information exchange rather than directives from above, authority based on expertise and knowledge rather than on authority of position, conflict resolution by interaction rather than by superiors, and loosely rather than tightly defined responsibilities. To facilitate these processes, a structure characterized by low vertical differentiation, low formalization and decentralization was found to be optimal[14].

Lawrence and Lorsch: Sub-unit Environment and Design Complexity. To cope with a complex task environment, organizations typically develop specialized sub-units to deal with specific parts of the environment. Lawrence and Lorsch conducted a series of field studies to determine what type of organizational design was best for coping with different economic and market environments. They studied six plastics firms and, later, investigated a highly effective and a relatively ineffective firm in three different industries: plastics, food, and containers. These industries varied considerably in their degree of environmental uncertainty. From their study of these organizations, Lawrence and Lorsch identified five major variables that can be assessed regarding sub-unit environments to determine the optimal degree of horizontal differentiation. These are (a) uncertainty of information (low, moderate, or high), (b) time span of feedback (short, medium, or long), (c) pattern of goal orientation (focus of task), (d) pattern of time orientation (short, medium, or long), and (e) pattern of interpersonal relationships (task or social). In general, the more disimilar functions are on one or more of these dimensions, the greater is the likelihood that they should be differentiated into separate sub-units for effective system operation15. A second major finding was that the greater the differentiation, the greater the need for integrating mechanisms. A third major finding was that the level of uncertainty of the external environment was of foremost importance in selecting the structure best suited for effective functioning. Effective sub-units with more stable environments (e.g., production) tended to have high formalization, whereas those having less predictable environments (e.g., research and development) had low formalization. In summarizing from Lawrence and Lorsch's findings, we can note than whenever an organization's design does not fit its mission, environment or resources, the organization is likely to suffer.

REFERENCES

1. Robbins, S.R. (1983). Organizatioonal theory: the structure and design of organizations. Englewood Cliffs, NJ Prentice-Hall.

2. Smith, A. (1970). The wealth of nations. London: Penguin (originally published 1776).

3. Mileti, D.S., Gillespie, D.F., and Haas, J.E. (1977). Size and structure in complex organizations. Social Forces, September, 208-217.

4. Hall, R.H., Haas, J.E., and Johnson, N.J. (1967). Organizational size, complexity and formalization. American Sociological Review, December, 905-912.

5. Emory, F.E., and Trist, E.L. (1960). Sociotechnical systems. In C.W. Churchman and M. Verhulst (Eds.), Management science: models and techniques (vol. 2). Oxford: Pergamon.

6. Woodward, J. (1965). Industrial organizatioon: theory and practice. London: Oxford University Press.

7. Thompson, J.D. (1967). Organizations in action. New York: McGraw-Hill.

8. Perrow, C. (1967). A framework for the comparative analysis of organizations. American Sociological Review, April, 194-208.

9. Harvey, O.J., Hunt, D.E. and Schroder, H.M. (1961). Conceptual systems and personality organization. New York: Wiley.

10. Harvey, O.J. (1963). System structure, flexibility and creativity. In O.J. Harvey (Ed.), Experience, structure and adaptability. New York: Springer.

11. Hendrick, H.W. (1981). Abstractness, conceptual systems, and the functioning of complex organizations. In G.W. England, A.R. Negandhi, and B. Wilpert (Eds.), The functioning of complex organizations (pp. 25-50). Cambridge, MS: Oelgeschiager, Gunn & Hain.

12. Hendrick, H.W. (1967). Differences in group problem solving as a function of cognitive complexity. Journal of Applied Psychology,64,518-525.

13. Duncan, R.B. (1972). Characteristics of organizational environments and perceived environmental uncertainty. Administrative Science Quarterly, September, 315.

14. Burns, T., and Stalker, G.M. (1961). The management of innovativation. London: Tavistock.

15. Lawrence, P.R. and Lorsch, J.W. (1969). Organization and Environment. Homewood, IL: Irwin.

Human Factors in Organizational Design and Management - II
O. Brown, Jr. and H.W. Hendrick (Editors)

SYSTEMS THEORY, MACROERGONOMICS, AND THE DESIGN OF ADAPTIVE ORGANIZATIONS

Kenyon B. De Greene

Institute of Safety and Systems Management
University of Southern California
Los Angeles, California
U.S.A.

The history of human factors/ergonomics search for more macro units of study is reviewed briefly and the role of systems theory in stimulating a major paradigm shift in behavioral/ social science noted. The coevolution of organization and environment as a fundamental organizing principle is stressed, and particular attention is paid to adaptations and maladaptations to the turbulent-field environment. Several systems theoretic interpretations of organization-environment dynamics are made. The necessity for the emergence of second-order organizations to integrate and steer the primary producers of goods and services is discussed. The importance of self-renewal and self-design in organizations is stressed. A number of macrolevel management functions are identified. The sociotechnical systems approach is reviewed and applied to emerging organizational and management problems. Particular attention is paid to microelectronics technology. A number of macroergonomics problems are identified. Some areas for further study are identified.

INTRODUCTION

This paper treats of a number of issues emergent at the complex systems level of organizational design. Systems theories and systems-theoretic constructs and assumptions are employed in order to focus on some limitations to traditional human factors/ergonomics contributions to organizational design. The paper extends the theory and practice of large-scale sociotechnical systems (cf. De Greene, 1973, 1982). The focus of the paper is holistic and top-down. It is argued that bottom-up approaches to the design of person-machine and sociotechnical systems, and attempts to integrate systems parts excised for analytic study, is improper and leads to faulty designs.

These limitations present continuing difficulties for the academic and professional fields of human factors and ergonomics. Indeed, researchers have been concerned for some two decades with finding the proper units or levels of analysis and with building predictive models that are not constrained by the aggregation of system elements from successively lower levels. As long ago as the mid-1960s, Altman (cited in De Greene, 1970) was concerned with the atomistic nature of behavioral units and with the need for more "molar" as opposed to "molecular" units of analysis. At the same time (De Greene, 1970), several authors noted deficiencies to the elementaristic, combinatorial approach to human-reliability modeling. Pew, Baron, Feehrer, and

Miller (1977) reviewed about 40 human-performance models or modelling approaches applicable to the design and evaluation of large-scale command and control systems. Focus was on control-theoretic, decision-theoretic, and information-processing models. They concluded that integrative models did not exist and that attempts to assemble component models will require a major effort and many assumptions. De Greene (1977) criticized the analytic/reductionistic legacy of experimental psychology in human factors research and called for the development of basic theory, particularly at the collective level of ensembles of people and machines. Hendrick (1985) discusses a third generation of human factors efforts, the macroergonomic, concerned with human factors at the level of the organization as opposed to the focus of earlier generations on the physical, perceptual, and cognitive characteristics of people in specific jobs and work groups. Macroergonomics is presented as a top-down approach, prerequisite to microergonomic design of jobs, work groups, and person-machine interfaces. De Montmollin (1985) suggests that this macroergonomics may represent a fourth generation, with a third generation oriented to the workers' actual activities as opposed to the human capabilities and limitations brought to the work.

Many more questions are raised than answers provided by such an evolution in thinking, however. The most profound questions from the perspective of systems science are those of direction of causality and of the emergence of qualitatively new properties at successively higher hierarchical levels of the organization of matter/energy and information. Do the theoretical and research approach to the study of and experience with microsystems provide a basis for designing macrosystems? And, at a rather more trivial level, how does the macroergonomics approach differ from the sociotechnical approach to the design of information-intensive organizations? How can the two approaches complement one another?

It is the belief of this author that progress in the theory and design of person-machine systems, organizations, and supraorganizational institutions will continue to lag as long as there is no fundamental change in the parent behavioral and social sciences. Revolutions in integrative conceptualization that have led to major restructuring have taken place within the past one to two decades in basic physics, cosmology, geology, and molecular biology. However, psychology, sociology, cultural anthropology, economics, and political science appear, for the most part, to be stalled in the normal science stage defined by Kuhn (1970). The use of computers by these disciplines does not represent a paradigm shift. Substitutions such as "information-processing" for "learning" are misleading, and, worse, are delimiting. Methodologies, for example, experimental design in psychology and regression analysis in sociology and economics applied to new problems may be simply old wine in new glasses. It is the belief of this author that revolution in behavioral/social science will be triggered by developments in systems theory. Although the development of integrated theory of sociotechnical systems evolution and structural change is beyond the scope of this paper, it is the purpose of the paper to examine some systems-theoretic constructs applied to the design of organizations.

THE COEVOLUTION OF ORGANIZATION AND ENVIRONMENT

Human evolution is part of organic evolution, and sociotechnical evolution shows many of the patterns displayed by biological evolution. Organisms evolve in relationship to their external environments, and organisms have always been able to transform these environments. Even in the Archean Eon blue-green algae had begun through photosynthesis to generate Earth's atmospheric oxygen (Jantsch, 1980). The new environment in turn encouraged the acceleration of evolution of ever more complex life forms. Many interesting examples of organism-environment coevolution can be seen in the natural world today. The human ability to transform nature in both obvious and subtle ways presents one of the most formidable problems of our times.

Human organizations must evolve with and adapt to changing environments, whereby change in the environment itself results from continuing system-environment interactions. Human organizations that cannot adapt undergo structural change, perhaps to the extent of extinction. This definition is made here. An adaptive organization is an organization that dynamically modifies its internal structure, function, and behavior so as to maintain congruence or fit with its dynamically evolving external environments.

The issues of organizational adaptation and modification of the external environment can be pursued further using the seminal theories of Emery and Trist (1965, 1973) on causal texture of organizational environments. Briefly, in order of increasing complexity, lawful connections evolve within the organization, between the organization and its environment, between the environment and the organization, and within the environment itself. Each evolutionary stage incorporates the lawful connections evolved during earlier stages. The fourth stage is called the turbulent-field environment. Emery proposed that the turbulent-field environment emerged about 1967-1969 concident with the acceleration of high-speed, long-distance communications, transportation, and data-processing. Many organization theorists believe that present environments are turbulent, and there is considerable concern with rampant failures to adapt. Failure to adapt to the turbulent field of forces may precipitate the emergence of a fifth kind of causal texture, barely alluded to by Emery and Trist (1973), the vortical or anarchic environment. There is no possible adaptation to this environment.

Metcalfe (1974) has examined some of the consequences of failure to adapt to the turbulent-field environment. Adaptation must take place at the level of the system of interrelated organizations as a whole. Constituent organizations must interact, and cooperation, not competition, must lead to the development of macropolicies. The attempts of individual organizations to adapt can generate further turbulence by triggering uncontrolled positive-feedback processes. A macroorganization theory must be developed at the level of the system/environment field. Metcalfe indicates that conventional organization theory is an unpromising place to start. In particular, macrotheory must explain (1) deviation-amplifying, mutually-causal, positive-feedback processes, whether deliberately managed multiplier effects or uncontrolled turbulence, and (2) vicious circles like arms races and international economic instability.

Metcalfe (1974) stresses the need for second-order organizations that would function in terms of the integrative needs of the system as a whole by steering that system and managing the environments of the primary organizations that do things like making cars and supplying electricity. However, pluralistic systems of organizations presently are undermanaged, fail to make major structural changes, fail to resolve emergent problems, and make only specific incremental or *ad hoc* adjustments. The result is drift into crisis and unmanageable turbulence.

Crisis is generated by the pressure of environmental forces and the complexity of environmental inputs. Decisionmakers rely mostly on internally generated information applied to safeguarding immediate organizational interests. Interorganizational cooperation is harmed as is individual organizational performance that has been weakened by the internal focus and attenuation of interorganizational interrelationships. Such organizations tend to develop rigidly defensive behaviors that transmit stress back into the external environment where it is amplified and fed back. Irreversible structural changes emerge in the environment itself following the triggering effects of organizational actions (see also De Greene, in press).

Metcalfe (1974) discusses the distinct kinds of change that social systems experience, incremental and structural. Structural transformation of established patterns is not the cumulative effect of continuous incremental changes. Rather, major discontinuities (e.g., step-function triggers) are involved.

Macroorganizational management requires both interaction among the networks of second-order organization and between them and the primary organizations. But most organizations have had much more experience with conflict, a rational stretegy in keeping with previous causal textures, than with consensus-building and cooperation. The new requirements for adaptation are likely to threaten the power, status, and legitimacy of individual organizations, leading to defensive maladaptations. Baburoglu (1986) discusses such maladaptive responses of organizations--stalemate, polarization, and dogmatism--that lead to dissolution of all links between organization and environment and to emergence of the vortical environment.

In practice lack or erosion of adaptive fit may be difficult to detect because of the effects of time lags in the dynamics of the system and because large organizations possess vast resources and large momentum. Nevertheless, early-warning signs can almost always be identified (cf. Argenti and SRI International referenced in De Greene [1982]).

Systems Theoretic Interpretations

The development of macroorganization theory and of macroergonomics can be hastened and improved by integration into the framework of systems theory and in particular field theory (e.g., De Greene, 1982, in press). For instance, dissipative-structure theory (e.g., Nicolis and Prigogine, 1977) can explain the deviation-amplifying, mutually-causal processes, and catastrophe theory (e.g., Zeeman, 1977) the discontinuities identified by Metcalfe (1974).

Briefly, dissipative-structure theory deals with the interaction of stochastic/probabilistic events and deterministic processes in the evolution of systems. A desirable or undesirable fluctuation or mutation (chemical, genetic, social, or technological) may arise spontaneously. Surface effects at the system-environment boundary tend to damp the qualitatively or quantitatively different mutations. However, the system-environment state becomes progressively less stable as it evolves or is perturbed far from (thermodynamic) equilibrium. Under such nonequilibrium (not just reversible disequilibrium) conditions, a fluctuation and ensuing nucleation may grow through autocatalytic or deviation-amplifying processes. The new self-organizing structure may now defeat the old regime, producing a new, usually more complex system-environment state.

Catastrophe theory deals with systems in topological space when slow, continuous changes suddenly produce fast, discontinuous behaviors. Both dissipative-structure theory and catastrophe theory depend on bifurcation theory in that structural change follows critical bifurcation points or sets (i.e., threshold points or regions) that mark the onset of system instability. Catastrophe theory might best be viewed as explaining some terminal changes originating in fields of force better approximated by dissipative-structure theory. Technically, catastrophe theory is limited to systems describable as a potential (or equivalent) function and to systems that are smoothly differentiable in all orders. The forces in catastrophe models are exogenous. Nevertheless, catastrophe theory is unexcelled in explaining powerful driving forces to equilibrium and flips or jumps in multi-equilibrium systems.

Dissipative-structure theory, catastrophe theory and other field theories help to explain many otherwise baffling features of the turbulent-field environment, for example, persistent structural unemployment, international financial instability, ecosystem instability, the failure of arms races and of "tough" international diplomacy, the rise and spread of crime and terrorism, and the emergence of autonomous structures in the environment itself (De Greene, in press).

To conclude this section: The coevolution of system and environment is the primary driving force shaping adaptive organizational design. The turbulent environment precludes rigid, overly formalized, and enduring forms. At present there is an unhealthy evolution toward larger and more powerful organizations. Acquisition and merger partly account for this condition, as do the falling by the wayside of many competing organizations through condensation effects and the growth of government megabureaucracies. These developments lead to the very richly joined system and the anarchic environment of Ashby to which no adaptation is possible (see Metcalfe, 1974). Alternative organization designs, both at the primary and secondary levels, must emerge. Innovative organization design in turn establishes the framework for interactive macroergonomic design.

SELF-RENEWAL -- THE BRIDGE BETWEEN ENVIRONMENT AND SYSTEM

The importance of self-organization and self-renewal in the development and maintenance of system structure has been stressed by a number of systems thinkers including Ashby (cited in De Greene, 1982; Metcalfe, 1974), Jantsch (1980), and Nicolis and Prigogine (1977).

The concept of autopoiesis is an outgrowth of studies of the self-organizing properties of neural networks. Autopoietic systems--ideal cells, organisms, organizations, societies--are self-renewing, self-repairing, unity-maintaining, and self-perpetuating (Jantsch, 1980; Zeleny and Pierre, 1976). An autopoietic organization through the continuous processing of matter/energy and information renews that very organization. The components of the organization interact and generate recursively the same network of processes that produced them. An autopoietic organization is semi-autonomous. The bottom line of autopoiesis is self-renewal.

The manager functions as a catalyst rather than as a director or controller in an autopoietic organization. The function of management is to stimulate the growth of systems of programs and decision processes that operate recursively throughout the organization and lead to attaining goals and objectives.

In contrast, an allopoietic organization is not self-renewing although it may well be self-perpetuating. Structure, function, and behavior are imposed from without or above, and the result is typically a static, mechanistic, and rigid hierarchy. Programs, processes, and decisions are directed toward reinforcing existing structure rather than stimulating the development of new structures more in keeping with changing environmental conditions. Because of mounting environmental pressures and the development of internal schisms and the concentration of effort on maintaining control, allopoietic organizations eventually lose the capability for self-repair and maintenance of unity.

The concepts of autopoiesis, the semi-autonomous working groups and the role of human catalysts stressed in sociotechnical systems theory, and the Adaptivity Field developed by De Greene (1982) are highly compatible. As already discussed, many organization theorists, particularly contingency theorists, have long noted the need for organizations to maintain some kind and degree of fit with their external environments. Typically, this need has been expressed in terms of internal differentiation to match somewhat the complexity of the external environment and internal integration to tie together the differentiated units. For example, large organizations may differentiate according to product lines, and there may be specialized boundary-spanning units of lawyers or ex-government employees to deal with environmental challenges and opportunities.

These strategies appear to be more responsive to the past disturbed-reactive causal texture than to the turbulent field. To maintain internal self-renewal and continuous adaptive fit to the organization/external environment coevolving system, several grand design strategies and macro-policies must be developed. First, design cannot be based on structure that is rigid and that takes long times to develop and consequently long times to change. Hierarchical control and product groups, for example, are inappropriate starting points for designing adaptive organizations. Second, design must be based on the interrelated dimensions of the internal environment (i.e., the psychological, social, technological, and political) and on the forces of the external environment (e.g., economic, ecological, technological). Accordingly, third, differentiation must proceed to the ultimate unit of the organization, the individual human being, and integration must be based on semi-autonomous working groups in the broadest sense of the concept and problemsolving teams that function recursively both upwards and downwards. The entire organization will self-design as a learning organization and will adapt throughout in the face of environmental turbulence.

These principles of design are still incomplete, however. Features of environmental turbulence such as escalating competition and conflict, structural increases in unemployment, social dissolution, and international economic, ecological, and political instability must be recognized and managed throughout the organization. At each recursive level, actions must be taken to avoid triggering further turbulence. Nevertheless, each semi-autonomous unit cannot interact completely and independently with the environment external to the organization as a whole. A top management, that includes representatives from the various working and problemsolving groups and that reflects a network of lateral, bottom-up, and top-down communications, can serve as a primary interface with other primary organizations. Similarly, representatives of the top management of cooperating primary organizations would be the main interface with the second-order organizations still to emerge.

This design framework sets the stage for defining the functions of management of organizations in turbulent-field environments and, accordingly, the requirements for macroergonomics. In these organizations, management functions are distributed throughout and are not the sole prerogative of those well up toward the top of a pyramid. The organizational profile is no longer pyramidal; it may be ovoid. Most of the traditionally defined functions of management--organizing, staffing, directing, controlling, decision-making--are absorbed within the semi-autonomous groups. The organization self-designs with problemsolving the major function applied to the meeting of goals and objectives, and planning is also distributed throughout the organization. For our purposes here, however, the main functions of the top management are long-range and strategic planning, the solving of complex systems problems, the making of macropolicies, and the integrating of system and environmental components.

To conclude this section, a few words are in order about the nature of the sociotechnical systems approach. The sociotechnical systems approach is not simply the recognition that there are interacting social and technical factors in organizations. The sociotechnical systems approach does not interface well with mechanistic organizations. Many, perhaps most, sociotechnical interventions have eventually failed in these settings. As discussed earlier, large systems usually damp such nucleations or swallow them up as foreign bodies. From the very beginning sociotechnical systems theory has had an important philosophical orientation. It represents a reaction against the extremes of the Industrial-Production Revolution, the Scientific Management of Frederick W. Taylor, and the modern-day Taylorism of industrial engineering, operations research/management science, and psychotechnology. Sociotechnical systems theory has been strongly influenced by the holistic personality theories of Angyal and by studies in Norway on the democratization of work. The approach is strongly committed to the practice of industrial democracy. Designs that utilize alternatives to hierarchy, experts as catalysts, matrix organizations, semi-autonomous working groups, and quality of work life reflect the philosophical orientation. See De Greene (1982) for further discussion and citations of the writings of the major theorists.

Unfortunately, mechanistic organizations often try to take over the latest ideas at the level of slogans. An ergonomics program or a quality-of-work-life program in some of these organizations becomes a travesty.

SOCIOTECHNICAL AND HUMAN FACTORS STUDIES OF INFORMATION - INTENSIVE ORGANIZATIONS

Sociotechnical, ergonomics, and human factors studies at the level of the organization as a whole are still few and far between. Modern organizations near the forefront of social, technological, and environmental change must deal with increasing amounts, varieties, and complex combinations of information and knowledge. Indeed, it is often said that the world is experiencing an Information Revolution of at least the scope of the Industrial-Production Revolution. As we have seen this revolution has triggered the emergence of the turbulent-field environment. And theory, design, and management must be in accord with the dynamics of organization-environment interaction.

Pava (1983) stresses the predominantly nonroutine nature of the work of executives and professionals in the modern office. Rational and linear problemsolving are irrelevant or wrong (see also De Greene [1985b] for a discussion of limitations to management thinking with regard to nonlinearity, equilibrium, complexity, and structural change). Inputs, outputs, and input/output interrelationships may be uncertain. Modifications of the traditional sociotechnical approach to design in production organizations is necessary. Pava indicates that tracing the deliberations in which people engage is a useful new approach. Deliberations are defined as "sequences of reflective and communicative acts employed to resolve problematic issues" (Pava, 1983, p. 177). Deliberations involve topics, forums, and participants.

Discretionary coalitions are to nonroutine office work what work groups are to more routine work. The principal goal of sociotechnical design becomes to obtain the best match (not as previously thought joint optimization that even mathematically is intractable if not impossible) between the social and technological subsystem. This is accomplished by creating a variety-increasing system based on redundancy of functions. Deliberations and coalitions are the units of analysis in the approach for the technological and social subsystems, respectively. Organizational structure as a whole becomes reticular, that is, there is more than one conversion process at the same time, and information and authority are fluid and shift as required. Organization is that of dynamic coalitions dependent on debate and collaboration to generate overall understanding and to make trade-offs.

Reticular organizations stress mutual complementarity among differing perspectives in the coalitions. Concern in this paper is at this level and with the highest degree of nonroutine work. Pava (1983) limits the term "work group" to traditional sociotechnical applications involving mostly nonroutine work. In this sense a modern office could include mixtures of work group, product line/market segment, and reticular organizations.

Beyond the general guidelines for the design of adaptive organizations that have been discussed so far, for example, fit with the turbulent-field environment, organicity, capability for self-renewal, emphasis on problem-solving and learning, and member participation, it is difficult, and probably undesirable, to specify in advance the detailed form an organization should assume. The mix of routine and nonroutine jobs could be one factor, but jobs can and should evolve with experience and learning. A number of other organization and societal developments, the mention of which is no longer profound and indeed is almost banal, will help shape the organization. These include the distribution of work of all levels of complexity throughout the organization, distribution of work between home and office, merging of the functions of office and factory (De Greene, 1985a), and capability to "farm out" services to vendors.

These evolutionary processes and structural changes are closely associated with the technological development of microelectronics (future developments may be based on complex organic chemical structures as well as on electronics). The microelectronics revolution differs from earlier industrial revolutions in that it substitutes information capital for human labor, leading to decreases in the need for workers ranging from clerks through middle managers to top executives (Tapscott, 1982). This revolution is seen to have begun about 1970, the time of emergence of the turbulent-field environment as interpreted by sociotechnical systems theorists and the time of the beginning of the downturn of the latest Kondratieff cycle or socio-economic/sociotechnical long wave (De Greene, 1986). Organizational design will have to reflect changes in employment levels and probably massive unemployment, limits to the amount of paid work and therefore major structural change in the very meaning of work, and the way organizations interrelate. From both the organization design and macroergonomics perspectives, microelectronics technology presents a critical choice point. The technology can be, and is being, used to strengthen the rigid, authoritarian, bureaucratic, and hierarchical, that is, mechanistic designs. Consider bullpens of word-processor operators, whose keystrokes are measured by remote supervisors! Alternatively, microelectronics-based systems can be used to enhance organizational self-renewal, personal growth, variety, quality of work life, and the capability to handle uncertainty.

Microelectronics technology may have an even more subtle effect on the nature of work and productivity at both the micro- and macroergonomics levels. Sociotechnical systems theorists have long recognized excessive television-viewing as one form of oversimplification and retreat from the complexities and pace of environmental change. Dogmatic beliefs and highly rationalized decision models are other examples. Pava (1983) suggests that increasing interfacing with equipment, associated with reduced contact with people and tangible reality, may contribute to people becoming more passive, docile, and complacent. So-called knowledge work may actually discourage intelligence! Further, this author observes, shifts to new technologies like automatic bank-teller machines may reflect a growing depersonalization of society more than the convenience involved. Comfort with electronic machines will probably increase as the present generation, growing up with video games and personal computers, matures. This may represent a very disturbing ergonomics trend in the European sense of the term, that is, the meaning of work as opposed to the analysis of work components.

In keeping with the emphasis of this paper on the dynamic interaction between the organization and its external environment and the coevolution of the two, a further important area of management and macroergonomics can be identified. This is the area of boundary-spanning functions associated with requirements to interact with other organizations and with emerging new groups of stakeholders. This design problem has variously been discussed by De Greene (1982), Pava (1983), and Tapscott (1982). Pava, for example, sees the formation of collectivities outside the organization as describable along the dimensions of self-awareness and public recognition. These collectivities evolve through the three stages of interest groups, claimants, and stakeholders. The last stage is the most concretized. Hence, management can best anticipate future confrontation and conflict by seeking clues on emergent groups early in their formation. Dissipative-structure theory and the design proposed by De Greene (1982) might provide the necessary framework.

The present trajectory of sociotechnical systems evolution just about guarantees continued and escalating destructive competition, confrontation, and conflict; dissolution of the world society, economy, and natural environment; and even major war (see, e.g., De Greene, 1982, 1985b, 1986, in press). Consider again the organizational and institutional behaviors, maladaptive vis-a-vis the turbulent-field environment, that were identified by Baburoglu (1986) and Metcalfe (1974) and discussed earlier. Cooperative understanding, consensus-building, and the development of shared values and goals are clearly now among the most important organizational-design and management requirements. It is the responsibility of all organizations to help build a world trajectory alternative to the present one and to replace eventually the present one.

In the design of the second-order organizations discussed by Metcalfe (1974), a good place to begin would be the control and direction of advanced technology. What will be the effects on the nature of work and on employment and unemployment? How will everybody share in the benefits? How can destructive side effects be minimized? What will be the impacts on different social and cultural groups? What will be the impacts internationally and differentially in the so-called developed and underdeveloped worlds? Who owns technology?

Other second-order organizations could be designed around urban, financial, and environmental problem areas. It is clear that individual organizations and managements can no longer abdicate their responsibilities for actions that affect the world outside their immediate boundaries, actions that trigger further turbulence and instability in the world field of forces. If second-order organizations are designed according to shared values and macrogoals, then the deficiencies of present, say, regulatory, organizations that are based on confrontation might be eliminated. Likewise, cooperating organizations would have to do precisely that--cooperate--and not seek opportunities for cooptation of the higher-order organizations.

Fortunately, there does appear to be a maturing in attempts to design organizations, not only as represented by the sociotechnical systems approach but also from the perspective of designing modern computer systems. For example, there has been some shift from "user-friendly" design by experts to active participation of users in design. There is at least some consideration of the organization as a whole and impacts of computer technology outside the organization (e.g., Norman and Draper, 1986; Tapscott, 1982). This can be interpreted as a shift too from micro- to macroergonomics. The teleconferencing techniques, computer conferencing and video conferencing, are also of interest here. Hiltz (1984) studied the impacts of a computer-conferencing system on the communications patterns and productivity of scientifc researchers. As the study progressed her emphasis shifted from purely objective to subjective concerns. The process of communication by computer, initially only an intervening variable, became worthy of study in itself. The experiences with working with the computer were felt to have far-reaching effects for society in general. Hiltz's observations and those of Pava (1983) on the passivity encouraged by too much person-computer-interface time appear to highlight the distinctions between European ergonomics and analytical human factors made by de Montmollin (1985).

The evolution of complex systems presents many choice or bifurcation points where there is a great deal of initial indeterminacy. Specifically, here, organizational design, the use of advanced technology, and macroergonomics can take advantage of the challenges and opportunities offered by the turbulent-field environment or they can assume forms more consistent with past worlds. Hendrick (1985) has warned of the microergonomic tail's wagging the macroergonomic dog, and Metcalfe (1974) has cautioned that traditional organization theory, with its microapproach and inward direction, has little to contribute to the development of macroorganization theory. Moreover, in the face of systems complexity, managers maladapt and organizations become managed by slogan--consider PRIDE, Zero Defects, management-by-objectives, Program Planning and Budgeting System, quality circles, the one-minute manager.

It can be noted that artificial intelligence and expert systems may be just the latest of quick fixes and slogans at the macrosystemic level. Truly, advances can be expected in robot capabilities such as better vision and grasp. Truly, new traces will be developed of the methods followed by individuals in diagnosis in science, engineering, and medicine. But whether artificial intelligence and expert systems will apply in the near future to novel and emergent situations is highly questionable. The intelligence of human beings is expressable along many dimensions, and machines can equal or exceed people in many aspects of memory, computation, combination, and comparison. However, the most unique and important aspect of human intelligence is creative and insight learning and the storage in and retrieval from memory of models or patterns, not bits, channels, externally reinforced pieces, or streams. Following the original Gestalt psychologists, and in agreement with modern field theories like dissipative-structure theory and catastrophe theory, insight learning represents a sudden restructuring of the cognitive field.

Sternberg (1986) discusses recent cognitive, as opposed to psychometric, research on intelligence including several kinds of insight. He distinguishes between cognitive and metacognitive processes, and the latter appear to be about the same as problemsolving as used in this paper and earlier (De Greene, 1982). Sternberg further found that the more successful business executives differ from the less successful ones in the possesion of tacit knowledge, that is, knowledge that is not explicitly taught or even verbalized but is needed for a given field. This appears to be another important macroergonomics observation.

Denning (1986, p. 18) provides this definition. "An expert system is a computer system designed to simulate the problem-solving behavior of a human who is expert in a narrow domain." Examples include classifying chemical compounds, prescribing antibiotics, and diagnosing equipment failures. Expert systems are limited by the information in their databases and the processes by which the information got there. They cannot give results that are not already implicit in the databases.

Finally, macroergonomics principles and findings must be embodied in learning organizations. Considering present organizations this may be more difficult than would at first appear. The nuclear reactor failure at Three Mile Island generated an immense amount of attention and study. Nevertheless, on 26 December 1985, the Rancho Seco reactor near Sacramento underwent a four-hour crisis characterized by confusion in the control room, poor grasp by operators as to the state of the system, misfunction of alarms and controlled systems, and a chaos of noise (Marshall, 1986).

CONCLUDING REMARKS

The limited scope of this paper precludes further analysis and discussion. Clearly, the topics considered represent a whole new area for theory-building, discovery, invention, and design related to how organizations and societies both adapt and contribute to turbulence. Further work should include (1) how advanced technology can help problemsolving involving fields of forces; (2) how organizations can identify and monitor and control behaviors that trigger turbulence, much as the traditional sociotechnical approach monitors and controls internal variances; (3) specifically how organizations and societies can manage the present Kondratieff recession/depression.

REFERENCES

[1] Baburoglu, O., Is the end of free fall, free fall? Emergence of vortical environments, in: Proceedings of the 30th Annual Meeting of the Society for General Systems Research (University of Louisville, Louisville, Kentucky, 1986).

[2] De Greene, K.B., Systems Psychology (McGraw-Hill, New York, 1970).

[3] De Greene, K.B., Sociotechnical Systems: Factors in Analysis, Design, and Management (Prentice-Hall, Englewood Cliffs, New Jersey, 1973).

[4] De Greene, K.B., Has human factors come of age? in: Neal, A.S. and Palasek, R.F. (eds.), Proceedings of the 21st Annual Human Factors Society Convention (Human Factors Society, Santa Monica, California, 1977).

[5] De Greene, K.B., The Adaptive Organization: Anticipation and Management of Crisis (Wiley, New York, 1982).

[6] De Greene, K.B., Integrated Sociotechnical Systems - Human Factors Research on Advanced Factory Automation, a Research Proposal (proprietary) (University of Southern California, Los Angeles, 1985a).

[7] De Greene, K.B., International stability and policy-related thinking about equilibrium, complexity, and structural change, in: Banathy, B.H. (ed.-in-chief), Systems Inquiring II (Intersystems, Seaside, California, 1985b).

[8] De Greene, K.B., Dimension, evolution, and structural change in sociotechnical macrosystems, in: Proceedings of the 30th Annual Meeting of the Society for General Systems Research (University of Louisville, Louisville, Kentucky, 1986).

[9] De Greene, K.B., US and USSR adaptation to a turbulent-environmental field of forces: Avoidance of catastrophic flip to the EMERGENCY state, in: Contributions of Technology to International Conflict Resolution (Pergamon, New York, in press).

[10] De Montmollin, M., Macroergonomics? - yes, but don't miss the work itself! ODAM Bulletin, 4 (1985) 4-7.

[11] Denning, P.T., The science of computing: Expert systems, American Scientist 74 (1986) 18-20.

[12] Emery, F.E., and Trist, E.L., Causal texture of organizational environments, Human Relations, 18 (1965) 21-32.

[13] Emery, F.E., and Trist, E.L., Towards a Social Ecology: Contextual Appreciations of the Future in the Present (Plenum, New York, 1973).

[14] Hendrick, H.W., Macroergonomics, a concept whose time has come, in: Proceedings of the Human Factors Society, Los Angeles Chapter Eighth Annual Meeting (Human Factors Society, Santa Monica, California, 1985).

[15] Hiltz, S.R., Online Communities: A Case Study of the Office of the Future (Ablex, Norwood, New Jersey, 1984).

[16] Jantsch, E., The Self-Organizing Universe: Scientific and Human Implications of the Emerging Paradigm of Evolution (Pergamon, New York, 1980).

[17] Kuhn, T.S., The Structure of Scientific Revolutions, 2nd Ed. (University of Chicago Press, Chicago, 1970).

[18] Marshall, E., Rancho Seco reactor suffers another mishap, Science, 231 (1986) 334.

[19] Metcalfe, J.L., Systems models, economic models and the causal texture of organizational environments: An approach to macro-organization theory, Human Relations, 27 (1974) 639-663.

[20] Nicolis, G. and Prigogine, I., Self-Organization in Nonequilibrium Systems: From Dissipative Structures to Order Through Fluctuations (Wiley, New York, 1977).

[21] Norman, D.A. and Draper, S.W. (eds.), User Centered System Design: New Perspectives on Human-Computer Interaction (Erlbaum, Hillsdale, New Jersey, 1986).

[22] Pava, C., Managing New Office Technology: An Organizational Strategy (Free Press, New York, 1983).

[23] Pew, R.W., Baron, S., Feehrer, C.D., and Miller, D.C. Critical Review and Analysis of Performance Models Applicable to Man-Machine Systems Evaluation (Bolt Beranek and Newman Inc., Boston, March 1977).

[24] Sternberg, R.J., Inside intelligence, American Scientist 74 (1986) 137-143.

[25] Tapscott, D., Office Automation: A User-Driven Method (Plenum, New York, 1982).

[26] Zeeman, E.C., Catastrophe Theory: Selected Papers 1972-1977 (Addison-Wesley, Reading, Massachusetts, 1977).

[27] Zeleny, M. and Pierre, N.A., Simulation in self-renewing systems, in: Jantsch, E. and Waddington, C.H. (eds.), Evolution and Consciousness: Human Systems in Transition (Addison-Wesley, Reading, Massachusetts, 1976).

Human Factors in Organizational Design and Management - II .
O. Brown, Jr. and H.W. Hendrick (Editors)

HUMAN SCIENCES AND THE FUTURE ROLES OF HUMAN FACTORS

Sadao Sugiyama

School of Sociology
Kwansei Gakuin University
Nishinomiya, Hyogo
Japan

The aim of this paper is to analyze the structural relation of various sectors of human sciences and to suggest the ways to synthesize information obtained from various independent levels of our scope toward human being viewed both from molecular and molar aspects, so that the technology of human factors can widen its scope for accomplishing future human welfare.

INTRODUCTION

According to scientific information on human functions already provided in handbooks, dictionaries, research articles and proceedings of various human sciences, such as life science, medical sciences, physiological and psychological sciences and, of course, ergonomics and human factors, we can easily be possessed by the illusion that we have already acquired almost sufficient knowledges of human function. In reality, however, we are not so self-confident yet as to use such scientific data that will enable us to design human life by simple application of necessary information, and consequently we are still producing more and more information in our daily research life.

Indeed, the scope of human sciences has been widening and deepening with the progress of sciences and technologies, and our scientific viewpoint toward human being has become more analytical as well as synthetic. We have been accumulating information ranging from molecular biological studies on behavior of human cells to macroscopic behavioral studies of human being as a unit in a vast environmental setting in human organization, in total society and in such cultural settings that may determine human behavior as a whole. No doubt, human science is a complex of various sciences on human being, which includes the one using microscopic viewpoints and methodologies to the one with vast macroscopic viewpoints and methodologies whose target is human being as a unit in human society.

CERTAINTY AND UNCERTAINTY OF CAUSAL RELATION

Regarding scientific analysis of human function and synthesis of all available information, materials and energy for materializing the well-being of human life by means of technological endevors, mechanism of certainty and uncertainty of causal relation has to be carefully considered, namely the relation between a set of causes and effects whose hypothetical location at all levels of human function is not so simply and directly connected. This must be considered more carefully in case of technological synthesis than that of scientific analysis, because the final technological outcome may create a chance to do more harm than good for human function at some unexpected level which engineer has never thought about.

For example, the phenomenon called "fatigue" which has been and will be an everlasting human issue could roughly be recognized as the complex of 'molecular, physiological, psychological, organizational and socio-cultural factors which affect the total well-being of individual's working life. And we have known that the maintenance of absolute functional stability of each set of factors is not a realistic matter, thus is considered as an impossible one, because of the fact that human being is dynamically changing at all levels all the time when he is at work. If this is an acceptable assumption, we can easily come to assume that each set of causal relation, if any, is not a rigid and tight one but is a dynamically changing and floating soft one, and thus we can obtain the data from each level of human function as a random variable. In reality, we can notice the fact that man can cope with difficult situation, even if his physiological condition is heavily depressed, by being highly motivated or by socio-organizational demand, such as time pressure given to him. Since "fatigue" is the effect, we can control any possible causes so that the effect could be eliminated. However, it is obvious that this causal relation is not a linear one, because the cause affects the functions hypothetically located at various levels of human functional structure. It may be more logical to assume that the relationship between the cause and the effect not as point-to-point relation but as network-to-network relation.

From our daily experience, the causal relation seems to be more direct and linear (a high degree of certainty could be expected) on the molecular or microscopic level than on the molar or macroscopic level (more uncertainty could be expected). Therefore, logic at one particular level is independent of the one at other levels. In other words, logic used in physiological sciences is not the same as the one used in social scientific analysis. However, the logics of mathematics and physics seem to be exceptions. Those two have been accepted as the universal logic, because of the fact that the human is a materialistic entity which obeys the laws of nature. In today's sciences, two other logics might be added, namely those of chemistry and of information.

Presentday logic of engineering has mainly been drawn from mathematics and physics. But more recently, it has become to be more influenced by logics of sciences of chemistry and electronics. One example of former case is the machine system. However, for some strange reason, almost the same logic is used in modern economic theory and its application, in science of human group and its application to our daily group operation, and even in operation research and its application to organizational and social design. Objects seem to be more vast, ambiguous and complicated for the logics used in analysis and synthesis which might be too precise and microscopic. Accordingly, in our society, there are so many uncertain factors made by comparatively simple logic. But, this uncertainty will not be recognized until the time when something, such as an accident, happens. Fortunately or not, human tends to forget or neglect the real potentiality of failure of logic. It seems to me that we have to pay more attention to the mechanism of complexity of the object to be observed before we apply the engineering logic to the complicated system which we are going to build. Examples of such a complexed system are numerous. The most typical ones are the human system and social system which show uncertain characteristics of causal relation, that is, probabilistic occurrence of outcomes. Too many unknown and undetectable results are drawn from engineering application of various scientific knowledges. Accordingly, the occurrence of consequence has come to seem probabilistic for us, namely the certainty has been lost and the degree of uncertainty has gradually been going up. Even the logical machine system will come to have such characteristics, as shown in the case of unexpected failure during the most critical portion of flight operation when the total aircraft operation system becomes very complex. Increasing uncertainty degree of total man-machine system operation will definitely make other people distrustful of the system itself.

However, this suspicion is not a mere psychological response of the public. It is originated from the most unscientific human behavior based on over-estimation of the degree of certainty or of probability. History of mankind has taught us that adequate degree of uncertainty will make us think and behave carefully. Even if the machine is designed according to the absolute law of nature, man who designs it and uses it is full of uncertainty. As the result, it may be concluded that there is no absolutely certain system in which human or social system is involved, that is, the man-machine, man-environment, man-organization, and man-society systems.

NECESSITY OF VIEWPOINT TOWARD MULTI-LAYER STRUCTURE OF HUMAN FUNCTION WHICH CORRESPONDS WITH VARIOUS SECTORS OF HUMAN SCIENCES

Assuming an entity called human being is in front of us, physiological sciences aim to observe its inner structure and functions. Molecular studies such as human biology try to investigate about molecular behavior in the cell at the extreme end of microscopic level of human sciences. On the other hand, another end of our view is more or less macroscopic. For example, social sciences try to observe mechanisms and functions of human society which is a huge organization of human being. In spite of the fact that all human sciences treat human being, their purposes, objectives, methodologies and even language used differ, often widely. As it is, the logic of physiological analysis can not be compatible with the one used in sociological sciences. In disregard of such incompatibility usually found within the framework of human sciences, in reality, the effects caused by human social activities sometimes either acutely or mildly cause certain changes in human physiological functions. Thus, we can conclude that the causal relation may exist between one layer and another of human functions.

It has been our common understanding that physiology deals with physiological structures and functions of human body, psychology with human behavior and mind, and sociology with human society by meta-physical methods. And it has been also understood that these knowledges drawn from each level of human sciences can be directly applied to technologies corresponding to each level. However, it is obvious that, for example, man-machine system and other similar systems can not be established only by mere integration of corresponding system of knowledges drawn from a single sector of science. I think that human sciences ranging from molecular to molar have to be more wisely integrated in order to create more useful knowledges, so that the technology of the times can create more human-oriented outcomes. It is not, however, so simple a matter to do so.

As already indicated in the previous section, it is not so realistic to assume that the point-to-point type causal relation can be applied to all types of analysis. Therefore, I assume a hypothetical network whose components disperse to various levels of human functions (as a causal network) which correspond with a network of effects whose components are also scattered to various levels. This network-to-network relation might not be like point-to-point relation such as firing observed in neural transmission. It might be probabilistically connected. Another assumption is about the characteristics of the structure of the network which is assumed to be hard or soft (which correspond to a tightly formed cluster of components or fuzzily structured network of components) connectivity between the cause and effect, depending on the probabilistic order of connection. Thorndike's connectionism was presented under the assumption that connectivity between stimulus and response was strengthened by learning. In this paper, however, it is assumed that any potentially related matter can be the cause for certain effect and it depends on the degree of tightness of the order of the network. How tight it is depends on probabilistic order of the network. As explained, connectivity of cause and effect depends on probability of information transmitted, thus, if the connection is repetitively reinforced, the probability of connection increases.

Since the analysis of such a cluster as well as of fuzzy structure of a network whose components dispersed to various levels of human functions is not so simple a matter to do, my theory might be unrealistic. However, from our daily experiences, we have known the fact that an unexpected cause may elicit unexpected effect, as in the case of claim caused from VDT use which showed the importance of psycho-social stressor. However, without forecasting this causal-but-probabilistic relation, we can not analyze the effects arising from so-called psycho-social stressor which is now commonly used in various settings of human working situation. I think that the term "psycho-social" does not sufficiently explain the matter, and that any stressors have physico-chemical, physiological, psychological, organizational, socio-cultural characteristics.

Presentday technology which treats human being must establish such a elastic, adaptive, versatile logic that may be applicable for formation of the system. Otherwise, man-machine system can not be optimized by means of technological measures.

NECESSARY VIEWPOINTS FOR HUMAN FACTORS

As was already indicated, a dominant scientific logic which is particular for each sector of human sciences has been commonly used either for analysis or for synthesis. If we depend too much on such an independent logic having to do only with a single sector of science and technology, the future technology may be able only insufficiently to solve human technological problems given by future society, and this will do more harm than good, or do harm as well as good in the long run. Such inefficiency might be due to scientists' and engineers' insufficient knowledges and experiences in other sectors of human sciences and technologies than their own.

Again it is because of the fact that, even if the cause and effect were matched within a narrow range of each level of science and technology, such causal relation is only applicable within a narrow range and can not be the logical relation of human sciences as a whole. Of course, I do not deny either the necessity of technologizing narrow and definite knowledges created by an exact science by means of a specific precise logic applicable only to the limited range of technology, or an importance of precise analysis of pure object by means of absolute analysis of precise scientific technique. However, what I would like to add to them is the necessity of technologizing all possible scientific knowledges created by various sectors of natural, human and social sciences by means of more general and softly integrated logic, whatever it is even an immature one or the one to some extent roughly constructed.

Although we have developed some techniques of so-called technology assessment, it seems to me that the technology assessment has been discussed only in relation with characteristics of physical environment which in turn affects social welfare and that less attention was paid for necessary assessment of each individual characteristics, such as individual aptitude and capability, his particular value system and preference, etc. which might hypothetically locate somewhere between physical environment and social welfare.

Ergonomics or human factors has been developed with the progress of mass-production system which was commonly employed by the industrial society and still it can be accepted in this ages of post-industrial society. However, at this stage of technology, the science and technology of human factors must pay more attention to the individual not only from a merely psychological point of view but more widely from the viewpoints of other natural and social sciences, because the age requires to meet individual necessity and needs. It is also because of the fact that human being sometimes adapts himself to the technological advancement but at the same time is considered as an entity that escapes from it or ignores and disregards the advancement of science and technology.

CONCLUSION

It has been said that once knowledges are created by our scientific endevors, they will be technologized sometime in the future, and then industrialized before long, if the human society accepts them as the useful means to promote human and social welfare. During this process of development, the roles of the individuals as well as of the organization can not be neglected. Especially, in the latter part of developmental process, human organization will play more important roles than individuals. Organization does not deal directly with hardware, such as production machines, but individual worker or a group of workers in the organization will have to do with them. In this respect, it can be said that organization design and management does deal with human being and not with physical factors of hardware system. The role of organization design and management is not so simple to be accomplished easily. The aim of this discipline is not merely to promote productivity of an organization but is to adequately integrate all possibilities which human resources have, so that they can gain not only so-called job-satisfaction but also life-satisfaction. In this sense, its main target is not an organization itself but the human life of individuals.

As was already mentioned, human being cannot be seen only from a single viewpoint, such as physiology, psychology, etc. Since we have microscopic viewpoint toward human being, such as molecular biological one whose results may lead to change human function, we badly need outcomes from microscopic studies not only merely for understanding human behavior but for technologizing those outcomes from of point of view of human being. At the same time, rules and viewpoints drawn from knowledges and experiences of macroscopic sciences of human society, socio-cultural limitation and even of global aspect must be adequately and carefully utilized for the future progress of human factors. I think scientists, and then engineers who are interested in organization design and management have responsibility to organize knowledges on living human being both from microscopic and macroscopic points of view, so that human welfare could be achieved in the organization. If such an endevor is not paid in the future, man in the organization who is awakened to the stern realities of life will try to escape from an empty life in the organization having nothing to work for or live for. Therefore, I conclude that human science is the very means to understand humans as a whole and human technology is the method to promote welfare for human being so that both human science and technology can serve for future mankind. I sincerely hope to hear the words of future man saying "presentday mankind owed benefits to scientists, engineers and organizations in the past."

Human Factors in Organizational Design and Management - II
O. Brown, Jr. and H.W. Hendrick (Editors)

THE NEW "INTERNATIONALISM" IN MACROERGONOMICS

Susan M. Dray
Manager, Human-Technology Impacts
Corporate Information Management
Honeywell Inc.
Minneapolis, MN, U.S.A.

Elsbeth Monod
Directeur Bureautique et
Telecommunications Interns
Bull Inc.
Paris, France

As the field of Human Factors has shifted to consider the co-evoultion of organizations and technology ("Macroergonomics"), it has become necessary to redefine "internationalism", since organizations are cultural entities. We need to be aware of culture as a construct, and to use cultural variation (of both national and professional cultures) to arrive at synergistic solutions to the introduction of technology within organizations.

Three Generations of Human Factors

One way to conceptualize the growth in the field of Human Factors is to describe three "generations" of Human Factors (Dray, 1985). Briefly, the first generation studies the physical human (anthropometry, perception), the second generation studies the cognitive nature of the task and the ways in which cognitive functioning impacts system design, and the third generation, also dubbed "Macroergonomics", studies ways the organization impacts on technological design and vice versa. Each of these generations has been characterized not only by different content, but also by a fundamentally different way of approaching problem solving. Thus, they represent shifts to 3 differenct pardigms (Kuhn, 1962).

A new meaning of "Internationalism"

One manifestation of this paradigm shift is a change in the meaning of "internationalism" for the field. Historically, "internationalism" meant sharing results across cultures, because the results were assumed to be relatively culture-free. When Human Factors research concentrates on basically physical (anthropometry) or physiologically-based (perception, cognition) phenomena, there is a clear reason to assume that research results are largely generalizable to different national cultures. Although it is obvious that there are differences between cultures in terms of range, for instance, the relation between anthropometric measurement and design of technology is not influenced by culture.

However, in the context of Macroergonomics, "internationalism" takes on a new meaning because we are studying technology in regard to cultural variation itself. "Culture" is "a way of life of a group of people, the configuration of all of the more or less stereotyped patterns of learned behavior, including norms, mores, attitudes, values, etc., which are handed down from one generation to the next through the means of language and imitation" (Barnouw, as cited in Adler, 1983b). Organizations are the sum total of the interactions of individuals and their values, attitudes, mores, and norms. Clearly, therefore, culture is a construct which is very important to examine if we are to understand the ways in which technology impacts organizations.

The study of national cultures

There is increasing evidence that organizational behavior is influenced significantly by national culture. Child (1981), for instance, reviewed a large number of cross-cultural studies. He found that on macro level dimensions (such as organizational structure and type of technology in use), organizations are becoming more similar across cultures (convergence). At the same time, on a micro level (behavior of individuals within organizations), national and cultural differences are being maintained, or even exaggerated (divergence). Therefore, for example, while the technology being used by organizations in the U.S. and France is becoming increasingly similar, the way in which American and French people behave within those organizations, and even the ways in which they use that technology is different. Even within a single multinational company, these differences in national culture persist (Hofstede, 1980).

Although these and other studies indicate the appropriateness of a cross-cultural approach to the study of organizational behavior, this has not been the approach that most researchers and practitioners in that field have taken. One review of over 11,000 articles in management journals published between 1970 and 1980 found fewer than 5% which focused on cross-cultural issues (Adler, 1983a).

Perhaps this is not surprising, since many cultural patterns are "invisible" to a member of the culture because they are so ingrained and because they are often implicit, rather than explicit. It is very easy, therefore, to overlook them. Indeed, in a recent study, Adler (1983b) found exactly this. She studied Organizational Development (OD) consultants in Canada. The majority of the work was done in Montreal, which has both English-speaking and French-speaking populations. Given both the nature of OD work, and the fact that these professionals lived in an area with two distinct cultures and languages, one would expect that they would be very sensitive to the impact of cultural diversity on the organization. Adler interviewed and surveyed both English-speaking (Anglophone) and French-speaking (Francophone) OD consultants, with rather startling results. These consultants were interviewed with broad, open-ended questions which did not explicitly mention cultural diversity. 2/3 of these OD consultants did not spontaneously identify an impact of cultural diversity on either the organization or on their role as OD consultants. In the questionnaire portion of the study, however, when specifically queried about positive and negative impacts of culture, these consultants did see both positive and negative impacts of cultural diversity. Because her subjects needed to have culture explicitly mentioned before they could see its impact, Adler concludes that these differences in response are due to the invisibility of culture as a construct used to explain organizational behavior. She states, "People naturally 'assume similarity' even when differences exist. They do not 'see' culture as one of the possible explanations for variations in organizational functioning." (p. 17)

Extrapolating to professional cultures

This new "internationalism" will have something to teach those of us working in a single national culture. Any large organization is made up of specialized professional groups responsible for different domains. It is not only the tasks of these groups which vary. There is a great deal of evidence which suggests that these groups also differ in other ways. There are superficial differences in language, jargon, humor, dress, and so on, as well as more profound psychological and social differences in work style, personality, values, and attitudes among these groups. One appropriate role for the Macroergonomist is to understand the relation of these different groups to technology.

One example is the issue of how to use this understanding to facilitate the introduction of new technology and the change it brings within the organization. Management of technological change can be subdivided into three primary tasks:

- The "technical" task - what are the technical requirements and possibilities, how can the system be acquired/supported, etc.
- The "human" or "organizational design" task - what kind of organization is this, how are things "done", etc.
- The "business" task - what are the business goals, and direction, where is the marketplace headed, etc.

If all three tasks were neatly embodied in one person or group, the job would be significantly easier. Unfortunately, they are usually the domains of specialized groups. As companies become larger, the pressure towards increasing specialization and fragmentation has exacerbated the functional isolation of these gruops, one from another. As an example, management is responsible for accomplishing the "business" task, but may have limited awareness of the others, while the Human Relations/Resources department may be the primary "keepers" of the "human/organizational design" task, with little real knowledge of the others. Similarly, the technical community (Engineering or Information Services) will be most aware of the "technical" task, but not necessarily know a great deal about the others. Not only do these groups differ according to division of labor, however, they also differ on all those dimensions described above as cultural. For example, the Human Resources person is more likely to advocate reliance upon group problem solving while the engineer is apt to prefer more solitary problem solving. Bowman (1986) has discussed more extensively elsewhere some of the stereotypes these groups hold about each other. Experience has shown that technological change can be managed more effectively when these three groups work collaboratively. The challenge, though is to achieve collaboration in the face of cultural diversity.

Organizational responses to cultural diversity

In general, it appears that the most common response of organizations to cultural diversity is, as Adler (1983b) has put it, "parochial". People in parochial orgnizations do not see cultural diversity or its impact on the organization at all. They, therefore, ignore differences, and, if problems occur, they do not attribute them to culture. If such problems actually do result from such diversity, they are unlikely to be able to solve them directly.

The second most common response is "ethnocentric" (Adler, 1983b). In these organizations, culture is viewed as impacting the organization, but only in a negative way. Cultural diversity is seen as the cause of problems, and therefore, such organizations try to minimize such diversity. They may enforce norms from one culture, or attempt to make select a presumedly culturally homogeneous workforce. In ethnocentric organizations, problems are readily attributed to culture, whereas potential benefits of diversity will be ignored.

It is uncommon for organizations to respond "synergistically" to cultural diversity (Adler, 1983b). Cultural synergy, as originally described by Adler (1980), is a way of managing cross-cultural interactions to use cultural diversity as a resource to adapt and shape new, integrated solutions which might never have arisen from either culture independently. In synergistic organizations, cultural diversity is viewed as the source of both problems AND advantages for the organization. These advantages include increased flexibility, creativity, and organizational adaptiveness as a result of the multiple perspectives which can be brought together to solve a specific problem (Adler, 1983b, Moran and Harris, 1982). Because a synergistic organization recognizes both positive and negative impacts of diversity, they are able to manage them so as to maximize the positive and minimize the negative.

The role of Macroergonomics in promoting cultural synergy

It should by now be apparent that a number of groups need to be involved synergistically in technological implementation. The question is how and where does Macroergonomics fit?

It is clear from our experience (Bowman, 1986) that good organizational skills are required to facilitate productive interaction among the groups that must be involved in this task. However, this is not sufficient. Also critically important is an ability to understand the culture, the jargon, and the tasks of each group well enough to translate among them. This includes understanding the technical constraints, along with the organizational and business constraints, as well as understanding how each of these groups views and uses technology itself. Since this role requires an interdisciplinary perspective, it is a role which the Macroergonomist is particularly suited to fill. For most of us, our training and experience has included an appreciation for multiple perspectives and an ability to communicate between them. If we are able to incorporate new concepts, such as culture, and learn from other disciplines, such as cross-cultural management, we will be able to play a major role in promoting cultural synergy.

References:

Adler, N. J., Cultural Synergy: The Management of cross-cultural organizations. In W. Burke & L. D. Goldstein, Eds. Trends and Issues in OD: Current Theory and Practice, San Diego, Ca: University Associates, 1980.

Adler, N. J., Cross-cultural management research: The ostrich and the trend. Academy of Management Review, 1983a, 8, 226-232.

Adler, N. J., Organizational development in a multicultural environment. Journal of Applied Behavior Science, 1983b, 19, 349-365.

Bowman, B., Cross-functional collaboration: Teaming for technological chanage. Second International Symposium on Human Factors in Organizational Design and Management, Vancouver, B.C., August, 1986.

Child, J. D., Culture, contingency and capitalism in the cross-national study of organizations. In L. L. Cummings & B. M. Shaw, Eds., Research in Organizational Behavior, Vol. III. Greenwich, Ct: JAI Publisher, 1981. pp. 303-356.

Dray, S. M., Macroergonomics in organizations: An introduction. In I. D. Brown, R. Goldsmith, K. Coombes, & M. Sinclair, eds., Ergonomics International 85. (Proceedings of the Ninth Congress of the International Ergonomics Association, 2 - 7 September, 1985, Bournemouth, England). London: Taylor & Francis, 1985. pp. 520-522.

Hofstede, G., Culture's consequences: International differences in work related values. Beverly Hills, Ca; Sage Publications, 1980.

Kuhn, T. S., The Structure of Scientific Revolutions, Chicago: University of Chicago Press, 1962.

Moran, R. T. and Harris, P. R., Managing Cultural Synergy, Houston, Tx: Gulf Publishing Co., 1982.

Human Factors in Organizational Design and Management - II
O. Brown, Jr. and H.W. Hendrick (Editors)
© Elsevier Science Publishers B.V. (North-Holland), 1986

COMMUNICATION FACTORS AFFECTING LEADERSHIP: AN ORGANIZATIONAL CULTURE PERSPECTIVE

Shay Sayre
Menlo College

This paper is based upon a field study conducted at a $40M company in Ventura, CA for the purpose of exploring the communication patterns within an organizational culture. The study lends credibility to the belief that transformational leadership grounded in constant and collegial communication can increase human potential and create a productive working environment in organizations.

INTRODUCTION

Pioneer Mary Parker Follett's contribution to organizational literature was the recognition of the importance of the human factor in organizations. For Follett, leadership--the key element of organizational effectiveness--is a group process. She recognized that the leader "must know how to create a group power rather than express personal power," and (Metcalf & Urwich, 1941, p. 248) that "the best leader has not followers but men and women working with him" (p. 262).

Over the decades, Follett's emphasis on leadership has taken a back seat to concerns about management in a technological age when, in fact, the emphasis should have been on leadership for the future of humankind. For one California company, leadership as a group process is alive and well, and provides a dynamic example for academicians and practicioners of human factors study.

Thirty five year old Kris McDivitt built a floundering Ventura, California company of 40 people into a $40 million international enterprise of 250 employees by skillfully utilizing personal communication, collegial decision making and power sharing. Her accomplishment sets an example for executives and managers who wish to increase productivity without sacrificing worker satisfaction and commitment.

A field study was undertaken at Lost Arrow/Patagonia to investigate the communication patterns of this transformational leader (Burns, 1972) within a framework of the organization's culture. Patagonia was conceived in 1957 out of climber Yvon Chouinard's need for equipment not available on the open market. By 1965, Chouinard's operation expanded from the trunk of his car to a tin shed; in 1971 he opened a retail store and within four years began making specialized clothing. In 1980, Kris McDivitt took over as CEO and has run the entire operation for absent Chouinard, establishing Patagonia as a leader in "rugged individualist" sporting equipment and attire.

LITERATURE REVIEW

An important area of agreement in the literature is that leaders influence the behaviors of the led through effective communication. In a cultural sense, the leader has both executive and symbolic roles. Bringing special resources to the group, structuring group behavior and helping define reality for others are functions of those roles (Morley, 1984).

Quinn's (1984) review of the literature elaborates on the fit between leadership style and cultural type. He demonstrated that successful corporate leaders provide image patterns of thinking that have meaning for organizational members. Smircich and Morgan (1982, p. 261) elaborated upon the importance of the leader's control by arguing that the actions and utterances of the leader guide attention and shape the meaning of situations. If this is true, then effective leadership depends upon the extent to which the leader's communication serves as a basis for the actions of others. In their research on leader communications, Penley and Hawkins (1985) categorized two levels of communications for leaders, content and relational. Kantor (1983), who said that "leaders of change deal in symbols and visions and shared understanding" (p. 305), approached leadership from a communication /cultural perspective. In the above examples, authors credited the leader with controlling meaning to create, sustain and change the organizational culture.

As illustrated by management literature, the relationship between leaders and followers is changing (Maccoby, 1983). Heller and Van Tel (1982) said that leaders are now thinking of followers as potential leaders rather than only as subordinates. Some change has been brought about by organizational members who have demanded more collective communication (Lippett, 1982). Shapiro (1984) and Smith (1984) found that effective leaders communicate a sensitivity to follower needs and to the organization as a human entity. Most important, said House (1977) is that effective leaders articulate the corporate ideal to evoke the collective will of followers. If a leader articulates and plays the language game (Pondy, 1978) effectively as the culturalists suggest, the organization will reflect those efforts in its united purpose and strength.

FIELD STUDY AS A METHODOLOGY

The field study was an appropriate method for looking at this organizational culture because it enabled the researcher to explore relationships among the parts of the system from the member' viewpoints, and to see that social system as a working whole from within. Perrow (1972) argued that only the field study can adequately deal with systems phenomena.

By developing a total picture of the organization, a holistic sense of the company and how it operated was possible through a combination of data collection techniques such as interviews, participant observations, review of written material and surveying. A cultural analysis of organization required the researcher to focus on communication as performance.

For this field study, the researcher used a multimethod design in three stages. An abbreviated ethnographic immersion into the culture was accomplished in the first stage through a combination of observation and in-depth interviews. The second and third stages provided comparative quantitative data which determined the extent to which members shared the values and assumptions of the leader.

RESULTS

The qualitative data, which included extensive interviews, member and leader observations, evaluations of corporate publications, and attention to myths, stories, symbols and rituals, was verified by a survey instrument. According to data from both sources, members of this organization shared eight important assumptions about their company. They believed that Lost Arrow:

1. Maintains an image of quality where product form follows function.
2. Values the employee as its most important resource.
3. Encourages new ideas, risk taking and intuitive problem solving by employees.
4. Provides sincere responsiveness to customer concerns.
5. Makes decisions by informal, collegial processes at meetings.
6. Is a family composed of people who care for one another.
7. Is unique in its structure, operations, and concern for its members and the environment.
8. Believes in self education and holds as suspect people whose academic credentials certify them to be experts.

The data also revealed that communication in this organization was the key to effective leadership. The following four aspects of communication characterize the organization and its leader:

(1) Type. Communication did not follow the organization's structure-- it determined it. An open system existed, facilitating communication through, around, and across company groups and functions. Communication was informal in nature and personal in delivery. Verbal communication provided 90% of the company's message exchanges; nonverbal communication punctuated and gave dimension to those exchanges because they were given in person and not over telephones or through barriers. Communication was constant; it constantly circulated throughout the company, penetrating all levels, divisions and functions. Daily, meetings facilitated input, discussion and participation. A metaphorical use of language, jokes, asides, allegories and exaggeration punctuated communication and give its a style of its own. Meeting communication took an interrogative form as members asked questions for clarification and to elicit further response. Communication was direct, candid, and pregnant with meaning.

(2) Affect on behavior. Culture influenced behavior in two ways: indirectly by determining kinds of regulatory mechanisms (structure) used in the organization, and indirectly as that mechanism. By removing traditional structures, the leader removed traditional behavior patterns; idiosyncratic behavior patterns were exemplified through dress, conversation and rituals.

(3) Performance. Those forms of communication which can be directly related to performance are: oral communication, frequent meetings, honest and direct communication, decisive communication, and informal communication. Esprit de corps and comaraderie among members were expanded with humor and personalized conversation which symbolized the bond which existed between company values and personal goals.

(4) Ideology. The CEO communicated her ideology by believing in the ability of human persons to learn, to achieve, and to grow, and by demonstrating that belief in all of her daily activities. She espoused vulnerability, confessed to naivete, and permitted others to do likewise without penalty. Employees, like disciples, spread her ideology to new members, to families and to friends who came to know the company for its dedicated, productive workers.

DISCUSSION

Many of the company's values can be traced directly to the behavior of its founder and CEO. The founder's commitment to maintaining an image of quality reflects his concern that all the company products are made with function in mind. The products' success has always been how well they are received by users, and the underlying value of responsiveness to customer concerns is a direct result of his priority for meeting the needs of his customers. The collegial style of the CEO, one which stresses individual input and group consensus for decision-making, may be a result of her need to be connected to members of the organization who need to be connected to one another. Raised in a close-knit family, McDivitt has perpetuated a meeting forum for all levels of problems solving activities at Lost Arrow.

What makes this company both interesting and unique is its view of reality and the importance it places on values. Its philosophy is evident through aspects of how the leader and company employees view the environment, reality, space, and human nature. Lost Arrow finds itself in harmony with the natural world for which it designs products, yet simultaneously at odds with the bureaucratic world in which it finds itself. The company has no formalized hierarchy or bottom line, aspects which it considers to be the evils of bureaucracy and big business.

There is an obvious gender orientation at Lost Arrow, a direct result of its leader's ideology which is constantly and effectively communicated to the members of the organization.

Based upon data gathered, one can conclude that ideology of this leader directly impacts upon the culture by the organization and the behavior of the members within that culture in a transformational process. McDivitt's style can be most appropriately called "leadership by departure from the norm" of traditional, bottom-line business.

Lost Arrow is a high performing company in which the leader and managers are united in actualizing the goals of the leader and founder. Mutual goals build strong cultures, and cultural strength aids organizational performance. The mechanism for securing consistency and performance is McDivitt's skillful use of her ideology for transformative purposes, a skill which has maintained Lost Arrow in the best of organizational health.

CONCLUSION

This study clearly supports the value of studying leadership within organizational cultures for the purpose of improving knowledge in the area of human factors. While every study is unique and results from research in one setting may not apply to others, it is especially crucial to evaluate the importance of studying microcosims for a more complete understanding of the macro organizational world.

It is the hope of this author that leader communication in organizational cultures as it relates to human performance can be incorporated into future studies for human factors researchers. In order to make the workplace one which stimulates growth and curiosity rather than one which is structured to operate as a hierarchy, leaders must overcome the pressures of technology and bureaucracy through effective, personal, constant and collegial communication.

Also, research which is innovative and departs from traditional methods is encourages for students of human factors. Studying organizations as cultural groups by integrating qualitative and quantitative methods can provide researchers with clues for uncovering new insights into human behavior.

REFERENCES

(1) Burns, J. M. Leadership. (New York, 1972).

(2) Heller, T. & Van Tel, J. Leadership and followership: some summary propositions, Journal of Applied Behavior 18 (1982) 405-414.

(3) House, R. L., A 1976 theory of charismatic leadership, in: Hunt, J. G. and Larson, L. L. (eds.), Leadership: The cutting edge (New York, 1977).

(4) Kantor, R. M., Change Masters (New York, 1983).

(5) Lippett, R. The changing leader-follower relationship of the 80s, Journal of Applied Behavioral Science 18 (1982) 395-403.

(6) Maccoby, M. The leader, Sloan Management Review (1983, Fall) 56-64.

(7) Metcalf, H. C. & Urwich, L., (eds.), Dynamic Administration: the collected works of Mary Parker Follett (New York, 1941).

(8) Morley, E. I., On imagery and the cycling of decision making, in: Hunt, J. G., Hosking, D. M. , Schriesheim, C. a. and Stewart, R. (eds.), Leaders and Managers (New York, 1984).

(9) Penley, L. & Hawkins, B., Studying interpersonal communications in organizations: a leadership application, Academy of Management Journal 28 (1985) 309-326.

(10) Perrow, C., Complex Organizations, a critical essay (Glenville, IL, 1982).

(11) Pondy, L. R., Frost, P. J., Morgan, G. & Dandridge, T. C., (eds.), Organizational Symbolism (Greenwich, CT, 1983).

(12) Quinn, R. E., Applying the competing values approach to leadership, in: Hunt, J. G., Stewart, R., Schreisheim, C. & Hosking, D., (eds.), Managerial Work and Leadership: internal perspectives (New York, 1984).

(13) Shapiro, I. Managerial communication: the view from inside, California Management Review 27 (1984) 112-124.

(14) Smircich, L. & Morgan, G., Leadership: the management of meaning, Journal of Applied Behavioral Sciences 18 (1982) 257-273).

(15) Smith, A. , Present Need for Changing the Nature of Organizational Leadership, Unpublished Manuscript, University of San Diego (August, 1984).

Human Factors in Organizational Design and Management - II
O. Brown, Jr. and H.W. Hendrick (Editors)

CROSS-FUNCTIONAL COLLABORATION: TEAMING FOR TECHNOLOGICAL CHANGE

Betty L. Bowman
Human-Technology Impacts Specialist
Corporate Information Management
Honeywell, Inc.
Minneapolis, Minnesota
U.S.A.

Effective sociotechnical systems design requires collaboration among experts in technical systems, organizational systems, and the work unit involved. For the past year, Honeywell, Inc. has been examining the elements and dynamics of cross-functional collaboration on technical projects. This paper discusses the expertise that is needed, as well as the barriers to and mechanics of teamwork among representatives of three functional groups: Technical Systems, Human Resources, and the Work Unit.

PURPOSE:

The inclusion of sophisticated technology in the workplace is one of the most common organizational changes now occuring in our public and private institutions. Increasingly, those who analyze successful implementations of technical systems are recognizing the interdependence of the technology, the organization, and the people in that organization. However, our approach to managing technical projects has not changed significantly - it is usually managed as a technical process, and not as an organizational change.

This focus should not surpise us, given the significant capital outlay for technology. But there are also human and organizational costs which can escalate if not properly anticipated and managed. Proponents of sociotechnical systems design advocate blending technical systems development methodologies with organizational systems strategies; augmented by techniques for managing organizational change, all elements of successful systems implementation can be addressed. Central to connecting these disciplines on a given project is the collaboration of appropriate experts across functions.

In early 1985 some key Honeywell executives recognized this need for cross-functional collaboration between three major groups - technical or systems experts, human resource experts, and work unit management. The Human-Technology Impacts Department within Corporate Information Management was charged with finding ways to facilitate that process. The research in this area has focused on three major tasks: identifying the relevant expertise required of each group; identifying barriers to collaboration; and identifying the mechanics of collaboration.

SUMMARY OF METHODS:

Three principle methods have been utilized: a cross-functional task team of experts, individual interviews, and pilot studies. The cross-functional task team was composed of well-respected and experienced technical, human resources, human factors, and organizational development specialists. They pulled from their past professional experiences, as well as from their experience collaborating cross-functionally on this task team.

The initial findings of the task team were validated and augmented via individual unstructured interviews with additional specialists in these areas, as well as with work unit managers. The preliminary findings on cross-functional collaboration are being piloted in projects underway within Honeywell divisions; initial feedback will be available from pilot observers mid-summer.

SUMMARY OF FINDINGS:

Relevant Expertise

From the outset of even a potential project, involvement of all three groups - Technical Systems, Human Resources, and Work Group Management - is important, if methodologies and perspectives are to compliment one another. A three-person project management team with a representative from each functional group best assures that integration. If specialized expertise from a functional group is needed at some stage in the project, the appropriate resource can be called in.

Organizations vary in where specific expertise and knowledge resides. In addition, every project will be unique, as there may be employees available who have expertise but do not have the corresponding functional responsibility. In some areas, having functional responsibility is not necessary to serve as a resource for the project team; in others, such as employee compensation, authority is required along with expertise.

Technical Systems experience and knowledge is needed in technical needs assessment, user requirements, human factors, hardware and software evaluation, software design and development, prototyping and testing, and equipment installation.

Often residing in the Human Resource function are those with experience and knowledge in organization development, organization communications, organization/human needs assessment, organization design, job design, staffing, compensation, labor relations, training and education, user involvement, performance measurement and appraisal, career development, as well as individual adjustment to change.

Work Unit Management needs to provide information on business strategies and vision, as well as on organizational operations, politics, personnel/staff, organization traits, management style, and informal/formal norms.

Barriers to Collaboration

Merely having input from these three communities is not enough. They need to work together, each bringing their own perspectives and expertise to the tasks at hand. The interdependence of technical, business, organizational, and people decisions cannot be ignored; each system must work in concert with the others, because at any point, one can inadvertently affect the others.

However, effective collaboration across the three groups does not come easily. Although they all are part of the same larger organization, there are some inherent differences among these groups which can make collaboration initially difficult. Each member of the project management team may have a separate reporting chain with its own priorities and politics. Organizational purposes differ signficantly: Technical Systems experts are primarily focused on the technology, Human Resources on people, and Work Unit Managers on business objectives. Rewards systems are typically based on individual or on team performance within a function, not on team performance cross-functionally.

A significant barrier to this type of collaboration can be the failure to recognize the value of collaborating. This may be attributable to insufficient understanding about each other's disciplines and their role in the success of technical systems. For instance, quite typically, Human Resources is viewed strictly as a source for help with training; the fact that there is expertise in other areas which need addressing (i.e. job design) is overlooked due to lack of information.

Another potential inhibitor to collaboration is the stereotypes which result from past history among the functional groups. For instance, "Those organization development people just come in and stir things up!" "The techies are so enamored with the technology that they ignore the poor end users!" "Nobody bothers to really understand my business needs." Just as easily, though, history among groups can make collaboration very natural.

Even with a positive history, communicating cross-functionally can be frustrating. Each group has their own language and jargon which is often unfamiliar to the others. Familiarity with terms does not preclude miscommunication when there are incorrect assumptions of meaning. For instance, "processing" to a technical systems expert brings to mind the technical processing of data; "processing" to an organization development specialist connotes a psychological processing of an event or feelings.

Communication is especially problematic when a word has the same meaning across groups but carries with it different assumptions. For instance, a Technical Systems expert might assume "user involvement" to mean having the management of the user organization involved in a task, while the Human Resources expert might picture all potential end users having input, and the Work Unit Manager might envision having a team of representative users.

Based on the value system of each functional group and the type of individual attracted to that kind of work, there are often differences in management and decision-making styles between the Technical and Human Resource functions. Task/content and people/process orientations may come into conflict. These differences are clearly reflected in the methodologies which need to be blended together. Systems development methodologies are highly structured and detail-oriented. Organizational change methodologies on the other hand, are less structured, dynamic, and process-oriented. One is extremely analytical and factual, the other is analytical but highly intuitive. One is dealing with predictable technology, the other with unpredictable human beings.

The potential organizational and professional differences between these groups can be accentuated by individual differences in people who typically choose to work in these functional groups. Research on work and friendship has shown that people prefer to work with people they like - people with similar interests who conceptualize the world in a similar manner. These tendencies have even given rise to stereotypes about "data processing types" ("they like to be off by themselves with their data") or "personnel types" ("they like to be out working with people"). While there are always individuals who do not fit the pattern, there is growing evidence that these two functional groups typically think differently and prefer different environments. Recent research correlating Strong-Campbell Vocational Interest Preferences with Myers-Briggs Type Indicators emphasized this. It can help us understand why collaboration does not come easily to these groups.

Mechanics of Collaboration

Given these potential barriers, collaboration will not happen naturally and must be orchestrated. There are five distinct elements: Commitment from the management of the functional groups, selection of project management team members, organization of project management team, team facilitation, and team development.

Top management of each functional group must be commited to cross-functional collaboration, if it is going to be effective. Without that commitment, the collaborative effort is not assummed part of job accountabilities, and therefore is not allotted time nor is it rewarded. The approach to gaining that commitment varies with each functional group, depending upon its mission, current priorities, and politics. This type of support often takes time to build. A clearly-stated, compelling business reason and motivating vision are especially important in maintaining top management support in the face of constantly-changing priorities.

Once support is in place, specific individuals are selected to serve on the project management team. While there may be several experts from each functional area who assist with the project, the project management team members are responsible for setting the tone for collaboration, so their collaborative potential is important. Assuring some commonalities among team members can pave the way for easy communication (i.e. good analytical skills, "team players", focus on central business objectives). Some prior cross-functional experience is helpful, especially if it has been with the specific functional areas involved on the project. Most importantly, each member needs to feel as though he or she will gain from collaborating.

The Strong-Campbell and Myers-Briggs can be helpful in selecting appropriate members for a cross-functional team. Given the human and organizational content and the teamwork involved, preference and comfort in dealing with people can be beneficial; this would be reflected in Strong-Campbell with high scores in Social or Enterprising; in the Myers-Briggs, it is reflected by preference for Extraversion. In working with socio-technical systems design and organizational change, which are both logical processes requiring careful analysis, Strong-Campbell scores in Investigative or Conventional and a Myers-Briggs preference for Thinking would be appropriate.

The organization of the project management team will vary with the situation and the personalities involved. The leader may logically be the Work Unit Manager or it might rotate periodically. Each project management team member might have their own project teams to handle specific tasks or there could be a single project team with which the functional specialists consult.

Because of the difficulties inherent in collaboration of this type, a facilitator can be invaluable during the initial stages of project management team development. The facilitator has heightened awareness of the potential differences among members (language, values, decision-making styles, etc.) and can monitor potential miscommunication and conflict. He or she can help the team develop shared goals, clarify roles, and establish norms for effective team functioning. The facilitator role can be taken by one of the three project management teams, but it is difficult at the outset to pay attention to both content and process; an outside facilitator is free to monitor the process during this important startup phase.

The development of a well-functioning team takes time and experience together. Productivity is low and progress is oftentimes painstakingly slow. Task-oriented groups particularly have difficulty, because they frequently are unaware of the time it takes to establish rapport, learn about each others' knowledge and skills, develop respect, and build trust among team members. Once that has occurred, productivity rises sharply.

OCCUPATIONAL IMPLICATIONS:

If information and communication technologies are going to be successful in the workplace, the interdependence of technology, organizations, and people must be reflected in the way we implement technological change. Involving Technical Systems experts, Human Resource experts, and Work Unit Management experts is central to assuring that the essential elements are addressed, accommodating individual and organizational differences.

As professionals in Macroergonomics who want to change the way systems are designed and implemented, we have an important role in motivating these three segments to work together. We are uniquely qualified. Our field is interdisiplinary in nature, and we can therefore easily recognize the importance of this type of interdependence. We also know the difficulties in communicating across disciplines. Our challenge is to build the understanding in these three functional groups, so that effective collaboration actually occurs.

Human Factors in Organizational Design and Management - II
O. Brown, Jr. and H.W. Hendrick (Editors)

TECHNOLOGICAL DETERMINISM IN THE OFFICE

Christopher E. Moen
Human-Technology Impacts Specialist
Corporate Information Management
Honeywell, Inc.
Minneapolis, Minnesota
U.S.A.

The theory of technological determinism is reviewed and its ability to predict the impact of electronic information systems on individuals and organizations is found to be lacking. An alternate theory of the way in which computers affect organizations is described. This theory suggests that social trends shape how orgaizations use computers.

Technological determinism, the theory that society and its organizations are shaped by advances in technology, has been used to describe how computer's impact organizations. But more recent research has shown computers to be adaptable to a variety of organizational structures.

Technological determinism has been a powerful and well-used persective since the early 20th century, when it was first applied to American history. Under this paradigm, a common approach would be to observe the uses of a technical instrument and assume that the concurrent social change was caused by the use of that technology. For example, innovations such as the typewriter have been assigned a prominant role in the entrance of women into the workforce.

A variety of consequences to using computer's in organizations have been proposed within the paradigm of technological determinism. Equally convincing speculative arguments can be found describing the inevitability of computers paving the way to an utopic, egalitarian society or to an Orwellian nightmare of control.

The empirical evidence on computers impact on organizations, however, has not fit these speculations. Technology does not seem to determine the structure of jobs or organizations.

In a study by Gutek, Bikson, and Mankin (1984) employees and managers in over 55 offices which had recently installed computer systems were surveyed. Few consistant changes were reported in people's work with the introduction of information systems.

Although performance standards grew and the pace of work evened, there were only a small number of changes in pay level and job title.

In another survey, Kling (1978), correlated changes in job characteristics attributed to the introduction of computers with the same characteristics in the overall job. These correlations proved to be nonsignificant. Kling concluded that the use of computers had some effect on people's jobs, but not a dominant effect.

The studies where change did occur reveal even more graphically that computers will not result in a single type of organization. In a case study of how computers impact management power structure, Bjorn-Anderson and Pedersen (1980), found that management relationships changed as a result of an intentional upper management decision to transfer power from one position to another. The vechicle for this new power structure was a computer system. The new information system was constructed in such a way that responsibility for critical decisions were transferred from the production managers to the planners.

In another research effort by Hedberg and Mehlmann (1985), a comparison was conducted of two publications offices, each with a very similar text processing system. The two organizational units were also comparable with respect to company, professional status, task, and level of experience.

It was found that the same type of computer system had allowed two very different methods of completing the same set of tasks. In both offices the work was divided between two positions - the writers and the production assistants. Basically, the writers were responsible for the content of the documents and the production assistants constructed the appearance. Both systems were implemented to improve document quality and to speed up the process of creating the documents. Both were successful in meeting these goals.

The first work group installed their document production system but made no changes in the procedures in how the work was accomplished.

The second work group, on the other hand, took advantage of the increased flexibility produced by the use of the information systems to restructure the work flow and task assignments. Work teams of writers and production assistants were constructed and the delegation of tasks were negotiated by these teams. This arrangement of work complemented the information systems ability to integrate the entire process of document development.

The outcome of this new arrangement was an increase in status and job enrichment on the part of the production assistants and a productivity gain greater than that made by the first group.

The impact of information systems on organizations as a whole has also been examined. In a case study of eight organizations, Robey (1981), found that in five of the organizations no changes in formal structure accompanied the introduction of information systems. When organizational changes did occur, the existing structure of the orgaizations was usually reinforced. Robey found that computers seemed to equally supportive of matrix organizational forms and traditional bureaucratic organizations.

In a review of the literature, Attewell and Rule (1984), found studies reporting evidence for computers resulting in greater centralization and studies showing greater decentralization. They also reviewed a number of articles which suggest that the impact of computers cannot be explained simply in terms of a single impact on organizational structure. The reviewers conclude that computers are compatible with a range of organizational structures.

Contrary to the theory of technological determinism, Daniels (1970), argues that it is only when technological innovation compliments already existing societal trends that innovation appears to produce significant change. Thus when the typewriter is described as the historic gateway for women gaining access to white collar work, he points out that before its advent, women were already working in the office and they were beginning to do so in great numbers. The emancipation of women was a social process which took advantage of appropriate inventions when they appeared.

Although the widespread use of computers is too recent a phenomenon to say with any certainty, some evidence has been documented which adds validity to the claim that societal trends will act to shape the use of information systems in organizations.

Olson and Turner (1985), have described two such trends. For the first time in history, there are more white collar workers than blue collar workers. The number of office workers has doubled in the last decade. The need for information processing abilities has grown with the growth of office workers. The second trend identified is the increasing cost of housing employees in one centralized office. Telecommunication offers the flexability to station workers in less expensive locations.

Nasbit (1985), has observed that the number of families where both parents work is growing. At the same time, the desire for variable work hours to meet family needs will increase. Telecommuting and the ability of computers to be accessed at any time will aid in meeting this need.

Population demographics have been described showing that most of the baby boom generation has settled into middle management positions, leaving little room for promotion of the next generation (Drucker 1980). Competition to keep these professionals will grow as the population declines. Instead of rewarding them with promotions, companies will turn to offering greater decision-making power through the use of information systems.

Technological determinism is insufficient in describing the impact of computers on organizations. The research suggests that electronic information systems are compatable with a variety of task and organizational structures. Rather than technology prescribing the structure of jobs and organizations, it is the environment, including societal trends, which will shape the use of computers in organizations.

BIBLIOGRAPHY

Attewell, P. & Rule, J. (1984) Computing and organizations: What we know and what we don't know. Communications of the ACM, 27 (12), 1184-1192.

Bjorn-Anderson, N. & Pedersen P. (1970) Computer facilitated changes in the management power structure. Accounting, Organizations and Society, 5(2), 203-216.

Daniels, G. (1979) Technological change and social change. In Burke, J. & Eakin M. (eds) Technology and Change, San Francisco, Boyd & Fraser Pub Co.

Drucker, P. (1980) Managing in turbulent times. London, Pan Books Ltd.

Gutek, B.A., Bikson, T., & Mankin D. (1984) Individual and organizational consequences of computer-based office information technology. In S. Oskamp (Ed), Applied Social Psychology Annual, 5, Beverly Hills, Sage Publications.

Hedberg, B. & Mehlmann, M. (1985) Processing and communicating words: Can telematics turn word processing into work processing? Paper in the 7th EGOS Colloquium: "Challenges to Organizational Authority", Saltsjobaden June 12-14, 1985.

Kling, R. (1978) The impacts of computing on the work of managers, data analysts, and clerks. Irvine, CA: Public Policy Research Organization, University of California, Irvine.

Kling, R. (1980) Social analyses of computing: Theoretical perspectives in recent empirical research. Computing Surveys, 12 (1), 61-110.

Olson, M. H. & Turner, J. A. (1985) Rethinking office automation. Unpublished manuscript, New York University.

Robey, D. (1981) Computer Information Systems and Organization Structure. Communications of the ACM, 24 (10), 679-687.

Human Factors in Organizational Design and Management - II
O. Brown, Jr. and H.W. Hendrick (Editors)

ASSEMBLY CELLS AND PARALLELIZATION: TWO SWEDISH CASES

Harvey F. Kolodny

Faculty of Management Studies
University of Toronto
Toronto, Ontario
Canada

Product focussed or product oriented forms of organization have begun to proliferate in the manufacturing sector. These forms reduce throughput time considerably. As a result of computerization of the production equipment, they also offer considerable flexibility and can quickly accomodate changes in products, in volumes, in process routing, etc. This paper describes one set of these product focussed forms - assembly cells and parallelization - and uses two Swedish case studies as illustrations of the forms.

INTRODUCTION

As a response to turbulent environments, there is a need to (1) structure organizations more flexibly and (2) decrease the throughput time of goods produced and/or services rendered. Organization forms, in recognition of these pressures, have shifted in this direction. With relatively self-maintaining or autonomous work groups at their base, an hierarchy of 'product focussed forms' have arisen: group technology and/or manufacturing cells (Hyer, 1984), flexible manufacturing systems, product shops or focussed factories (Skinner, 1974), world product mandates, divisional organization. All have in common that they are oriented "with the grain" (Lindholm, 1975), i.e., with the flow of production. This has resulted in throughput time reductions of as much as 80 percent (e.g., from 16 weeks to three weeks) with commensurate other savings (e.g., work-in-process inventories). They also all have in common computerization at the base of their control systems, and that high programmability leads to set-up times approaching zero, batches of one as a reality and overall systems that are highly flexible (see Figure 1).

These technical innovations have been accompanied by equally dramatic organizational changes. Driven by explicit value positions (such as openness, trust, equity, growth, etc.), or in some European cases, by legislation (e.g.,

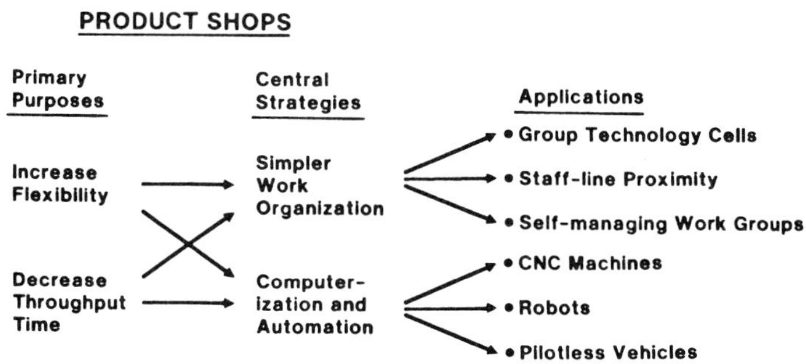

FIGURE 1

codetermination, self-management), new organizational philosophies have been developed to accompany the new technologies. The resulting work arrangements are designed and operated with the participation of the employees concerned, sometimes in the design stages, particularly in the case of 'greenfield' sites (Emery, 1980), more often in the ongoing operations. These new ways of working, variously referred to as sociotechnical systems (Trist, 1981), participative management, innovative work restructuring, employee involvement and high commitment work systems (Walton, 1985) result in an improved quality of working life for those involved (Kolodny and Stjernberg, 1986).

ASSEMBLY CELLS AND PARALLELIZATION

The technical innovations referred to above have focussed on machining operations (Kolodny, 1984). Another area of extensive social and technical innovation to date is continuous process operations (Cunningham and White, 1984). This paper describes cases in a third area: assembly operations. For want of a better name, these are referred to here as assembly cells. They subsume self-maintaining units (Davis, 1983) or semi-autonomous work groups (Sussman, 1976) in that they consider the technology of the assembly operations as well as the group process and structure. They differ from machining operations in that they are usually a stage beyond in the manufacturing process and the technology is often less computer controlled. Hence the flexibility objective is frequently derived from parallel arrangements; which accounts for the second half of the title phrase of this paper: 'parallelization'. These assembly cells or parallelization arrangements fit into a scheme of product focussed forms as illustrated in figure 2.

PRODUCT FOCUSSED FORMS

- Product Shops
  - focussed factories (but flexible)
  - factories-within-factories
- Parallelization
  - assembly cells
- Group Technology
  - machining cells
- Work Groups
  - semi-autonomous
  - self-maintaining

FIGURE 2

TWO SWEDISH CASES

The two cases presented here illustrate a level of product focussed hierarchy above the work group level. One case is from a consumer goods situation (lawn trimmers and mowers) and the other is from an industrial setting (submersible pumps). The background data for the case studies was collected during a research sabbatical by the author in Sweden in 1982-83 (Kolodny, 1985). Brief extracts of each of the cases were first discovered in a Swedish language publication of the Swedish Institute for Production Engineering Research (IVF) (Agervald, 1981). The data were collected through observation and interviews on site and with researchers at IVF.

Case 1

Stiga AB is a manufacturer of power lawn mowers, manual and rider driven, and garden trimmers. Sales in 1982 were approximately $50 million (Canadian). Employees numbered 540, 350 in production and 190 in support. Stiga's organizational innovations were directed at: (1) reducing throughput time, primarily by combining the locations for assembly and storage in order to reduce inventory administration and transportation costs (figure 3), and (2) improving the quality of working life by decoupling people from the assembly line and facilitating work team arrangements (figure 4). Figure 4 also illustrates the concept of parallelization in the two person, parallel work team cells.

Case 2

Flygt AB is a manufacturer of submersible pumps. In 1982, worldwide Flygt employment was 3,000 and sales were approximately $65 million (Canadian), 95 percent of which were exported. The 1,000 person Lindas plant, in southern Sweden, experimented with three successive organizational arrangements to assemble pumps: first by a driven assembly line with the operators at fixed locations, then by having the operators move in sequence through the fixed assembly stations, and finally by physically and organizationally redesigning the assembly area into eight parallel stations with two operators each per station. This brief vignette illustrates Flygt's evolution through different organizational arrangements to improve the assembly tasks in order to decrease throughput time and obtain the flexibility of parallel operations.

SUMMARY

The assembly cells and parallelization arrangements described in the brief case studies were both arrived at through considerable experimentation with different combinations of people and production technology. Nevertheless, both were successful in decreasing factory throughput time while remaining flexible enough to accomodate the variations that an increasingly turbulent world thrusts on organizations. At the same time, they were able to improve operators' jobs by providing individual operators with opportunities to work in groups with considerable autonomy about how they conducted their tasks, by enriching jobs to make them more meaingful, and by decoupling jobs from the demeaning features of driven assembly lines.

Assembly cells and parallelization are one set of organizational forms in a hierarchy of product focussed forms. They are an increasingly attractive alternative to traditional ways of organizing production.

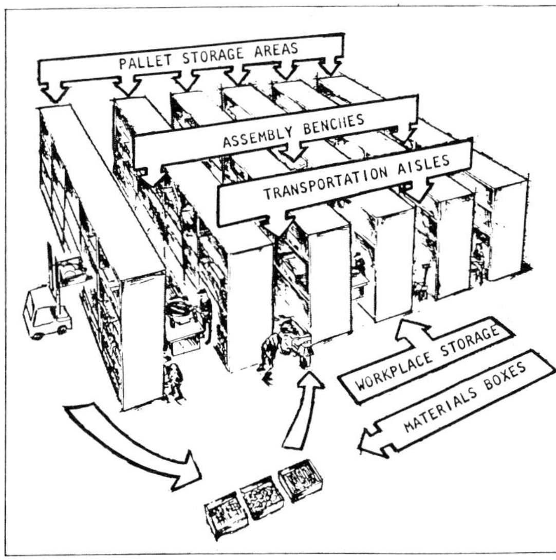

MATERIAL SUPPLY OF ASSEMBLY GROUPS AT STIGA

FIGURE 3

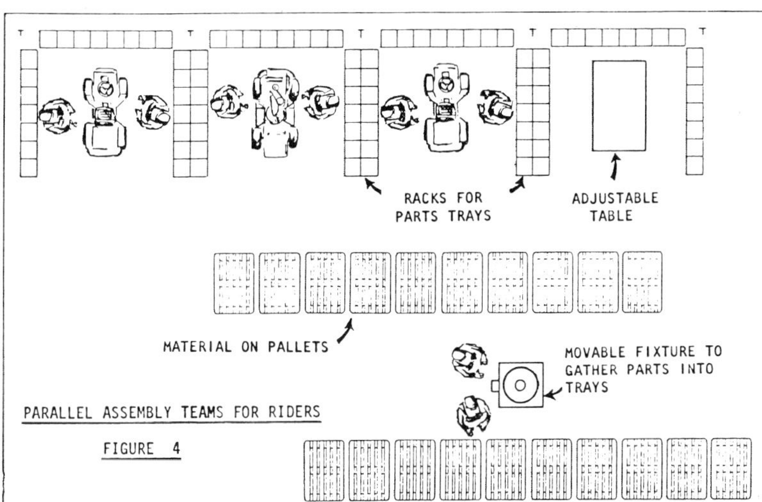

PARALLEL ASSEMBLY TEAMS FOR RIDERS

FIGURE 4

*Figures 3 and 4 are used with permission of the Swedish Institute for Production Engineering Research (Institutet for Verkstadsteknisk Forskning) of Sweden's Mechanical Association (Sveriges Mekanforbund).

References

Agervald, Olof, 1981. Exempel pa Monteringssystem for Mindre montage (Examples of small batch assembly systems), Stockholm: Swedish Institute for Production Engineering Research.

Cunningham, J.B. and T.H. White, 1984. Quality of Working Life: Contemporary Cases. Ottawa: Government Publishing Centre.

Davis, Louis E., 1983. Learnings from the Design of New Organizations. In H. Kolodny and H. van Beinum (Eds.), The Quality of Working Life in the 1980's. Praeger Publishers, pp. 65-86.

Emery, 1980. Designing Sociotechnical Systems for 'Greenfield' Sites. Journal of Occupational Behaviour, January, 1(1), 19-28.

Hyer, Nancy, 1984. Group Technology at Work, Society of Manufacturing Engineers, Publications Development Department.

Kolodny, H.F., 1984. Product Organization Structures Improve the Quality of Working Life. Paper presented at the 1984 SME World Congress on Human Aspects of Automation, Montreal, Sept. 16-19, 9 pages.

Kolodny, Harvey F., 1985. Work Organization Design in Sweden: Some Impressions from 1982-83. Human Systems Management, forthcoming.

Kolodny, H.F. and Stjernberg, T., 1986. The Change Process in Innovative Work Designs: New Design and Redesign in Sweden, Canada and the U.S.A. Journal of Applied Behavioural Science, forthcoming.

Lindholm, Rolf. 1975. Job Reform in Sweden. Stockholm: Swedish Employers' Conference.

Skinner, W., 1974, The Focussed Factory. Harvard Business Review. May-June, 113-121.

Sussman, G.I., 1976. Autonomy at Work, Praeger Publishers.

Trist, Eric L., 1981. The Evolution of Sociotechnical Systems. Occasional Paper No. 2. Toronto: Ontario Quality of Working Life Centre, June, 67 pages.

Walton, Richard E., 1985. From Control to Commitment in the Workplace. Harvard Business Review, March-April, 76-84.

Human Factors in Organizational Design and Management - II
O. Brown, Jr. and H.W. Hendrick (Editors)

R and D Management in Japanese Firms
-- A Study Based on the Sample Survey Results --

Ken'ichi Mori*, Norihiro Jidai**, Kanju Haneishi***, and Iwao Oohashi****

*Faculty of Economics, Shiga University, Hikone, Shiga, Japan
**School of Business Administration, Asahi University, Gifu, Japan
***School of Business Administration and Information, Setsunann University, Neyagawa, Osaka, Japan
****Faculty of Commerce, Hannan University, Matsubara, Osaka, Japan

This paper summerizes the study of 128 firms, the size of which distributed from the big firms to the smaller starting ups. The purpose of this study is to figure out the purspective view of R & D management in Japanese firms. As is well known in the production Management area, the Japanese kind of management has some differences from the Western counterparts Even in such intellectual activity as R&D and its management, we consider to characterize the Japanese way. This situation is also inferred to come from the differences in culture, work habits, science tradition, social environment and so on We think to point out successfully Japanese way here.

INTRODUCTION

Many of the Japanese firms have attained the successful innovation both in their produts and production systems. In this process, the R&D and its management have played the significant function and now are expected to have an incresing power. Originally, their R&D's were dependent on the seeds imported from the Western counteries. At the early catch up stages after WW II, this propensity has owned the more definitive place. But, nowadays after some big industrialization, they are required to get rid of this pattern to get more innovation.

Our main purpose of the study is to explore their contempolary state of R&D commitment. It is not an easy task to make clear the complete figure of this. This report, actually, is based on the questionare and interview survey results which cover the rather wide area. Here, we focused mainly on the organizational behavioral point of view. At first, we tried to clarify how they organize R&D to react to the social environmental changes. Secondly, the basic nature of their R&D is evaluated to have been incremental one based on the innovation by immitation. This kind of culture are deemed to be changing now, so we discuss on the direction of the change.

THE FRAMEWORK OF OUR SURVEY

Our quetionare survey covers the R&D organizations in 128 firms whose main business area distribution is shown in Table 1. As is clear from this, about 20 of them are from electric and electronics, and follow the mechanicals, chemicals, steels and so on. The electronics, mechanicals and chemicals are more innovative industrial areas in which many innovators exist.

As shown in Figure 1, our survey areas are categorized into three ones, i.e., the adaptation to the external business environment, the management process, and the internal administrative behavior. The R&D attainment is evaluated in relation with such performance values of firms as sales, profit and so on.

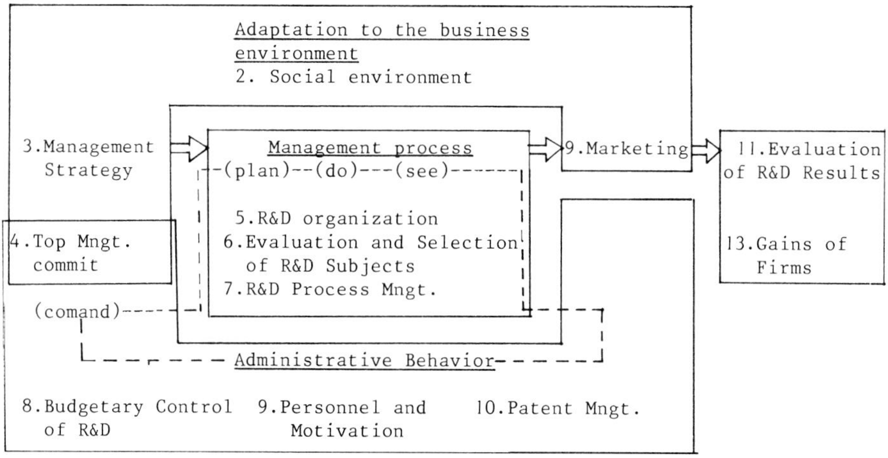

Figure 1
Category of Survey

THE DEFINITION OF THE INNOVATION BY FIRMS

We asked about the definition of new products and new production engineerings. This results are shown in Table 1 and both top answers are coincide as the first one for their firms even if same kinds are already in the markets or in the competitors' factories. These catch up attitude prove that the R&D of Japanese firms can be still in the immaturity stage. As for new products, they, in many cases, have introduced the basic seeds from outer organizations, and shown the excellence in polishing these ones up. In other words, they have devoted themselves to the incremental or improvement-oriented R&D, which object is not to publish the drastic invention. But some of the improvements were so sophisticated that those products brought about completely new utilities. In production engineering, new devices have been created in the framework of new production system concepts, which is represented by Toyota production system and FMS. The smaller improvements also have come from TQC activities.

Table 1 Definition of New Products and Production Engineering (in number of firms)

	Product	Production Engineering
First in Firm	102	91
First in Japan	85	87
First in World	83	83
Improvement	51	10

COMMITMENT OF TOP MANAGEMENT

Usually in Japanese firms, the age of presidents is rather high, and 55% of them belong to 60s and 25% to 50s. Some of them keep their position such long period as more than 30 years. About one fourth of them are from the production division and next follow sales, R&D and so on. Most of them have strong will for R&D and show the deep commitment as shown in Table 2. This inclination is notably high in firms with president from R&D area. In addition, about 80% of them have more than one member in the board of directors come from R&D area.

Many of presidents in the manufacturing industry firmly beleive that the technological power is the foundation for business, so it is their duty to strengthen it. It is, also ,taken as natural for the firms business strategy to involve the R&D perspectives.

Table 2 Commitment of Presidents and Their Initial Area

	strong	ordinary	weak	no ans.	T	
initial division	%	%	%	%	no.	%
Production	54.5	45.5	0.0	0.0	33	100.0
R&D	90.0	5.0	0.0	5.0	20	100.0
Sales	25.9	74.1	0.0	0.0	27	100.0
Others	39.5	55.8	4.7	0.0	43	100.0

THE STRUCTURE OF R&D ORGANIZATION AND ORGANIZATIONAL RELATIONSHIP WITH OTHER DIVISIONS

Almost all firms we investigated have some kinds of R&D organizations. More than 40% hold the several number of research divisions, and laboratories. A big car maker answered that they retain no independent R&D divisions but the groups are built in appropriate place of the production line. This kind of organization is estimated to be fitted for the incremantal R&D, which is also assumed by some others, especially in technologically mature industry. From the research point of view, about

30% are organized following product line, also about 30% engineering-wise, and 40% both type. So, the organizing policy can become more flexible and so-called matrix type.

About 75% of firms have the committe or specialized section for R&D coordination. The fact that three fourthes of them are of specialized section is worthy to notify as the much efforts to coordination, which may come from the increasing demands and complexity of R&D.

As for our survey, R&D division is inferred to have rather good relations with sales and production, in most firms. To be specific, the relation with production division is considered to be better. This may be nurtured through the tighter contacts between these two divisions, which include many kinds of new product testing, transfer and improvement, in addition, even the frequent personnel exchanges. On the other hand, the sales only transfer the customers order to them, so the contacts could be the one way.

Now, more firms have recognized the complexity of intradivisional relationship, so more than 60% of firms utilize committee or section to coordinate relations. Another and more agressive their intention looks like to control these relations following their policy and search for the efficient R&D processing. The leadership in the process of coordination is assumed by both coordination division and 3 divisions evenly.

THE PERSONNEL ADMINISTRATION AND MOTIVATION FOR RESEARCHERS

Well established work motivation system can create the favarable R&D environment, which directly is connected to the productive R&D activity. We investigated the 16 events from 4 categories, to which we refer in the following, with the application percentages. First category summerizes the work conditions, which contains home working system (0.8), flex time (7.0), special bonus (3.9), and specialist system (48.4). Second is the systemization information exchange, and consists of proposal system(64.8) , researchers salon (3.1), interdisciplinary exchange (19.5), and inter-department exchange (23.4). Thirdly, ability development is coming up, which constituents are the lecture by the outside specialists, study chance in other organization (60.9), special education system(19.5) and creation trainning(22.7). Fourth category can be named as the certification system which is of the publication of own R&D journal (14.8), own R&D conference (72.7), system of commendation (66.4), and patent encouragement (80.5). The systems in the fourth category, as clear from these results, are shown to get the most popularity of all.

Now, more firms have recognized the complexity of intradivisional relationship, so more than 60% of firms utilize committee or section to coordinate relations. Another and more agressive their intention looks like to control these relations following their policy and search for the efficient R&D processing. The leadership in the process of coordination is assumed by both coordination division and 3 divisions evenly.

THE PERSONNEL ADMINISTRATION AND MOTIVATION FOR RESEARCHERS

Well established work motivation system can create the favarable R&D

environment, which is directly connected to the productive R&D activity. We investigated the 16 events from 4 categories, to which we refer in the following, with the application percentages. First category summerizes the work conditions, which contains home working system (0.8), flex time (7.0), special bonus (3.9), and specialist system (48.4). Second is the systemization of information exchange, and consists of proposal system(64.8), researchers salon (3.1), interdisciplinary exchange (19.5), and inter-department exchange (23.4). Thirdly, ability development is coming up, which constituents are the lecture by the outside specialists, study chance in other organization (60.9), special education system(19.5) and creation trainning(22.7). Fourth category can be named as the qualification system which is of the publication of their own R&D journal (14.8), own R&D conference (72.7), system of commendation (66.4), and patent encouragement (80.5). The systems in the fourth category, as clear from these results, are shown to get the most popularity of all.

As the evaluation criteria for promotion, the potential ability of research and group management is preferred to the past achievements. This can be expected from the general climate for the personnel management of Japanese firms. The personal attainments are sometimes obscured in the groupism oriented work system which has power even in R&D work. Under this kind of work climate, the area of job rotation is of such wide range of sales engineering(60.9), management in R&D division(60.2), corporate planning(60.2), production engineering(36.7) and so on. We also witnessed a case that one research section has transferred to production division after developing a paticular new product.

CONCLUDING REMARKS

We can summerize as follows. Fistly, the innovation is inclined to be defined as the incremental one in comparison with the drastic invention orientation. Japanese R&D has found that way in nuturing or polishing up the initial seeds mostly imported from abroad. Secondly, the deep commitment by top management is in the certer of R&D management which also may be called gavernance. Thirdly, the climate of Japanese R&D organization is now changing and their efforts are also turnning to the more basic areas. Some big firms have established the sophisticated R&D organizations. Fourthly, they work the many kinds of motivation mechanisms, which pattern somewhat differet from Western counterparts.

REFERENCES

[1] R&D Management Research Group of JIMA, Report on the Sample Survey of R<D Management in Japanese Firms, (in Japanese), Japan Ind. Mngt. Ass. (1984)

[2] Mori, S., R&D Management, (in Japanese), (Yushindou, 1981)

[3] Pelz, D. C. and Andrews, F. M., Scientists in Organization (John Wiley, 1966)

[4] Allison, D., The R&D Game (MIT Press, 1969)

[5] Eto, H. and Matsui, K., ed., R&D Management Systems in Japanese Industry (North-Holland, 1984)

Human Factors in Organizational Design and Management - II
O. Brown, Jr. and H.W. Hendrick (Editors)

SUCCESSFUL IMPLEMENTATION OF CHANGE IN THE CORPORATE ENVIRONMENT: THE SOCIOTECHNICAL APPROACH

Ilan Hershman

Workers' Compensation Board
Alberta, Canada

INTRODUCTION:

The relationship between people, technology and change has in recent times received a great deal of attention. Under such diverse headings as *Resistance to Change*, *Pay for Performance*, *Back to Basics*, *Corporate Culture*, *Management by Objectives* and so forth, corporate society is attempting to deal with the consequences of a fractionized and mechanistic work environment. Reduced output and poor quality work are often correctly seen as by products of apathy, alienation and boredom.

In recent times quick fix methods have become the panacea for the ills of the corporation. Buzz words and clichés such as *Intrapreneurship*, *Matrix Management*, *Corporate Culture*, *Management By Walking About* proliferate. Yet, the undeniable truth is that these problems are too often created by the implementation of change without sensitivity to the requirements of the workforce.

The remainder of this presentation will offer a succinct overview of the sociotechnical system as it applies to the achievement of desired results, namely attainment of a higher level of productivity and gains in quality. Appropriate use of the work force in the day to day operations and, more importantly, in the intermediate and long term planning strategies of the company will be emphasized.

THE SOCIOTECHNICAL APPROACH:

Sociotechnical interventions are techniques that are systematically applied and attempt to,

(a) re-organize the work method

(b) re-structure the technology and/or redesign the organization's social environment.

The purpose of instituting these applications is to create a better fit between the people doing the work (human system) and the tools by which a product is produced or service rendered (technological system). In essence, sociotechnical interventions attempt to create organizations which function more effectively and more efficiently, produce a higher level of satisfaction within the workforce, and are adaptable to the contingencies created by change.

In reality, to maximize sociotechnical intervention requires the joint optimization of the social and technological system. This is a tremendously complex undertaking and necessitates the examination of many diverse aspects of the corporation. The analytical format must consider;

(a) the concept and underlying mechanics of organizational change,

(b) the social processes and climate of the organization, and

(c) the technological aspects of the organization and consequent interactions with the social system.

The recognition that the environment of each organization is vastly different will clearly demonstrate the inherent complexities faced by the change agent. This recognition offers the unique insight to the lack of wisdom in instituting change based on a cook book approach. Each change process is unique and must therefore be analyzed, developed, implemented and evaluated on the basis of that uniqueness.

The mere concept, let alone the process of change, is known to threaten people and therefore negate the positive impact successful change offers. New ideas within corporate society require for their introduction an implicit or explicit strategy. The threat and fear that change arouses must not be underestimated. This fear element can be minimized and possibly negated through the development of a corporate wide understanding of why change is necessary. Even more important is the clear dissemination of information on how the changes will be implemented throughout the corporation's environment. Without such beginnings the probability of reaping the rewards which successful change offers is minimized. With this in mind, I will review three essential assumptions required to optimize the probability of achieving successful change.

ASSUMPTION I:

The introduction of the change method should be instituted in a small segment of the corporation, to offer people the opportunity to see the process work. When experimenting with change under sanctioned and therefore protected conditions, the system has the opportunity to evaluate the impact that change produces. By using this approach the corporation may make informed decisions about the usefulness and general applicability of the method.

ASSUMPTION II:

To produce successful change by the sociotechnical approach, the organization must have the ability and willingness to accept risk. The staff must view senior personnel as persons of integrity and candor.

The idea that change affects the configuration of power, reward systems, status and the like, must be recognized at all levels of the corporation. It is imperative that these issues be dealt with openly and honestly prior to undertaking the sociotechnical approach. An organization which does not institute this procedure is destined to face an endless array of seemingly insoluble problems.

ASSUMPTION III:

The importance of involving those organizational members most affected by the proposed changes cannot be underestimated or overemphasized. Sociotechnical induced change emerges from inside of the work system. It is developed, implemented and evaluated by the workers who operate the technological process of the corporation. The sociotechnical approach requires the active participation of all the employees affected by the proposed changes.

The approval and support of the most senior management must always be clearly evident, to ensure a reasonable level of success.

Taken together, the previous assumptions provide the necessary pre-conditions required to institute successful change by use of the sociotechnological method.

SOCIOTECHNOLOGY AND THE ORGANIZATION:

Sociotechnical Systems Theory guides and directs the organization in analyzing itself. Sociotechnological strategy has evolved from the work done by Rice, Trist and Van Bienum at the Tavistock Institute of Human and Industrial Relations in London, England; and Einer Thorsurd at the Work Research Institute in Oslo, Norway during the last 30 years. In this period of time a number of propositions have been tested and observed to be stable, over time and in various environments. These propositions in brief specify that;

(a) organizations must be designed to fit their goals,

(b) variance needs to be controlled as close to its source as possible,

(c) subsystems need to be designed with a view to a recognizable whole task,

(d) support systems must be correlated to the design of the organization and its goals,

(e) those people most affected by the proposed changes must be actively involved in designing and developing the organizational structure,

(f) the organization must be sufficiently adaptable to accept change based on observable contingencies.

The strategy of applying Sociotechnology is based on these propositions, and follows a nine step sequence;

(1) define the experimental system,

(2) acquire appropriate sanctions for the experiment,

(3) form an "Action Group" to carry out the experiment,

(4) analyze the experimental system,

(5) develop hypotheses for redesign based on the analysis,

(6) evaluate the hypotheses and reduce them to manageable levels,

(7) implement and analyze the hypotheses under experimental conditions,

(8) revert from experimental to normal operating conditions,

(9) disseminate the result to the organization.

Prior to defining which area in the corporation is most suitable for the sociotechnological experiment, two questions need to be asked. The first is what constitutes an ideal unit and the second, does an ideal unit exist? The answer to the second question is obviously no, but there are conditions which make some units closer to the ideal than others. These conditions are:

(1) That the social and technological components are clearly differentiated from other sociotechnical units. In other words, the components must be more integrally related to each other within the unit than they are to external units. In technical terms this represents *Bounding*.

(2) That measurable input and output states be clearly defined. The logic for this is simple, if a distinction cannot be made between input and output, it is impossible to comprehend the system.

(3) That the unit has a high probability of achieving success. What constitutes a unit as having high probability of achieving success? The answer is two fold. A unit can be chosen that has reached rock bottom. Or on the other hand, a unit that is presently highly successful can be selected. Both of these criteria are extreme and pose inherent problems. Because they are extreme they will necessarily tend to statistically regress to their mean. In other words, there is a fair probability that the highly successful unit will, over time, become less successful. At the opposite extreme, the unit that is presently unsuccessful may begin orienting itself in the direction of success. So, what is the problem? These changes may occur not because of the sociotechnical interventions but simply due to a statistical artifact. The point to be emphasized is that there needs to be certainty that the highly successful or unsucessful unit is not operating in this fashion due to a statistical aberration but rather that it has operated in this manner for a long time. As can be imagined it becomes very embarrassing to explain why an intervention has taken a once highly productive unit and diminished its productivity.

(4) The fourth required component is that the unit has the essential characteristics which make the results generalizable to other units. Thus, ideally the results should be able to be diffused to other units with similar

characteristics. The point being stressed is that the experimental process should not be initiated on a totally atypical unit. If it is then the results are not generalizable to other units, and as such the dissemination of information phase becomes a much more difficult task.

(5) The last characteristic of the ideal unit is that the unit be open enough to allow its employees to be involved in designing the experimental process. It is easier to gain insight and understanding when experiencing a situation on a first hand basis. Change, and specifically change by the sociotechnological approach requires an experiential approach. This is especially so if it is to act upon the organization as a whole. An understanding of the sociotechnical process is most effectively and efficiently transmitted through direct involvement.

Having chosen the appropriate experimental unit, the acquisition of appropriate sanctions needs to be addressed.

It is of paramount importance that the employees receive management sanctions to experiment and innovate. If a labor union is represented, then a union-management sanction is required. What is a sanction? The answer is two fold. First, the experiment if it is to be successful, requires protection, to delve into how people do their work. The initial protection allows the workers to be shielded from the day to day operating requirements. It becomes patently obvious that you cannot innovate or properly experiment if the rate of production is an underlying issue.

Secondly, a psycho-social protective mechanism is required. This represents protection from remunerative changes, job security issues, changes in benefit structure, and so forth. If the sanctioning body cannot guarantee this protection, then the sociotechnical process will in all probability fail.

The sanctions should be obtained from the highest level of the organization. When acquired, the protectionist scheme can then be channeled to the organizational level having a direct bearing on the experiment. The ultimate aim of sociotechnology is to impact on the whole organization, and this is a crucial step in that direction. A final point on the sanctioning procedure. Past experience has shown that individuals are more accepting of the sociotechnical method when the ownership and ultimate decision making is theirs. Therefore, to succeed with this type of enterprise, the appropriate psychological state must be developed. Both groups, management and non management, must feel that they have a vested interest in the outcome.

When the appropriate sanctions are in place, an action group can be formed. The action group is composed of individuals who develop hypotheses, analyze procedures, collect data, evaluate the outcome, etc. In other words, these are the employees who carry out the experimental process. If the number of employees in the organizational unit is small, then there is no need to form an action group. In such a scenario, all staff can be incorporated into the planning process. If the unit is large then the sociotechnical experiment must be facilitated by an action group.

Classically, the action group is composed of management and non management personnel, all with direct ties to the outcome of the experiment. Numerically, the action group should never exceed seven people and if possible should be held to five. Observation of group dynamics reveals that the more people involved the more difficult it is to efficiently achieve the required goals. Variables such as politics and past internal grievances, militate against the efficient functioning of the group.

Although control over who participates in the action group is a discretionary issue, if at all possible certain types of individuals should be chosen. Ideally, individuals who are innovative, change oriented, and abreast of current information in their field of expertise should be selected. Those picked should be viewed as change rather than status quo oriented. Because sociotechnical information is imparted throughout the corporation, it is essential that the group contain influential individuals. The logic of this group construct is self explanatory.

Organizational analysis follows the formation of the action group. Essentially two types of data require analysis in a sociotechnical design. The first involves analyzing organizational data about technology, the second, about people. Based on the work done at the Tavistock Institute, the data is classified as hard or soft. When dealing with a technological organization, a hard analytical model is used. A service organization requires a soft approach. To use one data base and exclude the other is impractical and does not mirror corporate reality. If technology is operationally defined as the tools, techniques and methods used to produce a product, then it follows that even service corporations use technology to produce products or services. To do otherwise may jeopardize the acquisition of insight about the corporation.

Analytical tools are available which provide information about people, technology, the interaction of the two and the variance that is a natural by-product of that interaction. As well, inventories and questionnaires can be developed to analyze specific problems. The issue of validity and reliability must be overcome when designing such instruments in-house.

Once the data is acquired, the next step involves developing hypotheses for redesign. Every environment is unique and necessarily requires a unique design. Every unit will supply a different array of data and every redesign will reflect this difference. Another way of looking at this is that one is analyzing *WHAT IS* in order to come up with *WHAT WILL BE*.

This mode of intervention has a noteworthy feature, namely, that the employees developing the redesign paradigm are the same people who will benefit from the redesign process. When employees realize this, the ownership issue dissipates.

The action group is now ready to generate hypotheses. This is a creative and spontaneous process that should not be interrupted no matter how incredible the hypotheses seem. If at this juncture an evaluative component is interjected and, specifically, if it has negative connotations, there is a likelihood that the creative process will cease. This stage is the idea generating stage and is not amenable to an evaluative component. There is ample time to analyze and evaluate the quality of the hypotheses during the evaluative component.

If a large number of hypotheses are generated there is a need to reduce them to a manageable set. The hypotheses that affect change most directly are the ones that should be incorporated into the experimental situation. To class the hypotheses into workable sets it is helpful to categorize them as affecting changes in the social system, the technological system or both.

The action group is now able to categorize, analyze and evaluate the previously generated hypotheses. Those hypotheses which have a high probability of achieving the desired goals are implemented. Having knowledge of which hypothesis will be implemented allows the action group to ascertain the costs of implementing a specific hypothesis. A cost-benefit analysis of the difference between anticipated results and the status quo can now be derived. When this is finalized the action plan is ready to be implemented.

The logistics for implementing an action plan requires a great deal of analysis. The action group having knowledge of daily production or service demands are indispensable in evaluating the specific implementation requirements. Everything necessary for the implementation phase must be in place. The worst scenario possible is that the experiment begins and major unforeseen problems develop. If this segment of the sociotechnical process is not well developed the probability of the action program aborting is extremely high. A check list of requirements, do's and don'ts is of great benefit. Methodical consideration to detail is essential.

A word of warning is in order. In a number of sociotechnological change programs, it has been noted that employees anticipate immediate positive payoffs for instituting the action program. The employees involved should be made aware of the fact that a lag exists between initiation of the change process and goal attainment. This is specifically the case when social rather than technological changes occur. Machinery and tools can be easily shifted, humans are much more fragile. Social system change is a slow process and at times things seem to get worse before they get better. When the awareness exists that humans lag behind technology during stimulation of change, the expectation will correlate with reality. As well, when rapid change occurs regression in attitudinal measures should be anticipated.

Having implemented the action program, the next step involves evaluating the effectiveness of the change program. For the sake of brevity I assume that the experimental design is sensitive enough to evaluate the efficacy of the changes instituted. The type of experimental design, method of analysis and analytical instruments used are dependent on the changes being formulated. When changes encompass only technology then efficiency and economic results warrant analysis. If the intention is to create changes in both the technology and social perspective, then the measures should examine economic and social results. Valid and reliable designs are available (e.g. quasi-experimental designs) to assist with the evaluative component in the corporate environment. The action group can also play an instrumental role in assessing costs that are otherwise not easily discernible.

It is important that the design incorporate as many data points of the same events and of different events as realistically possible.

Everything from analyzing production records to various staff satisfaction measures should be considered. Once these sensory points are found effective, an added benefit is gained. Namely, that a feedback mechanism for the continual monitoring of adoption and innovation is now in place.

When the experimental change process is shown to be successful in a protected environment, the point is reached to return to normal operating conditions. Returning to normal operating conditions is similar to initiating the original change process. The exception being that experience will militate against false expectations. A lag and possible regression prior to the full benefits of change acting on the normal daily operation should be anticipated.

The sensory input is now functioning and is available to analyze the daily variations. The feedback mechanism previously established is vigilant and offers insight into further required changes. An organization using the results arising from the Sociotechnical approach is not easily accused of throwing good money after bad.

When the data analyzed is shown to be valid and reliable, the information acquired can be disseminated. Two issues warrant review prior to disseminating the sociotechnical change process to the corporation.

(1) What information needs to be disseminated?

(2) What is the best mechanism to disseminate the information throughout the corporation?

As with all of the nine steps representing the sociotechnological perspective, the dissemination needs to be designed and tailored to the specific corporate environment. Questions that assist in this endeavour include:

(1) Is the unit which has undergone the change process environmentally similar or dissimilar to other units within the corporation, and,

(2) do the employees in the experimental unit resemble others employees in socioeconomic status, sex, training, etc.?

Available options include face to face discussions with small groups. If great distances exist between units audio-visual packages can be developed. Many organizations have internal media departments which can assist in formulating an appropriate information dissemination package.

As with other aspects of sociotechnology a word of caution is in order. There is nothing worse than having conducted the previous eight steps successfully and falter at the stage of dissemination. Too often the sociotechnical experiment is successful, yet the information does not impact on the organization positively. It is imperative that the individual communicating the information be influential and credible enough to create acceptance. A number of successful sociotechnical designs have used the employees that were directly involved in the experiment to disseminate the information. The underlying point is that careful attention must be paid not only to what is said and how it is said, but just as importantly by whom.

The sociotechnical approach as it applies to successful change in the corporate environment has been reviewed. This by no means is an all encompassing treatise and many points represent the personal view of the author. Sociotechnology offers a powerful logic, strategy and procedure for positive intervention in the corporate environment. Fortunately or unfortunately as the case may be, sociotechnology does not lend itself to a cook book style, and as such cannot be easily consumed or clichéd. This should not be construed as a weakness but rather a strength. Niccolo Machiavelli stated "There is nothing more difficult to take in hand, more perilous to conduct, or more uncertain in its success, than to take the lead in the introduction of a new order of things." Anyone attempting to develop, implement and evaluate change, is well aware of the inner meaning of the above statement. Sociotechnology offers an effective and efficient means of introducing "a new order of things".

REFERENCES:

(1) Campbell, D.T. and Stanley, J.C., Experimental and Quasi-Experimental Designs for Research (Rand McNally and Company, Chicago, 1966).

(2) Conant, E.H. and Kilbridge, M.C., An Interdisplinary Analysis of Job Enlargement: Technology, Costs and Behavioral Implications, Industrial and Labour Relations Review 18 (1965)377-395.

(3) Cummings, T. G. and Scrivastva, S., Management of Work: A Socio-Technical Systems Approach (University Associates, San Diego, 1977).

(4) Harris, P.R., Management in Transition (Jossey-Bass Publishers, San Francisco, 1985).

(5) Kirkpatrick, D.L., How to Manage Change Effectively: Approaches, Methods, and Case Examples (Jossey-Bass Publishers, San Francisco, 1985).

(6) Martin, M., Operational Research: A New Discipline in, Tillet, A., Kempner, T. and Willis, G. (eds.) Management Thinkers (Penguin Books, Inc.,Baltimore, 1970, pp. 140-165).

(7) Naisbitt, J. and Aburdene, P., Re-inventing the Corporation (Warner Books, New York, 1985).

(8) Perrow, C., Complex Organizations: A Critical Essay (Random House, New York, 1986).

(9) Thorsurd, E., Industrial Democracy: Involvement, Commitment, Action: Some Observations During Field Research (Tavistock Institute of Human Relations, London, Doc. No. T. 886, 1966).

(10) Trist, E.L., Higgen, M.H. and Pollack, B.A. Organizational Choice (Tavistock Publications, London, 1963).

(11) Van Beinum, H.J.J. and deBel, P.D. Improving Attitudes Toward Work Especially By Participation (Tavistock Institute of Human Relations, London, Doc. No. HRC, 101, 1968).

Human Factors in Organizational Design and Management - II
O. Brown, Jr. and H.W. Hendrick (Editors)

MANAGING THE IMPACT OF COMPUTER INTEGRATED MANUFACTURING ON AN ORGANIZATION'S HUMAN RESOURCES

Daniel Levi
Charles Slem
California Polytechnic State University
San Luis Obispo

Andrew Young
Northern Telecom

This case study analyzes the impact of CIM on an organization's human resources. In order to increase commitment and reduce resistance to CIM, organizations need to focus on three key factors: employee involvement in the change process, job security, and job training and personnel development.

INTRODUCTION

How will the introduction of Computer Integrated Manufacturing (CIM) affect the people in an organization? How can the transition to CIM be used to increase commitment to the organization and improve the quality of work life? What can be done to minimize the negative social and psychological effects of CIM?

These are the questions we asked when we began our study of a large telecommunications manufacturing facility. This facility is in the planning and early implementation stages of a multi-year transition to CIM. Our task was to assess how they have dealt with change in the past and their anticipated beliefs about the impact of CIM. This information was integrated with the existing research on the social and psychological impact of innovation and technological change. Working with the Human Resources Department, our purpose was to provide advise about human resources management to the change agents responsible for developing and implementing CIM.

The methods used to study the facility included a review of existing research on the effects of CIM [3], a reanalysis of a company communications survey, a stress survey given to a sample of the managers, and in-depth interviews with thirty directors, managers and professionals in the company. After the initial analysis was completed, we re-interviewed a sample of the managers to check, clarify and expand our analysis.

The transition to CIM creates major changes in the operation of a company. These changes disrupt the existing social system, require that people change their habitual activities at work, and displace employees into new types of activities which will require a substantial amount of training. All of these changes put a large amount of stress on the employees at the same time that they are being asked for the extra commitment necessary to make the transition to CIM successful.

In order for the CIM transition process to be successful, it must increase the amount of commitment from the employees to the new systems and reduce or eliminate resistance to the changes. The following three issues seem to be key factors for doing this: employee involvement in the change process, job security, and job training and personnel development.

EMPLOYEE INVOLVEMENT

Every major study of successful organizational change or innovation addresses the issue of employee involvement [1,4,6]. People who are affected by organizational change need to be involved in the change process. Their involvement allows them to modify the changes to meet their specific needs. This improves the quality of the changes, and increases employee acceptance of change. Employee involvement is called a variety of things by different organizations such as distributative ownership, empowerment of others or user participation.

Employee involvement does not mean turning the change process over to the employees and letting them do what they want. Organizations need to find a suitable balance between the need of managers to have control and accountability over the innovation process, and the need of employees to have the power to develop the innovation to meet their specific needs.

Our interviews repeatedly demonstrated the importance of employee involvement. When asked to describe an example of change which failed or was not well received, almost every person stated that the worse thing that can happen is a change which was delivered directly from the corporate headquarters. There were numerous stories about trying to adapt to systems or practices which did not make sense under local conditions. On the other side, employees were very confident in their ability to work together to develop and implement all types of changes and innovations through local task forces even though these changes might come from corporate headquarters.

This manufacturing facility has had a substantial amount of experience using task forces to develop changes and innovations. Employees in these task forces generally reported that they had sufficient power to operate effectively. When there were complaints about the changes made by task forces, they were from people from related parts of the organization who had to adapt to a new system without being consulted.

Because one of the fears (and sources of resistance) which employees have about CIM is that it is a change which will be dumped upon them by corporate headquarters, the change agents responsible for the CIM program need to assure the employees that they will be involved in the development of CIM. It is important that all of the major changes recommended as part of CIM be made through task forces or other systems which allow employee involvement.

The change agents responsible for CIM need to publically define their role so that employees understand they will be involved in future changes. The role of the CIM change agents should be defined as a facilitator, catalyst or organizer for change and an information manager, parameter setter or contoller of the interfaces among the new systems.

The role of the CIM change agents should be to plan and facilitate change through the actions of the employees. They should set the context for change by identifying performance gaps and alternative solutions, and establishing task forces to develop the solutions. Once the task forces have been established, they should facilitate the activities of them by supplying resources and technical assistance, and identifying and promoting connections to other parts of the organization. The CIM change agents must also participate in the task forces as a member to make sure that the solutions which are developed can be integrated with other changes occuring in the organization.

The enactment of CIM will increase employee commitment to the organization if the employees feel that it is their system. One of the biggest problems facing the CIM change agents in using the task force approach is obtaining the time of good people in the organization so that they will be able to work on the development of CIM.

JOB SECURITY

One of the most threatening aspects of change and especially automation is its effect on job security [1,5]. It is a widely held belief that automation means job loss; however, this is often not the case. Automation often allows an organization to survive and grow. It is more likely to change the nature of jobs than to significantly reduce the number of jobs.

In a developed CIM organization, job security will play an important institutional role in facilitating internal innovation. Feeling secure about one's employment is a necessary precondition for encouraging people to think about ways to eliminate or modify their jobs. Without a sense job security, people will defend their existing jobs rather than trying to creatively deal with with new situations and opportunities.

Our interviews found that many employees view CIM and its related automation as leading to job layoffs. Everyone seemed to feel that their job was important, and the layoffs would occur elsewhere in the organization (usually at a level below them). Although they were aware of and concerned about job security, they did not seem to recognize that the nature of their jobs might change.

If CIM is viewed as a major threat to job security, it will be resisted. Organizations undergoing change need to develop a public policy stating how job security will be affected by CIM. When possible, employees should be retrained rather than being removed when CIM related changes occur.

Dealing with job security issues will be a difficult task for CIM change agents. Although an open policy about job security will be helpful, it suffers from a credibility problem which will be difficult to counteract. Identifying both who does and does not have job security will help to establish credibility. Keeping people informed about CIM activities and planned changes will also help because people can anticipate the changes. Connecting the organization's internal job bidding system to changes created by CIM will help people to view CIM as changing the nature of their job rather than displacing them. Finally, investing in the development of the employees through job training and other personnel development actions will demonstrate the organization's commitment to its current employees.

JOB TRAINING AND PERSONNEL DEVELOPMENT

Enactment of CIM will change the nature and activities of many jobs. The employees will need a variety of new skills to meet the demands of the new systems. They will also need a better understanding of the overall system because of the increased degree of interconnectedness created by CIM.

These widespread job changes will create insecurity and resistance. If people feel that their existing skills will not be needed by the organization, their sense of job security will be threatened. The CIM change agents need to develop a substantial job training and personnel development program to reduce this source of resistance, and to create the kind of skilled employees which are needed to operate the system [2,5].

Our interviews found that people were generally unaware of the potential impact of CIM on their jobs or the need for training. They tended to view CIM as affecting job security rather than changing the nature of their job. When they were aware of potential job changes, they were unsure about what skills would be needed in the future organization.

The people reponsible for developing CIM need to make an early and strong commitment to job training and personnel development. This will send an important signal to the employees about job security and the organization's commitment to developing its human resources. This training and development program should focus on three issues: technical skills and knowledge, overall system information, and group problem-solving skills.

The transition to CIM will force people to operate new kinds of equipment and to deal with new types of information. For example, some people will be required to operate computers for the first time which may be anxiety producing for them. Although employees are competent, it will take time for them to learn and become fully functional in their new roles. Sufficient time and resources need to be allocated so that this learning can occur.

In addition to providing specific technical training, employees need background information about how the equipment operates and why they are operating it in certain ways. Fully trained employees will be better able to fix problems with the system, modify the ways the system is used, and develop creative ways of using the system.

Although most training programs only provide specific job information, employees also need to have an increased understanding of the overall operation of the organization and the developments created by CIM. Over time, substantial improvements in the facility's operation will be created by CIM through the creative use of interconnections among the subsystems. This is the opportunity which "integrated" manufacturing creates. In order for this potential to be realized, employees must understand the boundaries of their functions and the operation of the overall system.

One of the misconceptions about CIM is that it will reduce the need for people to interact. "Everything will be on the computer so we won't have to go to meetings." In fact, CIM will probably have the opposite effect. The process of developing CIM will require substantial interaction within task forces and groups. Once the CIM system is operating, it will identify problems and opportunities which will require more interaction.

A CIM workforce will need people with both technical skills and knowledge, and leadership and group problem-solving skills. Personnel development programs need to be enacted to help increase the group problem-solving, communication and management skills of the employees. These programs should begin as soon as possible because they will be needed during the initial development stages of CIM.

CONCLUSION

How will CIM affect the people in a manufacturing organization? The answer to this question does not directly depend on the technology of CIM. The effect of CIM on people depends substantially on the initiation and implementation process used to adopt CIM. If the transition process to CIM is designed to deal with the legitimate concerns and issues of the people in the organization, CIM may help to increase the employees' commitment to the organization and to improve the quality of work life for them.

REFERENCES

[1] Kanter,R.M., The Change Masters (Simon and Schuster, New York, 1983).

[2] Peters,T. and Waterman,R., In Search of Excellence (Harper and Row, New York, 1982).

[3] Slem,C., Levi,D., and Young,A. The effects of technological change on the psycho-social characteristics of organizations, in: The Second International Symposium on Human Factors in Organizational Design and Management (North-Holland, Amsterdam, 1986).

[4] Snyder,R., To improve innovation, manage corporate culture, in: Bennis,W., Benne,K., and Chin,R. (eds.), The Planning of Change (Holt, Rinehart, and Winston, New York, 1985).

[5] Tornatzky,L.G., Technological change and the structure of work, paper presented at the American Psychological Association Conference (Los Angeles, 1985).

[6] Zaltman,G., Duncan,R., and Holbeck,J., Innovation and Organization (J. Wiley and Sons, New York, 1973).

Human Factors in Organizational Design and Management - II
O. Brown, Jr. and H.W. Hendrick (Editors)

PERCEIVED CHARACTERISTICS OF THE WORK ENVIRONMENT

JOHN R. WILSON and SUSAN M. GREY

Department of Production Engineering & Production Management
University of Nottingham,
Nottingham, NG7 2RD, U.K.

The physical environment at work can not only affect performance and worker well-being, but also can engender positive or negative attitudes towards all other aspects of work organisation. It is suggested that ergonomic models and instruments for assessing this are still at a rudimentary stage. The authors draw upon a series of loosely related case studies to expand upon their argument, within a wider examination of the relationships between the various perceived characteristics of work.

Background

The measurement and evaluation of peoples' attitudes, beliefs and values concerning their work has long underpinned many initiatives in job design and work organisation. Many measuring instruments have been constructed, and validated to varying extents, which assess aspects of the work experience from involvement to alienation, anxiety to self esteem. Some of these are short scales intended for independent use, some full instruments in their own right, and others are scales used as part of a broader instrument measuring a whole gamut of job attitudes'· Cook et al in an excellent review, have critically examined 249 such measures; many more have appeared since their work was completed [1].

This paper is concerned with the methodology of measuring peoples' feelings and beliefs about their work. It is concerned in the main with the content of instruments, particularly of questionnaires and various interviewing techniques, and especially with the incorporation of physical workplace and environment considerations in these. The interplay of such aspects with those of job design (task or job characteristics and work context factors), and with affective responses about work, is also discussed. There is no room here for deep examination of such matters as scale validity and reliability; this can be found in great detail in the text referred to above [1].

Some of the work of the authors has drawn in part upon elements of the Job Characteristics Model (JCM) and Job Diagnostic Survey (JDS) of Hackman and Oldham [2 and 3], although using elements of the latter in a different manner to that originally recommended. The model and the instruments used to tap its key elements have been criticised on many grounds, from the relations and linkages assumed (e.g. Roberts and Glick, 1981) to the impact of other variables (e.g. Loher et al, 1985); from the content of the scales (e.g. O'Brien, 1982) to the method of data collection - both job characteristics and affective responses from the same source, the job holder (e.g. Adler et al, 1985); and even to the statistics used (e.g. Watson, et al, 1985) [4-8]. Despite this however, there is much that is intuitively reasonable about the model. Consequently,in the early research stages we utilised some of the JDS scales in our instruments and in particular with regard to the influence of both job characteristics and work context factors.

Ergonomics and work attitude measurement

Two surprising gaps have struck the authors, both ergonomists, as they have absorbed the literature over time. The first arguably is the apparent paucity of use made of "job attitude" questionnaires by ergonomists, at least those working in the U.K. and possibly also those in Europe and the U.S.A. Such research has seemed largely the preserve of occupational psychologists and management specialists. For instance, an admittedly not totally exhaustive literature search revealed only Martin et al (1984) [9] and Long (1983) [10] (in addition to the authors - [11]), as U.K. ergonomist users of e.g. the JCM and JDS, around which alone there is a considerable psychological literature.

Secondly, there is an absence of a unifying theory, with appropriately validated instruments, regarding attitudes towards, perceptions of, and responses to the total physical work environment including technological factors. Of course, some work has been done. For instance, researchers have attempted to assess relationships between variables of technology and employee perceptions of job complexity [12], their satisfaction, performance and influence [13] and perceptual and affective responses [14]. As far as the physical environment is concerned, McCormick et al's (1969) Position Analysis Questionnaire (PAQ) does include nine items about the physical working environment and four on physical hazards [15]. In their UK version of this - the Job Structure Profile (JSP), Patrick and Moore include eleven and four such items respectively [16]. Campion and Thayer have developed a Multimethod Job Design Questionnaire, comprising four main scales labelled motivational, mechanistic, biological and perceptual/motor, the third of which contains several postural, workplace and environment related items, and the fourth some lighting, control and display items [17]. However, these three instruments are to be completed largely by job analysts and for the first two their questions are largely concerned with the time spent in certain environments. Schiro and Cox (1978), as ergonomists, have produced new measures of attitude accordance, in two directions [18]. They report a Work Environment Questionnaire and a parallel Work Attitude Questionnaire to produce a comprehensive analytical package. We are uncertain of its use and validation by other workers.

Overall though, there does seem to be under-use of job design models and assessment instruments by ergonomists, together with a lack of effort in the construction of these to tap attitudes to the physical setting of work. There are several possible reasons for this state of affairs. Many workers in U.K. and European ergonomics are also psychologists and we may be underestimating the ergonomic effort in the area of job satisfaction and so on. However, it is more probable that, at least historically, the fact that "the suggestion that ergonomists and human factors engineers consider measures of job satisfaction or engage in job restructuring projects has been criticised" [18] may have inhibited such an approach. It is however anticipated that such a regrettable state of affairs, if it exists, is changing now. It is furthermore possible that the impact of the physical environment has not been regarded as of importance, or of intellectual interest compared to such issues as mental workload, systems organisation and so on even amongst ergonomists themselves. A further reason is the fact that most physical aspects of work can be measured objectively using instrumentation, and thus the need for accompanying attitudinal measures is not seen as strong.

It is not that ergonomists have not jointly considered physical and social factors in their work. There is a long and honourable tradition of this, for instance in Scandanavia and France, and in inter-disciplinary teams of ergonomists and psychologists (see Klein, 1975 for instance [19]). The argument is that no rational model or approach for this has been proposed.Several ergonomics studies do of course assess workers' responses to their environment, especially now in VDU work. However, the present haphazard situation with each ergonomist seemingly devising and reporting a new instrument to tap relevant feelings is not really satisfactory any longer. We need some underlying models and scales of measurement for the influence of the work environment on attitudes,similar to those for job design research[2].

Research approach

The authors' approach to all their efforts in the widest area of work design can be typified by a number of key elements [20], amongst which three are prime.

a) An approach of joint investigation and subsequent optimisation, obviously akin to the socio-technical systems approach, is employed. Within ergonomics this means that we must recognise the impact of the physical and social environments (in their widest sense) upon each other, and their interacting effects upon the workers' attitudes. Ergonomics, as concerned with the total system, must take a broad perspective and concern itself intimately with issues of psycho-social conditions, job satisfaction, and work humanisation [21].

b) As a corollary to the above, and as discussed earlier, the physical environment (including the relevant technology) is regarded to be as important a determinant of "good jobs", and of peoples feelings about their jobs, as are the elements of job design. The two should thus be integral within any measurement and assessment philosophy and instruments. " The physical demands of a job, however, would seem to be an important factor in understanding such psychological states as fatigue, job satisfaction, and motivation ... current questionnaire measures do not index the working conditions or job hazards ..... . This omission is consistent with the bias in modern theories to classify working conditions arbitrarily as "extrinsic" to the job, and then to ignore them. It is quite possible that many workers consider such things as heat, fumes, oils and greases, and noise in the work place to be intrinsic characteristics of their jobs. Early studies in industrial psychology focused on the importance of working conditions, but they play almost no role in modern social science theories of work. Perhaps working conditions should be retrieved from the theoretical dustbin and examined more dispassionately as possibly important components of the job - whether intrinsic or extrinsic". (Taber et al, 1985 [22].

c) Despite some doubts as to whether we are enmeshed in wish-fulfillment, we believe strongly in participation as the key to all aspects of work design. This is so not only in the implementation of job design, interface, and work environment changes, but also in the gathering of information to support such initiatives. Within a generally multi-method approach, the authors have extensively used formal and informal interviews, informal debriefing sessions and design decision groups comprising workers and researchers.

We have recently been finding our clients initiating what we regard as only limited studies of work in their organisations. Typically, we are finding that the need for ergonomics in the specification and design of workplaces has penetrated U.K. industry and commerce to an increasing extent, but largely only as far as the physical environment is concerned. Employers are seeing the advantages, if not always the necessity, of well designed physical working environments and workspaces, not least because of increased union awareness. During negotiations and at the outset of projects we have attempted to persuade clients (with varying degrees of success) that one aspect of work cannot be examined in isolation; that work design studies should consider physical environment and organisational and job design issues also. Thus we have had opportunities to conduct a number of studies with a parallel or joint optimisation approach. These studies have considered, to varying extents, the actual states of, and workers' attitudes towards, five groups of factors, namely job characteristics, work context, work procedures (for instance training, consultative procedures etc.), physical work environment, and the technology itself, although not always simultaneously. As a consequence of these studies we have become interested in connections, particularly in terms of the attitudes and perceptions of the workforce, between the physical and organisational, or social, aspects of work. We are not necessarily hypothesising causal links in either direction, merely examining patterns and relationships. Of course, this must be linked to accepted wisdom in job design implementation, where a fundamental requirement for success has been said to be satisfactory working conditions.

The authors have explored a number of connections between these five fundamental work factors during a number of independent research studies. Each of these studies was structured and reported to meet the requirements of individual clients; however, interwoven have been the developing interests and expertise of the investigators. Consequently one unifying characteristic of the investigations has been an attempt to explore the connections between both physical and organisational factors as they manifested themselves in the varying research contexts. Figure 1 is put forward tentatively to describe in simple form the authors' approach to their work so far. The five types of factor, "objectively" measurable through instrumentation, direct observation or analyst or supervisor opinions etc., are perceived by workers and inform and colour their attitudes[3].The effects of these attitudes can be felt within worker well-being and worker performance, just as can the factors (or at least some of them) themselves directly. Factor perception and attitude formation will be moderated by individual physical and mental differences. A further link can be seen for much of the service sector where employee attitudes can have a marked direct effect upon customer satisfaction.

There is much criticism as noted earlier about the use of self-reports to measure both task characteristics and affective responses, "cause" and "effect". However, we have been interested in attitudes not merely as their measurement will tap feelings about what a job is and what someone wants it to be, but also as instrumental in their own right in the way they affect peoples' behaviour. Also, we are interested in attitudes to the physical environment and to the technology for two other reasons. First we have reason to believe that, because people find complaints about these factors as easier to conceptualise and verbalise, and they seem more "reasonable" attitudes to hold, then they are perhaps used to mirror dissatisfaction with a host of other, perhaps more intangible, characteristics of work. Second, we believe that the attitudes engendered by people's perceptions of the work environment and technology they are 'forced to work in by management' will have a marked effect upon their attitudes to other facets of their work, to their acceptance of change, and upon their affective responses in terms of motivation and performance. A conceptualisation of the key role of environmental conditions is presented in Figure 2.

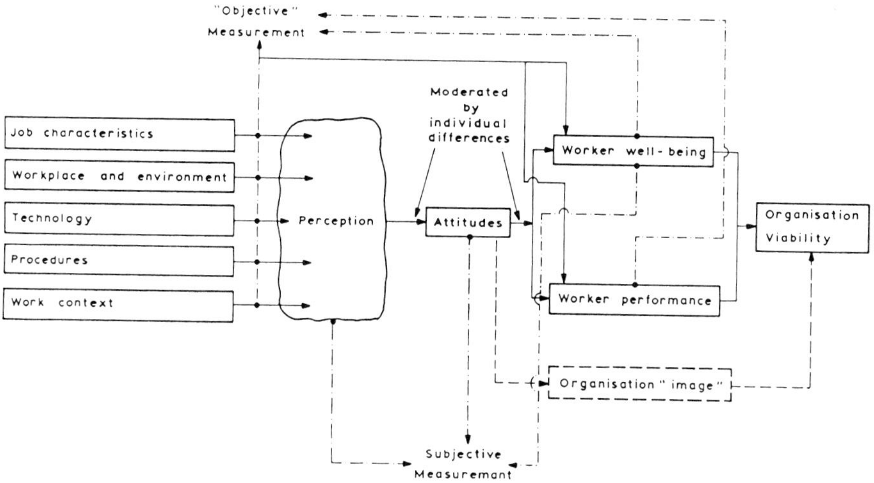

Figure 1. Relationships between perceived work characteristics

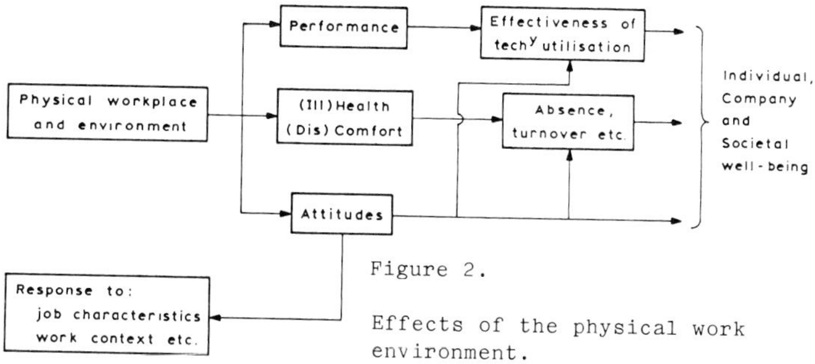

Figure 2.

Effects of the physical work environment.

Research Studies

The last part of this paper gives some brief details of a number of research studies undertaken by the authors, which illustrate some of the points made earlier. It must be emphasised that they have been conducted in response to specified practical needs of individual clients. Thus a rational series of methodological examinations has not been entirely possible, with consequences that direct comparisons between studies are difficult if not impossible. Also rigourous reliability analyses of the instruments used are difficult to conduct since the numbers of items employed per scale were limited due to respondent time constraints. To prevent any confusion it should be understood that we have not in general been exploring job characteristics as "causes" of affective responses. Job-holders' subjective assessments of these characteristics have been used to help provide a picture of how people view their actual and potential jobs and roles; thus criticisms of "same-source" data collection are not really applicable [7].

Four research studies have relevance to the arguments in this paper.Initial exploration began with a study of problems of workplace design and work organisation for point-of-sale operators in a number of different retail outlets. The methodology used here was developed further for a similar dual-purpose study of library issue desk services. A third study was of the coding function of a major U.K. telecommunications enterprise, particularly relating environmental design to performance, health and other job-related complaints. The fourth ongoing area of research, which is long term and involves a number of case studies, is for an automobile component manufacturer, with the focus of research on the semi-skilled shopfloor operator. The orientation of our research necessitates employing multiple methods to obtain and analyse data. The methods adopted within this approach have ranged from direct and participant observation to operator design groups and computer workspace modelling.

a. Retail Outlets

Work attitudes of 24 checkout operators using a laser scanner operation were compared with those of 24 operators at conventional electronic cash registers with "ergonomic" (manufacturer's description) features. The instrument used consisted of Section's 1, 2 and 4 of the Job Diagnostic Survey (JDS) [3] and Scales 5 (9 out of 10 items) and 6 (11 out of 16 items) from Warr et al's Work and Life Attitudes Survey (WLAS) [24]. This allowed comparison of two instruments with respect to job characteristics and work context measurements. Use of Campbell and Fiske's multi trait-multi method matrix (MTMM) [25] shouldconvergent validity between the two instruments, but only partial discriminant validity [full results in 11]. No significant between-groups differences were found for job characteristics scores; however the extremely low nature of these is shown by the Motivating Potential Scores of about 20 (range 2(!)-62), compared to U.S. means across jobs of 128 and means for "rank and file" problem jobs of about 50 [3]. As for work context, the laser scanner operators were less satisfied, and significantly so for pay, security and supervision, although few observable differences in these actually existed.

Increasingly apparent to the authors is the narrowing, and occasionally misleading, effect of questionnaire use, sometimes not allowing identification of some vitally important positive and negative attitudes. This has been strikingly demonstrated through the use of informal debriefing sessions following questionnaire completion. Use of this technique provided clarification of what might otherwise have been misleading results in this retail outlets study. Operator dissatisfaction with the laser scanning system seems to have been expressed through the work context factors relating to pay, supervision, co-workers and job security, rather than as might have been expected through responses regarding job characteristics. In general, dissatisfaction felt by laser-scanner operators with technical aspects such as defective codes and physical workplace aspects such as seating was most easily expressed in terms of work context factors, to which interviewees found it easier to relate their own feelings about their work. It is clear to the authors that in such an instance the use of questionnaires alone, however comprehensive and however suitably worded, cannot tap the interrelationships between such factors both in terms of the respondents' perception of them and their subsequent expression. We have recently commenced research with the specific purpose of comparing the use of questionnnaire and unstructured interviews. Our initial findings point towards advantages in the latter as a means of eliciting information regarding attitudes to intrinsic job factors, on the basis of which meaningful job design changes can be made, although inter-study comparisons are then almost impossible.

b) Libraries

Two aspects of the research in libraries are germane here. First, extensive pilot work, again including in-depth group and individual debriefings, revealed an apparent inadequacy of coverage of the job characteristics of Hackman and Oldham and others. Although factor analysis techniques were not applied we found good evidence that two further characteristics were relevant to the work of librarians. These were role clarity and skill utilisation (see also [23] and [6]). The need for the latter construct was emphasised by responses to the main study (n=112) where professional trained staff responded unfavourably to the amount of variety in their jobs. This was because they felt there was too much variety, too many distracting low-level public service tasks, thus contradicting one of the tenets of the JCM. The need to investigate the more pertinant issue in this case, of the degree to which they could use their expertise and training in complex tasks (skill utilisation) was therefore confirmed. Correlational analysis showed that variables of the workplace, technical system and procedures correlated more highly with expressions of satisfaction than did extrinsic or intrinsic job characteristics. This was taken further in the third study.

As part of research into library issue desks, we tested a technique to further supplement the information on workspace criteria already obtained. The technique involved operators in sessions of discussion, modelling and evaluation, with a view to participants designing and building a mock-up of their "ideal" workstation, given certain constraints. What was interesting was that although the focus of the technique was on physical workplace factors, the discussion turned at a very early stage in the sessions from purely physical workplace considerations to an examination of the work organisation involved together with its associated procedures. Respondents found it impossible to achieve the desired result without the parallel consideration of these factors.

c) Telecommunications enterprise

The advent of increased automation in a major U.K. telecommunications company has brought about a move from physically active but repetitive and mentally undemanding tasks to seated, keyboard/coding tasks', also repetitive and even less demanding. However, the surrounding environment has not changed and by-and-large workers are carrying out office-type tasks in a warehouse or older factory environment. Objective measurement of many variables showed especially the climate and noise levels as inappropriate for comfort and performance. Given certain worries about worker health (including "repetitive strain injury") and productivity, the employers initiated an ergonomics study involving 8 sites and, for subjective assessments, over 300 personnel.

Data from formal questionnaires, physical measurement, and informal de-briefings are presently being analysed. Initial indications are again of the strong influence of the physical environment (and importantly what it is believed that this says about management's attitude to the workforce), upon the attitudes of operators towards their work and the organisation. Worker comments typically are extremely cogent arguments about the personal and productivity effects of working conditions and procedures. Test of association and correlation are revealing connections between subjective experiences such as noise and temperature with some job design variables, for instance feedback. Workplace problems such as chair adjustability and coding desk design and ease of use correlate significantly (even assuming all orthogonal variables) with job design variables such as feedback, variety, pacing and initiative. Of all job design variables, communication with one's workmates seems to be that most closely correlated with reports of health problems. Further analysis is taking place to examine whether all the connections being made are empirically and conceptually relevant, as against them being merely spurious, random results.

d) Automobile component manufacturer

As part of a current case study one of the authors is working with a small group of semi-skilled shopfloor operators with the purpose of setting up a semi-autonomous work group (in a factory with a traditional functional layout of machinery), and also developing an appropriate group payment system. In this context the researcher is acting as a facilitator and the method of information gathering to date has largely been informal. Although the operators have paid lip service to the question of pay as their most important concern, it is very evident that critical industrial relations incidents tend to occur more frequently over other issues. More negative reactions tend to occur when responsibility for work organisation is undermined, when task variety is restricted, when the support services fail or when technology problems occur. Given the very good relations and insight built up by the investigator actually working full-time within the company for over a year, opportunity will be taken to attempt to provide formal evidence of some of the experiential observations made hitherto.

Conclusions

The "hidden agenda" of a number of ostensibly independent research studies has revealed the existence of some interdependencies between the five factors of job characteristics, work context, technology, work procedures, and the physical workplace. The nature of these interdependencies is not certain, although it is probable that this is variable and context-specific, and would not benefit from generalisation. What has been confirmed for the investigators is that it is artificial to examine one of the factors in isolation, although this is often difficult given constraints determined by the clients'. The use of more informal, participative methods, alongside formal instruments, in tapping worker perceptions is one approach which will broaden research orientation and encourage parallel consideration of the five core work factors. Most importantly, such methods reduce the likelihood of narrow, "automatic", and obvious attitude responses which may result from respondent expectations or investigator preconceptions.

Ergonomists are ideally placed to be the driving force behind such in-depth multi-factor investigations and subsequent job design initiatives. Moreover, they must extend explorations, and the means for making them, of the effects of physical working conditions upon people's attitudes to all other aspects of their work also, including their willingness to accept new technical systems. Such an approach, in somewhat different context was suggested by Corlett as long ago as 1975 [25]. Let us hope that todays climate is more favourable.

Footnote

1. Throughout this paper, if a particular form of scale, questionnaire or other instrument is considered then it will be referred to as measuring perceived job characteristics, affective responses, and so on as relevant. On the other hand, if all such instruments are referred to they will be called, for convenience, job attitude questionnaires.
2. It is accepted of course that certain particular facets of the environment have been associated with, often extremely detailed, subjective response assessments, for instance thermal comfort or noise annoyance.
3. The authors concur with Clegg's [26] distinction between task and role within the job, but for brevity only the embracing term "job characteristics" is used.

References

[1] Cook, J., Hepworth, S., Wall, T. and Warr, P., The Experience of Work (Academic Press, London, 1981).

[2] Hackman, J.R. and Oldham, G.R. Development of the job diagnostic survey, J. of App. Psych. 60 (1975) 159-170.

[3] Hackman, J.R. and Oldham, G.R., Work Redesign (Addison-Wesley, Reading, Mass., 1980).

[4] Roberts, K.H. and Glick, W. The job characteristics approach to task design: a critical review, J. of App. Psych., 66 (1981), 193-217.

[5] Loher, B., Noe, R., Moeller, N. and Fitzgerald, M., A meta-analysis of the relation of job characteristics to job satisfaction, J. of App. Psych. 70 (1985) 280-289.

[6] O'Brien, G.E., Evaluation of the job characteristics theory of work attitudes and performance, Aust. J. of Psych. 34 (1982), 383-401.

[7] Adler, S. Skov, R. and Salvemini, N., Job characteristics and job satisfaction: when cause becomes consequence, Org. Behav. & Human Dec. Processes, 35 (1985), 266-278.

[8] Watson, C., Watson, K. and Stowe, J., Univariate and Multivariate distributions of the job descriptive index's measures of job satisfaction. Org. Behav. & Human Dec. Processes, 35 (1985), 241-251.

[9] Martin, R., Simpson, G. and Talbot, C., Job design in colliery control rooms, I.Chem.E.Symp. Series No. 90, (1984) 201-210.

[10] Long, T., The response of staff to changing work - predictive assessments, in: Coombes, K. (Ed.)Proceedings of the Ergonomics Society's Annual Conference (1983).43-48.

[11] Wilson, J.R. and Grey, S.M., Reach requirements and job attitudes at laser-scanner checkout systems, Ergonomics, 27, (1984) 1247-1266.

[12] Pierce, J.L. Job design and technology: a sociotechnical systems perspective, J. of Occup. Behav., 5 (1984), 147-154.

[13] Brass, D.J., Technology and the structuring of jobs: employee satisfaction, performance, and influence, Org. Behav. and Human Dec. Proc., 35 (1985), 216-240.

[14] Roznowski, M. and Hulin, C.L., Influences of functional specialty and job technology on employees' perceptual and affective responses to their jobs, Org. Behav. and Human Dec. Proc., 36 (1985), 186-208.

[15] McCormick, E.J., Jeannert, P. and Mecham, R. , Position Analysis Questionnaire, Dept. Psych. Sci., Purdue Univ.(1969)

[16] Patrick, J., and Moore, A.K., Development and reliability of a job analysis technique, J. of Occup. Psych., 58 (1985) 149-158.

[17] Campion, M.A. and Thayer, P.W., Development and field evaluation of an interdisciplinary measure of job design, J. of App. Psych., 70 (1985), 28-43.

[18] Schiro, S.G. and Cox, E.H., A new approach to the evaluation of ergonomics changes, Proc. of the 22nd Human Factors Soc. Ann. Meeting, (1978) 479-484.

[19] Klein, L., New Form of Work Organisation (Univ. Press, Cambs., 1975).

[20] Wilson, J.R. and Grey, S.M., Design of Work in service industries, in: Proc. Int. Conf. in Occup. Ergs., Toronto (1984).

[21] Fraser, T.M., Job satisfaction and work humanisation: an expanding role for ergonomics, Ergonomics 21 (1978), 11-19

[22] Taber, T.D., Beehr, T. and Walsh, J., Relationships between job evaluation ratings and self-ratings of job characteristics, Organ. Behav. & Human Dec. Proc., 35, (1985) 27-45.

[23] Warr, P., Cook, J., and Wall, T., Scales for the measurement of some work attitudes and aspects psychological well-being. J. of Occup. Psych., 52 (1979), 129-148.

[24] Campbell, D.T. and Fiske, D.W., Convergent and discriminant validation by the multitrait-multimethod matrix. Psych. Bull., 56 (1959), 81-105.

[25] Corlett, E.N., The physiological and behavioural correlates of psychological well-being at work, Discussion paper, Dept. of Eng. Prod., Univ. of Birmingham (Sept. 1975).

[26] Clegg, C.W., The derivation of job designs, J. of Occup. Behav., 5 (1984), 131-146.

Human Factors in Organizational Design and Management - II
O. Brown, Jr. and H.W. Hendrick (Editors)

The Development of a Human Factors Satisfaction Questionnaire

Jim Carlopio
Department of Psychology
Old Dominion University, Norfolk, VA. 23508

INTRODUCTION

"Human Factors", born of engineering and experimental psychology, is a highly technical field. Technical studies of human perception, physical dimensions, and function have influenced its product and systems design (Dray, 1985). In recent decades, human factors has shifted from the study of the purely physical individual to the study of the thinking being and more recently, has expanded areas of interest to include the study of groups of interacting individuals (Hendrick, 1985). Researchers have begun to investigate organizations and organizational behavior from a human factors point of view. This view contains two major elements: 1) it shows a concern for measurement, experimentation, and control, 2) it shows a concern for technology, the environment, and health and safety.

Organizational, managerial, and social psychologists have been studying groups and organizations for decades. They share, with the human factors researchers, a concern for measurement, experimentation, and control in their studies. However, these researchers have not examined technology, the environment, or health and safety as these concepts are envisioned by human factors practitioners. March and Simon (1958) and Katz and Kahn (1978) have enlightened us regarding the effect of the environment on organizations and behavior in organizational settings. However, the study of technology, i.e., machines, tools, controls, and the processes by which these are interrelated with each other and the human operator, has been neglected. The environment, seen through the lens of the human factors practitioner, is the internal working mileu of the employee composed of physical facilities, environmental ambients, work systems design, and workload characteristics susceptable to individual stress, fatigue, and concerns for workers' safety. These areas of research are truely unique to the human factors point of view and have not been explored by organizational, managerial, or social psychologists for many years.

Earlier in this century, physiological organization theory, which included Taylor's scientific management and engineering notions such as motion economy, was very much concerned with the use of the human body, the arrangement of the work place, and the design of tools and equipment (cf. March & Simon, 1958). Organizational, managerial, and social psychologists in their zeal to develop purely psycho-organizational theories have discarded useful physiological and behavioral approaches. The predominantly psychological variables frequently studied in organizations (e.g., task and job characteristics, satisfaction, commitment), often leave behavioral and physiological variance unexplained and its effects on measures of performance, productivity, absenteeism, turnover, and even workers' satisfaction and commitment, unanalyzed.

A macroergonomic view of organizations is needed to encompass these issues traditionally discarded by organizational researchers and will enhance our understanding of organizations and of individuals' behavior within them. The purpose of this study was to develop a questionnaire which would be sensitive to more technical, individual level human factors/organizational issues and could aid in accounting for behavioral and attitudinal variance within organizations.

The questionnaire pre-supposes that employees know if their surroundings are uncomfortable or are not conducive to optimal performance. If the amount of light, noise or dust is a distraction, they can tell you. Workers will know if their tools are not efficient or if needed information and equipment is not available or poorly located. If workers fear the workplace is detrimental to their health or if they perceive that they are not provided with adequate facilities, it will affect their behavior. By addressing human factors concerns in the workplace which are explicitly individual level variables directly impacting on workers' quality of work life and their ability to perform their jobs, both the individuals' and the organizations' needs may be examined.

METHODS AND RESULTS

Six dimensions related to individual level environmental issues were identified and 109 items generated which were believed to measure these dimensions. The dimensions were: 1) environmental design, 2) plant facilities, 3) workload characteristics, 4) systems design, 5) equipment design, and 6) health and safety. After retranslation of the items back to dimensions, by three content experts (graduate students), the workload characteristics and systems design dimensions were combined into a single workload/system characteristics dimension.

The initial administration of the scales was conducted with 229 undergraduate students at Old Dominion University. Subjects were volunteers from eight psychology classes (sophomore and junior level) who received experimental credit for their participation in the study. Subjects were administered the 109 item Human Factors Satisfaction Questionnaire (HFSQ) along with the twenty item Minnessota Satisfaction Questionnaire (MSQ) short form and the five item "working conditions" sub-scale from the MSQ. The format and instructions for the HFSQ were taken directly from those used for the MSQ (Weiss, Dawis, England, & Lofquist, 1967).

The mean age of the sample was 22.7 and was composed of 44% males, 76% non-supervisors, 93% civilian, and had a mean job tenure of 2.5 years. Fifteen job environments were identified (e.g., office, warehouse, hospital, bank, etc.) from the job titles and descriptions provided by the subjects. Two independent raters initially classified the descriptions. The initial inter-rater agreement on classification was 94%. The 14 cases not initially agreed upon were discussed and recoded.

The scale reliabilities (Table 1) reveal high internal consistency, with reliabilities (coefficient alpha) ranging from .89 (environmental design) to .95 (health and safety). The scale inter-correlations (Table 2) show low scale independence, with inter-correlations ranging from .47 (environmental design with health and safety) to .64 (equipment design with work and system characteristics). Scale means, standard deviations, and the standard error of measurement are also presented in Table 1.

Table 1
Reliabilities, Means, Standard Deviations, and Standard Errors of Measurement of the Human Factors Satisfaction (HFSQ) Scales

HFSQ scale	Number of Items	Internal Consistency Reliability	Scale Means (N=229)	Standard Deviations	Standard Error of Measurement
Environmental Design	16	.89	53.82	11.23	3.72
Work Facilities	16	.91	53.87	12.62	3.79
Work and System Characteristics	27	.93	95.67	17.99	4.76
Equipment Design	11	.93	38.79	8.51	2.25
Health and Safety	39	.95	132.10	25.57	5.72
HFSQ total	110	.97	377.71	62.72	10.86

Table 2
Inter-Correlations for the
Human Factors Satisfaction Questionnaire (HFSQ) Scales

HFSQ scale	1	2	3	4	5	6
1. Environmental Design	--					
2. Work Facilities	.64	--				
3. Work & System Char.	.59	.56	--			
4. Equipment Design	.54	.50	.64	--		
5. Health and Safety	.47	.62	.62	.49	--	
6. HFSQ Total Score	.75	.80	.85	.72	.87	--

All correlations are significant (p<.001 two-tailed.) N=229

A oneway analysis of variance (ANOVA) was conducted on each scale mean over the eight job environments which had ten or more respondents (Table 3).

Table 3
Means, Variances, and Analysis of Variance for
Human Factors Satisfaction Questionnaire (HFSQ) Scores.

	Analysis of variance across job environments.			
HFSQ scales	Mean (N=203)	Variance Within Job Environment	Variance Between Job Environment	F(7,196)
Environmental Design	54.20	122.06	253.27	2.08**
Facilities	53.78	165.01	297.60	1.80*
Work and System Characteristics	95.94	309.23	857.82	2.77***
Equipment Design	38.87	72.64	104.63	1.44
Health and Safety	132.45	672.99	759.45	1.13
HFSQ total	374.63	3838.01	8028.57	2.09**

* p<.10 ** p<.05 *** p<.01

Although between-job differences were statistically significant for only two of the five scales ($p<.05$), the total HFSQ score was significant as expected. The eight job environments used in the above analysis were office (n=45), warehouse (n=15), retail (n=31), restaurant (n=30), hospital (n=32), outdoors (N=16), miscellaneous general labor (n=22), and school (n=13).

The results of a factor analysis revealed twelve items which did not load (.5 or above) on a factor. One complex item was found. The scales, their factor identifiers, and the value of the highest loading items are presented in Table 4. These results should be cautiously interpreted in light of the small number of respondents (n=229) in relation to the number of HFSQ items (n=110).

Table 4
Factor Identifiers and Highest Factor Loadings of Items on The Human Factors Satisfaction Questionnaire (HFSQ)

HFSQ SCALES Factor Identifiers	Highest Loading	% Variance Accounted For
ENVIRONMENTAL DESIGN		55%
The general design of the work environment.	.76	39%
Air movement and temperature.	.77	10%
Level of vibration and glare.	.76	7%
WORK FACILITIES		76%
Work area pleasantness and cleanliness.	.84	43%
Eating Facilities.	.89	15%
Recreation factilities.	.92	10%
Restroom facilities.	.89	8%
WORK AND SYSTEM CHARACTERISTICS		68%
Information quality and storage.	.85	35%
Number and length of restbreaks.	.81	10%
Control of work pace; workflow speed.	.71	6%
Attention level and challenge of work.	.83	5%
Error tolerance and importance.	.83	5%
Depertmental and organizational goals.	.77	4%
Amount of work able to produce.	.76	4%
EQUIPMENT DESIGN		69%
Machine and tool effectiveness.	.83	59%
Material storage and handling.	.89	10%
HEALTH AND SAFETY		71%
Safety training and near-accident study.	.82	35%
Mental and emotional stress.	.88	14%
Eye and hearing protection quality/comfort.	.91	7%
Physical and muscular fatigue and strain.	.78	4%
Training on machine use and safety.	.78	4%
Hazardous material handling and storage.	.75	4%
Smoking.	.88	3%

The five scale scores were then treated as items and subject to a second order factor analysis (Hunter and Gerbing, 1982.) A single factor was extracted with the scales loading between .78 (Equipment Design) and .85 (Work and System Characteristics) accounting for 65% of the variance.

The correlations among the HFSQ total score, the MSQ short form and the MSQ working conditions sub-scale were used to examine the relations among these scales. The HFSQ total score correlated .5 with the MSQ total score, and .54 with the working conditions sub-scale. The MSQ total score and sub-scale were correlated .79 .

The item-total correlations and internal consistency reliabilities were examined and used in an attempt to reduce the number of scale items to approximately three items per factor within each of the original scales. The shortened version of the HFSQ is presented in Table 5 with the revised scale reliabilities.

CONCLUSIONS

The findings from this initial administration of the HFSQ are encouraging. The scales and total score show extremely high internal consistency as well as some ability to discriminate among the job environments in this limited sample. (The shortened version of the HFSQ also displays high overall internal consistency, alpha=.94 .) The high sub-scale intercorrelations and the results of the second order factor analysis suggest that the HFSQ has a rather narrow focus. The correlations among the HFSQ total score, the MSQ short form and the working conditions sub-scale, illustrate a moderate positive relationship among them. Further study is needed (and planned) to determine if the HFSQ and the MSQ are measuring the same, or different constructs by examining their abilities to predict various organization outcome measures (e.g., intent to turnover, commitment to the organization, absenteeism, and production).

It is hoped that versions of this questionnaire will provide organizational researchers with tools to examine potentially powerful sources of behavioral and attitudinal variance in organizations. Human factors practitioners can use the scales to illustrate the impact human factors have within organizations. The scales may also be used to monitor variables which are likely to be affected by the introduction of new technologies (e.g., computer assisted manufacture and computerized word processing) within organizations.

Table 5
Human Factors Satisfaction
Questionnaire Short Form and Reliabilities

Environmental Design (alpha=.86)
The lighting in your work area.
The air quality in your work area.
The surfaces on which you usually walk.
The surfaces on which you usually work.
The direction of the light which enters your working area.
The general atmosphere in your work area.
The colors used in your work area (walls, furnishings, etc.).

Work Facilities (alpha=.85)
The cleanliness of the facilities at work.
In general, the type of facilities provided at work.
The pleasantness of the restrooms you use.
The cleanliness of the recreation facilities you use.
The pleasantness of the recreation facilities.
The size of the eating facilities (e.g., lunch room) provided.
The cleanliness of the eating facilities.
The pleasantness of your eating facilities.

Work and System Characteristics (alpha=.86)
How your time at work is scheduled.
The length of the rest breaks you receive.
The amount of work you are required to do.
The amount of activity/movement needed to perform your job.
The flexibility of your work pace.
The general design of your work system.
The amount of time you are given to complete your work.
The quality of information you receive to do your work.
How information is handled at work.

Equipment Design (alpha=.92)
The number of tools with which you have to work.
The effectiveness of the machines with which you have to work.
The effectiveness of the tools with which you have to work.
The effeciency of the machines with which you have to work.
The effeciency of the tools with which you have to work.
The quality of materials you are given to do your job.

Health and Safety (alpha=.95)
How accidents are avoided at work.
The safety preqautions taken in your work place.
The warnings you are given regarding work place hazards.
The safety training you have received.
How hazards are controled in your work place.
The safety training available to you through work.
How hazardous materials or products are handled/moved.
The safety training other workers receive.
The way accidents are reported at work.
The way accidents are investigated at work.
The fire prevention system(s) you have at work.

REFERENCES

Dray, S. M. (1985). Building bridges: Implementing computer technology in organizations. Proceedings of the Human Factors Society 29th Annual Meeting, Baltimore, Maryland.

Hendrick, H. W. (1985). Macroergonomics, a concept whose time has come. Proceedings of the Los Angeles Chapter Human Factors Society 8th. Annual Meeting, Downey, CA.

Hunter, J. E. and Gerbing, D. W. (1982). Unidimentional measurement, second order factor analysis, and causal models. In B. Staw & L. Cummings (Eds.), _Research in Organizational Behavior_, _4_, 267-320.

Katz, D. & Kahn, R. L. (1978). _The Social Psychology of Organizations_. New York: John Wiley & Sons.

March, J. G. & Simon, H. A. (1958). _Organizations_. New York: Wiley & Sons.

Weiss, D. J., Dawis, R. V., England, G. W., & Lofquist, L. H. (1967). The manual for the Minnesota Satisfaction Questionnaire. Work adjustment project, Industrial Relations Center, University of Minnesota.

Human Factors in Organizational Design and Management - II
O. Brown, Jr. and H.W. Hendrick (Editors)

The Effects of Technological Change on the Psycho-Social Characteristics of Organizations

Charles M. Slem
Daniel Levi

California Polytechnic State University
San Luis Obispo
USA

Andrew Young

Northern Telecom

To effectively use ergonomics in organizations undergoing technological change, the psycho-social effects of that change on the following dimensions must be considered: Task; roles; formal and informal group relationships; organizational culture; communication patterns; decision-making and problem solving; and career development. Using Computer Integrated Manufacturing as an example of change, potential positive and negative effects are reviewed.

The effects of rapid technological change on ergonomics has been the focus of recent attention [1,2]. Hendricks [2] has pointed out that the introduction of microelectronics and related automation into the office and factory affect organizational design far more than traditional labor-intensive human-machine systems. To continue to effectively use ergonomic design in the face of pervasive organizational change, Hendricks proposes a "macroergonomics" approach which considers the design of the system's organizational structure before proceeding with "microergonomic" design.

Within the macroergonomic approach, the psycho-social characteristics of the organization represent a critical component in effective organizational design. An implicit assumption in a psycho-social perspective is that there is little inherent in technology itself that will enhance or degrade organizational effectiveness or quality of work life. It is rather the specific design, application and implementation of technology that is critical to its success or failure.

A model of the effects of technological change (using Computer Integrated Manufacturing, CIM, as an example of pervasive change) on the psycho-social properties of an organization is presented in Figure 1. This model synthesizes much of the relevant work in this area and offers a series of

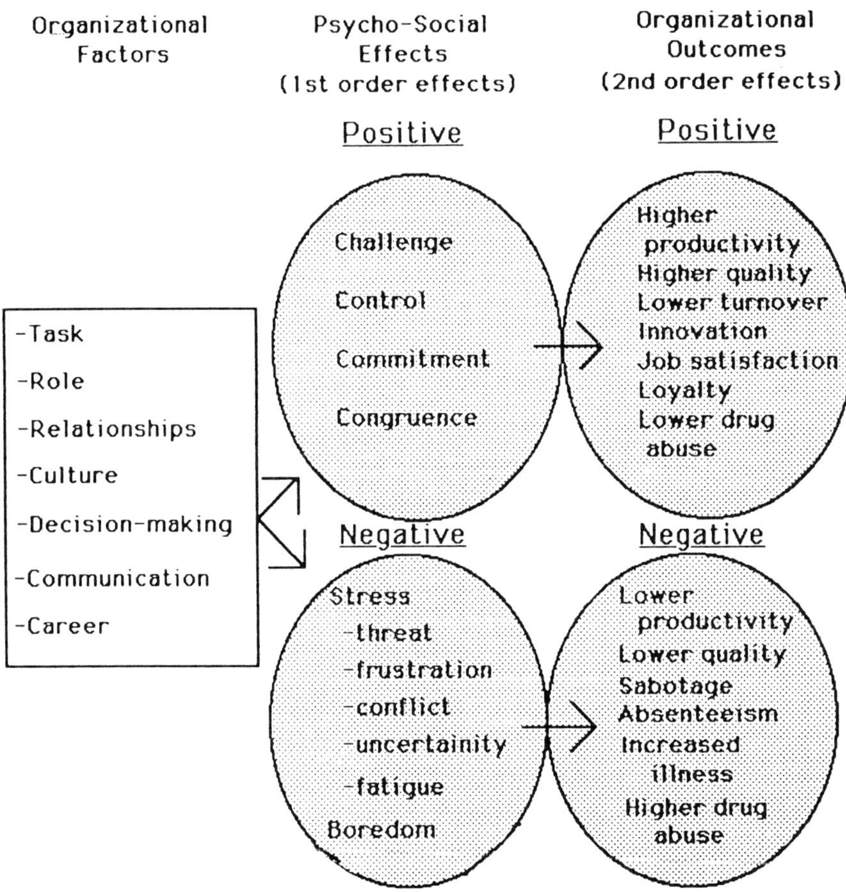

Figure 1

Effects of Technological Change on Psycho-Social Characteristics

hypotheses concerning the effects of successful and unsuccessful implementation of new technological systems like CIM.

The characteristics of an organization which have important implications for the psycho-social environment include: (1) the task itself, (2) formal and informal group role relationships, (3) organizational culture, (4) communication patterns, (5) decision making/problem solving strategies, and (6) career development issues such as job security and advancement. Pervasive technological change will have significant impact on each of these factors, and both workers and managers are likely to experience both positive or negative reactions to technological change.

In assessing the positive impact of change on human resources (often overlooked by traditional industrial/organizational models), a properly designed and implemented technological change can enhance the worker's or manager's sense of control, challenge and commitment to the organization. The goals of the individual may become more congruent with the goals of the organization. When these positive psycho-social effects occur, they are likely to influence organizational functioning in terms of higher productivity, product quality, innovation, loyalty, and lower absences and turnover. When technological change is viewed positively, managers and workers are likely to compensate for even suboptimal ergonomic designs and productively use the new technology.

On the negative side, technological change can lead to fatigue, stress and boredom. Technological changes are often perceived by managers and workers as threatening, frustrating, conflicting, or making their work environment uncertain. This negative psycho-social state can lead to reduced organizational functioning in terms of lower productivity, lower product quality, sabotage, substance and alcohol use, absenteeism and turnover. Under these circumstances, erogonomically optimal designs of work stations and primary job groups may not have the anticipated effect on productivity and quality of work life.

Computer Integrated Manufacturing (CIM) epitomizes a pervasive technology driven organizational change which can serve as a model for exploring the positive and negative consequences of technological change on the psycho-social characteristics of an organization. CIM embodies a strategy for integrating the manufacturing process from product design to sales and service, and includes all the technology that facilitates the integration of the manufacturing process [3]. Terms like automation, robotics, flexible manufacturing systems, automated warehousing, computer assisted design, materials resource planning, and "just in time" are typical components of CIM systems.

Given this brief description of CIM, the potential positive and negative effects of technological change (embodied by the introduction if CIM) on psycho-social factors can be reviewed. It must be kept in mind that the occurrence of positive or negative consequences is dependent upon the specific design and implementation of the new technology rather than anything inherent in the technology itself.

I. TASK CHARACTERISTICS

This dimension is the major focus of the traditional ergonomic approach [4] and is only briefly reviewed here.

A. Task Size, Complexity and Variety

In the traditional implementation of automation, tasks have been reduced into simpler sub-routines which can be more easily managed by automated machinery. From the psycho-social perspective, the fragmented sub-routines quickly become repetitive, boring and fatiguing. It is difficult for the worker to see how the sub-routine fits into the finished product [5,9]. Task challenge may disappear as well as a sense of commitment to the product and company.

Although automation can be designed to reduce complexity and size of a specific task, it can also be designed to increase task variety and complexity for the worker. Through the simplification of automation, the worker may be able to work with a larger variety of tasks, rotating from task to task.

Automation can also create more challenging tasks for the worker. Instead of performing a rote multi-step task which a machine can perform, the worker is in charge of or "supervising" the machine's performance. This requires additional training and could easily be experienced as challenging. This new skill in turn leads to greater mastery and potential for advancement within the organization.

Automation at the managerial level can also free valuable human resources by having a computer perform the routine and often boring tasks. Managers and professionals are then free to pursue more complex and challenging tasks.

B. Physical Characteristics of the Job

On the plant floor, one of the greatest benefits of automation is that it can remove humans from environments which might be dangerous, dirty, physically difficult and unpleasant. On the managerial level, the addition of computer terminals gives the individual almost immediate access to information needed to coordinate and manage complex tasks. This increases the manager's sense of personal control over the work environment.

Unfortunately, automated machines can also create their own hazardous environments on the plant floor. In addition, there is research which suggests that working long hours at a video display terminal may have some negative physical consequences--- e.g. fatigue, eye strain, muscle aches and headaches [7]. While working long hours at a VDT is not as likely for the managerial personnel as for the clerical, it appears that there is much more resistance by managerial level people to using the VDT. While much of this resistance appears to be based on a fear of the machine itself [8], the use of computers may threaten the manager's perception of status and appropriate role behavior.

C. Autonomy, Control and Supervision

Having been replaced by a machine and then being required to simply monitor its activities can make the individual feel as though he or she is a dehumanized extension of the machine with little autonomy and control [5,6]. Work tasks dominated by machine pacing have generally been associated with higher levels of stress [9].

Automation also creates the possibility of centralized control through monitoring via sophisticated tracking systems and close supervision for all levels of personnel. To achieve the goals of the organization, a delicate balance between the organizational need for control and individual autonomy is required. In the past, individual autonomy was structurally supported through inefficiencies in information and reporting systems regardless of wishes of top management. CIM technology offers an opportunity for direct monitoring of all phases of task and project action. Such close supervision could easily be experienced as a threat or frustration with resulting attempts to regain a sense of security. Implementation sabotage, higher rates of illness, absenteeism and turnover could be expected. In instances where close supervision and monitoring have been the consequence of a new technology, efforts are made by the individual (worker, professional or manager) to compromise the monitoring system rather than attempt to meet the corporate goals of the monitoring system [7].

On the other hand, the ability to have direct access to information required to do his/her job offers the possibility of an increased sense of autonomy and control over those activities. In addition, complex technologies like CIM create new channels of communication. Many clerical, professional and managerial personnel believe they have greater influence in decisions irrespective of an "objective" estimate of their actual control or influence [10,11].

D. Time/Scheduling Factors

CIM offers methods to greatly reduce wasted or inefficiently used time (e.g. CAD versus drafting). New technology makes the timely transmission of information more likely and shortens the cycle between design and manufacturing. Delays resulting in untimely and often inaccurate information are minimized. As a result, the individual experiences a greater sense of control over the situation which should be directly related to higher quality and faster speed.

On the negative side, people may feel pressured to produce quickly under this technology driven system. CIM also requires the timely coordination of subsystems or an entire and very costly automated line may have to be shut down. The effect of this kind of time pressure and concern about disrupting an entire system is a sense of uncertainty, frustration or conflict [5,6,10].

II. ORGANIZATIONAL ROLE FACTORS

A. Role Ambiguity

Role ambiguity can be defined as a lack of clarity about one's responsibilities, job objectives and expectations from others [12]. If role ambiguity is a major chronic organizational problem, it is stress producing and has been found to lead to lower individual performance and reduced organizational effectiveness [13].

The logic of CIM should produce a system which more clearly specifies the details of tasks and provides feedback to monitor and evaluate task behavior. This should produce a greater sense of control and role congruence, and result in more accurate and timely decisions. Instances of chronic role ambiguity are indicators that the CIM system was not properly integrated.

This potential reduction of a whole host of problems associated with understanding expected role behaviors points out one of the major paradoxes of complex and integrated systems. Contrary to the expectation that less human interaction will be needed in technologically advanced systems, it appears that more interaction is required during the design, implementation and maintenance of technologically advanced systems like CIM. It may well be that there will be a smaller total number of people involved in the process, but those who are in the process will have greater interpersonal demands placed upon them, especially at the managerial level.

B. Role Conflict

Role conflict occurs when the individual receives two sets of mutually contradictory expectations about role behavior. That is, to comply with one set of expectations would make it difficult to comply with the other expectations [12].

As with role ambiguity, the psychological benefit of an implemented CIM is that role conflicts should be substantially reduced or eliminated because a certain level of clarity has been built into the CIM system. Pre-existing levels of stress related frustration, threat and uncertainty should be reduced. Role conflict may still occur when the needs of an individual's unit are in conflict with system constraints generated by other units.

C. Role Overload

There are two types of role overload. Quantitative role overload occurs when the individual has too many activities to do in a prescribed period of time. Qualitative role overload occurs when the individual is confronted with tasks for which he or she feels that they are not qualified to complete [12].

Quantitative Role Overload: The ability of advanced technology to coordinate activities should reduce quantitative role overload. Role overload could be eliminated by coordinated scheduling schemes available through CIM. Since the technology provides endless feedback loops, real time spent can be compared with initial estimates. Over succeeding cycles the process of scheduling becomes more realistic.

Since role overload seems to be a widespread corporate characteristic, this capability to reduce overload is potentially very helpful. However, whether or not a corporation will make use of this opportunity depends upon the organization's cultural beliefs about the impact of unrealistic and demanding schedules on individual performance (especially at the managerial level). Many corporations value unrealistically demanding schedules and are likely to subvert this new scheduling capability.

In addition, since CIM eliminates a great deal of time delay inefficiency, it may initially appear that a much greater number of tasks can be effectively completed in the same period of time, resulting in greater role overload. At the psycho-social level, role overload will initially increase feelings of frustration and conflict, giving way to general feelings of threat, and then "burnout" [13].

Qualitative Role Overload: The implementation of a CIM system may initially provide a challenge which the person experiences as being "too difficult". The logic of the system, the potential changes in job function, and the necessity of learning how to use new machines (e.g. robots or computers) may be perceived as a threat to the individual. On the other hand, since CIM connects all users with a common integrating system, it theoretically supplies each new project manager with a potential list of expert resources. At the psycho-social level, a greater sense of control and resource availability would likely change the experience of threat to one of challenge.

III. RELATIONSHIPS WITHIN ORGANIZATIONS

Relationships among people in an organization have a direct effect on psycho-social outcomes because the impact of almost all of the other factors are mediated through personal relationships among people. The two primary types of relationships are formal role relations (supervisor-subordinate interactions) and informal relationships (social group interactions). The formal/informal distinction blurs in daily organizational interactions because individuals often shift continuously between their formal and informal relationship with another person.

A. Supervisor-Subordinate Relationship

Because CIM allows for the opportunity to more closely monitor activities and exercise greater control, the negative effects listed earlier in discussing control and close supervision apply here as well. The new ability to more effectively schedule activities may also inadvertently produce time and scheduling pressures. During implementation of CIM, supervisors may increase role ambiguity and role conflict. With task clarity enhanced by CIM, supervisors may initially underestimate the need for human relations skills which will actually increase in a CIM environment. The increased importance of interpersonal management skills may produce qualitative role overload, threat, frustration and stress for managers.

Despite these possible negative consequences, the supervisor-subordinate relationship could actually be strengthened through CIM because the supervisor will be freed from many routine operations. This will allow the supervisor to focus on planning, coordination, problem solving and innovation. The effect of this should have a direct impact on organizational productivity, product quality and innovation. Supervisors may have additional time to develop human relations skills which strengthen referent as well as expert power over subordinates, and offers the opportunity to focus on personnel issues such as stress and boredom.

B. Informal Group Relationships

In almost all organizational settings, informal group relations emerge in addition to formal group arrangements. The powerful effect of informal group norms on worker behavior, especially in restricting productivity, is well documented. This has mistakenly led to the belief that every effort should be taken to eliminate informal groups.

What has recently become more apparent is that informal group interactions are also critical in promoting organizationally desirable behavior by satisfying human needs which are not readily satisfied through a formal organization [12]. For example, informal groups satisfy human needs for personal interaction (affiliative needs), needs for recognition as a valued individual (esteem needs), and needs for social support to deal with stress. In fact, highly cohesive informal groups promote group pride and esprit which can translate into production competitiveness to demonstrate group superiority and facilitate cooperative helping to get a task completed under difficult circumstances.

If poorly designed and implemented, CIM can potentially isolate individuals and make informal group formation and function difficult. Without some sort of informal group interaction, boredom, isolation and stress are likely. To the extent that supportive adjustments are not made to ensure the development of new social arrangements, organizational change may promote feelings of threat and frustration [14]. At the level of the group itself, threat of disbandment may encourage resistance to maintain the group's integrity.

A properly functioning informal group should merge informal and formal goals, increase cohesiveness, generate a sense of control over situations, and produce commitment to projects. In addition, effective informal group interaction will facilitate the process of planning, integration and problem solving required in a CIM environment.

IV. SYSTEM CHARACTERISTICS

A. Culture

Culture can be defined as the taken-for-granted and shared assumptions, beliefs and norms about the way in which an organization operates. Organizational culture implicitly makes certain experiences, interpretations and organizational behaviors more likely to occur.

There is little research on the effects of technological change on corporate culture, although much has been written on changing culture to improve corporate technological innovativeness [15]. It is likely that corporate culture will affect the acceptance of technological change [16,17], and in turn technological change will eventually affect culture (e.g. the long term effects of the automobile on the U.S. culture). Innovative corporations will find CIM consistent with prevailing corporate culture. Corporations with non-innovative cultures will foster psycho-social experiences of threat, frustration and uncertainty which would present formidable sources of resistance.

Although technological determinism has not necessarily happened as change is introduced at the societal level [18], it is likely that the kind of pervasive technological change represented by CIM would have a major impact on cultural beliefs and assumptions. Through the underlying logic of integrated systems like CIM, the corporate culture should move towards innovativeness and continued technological change.

B. Communication

The specific effects of computer-mediated-communication are not well known at this time [19,20,21]. It appears, however, that new technologies like CIM will replace a centralized communication pattern with a flexible combination pattern which mimics aspects of the pinwheel communication pattern, (a decentralized pattern where every unit appears capable of "directly" communicating with every other unit). CIM will simultaneously be operating via a centralized wheel communication pattern (common centralized data base at the hub of functional unit reporting). This communication combination can optimize two of the prime goals of CIM: (1) taking advantage of automated tasks, i.e. component job tasks which are relatively simple; and (2) enhancing the ability to integrate a wide number of tasks and make relatively quick and timely modifications.

In addition to these benefits, having access to pertinent information fosters a sense of predictability and control in organizations. The individual unit is potentially aware of all the factors which affect its operation. The centralized aspect of CIM offers an opportunity for overall coordination and consistency which should significantly reduce role ambiguity and role conflict when the system is fully implemented.

Because of the tremendous computing power available via new technologies, a complex CIM produced communication system could produce information overload and qualitative role overload. Vast amounts of data will have to be interpreted by each unit to determine how the information relates to the unit's internal functioning. Instead of giving each unit a sense of control over its work environment, unpredictability and stress would be experienced.

Organizational power and job satisfaction are often associated with control over information [23]. CIM provides opportunities for many more units to monitor information which one area or person has controlled in the past. The anticipated loss of the power associated with the control of this information may be experienced as stress (threat) which would likely precipitate resistance, e.g. sabotage.

C. Problem-Solving/Decision-Making

Organizational problems are often experienced as a source of threat or frustration to the individual, and problem solving acts to eliminate the threatening stressor or frustrating block. The individual's experience of the problem solving/decision-making process has powerful psycho-social effects. Most organizational stress research focuses on organizational problem-solving and decision making [13]. CIM can be seen as an important catalyst in effective problem solving. The design of CIM allows for earlier problem recognition, supplies specific and detailed data about the nature of the problem [11], identifies those units which are important players in the problem solving process, allows for 'what if' modeling and provides timely feedback. A greater sense of participation and control are likely if the system embodies these characteristics.

If psycho-social factors are not carefully considered, critical information may not be accessible. This generates a great deal of frustration and erodes any sense of control one may have developed. There is evidence that early CIM systems did not include these capabilities and had a negative impact on supervisors and workers [6]. A related issue focuses on the organization's ability to go beyond the built-in biases of the system [24]. CIM will be based on technical assumptions and cultural values which may not be applicable in all situations. For example, reduced product quality may be the result of the social isolation of operators which could easily be resolved, but a CIM driven problem analysis might produce as a solution an additional and costly piece of automated equipment. In the maritime industry, this kind of technical solution to marine accidents by building increasingly sophisticated navigation and collision-avoidance aids did not reduce marine accidents [14] and a psycho-social analysis had to be employed.

V. Career Development

A. Job Security

The introduction of technological change questions the need for positions which have been transformed into automated machine driven processes. Although this issue has been more keenly appreciated on the production floor, CIM makes this also true for middle management and technical professionals. Because of the volatile effect of job insecurity, personnel policies play a dominating role in whether or not the implementation of technological change is ultimately successful.

CIM is likely to be perceived as a intense source of stress through threat of job loss or uncertainty. Feelings of uncertainty are often translated into worse case scenarios because they enable the person to prepare and exercise personal control over the most threatening potential outcome. Unfortunately, individuals begin to take action as if the worse case was reality. Organizational effects can be seen in power struggles to maintain relative positions, sabotage, resistance, absenteeism and illness. Additionally, these strategies may lead to the unwanted turnover of valued employees who in fact would not have been replaced. Since successful implementation depends upon the cooperation of the employees, job insecurity poses a major risk to the implementation of new technologies.

If an individual believes that he or she will have a position regardless of the technological impact and that the technology is being implemented as a tool to enhance human functioning rather than replacing it, there is a much greater chance that change will be seen as a challenge rather than a threat [17]. This has an important impact during implementation. Motivated, challenged and loyal workers are likely to facilitate implementation and improve the system design because of their intimate knowledge of the "real" requirements of the task which is being automated. If the perceived threats to individuals (and groups) are minimized, adaptive modifications during implementation are likely, but with a sense of threat predominating, there is very little motivation to be helpful.

B. Career Development Opportunities

The introduction of "state of the art" technology provides personal and professional growth enhancing opportunities. As a result, CIM offers additional career development opportunities because new technology requires higher skilled positions. Psycho-social outcomes of challenge, commitment and congruence will affect company loyalty, job satisfaction and innovation.

If the individual's career aspirations have been thwarted because technological innovation eliminates jobs which were advancement steps within the organization, frustration is also likely.

CONCLUSION

A review of the potential impact of technological change on the psycho-social aspects of an organization highlights the importance of these factors in successfully implementing ergonomic designs. Careful consideration at the macro level will ensure that specific ergonomic applications will be used as intended.

REFERENCES

[1] Brown, Ivan, Ergonomics and technological change, Ergonomics, 28 (1985) 1303-1309.

[2] Hendricks, Hal, Wagging the tail with the dog: Organizational design considerations in ergonomics, Organizational Design and Management Bulletin, 3 (1985) 5-13.

[3] Harrington, Joeseph, Computer Integrated Manufacturing (Industrial Press, New York, 1973).

[4] Rohmert, Walter, Ergonomics and manufacturing industry, Ergonomics 28 (1985) 1115-1134.

[5] Blumenthal, Marjory, Computerized Manufacturing Automation: Employment, Education and the Workplace (Office of Technology Assessment, U.S. Congress, Washington, D.C. OTA-CIT-235 1984).

[6] Blumberg, Melvin & Gerwin, Don, Coping with advanced manufacturing technology, Journal of Occupational Behavior, 5 (1984) 113-130.

[7] Westin, Alan, Heather Schweder, Michael Baker, and Sheila Lehman, The Changing Workplace (Knowledge Industry Publications, White Plains, NY, 1985).

[8] Meier, Scott, Computer aversion, Computers in Human Behavior, (1986) in press.

[9] Salvendy, Gavrial & M. J. Smith, (eds.) Machine Pacing and Occupational Stress (Taylor and Francis, Ltd., London, 1981)

[10] Kraemer, Kenneth, & James Danziger, Computers and control in the work environment, Public Administration Review, (1984) 32-42.

[11] Fleischer, Mitchell, and Jonathan Morell, The organizational and managerial consequences of computer technology, Computers in Human Behavior, 1 (1985) 3-93.

[12] Katz, Daniel & Kahn, Robert, The Social Psychology of Organizations (Wiley, New York, 1978).

[13] Cooper, Cary & Payne, Roy (eds.), Stress at Work (Wiley, New York, 1978).

[14] Perrow, Charles, The organizational context of human factors engineering, Administrative Science Quarterly, 28 (1983) 521-541.

[15] Snyder, Richard, To improve innovation, manage corporate culture, in Bennis, Warren, Kenneth Benne, & Robert Chin (eds.), The Planning of Change (Holt, Rinehart & Winston, 1985).

[16] Levi, Daniel, Charles Slem & Andrew Young, Evaluating an organization's potential for adopting computer integrated manufacturing, in: Proceedings of the Second International Symposium on Human Factors in Organizational Design & Management (North Holland, Amsterdam, in press).

[17] Bikson, Tora, Cathleen Stasz & Donald Mankin, Computer Mediated Work (Rand, Santa Monica, CA, 1985).

[18] Brannigan, Augustine & Sheldon Goldenberg, Social Responses to Technological Change (Greenwood Press, Westport, Conn., 1985).

[19] Olson, Margrethe, The impact of office automation on the organization: Some implications for research and practice, Communications of the ACM, 25 (1982) 838-847.

[20] Szlichcinski, K.P., Designing for the day after tomorrow: I. The interaction between communciation systems and social change, Behavior and Information Technology, 2 (1983) 253-261.

[21] Kiesler, Sara, Jane Siegel & Timothy McGuire, Social psychological aspects of computer mediated communication, American Psychologist, 39 (1984) 1123-1134.

[22] Shaiken, Harley, Work Transformed: Automation and Labor in the Computer Age (Holt, Rinehart and Wintson, New York, 1984).

[23] Markus, M. Lynne, Implementation politics: Top management support & user involvement, Systems, Objectives & Solutions 1 (1981) 203-215.

[24] Hirschhorn, Larry, Beyond Mechanization (MIT: Press, Cambridge, Mass, 1984).

Human Factors in Organizational Design and Management - II
O. Brown, Jr. and H.W. Hendrick (Editors)

Macroergonomics: A New Approach to the Implementation of Advanced Technology

Jim Carlopio
Department of Psychology
Old Dominion University, Norfolk, VA. 23508

INTRODUCTION

With the ever increasing adoption of new technology, the changing demographic composition, values, and attitudes of the work force, and the emphasis on both productivity and quality of work life, has come the need for more complex, interdisciplinary approaches to the implementation, design, and use of organizational systems (Hendrick & Brown, 1984). One such approach has been dubbed "Macroergonomics" by Dray (1985) and Hendrick (1985). The macroergonomic view considers multiple levels of analysis (e.g., individual and organization) as well as the importance of the joint optimization of the social and technological organization systems. The goal of the macroergonomist is to maximize the acceptance and effective use of technology and to minimize its potential negative effects (Dray, 1985.)

In this paper, technology refers to computer-assisted machines and equipment. Computer-Assisted Design (CAD), Numerically Controlled Machining, Programabale Robotics, and Office Automation are four examples of "new technologies" human factors practitioners will continue to be concerned with in the furure. The true benefits of these technologies will be realized as they are electronically linked together for the execution of work forming Computer Integrated Systems (CIS) which begin to take on features of continuous process production systems (Tornatzky, 1985.) It is this conceptualization of CIS, as opposed to the "islands of automation" which are most often encountered (Helfgott, 1985), which has led to the growth of the need for Macroergonomics. Human factors engineers have realized that all organization systems are inter-related and that an organization's technological sub-systems have an equally important counterpart, the social sub-system.

This paper presents a review of the relevant human factors literature, applies principles from organizational psychology, and examines the state of the newly born art within the framework traditionally applied to the implementation of training technologies. Following this framework, in order to reach the goal of macroergonomics we must 1) adequately assess an organization's needs for technology, 2) design a suitable intervention, 3) implement the processes, and 4) measure its effects in order to feedback results to the users and to the designers of the innovation.

ORGANIZATION NEEDS ANALYSIS

Inadequate organizational systems analysis may lead to inappropriate design of new technology, difficult and ineffecient implementation of new technology, or poor results and rejection of new technology. The need for a more macroergonomic view of organizations is born of a dissatisfaction with both the current methods of organizational needs analysis and interventions, and with the current conceptualizations of organizations and organizational theories which drive needs analyses and interventions. Several researchers from the University of Texas (Carsrud, 1984; Glick & Beekun, 1984; Krus, Carsrud, & Glick, 1984) in conjunction with IBM (Brady, 1984) have been moved by their dissatisfaction with the current state of affairs to develop User System Analysis which attempts a "systematic, analytic, and emperical integration of a wide variety of organziational factors" (Carsrud, 1984, p. 156).

These researchers argue that current theoretical conceptualizations of organizations are either too micro-oriented (e.g., Open Systems or Control Theory) producing a wealth of data which is greatly limited by its narrow focus, or properly macro-oriented yet limited because of weaker methodologies which produce less convincing empirical data. Therefore, what is needed is a multi-theoretical approach, macro enough to consider complex inter-relationships from different theoretical and applied disciplines across multiple organizational levels and system boundries, and micro enough to provide a useful methodology which may produce valid, generalizable data. User Systems Analysis is proposed to provide both the needed conceptualization and the necessary methodology capable of accomplishing these goals.

According to Glick and Beekun (1984), User Systems Analysis attempts to integrate the strengths of socio-technical systems theory/analysis (i.e., an open systems perspective leading to multiple levels of analysis, the modeling of systems and processes, and consideration of both the technological and social factors), Thompson's theory of technological interdependencies (i.e., need to match patterns of technological interdependence with the method of coordination), and Burt's network analysis (i.e., providing a flexible, precise, analytical description of patterns of relations within a social system). Thus, User Systems Analysis theoretically considers both the technological and social subsystems and uses a methodolgy designed to quantify, measure and represent them. User Systems Analysis improves upon previous conceptualizations and assessment methodologies by collecting data from numerous sources, summarizing and presenting data in work and information flow diagrams, while providing a more generalizable, systematic theoretical approach (Glick and Beekun, 1984). This method has been successfully applied in the personnel office of a large southwestern school district by Kraus, et. al. (1984).

The User Systems Analysis process is extremely time consuming and difficult. A comprehensive information and work flow diagram developed by Kraus et. al. (1984) took over 140 hours of the researchers' time, and resulted in a six feet by twenty feet diagram covered in small print. The researchers are now involved in developing a CAD program to overcome many of the difficulties.

Although User Systems Analysis is a step in the right direction (from no systems analysis prior to technology implementation to some system, function, and task analysis), it fails to recognize the organizations truely systemic/integrated nature. User Systems Analysis embrases a conceptualization of an organization, and a method of organization analysis, which is little better than the ones it was proposed to replace. User Systems Analysis still views an organization as a static, physical entity which may be adequately represented by information and work flow diagrams. Organizations are processes. They are interrelated actions (acts of people) which continually and cyclicly effect each other. Therefore, an analysis of an organization must consider more than current conditions and cannot be represented by a single-state picture, no matter how complex that representation is. Methods for organization analysis should go beyond current state concerns and examine the effects of business plans and priorities, possible future changes in systems, functions, tasks, and department members, as well as the inter-relatedness of technological developments, strategic direction, company values and philosophy, managers' roles and activities, pervasive management style, and corporate culture (Albert, 1985; Kanter & Buck, 1985.) We should conduct "future organization needs analyses," not (current) organization needs analyses. We must discover what needs the organization with a new technology will have in the future. The technological system can then be designed to better fit into the organization which will exist after it is implemented. This will be discussed further in the next section.

DESIGN OF TECHNOLOGY

In addition to poor assessment methodologies or poor conceptualizations of organizations, another source of difficulties regarding the adoption of automation is the design of the technology. Zakay (1984) reviews several factors involved in why automation is not generally well accepted in organizations, one of which is that users are usually not included in the design of the new system. This often produces a strong "not-invented-here" attitude and resistance to its use. Zakay (1984) suggests that system designers should involve the user in system design as well as consider both individual level factors (e.g., cognitive style) and organization level factors (e.g., user requirements and organization dynamics) in their design. Thus, he proposes a clear macroergonomic view which he calls "Organizational Human Engineering."

Imada (1985) similarly argues that while implementating (hard) technology top-down (e.g., FMS, CAD/CAM) can improve existing systems, it frequently ignores individual level concerns. Conversly, implementing (soft) bottom-up interventions (e.g., quality circles) shows great promise for meeting both individual and system needs, yet will not solve many of the complex technological problems of today's systems. Imada, therefore, argues for a multidimensional view integrating both technological (hard) and participative (soft) approaches systematically. Imada and Shoquist (1984) again argue for a more macroergonomic point of view in systems design. They illustrate how management, engineering, human resources, and ergonomic input should be integrated in order to properly implement technology change in organizations and to overcome resistance to the changes. They suggest that "impersonal human factors" designs efficiency at the expense of human dynamics. They conclude that the climate of an organization must encourage and nurture this more integrative macro approach. If automatic systems designers hold these macroergonomic views in mind during the stages of system development many of the current difficulties may be eliminated. The design of new technology should incorporate the results of thorough current and future organizational analysis. System designers must be made aware of the far-reaching effects that their technology will have throughout the organization, the workers, their jobs and their lives.

As alluded to in the previous section, an interesting paradox occurs when we presently design and implement new technological systems in organizations. An organization has a current technology, a current social system, current job designs, current staff, current work flows, etc. When new technology is installed in an organization it likely results in a new social system, new job designs, new staff, new work flows, etc. The new technology, which was designed for the original organization which no longer exists once the technology is implemented, must operate within a system for which it was not designed. Therefore, the results of a "future organization needs analysis" should be critical to the successful, effective design and implementation of new technological systems.

IMPLEMENTATION

Ideally, by this point in the process the results of a current and future organization needs analysis have been incorporated into the design of a technological system for an organization. More realistically, however, if professionals are employed, it is likely to attempt to ensure that existing technologies are properly implemented (Dray, 1985.) The existing macroergonomic literature provides the practitioner with several suggestions regarding the implementation of new technology.

Many researchers suggest using participative ergonomics to combat several of the problems usually associated with technological change (Endsley, 1985a & b; Freely and Freely, 1984; Imada, 1985; Sell 1984a & b.) These researchers suggest that by involving workers in the design and implementation of new technology, users' needs will be better met, and user attitudes will be more positive regarding the implementation of the technology. However, the introduction of automation into the work place may pose special problems regarding the use of participation. Baitsch and Frii (1985) state that while participation is essential in successful work re-design efforts, because the aim of automation is to replace workers, the social system to be re-designed may change. They also suggest that the participation efforts may fail without explicit guarantees regarding workers' job security. Diani, Bagnara, and Misiti (1984) provide a case example where participation in adoption of office automation led to increases in workers stress and mental workload. In their review of the literature on the implementation of information technology in office settings, Bickson, Guteck, and Mankin (1981) relate that although user participation in the implementation process frequently emerges as a determinant of success, if the technology is not comprehensible to the users (e.g., robotics to most line workers) their participation may be problematic. In light of these arguments, participative ergonomics should not be seen as a panacea capable of solving all of the problems associated with the implementation of technological innovations.

Gustafson (1984) and Ficks and Suzansky (1984), of AT & T, both present an alternative implementation methodology called "Project Management". They suggest that one person be in charge of, and accountable for, bringing the implementation about. This person needs various implementation aids (e.g., advisory, steering, and user committees), a power base, and institutionalization of the change process. According to these authors, this hard-nosed approach is nothing new. Importantly, it is compatable with more human relations oriented methods (e.g., participation) which could be used as implementation aids. The "project management" notions (also referred to as "facilitator of technological change" by Conner (1985), "technology entrepreneur" by Bickson, Gutek, and Mankin (1984), and "change agent" or "process champion" by Tornatzky (1985)) illustrate a solid macroergonomic approach to the implementation of new technology. Through its systematic, planned, coordination of activities, it conceptualizes the implementation of technology as an innovation process constituting major changes in both technical and social organizational systems.

The organizational psychology literature provides numerous additional examples of change strategies and variables which may be considered separately or in combination to aid implementation of new technology: 1. focus on supervisory roles, key players (e.g., gate keepers), education, training, and networking; 2. consider reasons for adoption; 3. actively pursue adaptive planning; 4. conduct specific pilot projects (manageable size, allows experimentation); 5. consider basic assumptions making up organizational culture; 6. use task forces or project steering committees to provide additional temporary organizational structure; and 7. be aware of incentives and counterincentives for change, new learning, and increased flexibility (Bikson, Gutek, and Mankin, 1981; Poza, 1985.)

The implementation of new technology affects an organization at all levels. Recognizing the complexity of organizations and the interdependencies within them, suggests that the implementation of new technology will affect a wide array of organizational variables. Sell (1984a), suggests that the implementation of technology may affect coordination and flow of information within organizations. Information can be kept by select individuals (information is power) or be more timely and appropriately delivered when and where it is needed. Communication, management and organizational structure, as well as working hours and occupations, can all be similarly affected. Robotization, for example, can be seen as de-skilling or as job-enriching by freeing up workers' time and elevating them to more coordinating and supervisory roles. Amick, Weisman, and Smith (1984) explored the affects of technology on workers' health. Their results suggest that workers' job satisfaction is affected by the job characteristics associated with a particular technology. Endsley (1985 a,b) reports that job commitment is reduced as a result of technological change. Tornatzky (1985) suggests computer assisted manufacturing technology will disrupt functional and hierarchical organization structure, and affect staffing, process monitoring systems, production flexibility and quality control. Helfgott (1985) examined the changes produced by robotics on human resource concerns such as compensation and training. He found that the introduction of robotic technology was followed by revisions in the compensation system to reward workers for broadening their knowledge base, the expansion, creation and deletion of jobs and job requirements, and the training of employees at all organization levels. Eason and Gower (1984) suggest that many organizations introduce word processing technology without intending any other changes in the system or considering that any changes might occur. These examples illustrate the importance of assessing the effects of the implementation of technological innovations.

MEASUREMENT

It is important that we not only consider possible organization dynamics in the analysis, design, and implementation phases, but that we recognize the changes in an organization resulting from technology implementation must be measured in order to feed these results back to both the system's user and designer. In this model, technology will be designed with current and future organization needs in mind. Technology will be implemented and will create changes in the organization. The organization will be measured and the results fed back to both users, who modify the organization to finetune the system, and to designers who will start the process over again. However, there is presently a lack of sound measures for assessing the effects of new technology (Goodman and Argote, 1984) which effectively thwarts the above process. Solid measures, sensitive to factors at different organizational levels, which will be affected by the introduction of various technologies, must be developed.

We know from previous research that communication, management, and functional structures, processes, and roles, as well as human resource and quality of work life issues and perceptions, and product flexibility and quality, all may be affected by the implementation of advanced technology. The questions we should be asking now are those which may lead us to discovering which of the advanced technologies affect which organizations and what are the effects at different levels within the systems? We do not know, for example, how CAD alters the designers' job, the human resource concerns of compensation, performance appraisal, and training, or the system configuration as CAD is slowly linked to the production function to form a flexible manufacturing system. We do not know how word-processors or personal computers alter workers' perceptions or the coordination and flow of information through the group, the department, and the organization. In response to these issues, this author has begun the development of a Human Factors Satisfaction Questionnaire (described in more detail elsewhere in this volume) which is designed to be sensitive to individual level factors within organizations likely to change when new technology is implemented. The HFSQ assesses the areas of environmental design, plant facilities, workload and system characteristics, equipment design, health and safety. It is not yet known if the HFSQ will be useful across various industries and technologies, or more useful for specific situations (e.g., robotic manufacturing technology.)

Gutek, Bikson, and Mankin (1984) suggest that the impact of advanced office technology will be mainly at the "office level," as compared to the individual level, due to the technologies' effects on work group and unit communication. Their research procedures included questionnaires to all workers, interviews with the office managers, observation of equipment and physical surroundings, and examination of archival data (e.g., staffing and budget patterns.)

These are two examples of research which is sensitive to the critical issues of organization and technology type, and levels of analysis. Continued research will produce useful results which may systematically be compiled into more generalizable frameworks and theories regarding technology and organization design and management. With further research and instrument development we may derive the benefits of receiving feedback from our implementation efforts.

CONCLUSIONS

The study of macroergonomics is in its infancy. It was born of solid interdisciplinary stock, and should grow well if nurtured and properly guided. Adequate organization needs assessment methods are available. With an appropriate "future needs" conceptualization, needs analyses may fruitfully be employed. Engineers and systems designers must become more cognizant of the inter-relatedness of organizational systems, and more open to consideration of, and cooperation with workers. Implementation efforts will benefit from better needs analyses and system design, as well as from more advanced planning, the use of project management, and a multitude of other suggested skills and techniques. The realization, by all, of the necessity for adequate measurement of the effects due to technological innovation is critical. However, this awareness needs to drive increased efforts toward the development and use of measurement instruments to enable the benefits of such information to be realized.

Even in its early stages of development, the macroergonomic literature and a macroergonomic perspective have a great deal to offer managers, psychologists, engineers, and system designers. With its interdisciplinary, multi-level, macro conceptualization of the processes of organizing, designing, implementing, and measuring the adoption of technology, macroergonomics offers a unique approach to the study of the design and implementation of advanced technology.

REFERENCES

Albert, M. (1985). Assessing cultural change needs. Training and Development Journal, May, 94-98.

Amick, B.C., Weisman, C.S., & Smith, M. J., (1984). The implication of technological change for workers' health: The role of the job redesign model in public health. In Human Factors in Organizational Design and Management, H. W. Hendrick & O. Brown, Jr. (Eds.), Amsterdam: North-Holland.

Baitsch, C. & Frii, F. (1985). A case study in worker participation in work redesign: Some suppositions, results, and pitfalls. Proceedings of the Human Factors Society 29th. Annual Meeting, Baltimore, Maryland.

Bikson, T. K., Gutek, B. A., & Mankin, D. A. (1981). Implementation of information technology in office settings: Review of relevant literature. Rand Corporation, Santa Monica, California.

Brady, L. (1984). User System Analysis. In Human Factors in Organizational Design and Management, H. W. Hendrick & O. Brown, Jr. (Eds.), Amsterdam: North-Holland.

Brown, O. (1985). Some perspectives on participatory ergonomics in high technology organizations. Proceedings of the Human Factors Society 29th. Annual Meeting, Baltimore, Maryland.

Carsrud, A. L. (1984). Organizational Analysis: Idiosyncratic methodologies searching for holistic solutions, or what's wrong with the way they are done? In Human Factors in Organizational Design and Management, H. W. Hendrick & O. Brown, Jr. (Eds.), Amsterdam: North-Holland.

Conner, D. R. (1985). Introducing new technology humanely. Training and Development Journal, May, 33-36.

Diani, M., Bagnara, S., & Misiti, R. (1984). Participation in technological change: Are there only enexpected consequences? The case of office automation. In Human Factors in Organizational Design and Management, H. W. Hendrick & O. Brown, Jr. (Eds.), Amsterdam: North-Holland.

Dray, S. M. (1985). Building Bridges: Implementing computer technology in organizations. Proceedings of the Human Factors Society 29th. Annual Meeting, Baltimore, Maryland.

Eason, K. D. & Gower, J. C. (1984). Implications of new technology on work organizations: A case study. In Human Factors in Organizational Design and Management, H. W. Hendrick & O. Brown, Jr. (Eds.), Amsterdam: North-Holland.

Endsley, M. R. (1985a). Technological change and individual adjustment. Proceedings of the Human Factors Society 29th. Annual Meeting, Baltimore, Maryland.

Endsley, M. R. (1985b). The toll of technology on job commitment and an implementation process to prevent it. Proceedings of the Human Factors Society 29th. Annual Meeting, Baltimore, Maryland.

Ficks, F. L. & Suzansky, J. W. (1984). Work systems as a large scale system design process. In Human Factors in Organizational Design and Management, H. W. Hendrick & O. Brown, Jr. (Eds.), Amsterdam: North-Holland.

Freeley, J. L. & Freeley, M. E. (1984). Increasing white collar productivity by humanizing the work place. In Human Factors in Organizational Design and Management, H. W. Hendrick & O. Brown, Jr. (Eds.), Amsterdam: North-Holland.

Glick, W. H. & Beekun, R. I. (1984). A theoretical overview of User System Analysis. In Human Factors in Organizational Design and Management, H. W. Hendrick & O. Brown, Jr. (Eds.), Amsterdam: North-Holland.

Goodman, P. S. & Argote, L. (1984). Research on the social impact of robotics: Issues and some evidence. In, Applied Social Psychology Annual (5), Stuart Oskamp (Ed.), Sage: CA.

Gustafson, H. W. (1984). Implementing work system change. In Human Factors in Organizational Design and Management, H. W. Hendrick & O. Brown, Jr. (Eds.), Amsterdam: North-Holland.

Gutek, B. A., Bikson, T. K., & Mankin, D. (1984). Individual and organizational consequences of computer-based office information technology. In, Applied Social Psychology Annual (5), Stuart Oskamp (Ed.), Sage: CA.

Helfgott, R. B. (1985). The impact of robotics on human resources and employee relations. Industrial Relations Counselors, New York.

Hendrick, H. W. (1985). Macroergonomics, a concept whose time has come. Proceedings of the Los Angeles Chapter Human Factors Society 8th. Annual Meeting, Downey, CA.

Hendrick, H. W. & Brown, O. (1984). Preface. In Human Factors in Organizational Design and Management, H. W. Hendrick & O. Brown, Jr. (Eds.), Amsterdam: North-Holland.

Imada, A. S. (1985). Integrating participatory ergonomics in to an organizational culture: A macroergonomic approach. Proceedings of the Human Factors Society 29th. Annual Meeting, Baltimore, Maryland.

Imada, A. S. & Shoquist, D. (1984). Managing ergonomic and process design changes. In Human Factors in Organizational Design and Management, H. W. Hendrick & O. Brown, Jr. (Eds.), Amsterdam: North-Holland.

Kanter, R. M. & Buck, J. D. (1985). Reorganizing part of Honnywell: From strategy to structure. Organizational Dynamics, Winter, 5-25.

Kraus, E. J., Carsrud, A. L., & Glick, W. H. (1984). Interactive organization workflow analysis: An approach to problem solving in organizations. In Human Factors in Organizational Design and Management, H. W. Hendrick & O. Brown, Jr. (Eds.), Amsterdam: North-Holland.

Poza, E. J. (1985). Comprehensive change-making. Training and Development Journal, Febuary, 81-85.

Sell, R. G. (1984). New technology and effects on jobs. In Human Factors in Organizational Design and Management, H. W. Hendrick & O. Brown, Jr. (Eds.), Amsterdam: North-Holland.

Tornatzky, L. G. (1985). Technological change and the structure of work. Paper presented at the American Psychological Association Meetings, Los Angeles, CA.

Zakay, D. (1984). "Organizational Human Engineering:" The design and implmentation of computer systems. In *Human Factors in Organizational Design and Management*, H. W. Hendrick & O. Brown, Jr. (Eds.), Amsterdam: North-Holland.

Human Factors in Organizational Design and Management - II
O. Brown, Jr. and H.W. Hendrick (Editors)
Elsevier Science Publishers B.V. (North-Holland), 1986

PERFORMANCE MEASUREMENT IN THE CONTEXT OF ORGANIZATIONAL VALUES SYSTEMS

Paul R. Seesing

Project Management Systems
Battelle Pacific Northwest Laboratories
Richland, Washington
U.S.A.

James L. Eubanks

Department of Psychology
Central Washington University
Ellensburg, Washington
U.S.A.

Current literature on organizational excellence suggests the presence of an employee-internalized value system as a key feature of successful companies. This paper presents a model of values assimilation in an organizational context based on previous behavioral research. A case study is used to illustrate how a company can develop behavioral measures of organizational value assimilation. Preliminary results suggest that organizational structures and management policies can encourage value assimilation. Organizational values can thus provide the basis for a comprehensive organizational performance measurement system.

INTRODUCTION

Organizational value systems, until recently, have received very little attention by "rational" managers and systems designers. Part of this inattention is due to the fact that values have not been readily amenable to quantification or specification, like "harder" elements such as budgets, policies and procedures, organizational structures, and strategies. Even with the advent of socio-technical systems theory (Emory and Trist, 1965) and consideration of various psychosocial characteristics of the workforce, little attention has been directed at organizational value systems.

Given the pervasiveness of organizational values in current management thinking (cf. Deal and Kennedy, 1982; Peters and Waterman, 1982), and the lack of their explicit incorporation into current socio-technical systems theory, how can an organization measure its commitment to those values in a reliable and valid manner? The present paper provides some insights into this issue, by discussing the nature of values and values systems, developing a model of organizational values assimilation, presenting some preliminary work in translating values into measurable behaviors, and highlighting the implications for organization design and management.

VALUES AND VALUES SYSTEMS

Considerable attention has been directed toward the study of values in individuals and societies. The present paper suggests that the concepts and theories that have been developed in these contexts can be generalized to the organizational environment. Previous work by Rokeach (1973, 1979) provides a substantial research base for values system development.

Values are defined as the standards or criteria for guiding action, evaluating or comparing oneself and others, and judging the preferability of competing end-states or results. Standards of preference are learned and internalized through the individual's interaction with society and its institutions. Values have a cultural content and are shaped by the constraints and opportunities of the social systems in which the individual is involved. Values also guide the individual's sense of identity and self image as a competent and moral member of affiliation groups.

A value system is an organized set of preferential standards that are used in making selections of objects and actions, resolving conflicts, imposing rewards and sanctions, and defending the choices made or proposed. Values are not concrete rules that are invariant in their application; rather, they tend to interact with each other, amplifying or attenuating the other values which constitute the system. Value systems are arranged in a hierarchy or network that guides the individual in prioritizing and choosing between alternative goals and actions, and in conflict resolution.

Value systems allow individuals to build a sense of their own identity, but at the same time provide a commonality or point of sharing that fosters their affiliation with social institutions. Just as individual values may be considered to be cognitive representations of personal needs and the means for satisfying them, institutional values are socially shared cognitive representations of institutional goals and demands. Institutions compete with one another for priority in drawing upon society's human resources, and for influence and control over those individuals.

Sociological data suggest that the most distinctive function of institutions, both social and organizational, is the framework they provide for value specialization. Institutions are structured to reinforce the desired values in the individuals' value systems through macrosocial arrangements. Both stability and change in institutional value systems are affected by the degree of consensus among the affiliated individuals' value systems, value systems of the society at large, and intrinsic characteristics of the values themselves.

A VALUES ASSIMILATION MODEL

Figure 1 presents a systems model of organizational values assimilation. This model also draws from the model of corporate values expressed by Paul (1982), and on a model of organizational control systems (Flamholtz, 1979).

In the model, values are used in the context discussed previously. Goals provide the specific objectives, driven by internalized values, toward which an individual or organization directs its resources. Once the organizational values (1) and goals and critical success factors (2) have been established and communicated to the members, individual behavior (3) can be measured (4) against those expectations. Evaluations (5) are performed using data-based methods to ensure fairness and consistency, and become part of the formal rewards system (6) of the organization. Rewards and sanctions, of course, are the traditional mechanism for behavior modification. These modifications result from a reevaluation of the organization member's personal values (7) and goals (8) and their compatibility with organizational values and goals. Measured member behavior also provides an indication of societal norms and values (9) which tend to shape organizational values.

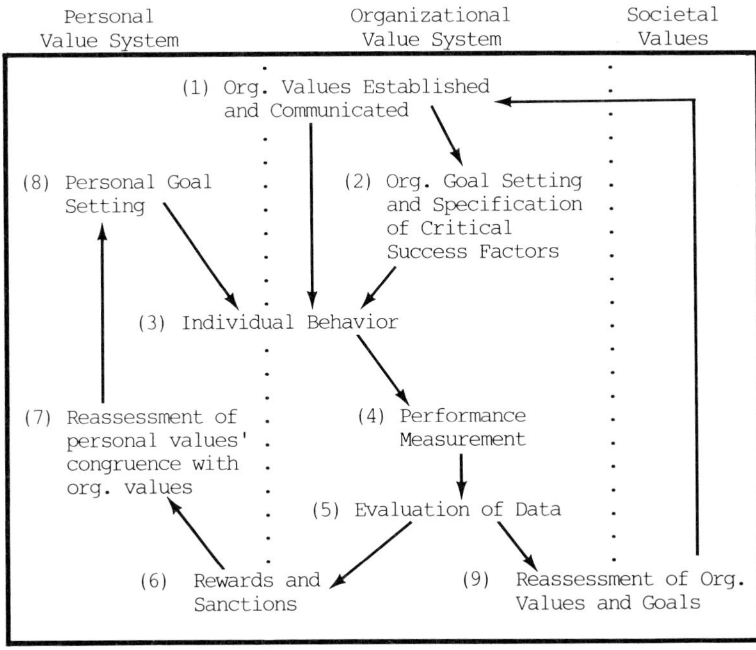

Figure 1
A Systems Model of Organizational Values Assimilation

The model depicted in Figure 1 describes the translation of corporate values into measurable individual behaviors. Such translation may occur subliminally for the individual, but needs to occur explicitly for organization managers engaged in performance measurement. This process is aided by organizational goal setting, which translates general corporate values into specific actions or conditions that provide the context for performance measurement.

Following is an example of how a set of corporate values can be translated into measurable behaviors using Battelle Memorial Institute (BMI) as a case study. Founded in 1929, BMI is an international multi-component research institution.

Institutions leave behavioral traces of their distinctive values in institutional documents and in the personal values of their leaders. Paul (1982) reviewed the writings of all prior CEOs and developed a set of eight perceived values that capture BMI's heritage and provide a leadership guide for future corporate challenges. These values are: a) benefit to humanity, b) innovation, c) integrity, d) quality, e) teamwork, f) corporate citizenship, g) growth, and h) profitability.

Leadership development is fostered at Battelle through a management orientation program conducted by the organization's senior executives. For each orientation, twenty managers from all components are invited to participate. One of the exercises in the program presents the eight values, expressed as organizational goals. The managers, working in teams, identify the critical success factors and an implementation plan for each of the goals/values. Critical success factors are the areas in which performance is necessary to ensure attainment of the organization's goals.

Initially, we examined the notes from this exercise for two orientations. Follow-up interviews were then conducted with some of the participants to amplify the notes. These notes provided lists of individual and group behaviors identified by Battelle managers as representing the translation of BMI's corporate values into measurable behaviors. Critical success factors delineated the "what" of the corporate values, the implementation plans indicated the "how." From these lists, we synthesized sets of objective measures that would reflect the implementation of the Battelle values. Figures 2 and 3 demonstrate the progression from value through goals and critical success factors to behavioral indicators. Though many of these indicators were expressed in organizational terms by the managers, they could be used as yardsticks for measuring individual behaviors.

Development of the eight values as an explicit part of BMI's corporate culture is at a very preliminary stage. As more management orientations are held, and more organization members are exposed to the values, support structures must be developed to reinforce the translation of this management philosophy into individual behavior. The assimilation model may

Value: QUALITY
Goal: To provide services and products of the highest quality commensurate with the needs and resources of those who pay for them.

Critical Success Factors	Performance Measures
Quality-minded staff	Results of internal and external peer reviews
Up-to-date equipment	Frequency of sponsor expressions of satisfaction and dissatisfaction
Formal quality assurance system	Percentage of repeat business clients
Careful review of proposals and reports	Existence of quality assurance (QA) plans
Involve client in project	Frequency of employee reference to QA plans
	Results of QA audits

Figure 2
Value Measurement Examples for Quality

Value: INNOVATION
Goal: To be an institutional leader in scientific discovery, technical inventiveness, and technological innovation.

Critical Success Factors	Performance Measures
Visionary, future-directed outlook	Number of publications in technical journals
Innovative staff members	Number of patents granted
An environment which fosters innovation	Percentage of R&D contracts which result in intellectual properties
Willingness to risk	Number of technical awards received by lab and staff
Willingness to invest	Existence and number of "skunk works" and "product champions" (cf. Peters and Waterman, 1982)

Figure 3
Value Measurement Examples for Innovation

be viewed as a blueprint for the organizational processes to be supported by such structures.

ORGANIZATION DESIGN AND MANAGEMENT IMPLICATIONS

Current organizational management literature frequently attributes the success of America's well-run companies to employee-internalized value systems. This literature suggests that employees who know what their company stands for, and know what standards they are to uphold, are more likely to make decisions that are consistent with these standards. It is also suggested that shaping and enhancing organizational values that are shared by its members can become the most important job a manager can do (Deal and Kennedy, 1982).

Although much of the current "trendiness" of values and culture in the organizational excellence literature is, in fact, based on anecdotal evidence, even human factors concerns are not without values-driven underpinnings. Take, for instance, the issue of user involvement in the design of management information systems (MIS). Few systems designers today would advocate the development of a comprehensive information system without user involvement in the design process, even though the empirical literature supporting such involvement is equivocal, at best (Ives and Olson, 1984). Thus, user involvement, like participative management, has become an underlying value, often more implicit than explicit, that drives much of the thinking of MIS designers today.

The importance of macroergonomic factors as necessary concerns for organization designers and managers is underscored by Hendrick (1985).

Psychosocial and cultural characteristics of the workforce are as much a part of the "large picture" as task and structural concerns. Since these characteristics are manipulable through the transmission and implementation of an organization's value system with its attendant implications for the welfare of the individual member, such values are clearly macroergonomic design elements.

Identification of organizational value systems is an important concern for human factors professionals. Not all organizations have explicit statements of their guiding values. Even when such statements exist, there are bound to be discrepancies between the ideal and the actual. The underlying value system in practice may vary with different contextual variables (such as hierarchy, formalization, education, technology) and is sure to affect organizational processes, managerial action, and ultimately, organizational performance. Procedures are required for converting implicit organizational values into explicit behavioral objectives and corresponding performance measures. The extensive human factors literature related to task analysis may be a useful tool for these efforts.

The organizational values assimilation model presented in this paper focuses on changing individual values through reinforcement contingencies that provide for reassessment by individuals whose personal value systems are incongruent with the organization's values. Organizations establish mechanisms to create and enforce institutional norms (and thereby standardize behavior) through assimilation of the individual into the organization's culture. These mechanisms need to be designed with a human factors approach taking into account the value systems perspective.

REFERENCES

[1] Deal, T. E., and Kennedy, A. A., Corporate Cultures: the rites and rituals of corporate life (Addison-Wesley, 1982).

[2] Emery, F. E., & Trist, E. L., Sociotechnical systems, Proceedings of the 6th Annual International Meeting of the Institute of Management Sciences 6 (1965) 92-93.

[3] Flamholtz, E., Organizational control systems as a managerial tool, California Management Review 22,2 (1979) 50-59.

[4] Hendrick, H. W., Wagging the tail with the dog: organizational design considerations in ergonomics, Organizational Design and Management Bulletin 3,4 (1985) 5-13.

[5] Ives, B., & Olson, M. H., User involvement and MIS success: a review of research, Management Science 30 (1984) 586-603.

[6] Paul, R. S., Battelle Values, BMI Bulletin, March 1983.

[7] Peters, T. J., and Waterman, R. H., Jr., In Search of Excellence: lessons from America's best-run companies (Harper & Row, 1982).

[8] Rokeach, M., The Nature of Human Values (Free Press, New York, 1973).

[9] Rokeach, M., Understanding Human Values (Free Press, New York, 1979).

ORGANIZATIONAL DESIGN, METHODOLOGY AND ANALYSIS

Human Factors in Organizational Design and Management - II
O. Brown, Jr. and H.W. Hendrick (Editors)
© Elsevier Science Publishers B.V. (North-Holland), 1986

ORGANIZATIONAL DESIGN STRATEGIES IN JAPANESE COMPANIES

Yoshio Matsumoto

Department of Business Administration, College of Commerce
Nihon University
5-2-1 Kinuta, Setagaya-ku, Tokyo 157
Japan

The purpose of this study is to examine organizational design strategies used by Japanese companies through data on their reorganizations. First, reorganizations taken place in Japanese companies in 1976 and 1982 are analyzed to explain their characteristics with regard to their environments and management strategies in each year. Then, based on this analysis, some of the organizational design problems facing Japanese companies today are explained especially in terms of human factors.

INTRODUCTION

After the first oil crisis in 1973, Japanese companies suffered a great decrease in profit and were obliged to take urgent measures such as assets disposal and layoffs. In addition to rises in prices of energy and raw materials, Japanese companies faced such problems as the strengthening of various regulations on environmental preservation. This was a general condition that existed in 1976. Recognizing a shift to a low growth economy, Japanese companies began trying the so-called "genryo-keiei" (weight reducing management) by reducing borrowing or employment, saving energy or raw materials, and simplifying organizational structures. Thanks to the "weight reducing management", the profits of these companies steadily increased through 1979, but after the second oil crisis in 1979, their profits started to decrease again. Under such conditions as weak domestic demand and decreasing export, it was feared that the economic stagnation would long continue. To cope with these conditions, some of the Japanese companies carried out strategies that induced the restructuring of their organizations. These reorganizations are analyzed next.

METHOD

<u>The Nikkei Sangyo Shinbun</u>, a newspaper for businessmen, reports daily the details of organizational changes made public by corporations. There are five or ten articles in a paper and each article usually contains from three to ten organizational changes carried out by a corporation. These articles on organizational changes in 1976 and in 1982 will be analyzed. The num-

bers of articles used for analysis are 836 in 1976 and 1,511 in 1982. Types of reorganizations are classified into 35 types, and occurences in each type is summed up by year and industry.

FINDINGS AND THEIR IMPLICATIONS

In 1976 and 1982, respectively after the first and second oil shocks, Japanese companies faced the same hard business condition. However, as is shown in Fig. 1, there is a difference between the goals given the highest priority by management in the two years.

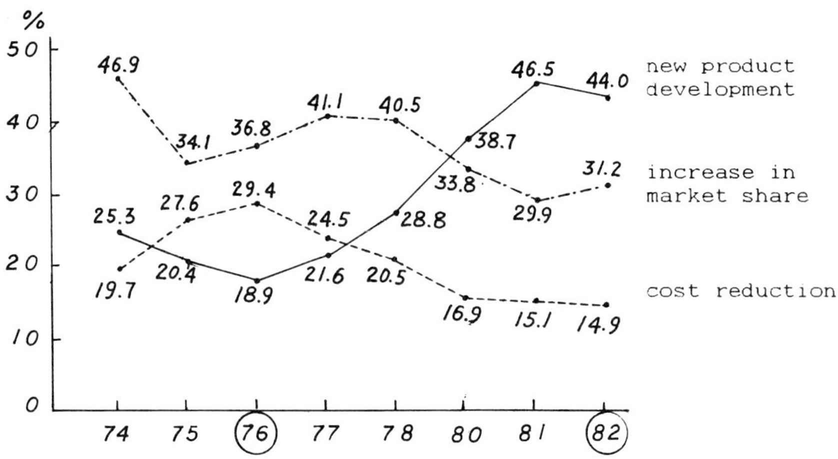

Fig.1 Change of goals given the highest priority by management

(Source : Ministry of International trade and Industry, New Management Indicators, 1983 edition)

type of reorganization	1976		1982
* Integration or abolition of functional divisions or sections	1.(267)	↘↗	1.(355)
* Establishment or expansion of a sales division	2.(148)	↗↘	2.(231)
* Establishment or expansion of an overseas affairs division	3.(96)	↘↗	3.(202)
* Establishment or expansion of a technology development division or a design division	4.(87)	↗↘	4.(172)
* Establishment of a new business division	5.(79)	⟶	5.(167)

Table 1 The most frequent types of reorganization in 1976 and in 1982

Table 1 shows the most frequent types of reorganization in 1976 and in 1982. In Table 1, the same types of reorganization are found among the most frequent types in 1976 and in 1982, but their sequences are different between the two years. Thus, Japanese companies tried to cope with the harsh business climate of those days by executing organizational design strategies such as simplification of organization, strengthening of sales, export or technology-development division, and establishing a new business division.

On the other hand, it is interesting to note that among the five most frequent types of reorganization, the first four changed their ranks between 1976 and 1982 (see Table 1). This suggests differences in priorities of strategies between the two years. Tese differences in ranking of strategies are more clearly shown in Table 2.

(1976)

* Integration or abolition of divisions	(41: 38)
* Integration or abolition of plants or prduction divisions	(30: 14)
* Integration or abolition of functional divisions or sections	(267:231)
* Establishment or expansion of an organizing div.	(10: 8)
* Establishment of a subsidiary or splitting off a division	(34: 21)
* Establishment or expansion of an environmental protection division or a safety control division	(22: 8)
* Establishment or expansion of a legal affairs division	(14: 14)

(1982)

* Introduction of a profit center system	(6: 26)
* Establishment of a new business division	(79:167)
* Establishment of a new business search or development group	(8: 59)
* Merger or absorption	(5: 20)
* Strengthening of a top management organization	(2: 15)
* Establishment or expansion of a laboratory or a research division	(41: 92)
* Establishment or expansion of a technology-development or a design division	(87:202)
* Establishment or expansion of a product planning division or a research division	(51:131)
* Establishment or expansion of a sales division	(148:355)
* Establishment or expansion of an information processing division	(23: 74)

The number within a parenthesis shows the number of each type of reorganization in 1976(left) and in 1982(right).

Table 2 Prominent reorganizations in 1976 and in 1982

Table 2 shows the types of reorganization which were large in number relatively either in 1976 or in 1982. According to Table 2, first, main characteristics of reorganization in 1976 are : 1) integration or abolishment of divisions, plants, production or other functional divisions or sections; and 2) establishment of subsidiaries by splitting off certain divisions. These organizational rearrangement are typical "weight reducing management" techniques used by companies trying to reduce organizational slack and trim the size of organization in coping with the hard business climate. In contrast, an organizing division or a management system division in charge of such organizational rearrangement was either established or expanded. Another characteristic of the reorganization in 1976 is that, reflecting rises of anti-pollution campaigns or consumerism in those days, an environmental protection or safety control division and a legal affairs division in charge of such problems were either established or expanded. In short, it can be said that the main characteristic of reorganizations in 1976 was a passive, defensive attempt to cope with the environment.

Then, main characteristics of reorganizations in 1982 may be summarized as follows :

1) The introduction of profit center system, establishment of new business divisions and mergers were greater in number than in 1976. This suggests that companies moved beyond the "weight reducing management" stage and began to take positive actions in some industries in spite of the severe business climate. To cope with the increased importance of these strategic decisions, top management was strengthened ;
2) The establishment of a "new business search group" was conspicuous. Its aims were---a) to pursue promising markets where high growth or high added value was expected, or b) to utilize employees who tended to be idle because of their old ages, the extension of retirement age or labor saving by microelectronics technologies ;
3) R&D related divisions were strengthened remarkably. This indicates that Japanese companies began to stress product oriented strategies rather than operation oriented strategies to cope with rapid technological developments or changes of industrial structure ;
4) Reorganizations to establish more business offices or sales divisions or to shift personnel from production, service or staff divisions to sales divisions were most noticeable. This indicates the companies' efforts to overcome stagnant demand by strengthening sales forces with idle personnel in their organizations ;
5) The information processing divisions were strengthened, which means that microelectronics technologies began to be introduced in plants and offices widely.

To sum up reorganizations in 1982, while taking such passive measures as simplification of organization or labor saving by microelectronics technologies, companies began to make positive efforts to strengthen their R&D organizations and sales forces.

The articles used for this analysis contains many important messages on organizational design problems which Japanese companies must cope with today. Based on these data, organizational design problems that are particularly important with regard to human factors are examined in the next section.

CURRENT ORGANIZATIONAL DESIGN PROBLEMS

1. Encouraging individuals to give a full play in their creativity

As a result of the fact that the level of technology in many fields in Japan has caught up with that in Western industrial nations, Japanese companies must change their R&D policy from "introduction of foreign technologies" or "stress on application" to "self development of original technologies". Many Japanese companies began to increase their investments in R&D and establish or expand their laboratories to strengthen their basic research potencials. Therefore, how to secure fields where individuals can fully express their creativity has become an important question in organizational design for Japanese companies which have traditionally used the "plinciple of organizing by group". The Japanese organization with strong Groupism has advantages in its product development stage where cooperation among individuals and groups is essential. But it tends to hinder the creativity of individuals engaged in basic research work. While it would be necessary to gain the advantages of efficient development by teamwork, the organization would also need to design an organization where individuals can restrain the tendency of over-tuning and fully express their originality and creativity. To achieve this dichotomy, some companies have tried following methods :

1) Organizational or geographical separation of the basic research laboratory from the development laboratory;
2) Establishment of a cooperative research system with universities, and public or overseas institutions;
3) Tacit approval of "under the table" research with certain limitations.

2. Taking measures against aging employees

The aging of population is taking place in Japan faster than in any other country. A retirement age has been extended to 60 in many big companies. On the other hand, the employment of young workers is not expected to increase much because of the low growth economy. Thus the average age of employees of Japanese companies tends to rise. As a result, problems such as increase in labor cost, shortage in managerial posts, and stagnation of organization are likely to occur. To cope with these problems and vitalize the organization, various methods have been tried in designing an organization. Some of the methods are as follows :

1) Modification of the seniority-oriented wage structure;
2) Incentives given to early retirement;

3) Leasing aged workers to subsidiaries;
4) Establishing subsidiaries for aged workers;
5) Counseling and relocation services for aged workers;
6) Specialist system;
7) Retirement system for managers;
8) Retraining of aged workers;
9) Redesigning jobs to fit aged workers;
10) Job rotation.

3. Coping with increased heterogeneity among employees

The so-called "Japanese Management" has functioned well based on homogeneity of members. But, recently, the heterogeneity of members has been increasing in Japanese companies in the following ways :

1) Increase in female employees;
2) Increase in temporary workers from other companies;
3) Change in values of young workers---a) weakening needs for promotion, b) incomplete commitment to an organization;
4) Increase in foreign employees through internationalization of business.

As a result, many Japanese managers are in great anxiety that the organizational vitality of their companies will wave. To avoid this to happen, it would be necessary to develop a system that is not heavily dependent on homogeneity of members.

Human Factors in Organizational Design and Management - II
O. Brown, Jr. and H.W. Hendrick (Editors)

SYSTEMATIC ORGANIZATIONAL DESIGN (SORD) METHODOLOGY

Joseph Conroy III

Human Performance Technology Division
Science Applications International Corporation [1]
McLean, Virginia 22102

The SORD methodology for designing U.S. Army organizational units is described. SORD is a step-by-step set of computer assisted procedures for the comprehensive, systematic, integrative and reliable designing of Army units. The methodology assists in deriving unit structures from the missions assigned to units. SORD consists of four subsystems: one for elaborating and quantifying missions; one for developing organizational designs; another for evaluating designs; and a fourth for designing team or crew structures.

INTRODUCTION

The focus of this paper is the development of organizational designs, and in particular organizational designs for the Army. In most instances organizational designs are the products of emulation. Designers copy wholesale or bits and pieces from other organizations. Often this is done without regard to the match between the purpose and function of the organization being designed and the one being copied (Davis, 1982).

In the Army, organizations which are known as units, are designed or redesigned each year by the hundreds. A review of the Army's publications on their unit design process revealed that each unit's organizational design is supposed to be mission-driven; to be designed specifically to accomplish its mission. Also, unit designs are supposed to be in accord with the Army's general doctrine and broad concepts about how it should fight.

Interviews with 40 personnel in the Army's unit design community revealed that in practice, unit designs are developed much as are those for civilian organizations - copied from other units (Conroy et al., 1986). Typically unit missions on which designs are to be based are terse and vague. The designs are not the product of a systematic and comprehensive set of procedures. Thus the Army's design process is not reliable nor documented so as to be easily audited. These two flaws in the process often result in a unit design including superflous elements whose justification for inclusion is nebulous. Futhermore, the process results in units whose capabilities are not well defined.

The Army needs a unit design process that is mission-driven. Also, as Davis (1982) points out for organization designing in general, the Army needs a process that is structured, comprehensive, integrative, and that proceeds from the general to the detailed. Thus it should involve a series of gradual steps. Since the Army frequently redesigns units, it needs a process that also includes the evaluation of unit designs.

Unit designers in the Army, as compared to designers in less structured organizations, must design using the Army's standardized components. These include individual personnel and teams and crews such as those for howitzers or tanks. The Army recruits and trains its enlisted personnel, the bulk of it's members, to be proficient at usually only one military occupational speciality (MOS). Also, the members of the Army's crews engage in specialized and formal training together and are part of very expensive systems the Army has purchased to address specific needs. Thus the Army's MOS and crews or teams are its standardized basic building blocks. As such they must typically be used unchanged in the design of units. Of course unit designs can incorporate some new components and sometimes units are designed for the sole purpose of incorporating new systems. Yet unit design in the Army is for the most part a process focused on organizational structures above the crew level, such as that of a company or battalion.

As with all new methodologies, a new unit design process for the Army must involve little if any additional work for the users. Moreover it must be a user-friendly system and be tailored to the Army.

METHODOLOGY

Given the previously mentioned general requirements for organizational design and in particular the needs of the Army, a unit design process was developed which has four subsystems (see figure below).

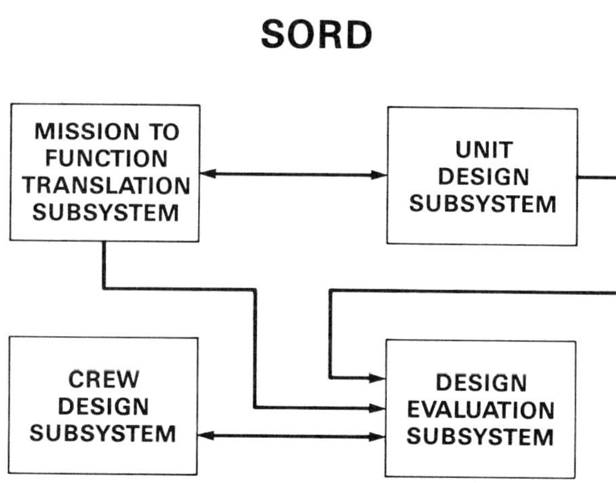

The first is called the "mission-to-function translation subsystem." Because unit designs are to be derived from their missions, this susbystem focuses first on elaborating and quantifying missions. Then the missions are distilled into the functions and subfunctions that must be performed. The second subsystem is called the "unit design subsystem." Its purpose is to match subfunctions with the capabilities of systems or personnel and to devise the managerial and organizational structure of the unit. The third subsystem is called the "design evaluation subsystem" and it focuses on evaluating the designs. The fourth, called the "crew design subsystem", focuses on designing and redesigning crews or teams.

Each of the four subsystems is designed to be independent so that users can enter them at any point. There is an introductory module for the whole methodology. It facilitates the independent use of each subsystem. It introduces the user to the system and provides guidance in the selection of modules and subsystems. Each subsystem has a tutor module containing specific instructions on the use of the subsystem's modules. The following describes each of the subsystems and modules.

The Mission to Function Translation Subsystem

This subsystem is a requirements generator. It consists of four modules designed to insure that each mission contains all the required elements and that they are quantified. The elements are distilled into functions and subfunctions with the aid of system prompts and a key-word accessed data base of functions and sub-functions. In the unit design sub-system the subfunctions are matched to the capabilities of systems and personnel.

The first module of this subsystem is called the "mission conditioning module." Since most missions are not sufficiently detailed enough to adequately drive the development of a unit design, the mission conditioning module assists the designer in the needed elaboration of missions. The module prompts the designer and provides formats and examples for developing a thoroughly articulated mission. Frequently this demands much more specification of the capabilities required of units than are implied in the concepts from which missions are derived. Thus it facilitates an early understanding and agreement on the exact capabilities to be expected from units to be designed.

Each module is designed to assist the user in taking a small step in the development of a unit design. In keeping with this, the second module, "the mission quantification module", is focused on only the quantification of the parts of missions. Conditions and standards are required to be supplied by the designers. The process is similar to that of the previous module in that the user is prompted by the system and supplied with a format and examples for each generic type of mission component.

The third module of this subsystem, the "function determination module", assists the user in taking the first step in the distillation of missions into characteristics that can be matched to the capabilities of the building blocks (the systems and MOS the Army trains). The module contains a data base of functions and subfunctions, many of which are tied to generic components of missions for ease of retrieval. Many of the functions and subfunctions were derived from Army doctrine and training literature.

The fourth module, the "function specification module", assists the user in the quantification of functions and subfunctions. This development derives from units' missions and the guidance of standard formats for each type of function plus examples. The result of the subsystem is functions and subfunctions with quantified conditions and standards.

The Unit Design Subsystem

This subsystem consists of five modules. In succession they assist the user in matching subfunctions to systems or personnel and then determining the numbers of the systems and personnel required. If the limits on unit size are exceeded, the unit is resized to conform to the limits. Finally the organizational and managerial structure is developed.

The first module is called the "major systems assignment module." With a data base of the capabilities of all the Army's major systems such as tanks and howitzers, it assists in the assignment of major systems to the performance of subfunctions. Subfunctions not assigned to major systems are by default assigned to be performed by personnel. The exact type of personnel are determined with the assistance of the next module.

An attempt is made to assign subfunctions to major systems prior to considering the assignment of them to personnel. The reason for this is the premise that the Army has developed and purchased major systems which are more suited to some subfunctions than are personnel. Thus the systems should be used for those subfunctions. For example, the Army has developed and purchased a remotely piloted airplane for forward area reconnaissance. Thus a designer should first consider assigning the subfunction of forward area reconnaissance to the remotely piloted vehicle prior to considering the assigment of that subfunction to personnel.

The second module of this subsystem is called the "MOS assignment module" and it assists in the assignment of MOS to subfunctions. The user has access to a data base of the capabilities of all MOS. The module also prompts the user to rely on the Army's reliability and maintainability data for the assignment of maintenance related personnel.

The next step is perfomed in the "unit sizing module." It helps determine how many of each type of personnel should be assigned to units. This is done by relying on the two data bases used in the two previous modules. The one for major systems details the capabilities of each system. The MOS data base does the same for personnel for which performance data is available, such as repair hours per type of item per day.

Probably all Army units which will be designed will have limitations imposed upon them in terms of the numbers of personnel and major systems that can be assigned to the unit. Thus, this module requires the user to compare assignments to constraints.

If the constraints were exceeded, the next module, the "function prioritization module", assists the user in prioritizing functions and subfunctions. The functions and subfunctions are prioritized on the basis of their paired comparisons, supplied by unit designers. The paired comparisons are converted into proportions and then into interval scale values using a table of normal deviates. Then the least important functions or subfunctions are used as guides to eliminate systems and personnel from the unit to comply with the constraints. If the functions or sub-functions assigned to those eliminated from the unit were critical, they must be performed by another unit. Otherwise, the functions or subfunctions will not be performed and thus the stated capabilities of the unit must be changed.

The last module, the "command, control and structure development module", assists the user in developing the organizational and managerial structure of the unit. It prompts the user to consider adopting the structures utilized by the component parts when they train. For example, if standard Army training for howitzer crews involves them being grouped into triads and managed by an additional managerial element, such practices should influence the assignment of a managerial structure to a unit that contains several howitzer crews.

The module also guides the user to consider span of control, information flow, the performance of common functions and typical geographic deployment configurations in the assignment of structure. The user has the assistance of a mouse-controlled graphics capability for conceptualizing and depicting the organizational structure. In addition, the user relies on a set of tables the Army has developed for the assignment of managerial personnel.

The Design Evaluation Subsystem

This subsystem evaluates the designs produced by the previous subsystem. The evaluations involve a comparison of the capability of units to the requirements specified in their missions. Rather than simply match the capabilities of systems or personnel to subfunctions, this subsystem evaluates all the elements of the unit interacting as they will most often be deployed and used. Thus it considers the sequences in which tasks will be performed and physical relationships between the elements of the unit.

The subsystem consists of three modules, two of which are used for preprocessing. The first is called the "conditioning module" and is used to set all the important mission and environmental parameters to be used in the evaluation process. Many of these have default values. The types of variables can be changed and typical examples include weather, terrain, physical deployment and level of fatigue.

The second module used for preprocessing is called the "functional organizational module." This is used to input the organizational structures to be evaluated.

The "activity and performance assessment module" performs the evaluation of unit designs. It takes the quantitative performance data and modifies them according to the previously specified parameters such as fatigue. For example, a high level of fatique may be represented as a 10 % reduction in all performance times or task completion probabilities. The module then sequences all task completion and productivity data and identifies the performances which do not meet their criteria.

The output of the module is the identification of the elements of units which are not productive or timely enough. These outputs are then used to redesign units. This can occur in the unit design subsystem or in the crew design subsystem.

The Crew Design Subsystem

This subsystem assists users in the design or redesign of crews. Typically the focus of this subsystem is crews for machines such as a howitzer. However, the subsystem can be used to design crews or teams which do not operate a machine such as a command headquarters team.

The subsystem consists of three modules: "a preprocessor"; an "allocation module" to allocate tasks to MOS; and "a postprocessor" to evaluate alternative allocations of tasks to MOS. The allocations of tasks to MOS is based on the optimization of any variables the user chooses. These may include task completion times, time and costs to train, and salary.

The first module of this subsystem, the preprocessor, assists the user in defining alternative candidate crews. Initial manpower, personnel and training constraints are based on the present crew or similar crews. Tasks and task completion times are required and are recorded in the system via the preprocessor.

The allocation module establishes the assignment of MOS to tasks and optimizes values for the variables of interest. Moreover, the variables on which optimization is to occur can be prioritized. The module uses goal programming to find the most optimal assignment of MOS to tasks. This is especially useful in the early engineering phases of system design. It allows systems designers to estimate the most efficient and least costly method of assigning MOS to tasks. It also provides indications of the costs and performance characteristics of the alternative MOS to task assignments.

The third module provides reporting detail to enable the user to compare alternative crews. It identifies critical tasks and MOS. It also has a graphics capability for displaying results.

APPLICATION

The SORD methodology is slated for application in the near future. The segment of the methodology to be applied first is the crew design subsystem. This subsystem has uses outside of organizational design. It could be an efficient tool for use in the early stages of the design of any man/machine system. It could be used to efficiently make human engineering design decisions. It's simultaneous optimization of many human factors could provide a designer with a capacity heretofore typically only approximated in an inexact way using fewer variables.

The remainder of the methodology is slated for application soon after the crew design subsystem. However, its efficient applicability in its current state appears to be limited to the military or a situation in which organizational designs are frequently developed and developed from standardized elements whose capabilities are defined. This includes most of the military and several industries such as the building construction industry.

FOOTNOTES:

[1] Portions of the SORD methodology were developed by Michael Fineberg, Brian McEnany and Robert Robinson.

REFERENCES:

Davis, L.E. Organization design, In Salvendy, G. (Ed.), Handbook of industrial engineering (John Wiley, NY, 1982).

Conroy III, J.A., Lewis, S.L., Escholz, T.S., Fineberg, M.L., McEnany, B.R., and Robinson, R. Description and model of the Army's unit design process, Technical Report, McLean, Va., Science Applications International Corporation (1986).

Human Factors in Organizational Design and Management - II
O. Brown, Jr. and H.W. Hendrick (Editors)

A CASE STUDY OF PRODUCTIVITY IMPROVEMENT RESULTING FROM ORGANIZATIONAL RESTRUCTURING

Henry E. Bender*
Bell Communications Research
Piscataway, New Jersey 08854

F.L. Ficks
F.L. Ficks Associates
Springfield, New Jersey 07081

Judy Bender
Concept Applications, Ltd.
Laurel Hollow, New York 11791

A major New York City commercial travel company was awarded a large contract during a period of expansion, low productivity and high turnover rates. External consultants performed an organizational analysis and an employee perception audit. Study methodology, findings and recommendations are presented.

INTRODUCTION

Many smaller and mid-sized organizations operate under the belief that they have optimized their organizational structure. The need to reexamine the organization does not become apparent until sufficient environmental change occurs, forcing an analytical/introspective analysis of resultant problems. Such a change occurred during the decline of employee productivity in the 70's (Business Week, 1980). It has been suggested that they are the result of 2 factors, the capitalization of American industry and employee morale (Scheer, 1969).

Larger business organizations often have in-place methods to measure productivity levels on an on-going basis. Such methods measure financial/production levels and, less frequently, employee attituded/opinions that might interact with productivity.

In the smaller closely held organizations, the reanalysis usually occurs with either a significant increase or decrease in business volume. To accommodate the change, the organization begins to institute change sometimes as a conscious effort and other times as a reaction to pressures. These changes result in the continued success of the organization, or its demise. These changes occur in one of many ways. For example:

1) a dynamic leader will emerge to restructure the organization,
2) a formal process can be started using either internal staff or outside consultants who have expertise in organization/personnel evaluation,
3) The employees change the work processes and process interaction without changing the organization.

---

*Dr. Bender participated in this survey as Principal Consultant to Concept Applications, Ltd. This work represents the views of the author and not those of Bell Communications Research.

The present case history describes a condition where after approach 1 & 3 were tried, as a last resort, outside help was obtained.

PRESENTING PROBLEM

A major independent travel organization in New York City, specializing in corporate travel, was faced with the desirable problem of having successfully competed for and won a major corporate travel account. If existing productivity levels remained constant, a 50% increase in reservationist staff would be required to service the client. Reservationist productivity was below industry levels and turnover was a persistant problem (See Exhibit 1).

Exhibit 1
Productivity and Turnover Rates

	In Company	Industry Standard
Productivity levels in $'s per Reservationist	$850,000	$1,000,000
Turnover rate per year	33%	12%

During the preceeding 60 days, considerable effort had been expended in recruiting new staff. This effort had met with little success owing to the agencies reputation for operating a "sweat shop" and the travel industry in general suffering from a shortage of a travel workforce. The environmental situation of having been awarded a new major contract requiring a 50% increase in staff, coupled with low productivity averages per reservationist and a high turnover rate created a situation that could easily jeopardize the organization. It was faced with a potential loss of existing business as well as the most probably loss of the significant new business acquired in the past year. To correct perceived "dangerous" situation, the organization decided to analyze its operations and to take corrective actions based on the findings.

METHOD

Corporate officers and internal staff members initiated several attempts to analyze the problem and to develop a course of action. One of the senior staff members on several occasions attempted to introduce change, but was continually undermined. Workers making suggestions or attempting to improve work flow were castigated or invited to leave and, in general, the staff had determined the best approach was to work long hours and search for a new job. The staff's joint effort was unfruitful due to the small size of the firm and their close working relationships with various members of management.

Study objectives, as agreed to by the management team, were to:

- analyze the existing organizational structure,
- describe the perceptions of the nature and reasons for existing problems, and
- identify employee attitudes to numerous variables impacting satisfaction, productivity levels and turnover rates (Conte, 1984).

Organizational structure analysis was to be performed by conducting in-depth interviews with top corporate management. These were conducted in an attempt to determine the existing structure and job responsibilities/ descriptions as well as written and unwritten policies and procedures. The management team, consisting of the corporate President, Vice-President of Marketing and Vice-President of Sales participated in these rounds of interviews. The structure that evolved and was agreed upon, as existing at that time, can be seen in Exhibit 2.

Exhibit 2
The Presidents Perception of the Organization Structure

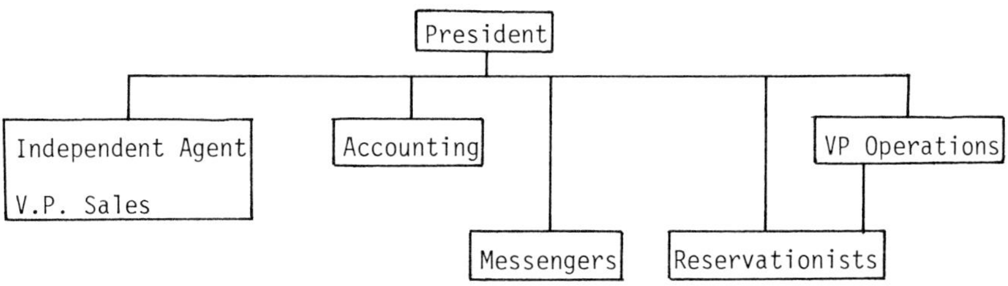

The organizational structure presented is the one described by the President of the corporation. He perceived the company as a closely run organization. Top management noted that policies and procedures were not documented and were, in fact, developed and changed on an individual-as-required basis.

Top management reviewed a draft form of an employee perception audit. This was necessary to ensure their support for the effort, and to provide them with guarantees that unnecessary or "embarrassing" categories of items would not be included. These discussions resulted in minor changes to the proposed instrument.

Members of the reservationist staff were informed that they would be participating in a study of the organization. Further, that any material/ or information supplied would remain confidential. Top management planned to use the findings as input to policy and practices development, therefore, the aggregated results would be presented to the management team. This was one of the better methods employees had to communicate their feelings/opinions to the decision makers.

Only the reservation workforce was surveyed. They completed the questionnaire on their own time and returned completed questionnaires in preposted envelopes directly to the authors. This was done in an attempt to assure respondent confidentality (Bender, Conte, 1986). There was a 60% response rate to the distributed questionnaires.

FINDINGS

A demographic analysis of respondents indicated they were representative of the organization as shown in Exhibit 3.

Exhibit 3
Demographic Characteristics of Respondents

Age: % of Respondents:	20-24 30%	25-29 30%	30-40 30%	40+ 10%
Salary % of Respondents:	$10,000-14,999 27%	$15,000-19,999 45%	$20,000-24,999 5%	$25,000-29,999 23%
Education level % of Respondents:	high school 0%	some college/post high school technical training 48%	college degree 41%	post college 12%
Length of Co. Employment % of Respondents:	less than 6 months 15%	6 mos.-1 year 19%	1 - 2 years 43%	3 or more years 23%

The data indicates that:

- o the employees represent a young workforce, with 60% being below the age of thirty,
- o they are moderately payed, at best. 70% $10,000 to $20,000 per year with a maximum of $30,000 per year,
- o they are fairly well educated. All have graduated high school and all have had some college or post high school technical training,
- o one-third of the employees have worked for the company less than one year. 15% have not completed a half year of employment.

Using the Hobbs (1947) satisfaction measure, nearly a third (29%) indicated a high level of satisfaction with their job. Half (50%) were moderately satisfied, while over a fifth (21%) responded that they derived low levels of satisfaction from their job.

Respondents rated 20 demensions as to level of importance. The recorded results are shown in Exhibit 4.

The ranking data indicates the importance placed on the subject area categories by respondents. It indicates a weighting to be placed on the category when attempting to resolve problem areas. It is interesting to note the apparent independence of importance rankings and satisfaction levels.

The extremes analysis findings indicate that the employees were favorably disposed to their own work quality and productivity, their immediate supervisor, the work itself, and the technology available for them to perform their jobs. On the other hand, low satisfaction levels were associated with communications, salary levels, suggestion handling (a subset of communications, along with a non-participative work environment), advancement possibilities, management, rewards and recognition and the benefit package. These findings resulted from an extremes analysis, in which comparisons were made between high satisfaction levels (70% or higher) and low satisfaction levels (30% or lower).

Exhibit 4
Order of Importance of Questionnaire Categories and Questionnaire Findings Using an Extremes Analysis

Subject Area	Satisfaction Level High	Low	Percent Difference	Rank
Your job	29%	21%	+08%	
Advancement	13%	52%	-39%	9
Benefits	21%	50%	-29%	10
Communications	08%	67%	-59%	8
Company	21%	46%	-25%	16.5
Suggestion handling	09%	73%	-64%	18
Personal future	23%	41%	-18%	5
Work Load	48%	37%	+11%	13
Job security	38%	21%	+17%	6.5
Productivity	62%	12%	+50%	6.5
Management	21%	50%	-29%	11-
Performance reviews	32%	55%	-23%	16.5
Problem solving	24%	48%	-24%	15-
Work quality	88%	0%	+88%	3
Rewards/recognition	26%	54%	-28%	1
Salary	13%	69%	-56%	-2
Technology used	40%	10%	+30%	14
Immediate supervisor	69%	19%	+50%	12
Work	61%	09%	+52%	4
Environment	29%	34%	-05%	

To validate the survey findings, top management and 20% of the staff participated in in-depth semi-structured interviews. Staff interview participants were selected in a stratified random manner, so that all aspects of the reservationists jobs could be sampled and represented. These interviews confirmed the reasonableness of these findings. A key word analysis confirmed the importance of the most critical "problem" areas. Additional findings, non-questionnaire related included:

- the initial organizational structure supported employees reporting to more than one boss,
- employees didn't have a source to reference for technical questions.

RECOMMENDATIONS

Revise the salary and benefit structure to be more consistent with industry standards. This, in fact, occurred and turnover was immediately reduced.

Interpersonal skills and communication training for the management team. This training was directed specifically to address the major problem areas in interpersonal relations between members of the top management team and members of the work force. Sessions were held using Synergistics tools emphasizing how others perceive you as a supervisor/manager and the value of working together.

Finally, a new organizational structure was recommended and implemented.

This is presented in Exhibit 5.

Exhibit 5
Proposed Organization Structure

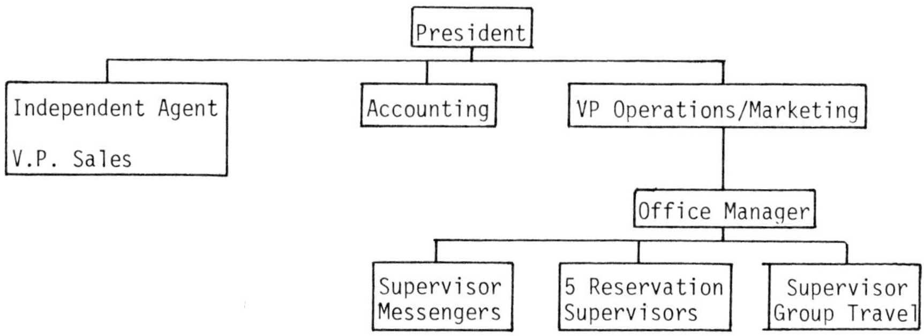

This structure reduced the span of control and added another level of management. Supporting the structure was a set of clear job title responsibilities and definitions. The proposed structure and clear job definition reduced the potential crossing of job responsibility boundaries by members of the top management team, reducing the possibility of issuing conflicting instructions to members of the work force.

Additional corrective measures included:

- A plan of recognizing the "Worker of the Month" and the "exceptional worker" were formulated in an attempt to increase reward/recognition.
- Additional supervisory positions, resulting from the organizational restructuring. It was believed this would increase satisfaction levels with career paths, personal future, advancement, communications, management, and suggestion handling items.
- A training program in interpersonal skills and communication improvement. Based on Synegistics tools, these sessions stressed the values of working together and how others perceive the individual as a supervisor/manager.

RESULTS

Management has reported that as a result of the changes in organization and the training program, there has been a 50% increase in productivity and a 20% reduction in turnover. A follow up on EPA is projected for 1986.

REFERENCES

(1) Bender, H.E., and Conte, V.A., The Human Factors/Applied Psychologists Role in the Corporate Culture Improvement Process, in: Second International Symposium on Human Factors in Organization Design and Management, 1986.

(2) "The Decline of U.S. Industry," Business Week, June 30, 1980, p. 59.

(3) Conte, V.A., A multivariate model of the turnover decision of data processing professionals. Unpublished dissertation, Hofstra U., 1984.

(4) Habbe, S., Job attitudes of life insurance agents. Journal of Applied Psychol., 1947, 31, 111-128.

(5) Scheer, W.E., The Dartnell Personnel Director's Handbook, The Dartnell Corp., Chicago, Ill., 1969.

Human Factors in Organizational Design and Management - II
O. Brown, Jr. and H.W. Hendrick (Editors)

INSERTING DESIGN INTO THE ORGANIZATION PROCESS

F. L. FICKS

F. L. Ficks Assoc.
Springfield, NJ 07081

ABSTRACT

This case study describes procedural modifications made by the workforce to accommodate a management-prescribed organization. These procedural modifications resulted in an increased rate of worker injury. Outside help was requested by both management and labor to improve working conditions that were components of an institutionalized structure and methods. Incorporating organization redesign into methods change resulted in changes that were nonthreatening to management and supported by the workforce.

INSERTING DESIGN INTO THE ORGANIZATION PROCESS

In this company, creating organizations (hierarchical structures) is viewed as management's responsibility. This responsibility is fulfilled by publication of hierarchical charts and manuals of organization purpose, signed by top managers, and displayed on walls or contained in company policy statements. This approach created an environment where the purveyor of organization design technology is viewed as an interloper or usurper of management's basic rights.

The organization described in this case study was designed by its present manager. Workers felt compelled to develop methods and/or modify prescribed work procedures that conformed to present organization structures. These procedures, perceived as simplifying the workflow, in some cases resulted in injury-inducing tasks. Workers believed that organization structures were the purview of management, and employees have to "work around" any problems the structure created.

BACKGROUND

The company Quality of Work Life Committee (QWLC) requested that management provide technical assistance selecting equipment and/or determining requirements for additional people in an organization responsible for operating laser printing machines. The committee felt that this would reduce the injury rate. Management's response was a request for assistance in the design of procedures or specialized training. The specific problem identified was back injuries that occurred during the loading and unloading of laser printing paper.

Both management and the workers, as represented by the QWLC, recognized the existence of a problem in the laser printing room. Each proposed solutions they felt were appropriate. Management proposed the development of new methods and training. The QWLC (workers) proposed the development of additional equipment or more people.

The literature is not clear and consistent on where the responsibility for organization design lies. Dunnette (ed.) (1983) refers to executives in authority, sponsors, managers, and change agents as interchangeable terms and responsible for organization design and implementation. This places authority for organization design squarely in the lap of management. Others such as Cherns and Davis (1975) assert that, "The analysis, preparation, and implementation of a sociotechnical design is ... the property of no individual or set of individuals; it belongs to the members of the organization whose working lives are being designed." In this case, neither of the proposals or groups recognized that the source of the problem could be the organizational structure itself although there are references to the numerous performance problems caused by poor organization and design (Kilmann, Pondy and Slevin, eds., 1976), (Grusky and Miller, eds., 1981).

THE PRESENT ENVIRONMENT AND WORKFLOW

A. Organization

Four years ago the unit was moved to its present location as a result of increased work loads. The present manager was responsible for moving the unit, designing the physical layout of the equipment, and defining the organizations and staffing required to print and mail bills.

Four work units (Figure 1) as well as the following responsibilities were identified:

Receiving - Receiving supplies (paper, envelopes, inserts, etc.), placing the supplies on storage racks close to each operator's work location, and moving the "stuffed" envelops to a mail pick-up location.
Printing - Laser printer operation and maintenance.
Bursting/Jogging - Selecting and sequencing the printed material and operation/maintenance of the bursting and jogging machines.
Insertion - Operation and maintenance of the insertion machine.

The organization structure is shown in Figure 1.

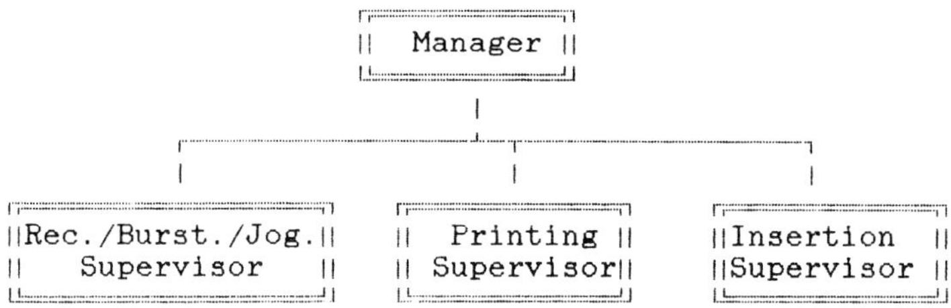

Figure 1. Original Organization Structure

B. Job descriptions

Warehousing - Receive palletized paper, place and stack the pallets, move the pallets to the beginning of the conveyer line, open the boxes, and place the boxes of paper on the conveyer. Move stock from the warehouse to the shelves of the insertion unit.

Laser printer operator - Lift paper from the conveyer to a dolly, push the dolly to the input side of the printer, thread the printer, queue the next printing job, remove the printed material from the printer, and place the stack on either a dolly, a work table, or a shelf. The input weight of supplies is 36 lbs and the removal weight varies between a few sheets and 40 lbs.

Bursting operator - Select printed material from shelves in the printing room, transport to bursting room, place lift from the transporting carts to the bursting machine feed tray, thread bursting machine, remove bursted media and place in trays, move to jogging area, jog material (by handful),and place trays of jogged material on carts or storage shelves for use by insertion group.

Insertion operator - Move bursted material from bursting area, pull supplies from shelves, load insertion machine, remove completed envelops from insertion machine, and place in postal boxes. Load postal boxes on carts.

C. Workflow

The flow of work from one organization to the other was initially prescribed. Over time the procedure evolved to the sequence described above. Changes to work methods were made by both workers and supervisors in response to complaints or "New" ideas. None of the workflows or reasons for change was documented. The QWLC indicated they had worked very hard to establish the present workflow and felt that it provided the best means of moving the large amount of material (approximately 7 tons per month) through the work units.

Figures 2, 3, and 4 show the workflow and weights lifted at each stage of the operation.

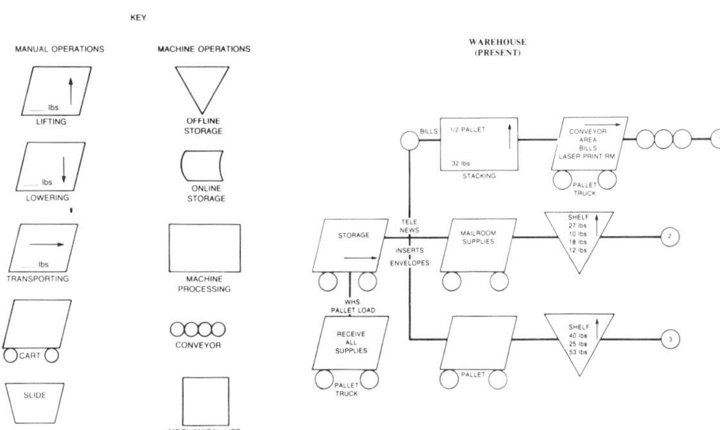

Figure 2 Original Work Flow and Weights Lifted

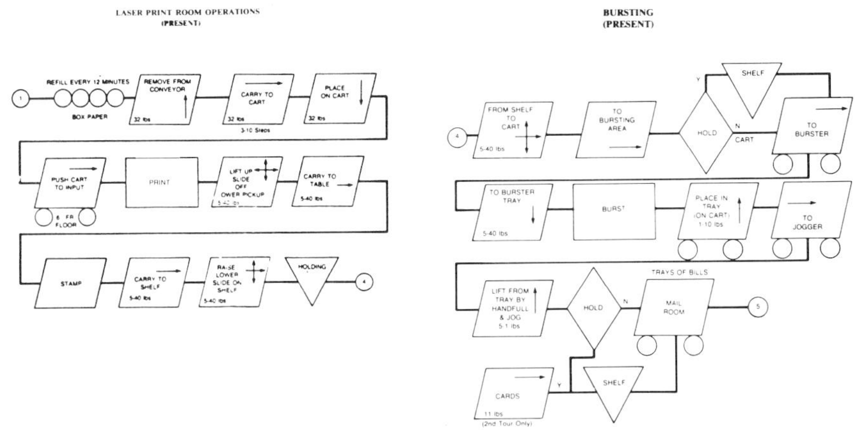

Figure 3 Original Work Flow and Weights Lifted (Cont.)

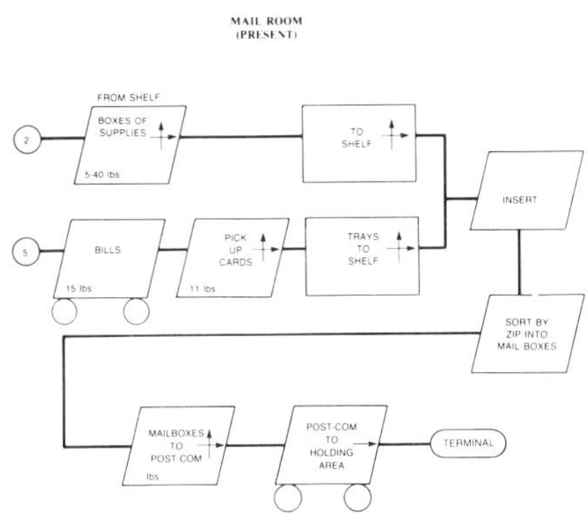

Figure 4 Original Work Flow and Weights Lifted (Cont.)

METHOD

Gustafson (1984) points out that acceptance and diffusion of organizational change under the prevailing doctrine results in "everybody - - or what amounts to the same, nobody, is supposed to be in charge." For this reason the Work Systems approach as described by Ficks and Suzansky (1984) which has a basic tenant, a project manager in charge, consultants, and subject matter experts as needed, was selected as most appropriate for this analysis.

The QWLC formed a project team with the chairperson of the QWLC as the project manager. (This person was a nonmanagement employee elected for a one-year term by members of the all volunteer QWLC.) Members of the project team included management's representative to the QWLC and one volunteer from each of the work units. The team was responsible for data input, working with the consultant, determining the most appropriate recommendations (in conjunction with the consultant), and implementing the finalized recommendations.

RESULTS AND RECOMMENDATIONS

The project team had little difficulty establishing what had to be done or in defining objectives. They resisted establishing or documenting how much was to be done as a union-negotiated item. When organization interfaces and job and organization designs were examined, any changes, except for those affecting only one unit, were documented as a management decision. Staffing and sizing determinations were made using mechanized programs, and reductions were accepted as "if that is what the computer says." In the case of physical layout, each of the team members had strong personal opinions. The team members refused to discuss work measurement, indicating that this was only negotiable between management and the union. The changes identified were primarily in the areas of interfaces and organizational structure. Additional training was determined as unnecessary. Detailed implementation plans were considered important because of the staffing and organizational changes recommended. The recommended organization chart can be seen in Figure 5. The recommended organization shows a reduction in the number of supervisory groups from three to two. In addition, one supervisory and nine working positions were eliminated. The new workflow shown in Figures 6 and 7 eliminated the need for lifting.

D. New Organization

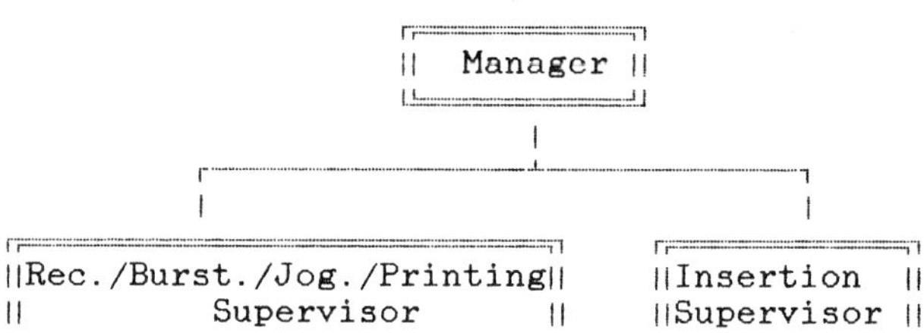

Figure 5. Recommended Organization

E. New Workflows

The new workflows are shown in Figures 6 & 7.

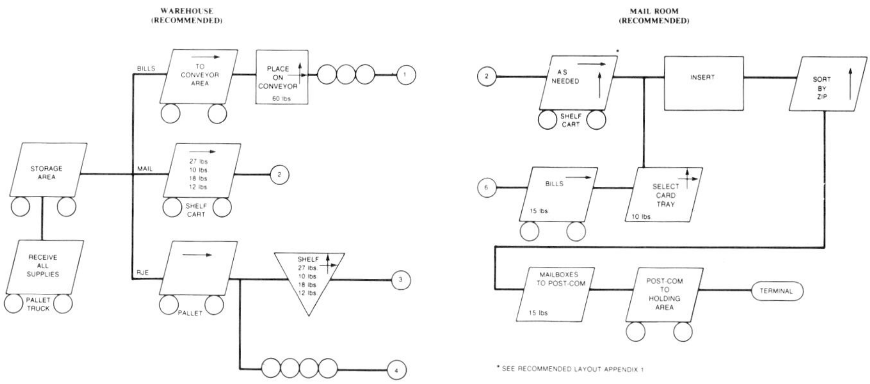

Figure 6. Recommended Workflow

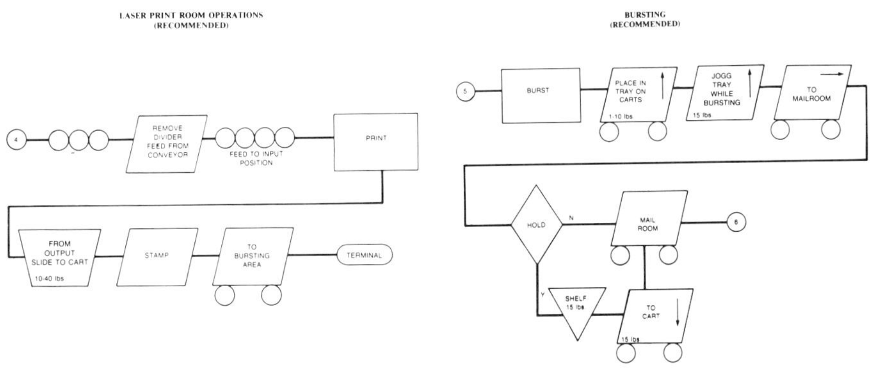

Figure 7. Recommended Workflow (Cont.)

DISCUSSION AND IMPLICATIONS

The impetus for this study was management's request for training and the workers' request for new mechanical aids. The study recommendations resulted in a more general use of available mechanical aids and a new organization structure. No requirements for training resulted from this effort. The most obvious question is, Did this experience in participative analysis result in an understanding of how organization design affects the working environment? The result seems to be a resounding No. Management and workers involved in this case were interviewed and asked to describe their beliefs about how organizations are typically designed and what effect the organization had. They reported:

1. Organizations are formed to meet management's need to control people.
2. Technical training or specialized knowledge is not required to design organizations.
3. Organization design tools are a pad and pencil. The process for creating or restructuring an organization is to estimate requirements based on past experiences and establish relationships based on personalities.
4. Power, politics, economics, and trade-offs in positioning, size, and function determine the hierarchical placement of the new organization.
5. Methods and training specialists are responsible for providing procedures and communicating the necessary skills -- good management will make it work (a management view). Workers believed that they do whatever is necessary to make it work.

The objective of this project was to change the working environment to reduce the number of back injuries. This objective was accomplished together with a major increase in productivity through redesign of the organization. Although successful in meeting this goal, workers and management continued in their belief that organizations were something outside successful operations.

BIBLIOGRAPHY

Cherns, L. D., and Davis, L. (1975) Quality of Working Life, Vol. 1 Problems, Prospects & the State of the Art. New York: Free Press.

Dunnette, M. D. (ed.) (1983) Handbook of Industrial and Organizational Psychology. New York: John Wiley & Sons.

Ficks, F. L., and Suzansky, J. W. (1984) Work Systems as a Large Scale Systems Design Process, Human Factors in Organization Design and Management, Proceedings of the First Symposium Held in Honolulu, 21-24 August, 1984. Hendrick, H. W., and Brown, O. Jr. (eds.), North-Holland 1984.

Grusky, O., and Miller G. A. (eds.) (1981) The Sociology of Organizations. New York: The Free Press.

Gustafson, H. W. (1984) Implementing Work System Change, Human Factors in Organization Design and Management, Proceedings of the First Symposium Held in Honolulu, 21-24 August, 1984. Hendrick, H. W., and Brown, O. Jr. (eds.), North-Holland 1984.

Kilmann, R. H., Pondy, L. R., and Slevin (eds.) (1976) The Management of Organization Design, Vol I. New York: North-Holland.

Human Factors in Organizational Design and Management - II
O. Brown, Jr. and H.W. Hendrick (Editors)

HUMAN FACTOR ISSUES IN REORGANIZING NTT

Akira Ishikawa
Aoyama Gakuin University
Shibuya, Tokyo 150, Japan

Muneya Sato
Yokohama City University
Kanazawa-ku, Yokohama 236, Japan

This paper is concerned with the changes of the human relations factors when a big organization undertakes an organizational reform. We will look into human factors, particularly, human behavior of many organizations and groups which are related to the new NTT (Nippon Telegraph and Telephone Corporation). We also intend to review its new strategy and make cost analysis in connection with productivity, organizational effectiveness and efficiency. Finally,several recommendations will be given to enhance them.

THE PURPOSES OF NTT REORGANIZATION

In Japan, Nippon Telegraph and Telephone Public Corporation was transformed into a new corporation named Nippon Telegraph and Telephone Corporation(NTT) in April,1985. This reorganization was carried out by the directive of the Japanese Government whichhas been promoting to reform many national enterprises. At the same time, telecommunications service market was changed to a competitive one. We must, at first, reconfirm the main purposes of these radical reform policy conducted by the Japanese Government. The basic ideas can be seen in a series of reports made by the DAINI-RINJI-CHOUSAKAI(Ad Hoc Public Administration Research Council) established by the Government. Following are the extracts from the Committee Report named "The Third Recommendation concerning Nation's Administration Reform Plan" in July, 1982.

The Third Recommendations (Extracts)

a) Under the control of the Congress and Govrnment, the existence of management responsibility is not clear and managerial efforts of the enterprise often are not fully made. Furthermore, referringto the management-labor relations, top managements of enterprises have no right to determine payroll-level by themselves, familiarizing themselves with easy and 'stick to formality' negotiations between management and labor union. On the labor side, union members would not build better and harmonious relationships because they are not granted labor rights and they become dull to go often on an illegal strike.

b) Each of three public corporations is too big, and NTT and Japan Tobacco Corp. enjoy monopoly. As such, it is difficult to manage their business properly and succesfully and to attain desirable performances.

c) To improve these problems, it is necessary to take such means as (1) giving freedom from external constraints, (2) giving them managerial autonomy, (3) promoting their self-consciousness on the part of workers, (4) making both managements and workers strive for high efficiency and diversifying into many new businesses. To accomplish these matters, it should be necessary to reform public enterprises perfectly and reorganize them to private corporations. At the same time, a new market structure must be built as active competition could take place.

INTER-ORGANIZATIONAL HUMAN RELATIONS AND CONFLICTS EMERGED

Let us sum up the main purposes of NTT reorganization and list them as follows (1) Acquisition of managerial autonomy through privatization. (2) Enhancement of a business spirit. (3) Improving organizational effectiveness through the introduction of competitive market principles.

These three purposes of reform this time were gradually approved by the public, and the interested groups such as NTT, the Ministry of Posts and Telecommunications (MPT), and Union of NTT's employees have agreed these purposes to be reasonable as well.

Top management of NTT was, of course, pleased to accept these purposes or aims of (1) and (2), but bureaucrats of MPT who wanted to sustain the regulatory powers raise a strong objection to it. NTT's management and Labor Union, at first, strongly opposed improving effectiveness through breaking up monopoly. The Government insisted on execution of (3) in exchange for (1) and(2).

Original plan of reforming NTT was to divide it into a headoffice and several regional companies. The Ad Hoc Public Administration Research Council, which was considered to speak from the nation's standpoints, had really an intention to deregulate the telecommunications industries and promote competition. This policy was warmly welcomed by many people concerned at home and abroad.

To sum up, since each of many interested groups has its aims and desire, each group wanted its own interests and power to be kept safe,and it is no use for us to accuse them. More important thing to be considered is that they reached a common consent of three purposes indicated above, although each party had different criteria of the reform at the time.

The figure 1 indicates the relations between interested groups of NTT reform. Now, we will list the conflicts emerged in NTT reorganization with bearing the related parties or groups in mind.

(a). MPT vs. Ministry of International Trade and Industry (MITI)---a discord of regulatory rights.

(b). MPT vs.NTT---regarding the new regulatory rights for telecommunications and it's scope.

(c). NTT management vs. Union (ZEN DEN TSU)---about the labor right of strike and the reducing plan of the number of workers.

(d). NTT vs.New Common Carrier (NCC)---regarding telecommunications market structure and access charge problem, as an example.

(e). NTT vs. Electronics and Telecom. manufacturing companies---NTT's new plan to launch new businesses, such as VAN (value-added network) business.

(f). Japan's Telecom. Manufacture vs.American Telecom. Manufacture---a big unbalance in trade of telecom. equipments between Japan and U.S.

In addition to the above, there occurred many conflicts inside of NTT.There were many discords in opinions between and among divisions.

For example, Engineering and R&D Division Managers, worring about the possibility to reduce the level of R&D activities, adhere to maintaining monopoly. On the other hand, the President and the other headquarter managers have thought much of the acquisition of management autonomy and living up to the spirit of entrepreneuership, dared to accept competitive doctrine which the Government have claimed. There were also discrepancies on the various matters between telephone division and data-communication division. The Government and MPT have an intention to separate the data-communication division from NTT as a subsidiary because of preventing the internal mutual assistance from the telephone division. Possibly, we think, this separation or reconstruction will take place in the near future.

These newly emerged conflicts in reform process were shown in Figure 1. We will classify these conflicts by various dimensions. At first, we can classify conflicts as to whether they have emerged inside of the organization(Internal Conflicts) or in relation to another external organization (External Conflicts). And concerning the hierarchy of organizational structure and industrial structure, we can make the distinction between Horizontal Conflict and Vertical Conflict. More important classification, we consider, is related to the nature of conflicts, and that is (1)remediable or remedied one (Remediable Conflicts) or (2) in progress and unresoluble one (Unresoluble Conflicts). We can define each conflict by using these classification multi-dimensionally. For example, we can define the conflict between NTT and Labor Union as an Internal Vertical Resoluble Conflict (IVRC), showing in the Figure. After considering the nature of conflicts in such a way, we can see what influences each conflict would give to the achievement of three purposes of NTT Reorganization.

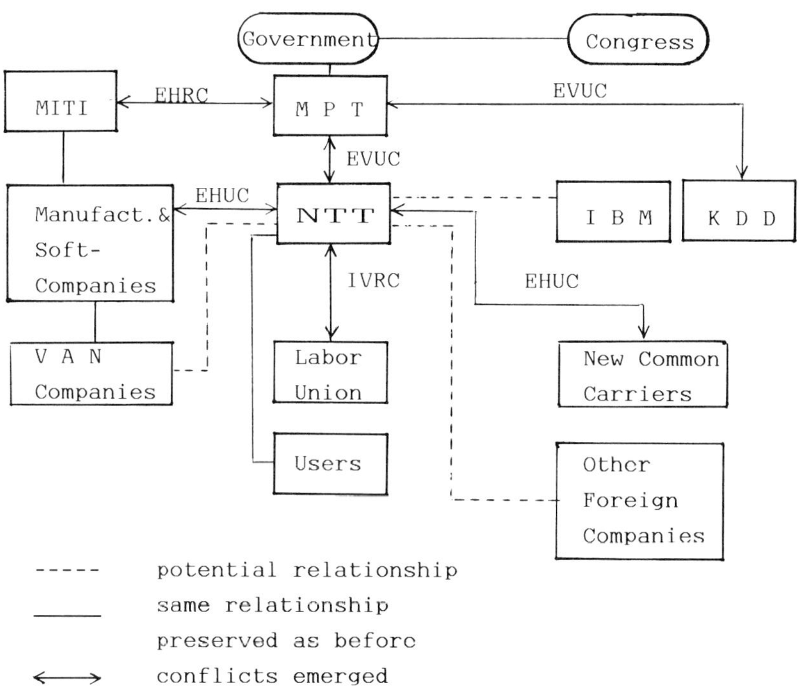

Figure 1

NTT and related groups

ORGANIZATIONAL STRATEGY OF NEW NTT

In December 1984, the new Telecommunication Utilities Regulation Act was passed and on April 1, 1985, new NTT has started as a private company. Its capital stock amounts to ¥780,000 billion (about $3,900 billion), gross asset amounts to ¥10 trillion-plus, and it has 310,000 employees. Mr.Hisashi Shinto, who had been a President of a big, private company for many years, was elected as President of the new NTT. How much has the deregulation policy exerted influence on a telecom. industry in Japan? This is clear through the following data in which about 146 companies have begun VAN businesses according to the report from Japanese magazine called "Computopia".

VAN business is a new area which relates closely with software and data-base business in Japan and it will also give strong impacts on the various services such as hardware maker, newspaper, advertizing, publication, and broadcasting businesses. This new business has made a fairly good performance, and this is due to the policy of opening the public and private lines of NTT to everyone. Furthermore, since the entry into the telephone services was permitted, many companies are planning their entries into the new market as a common carrier as well as several enhanced services including VAN, LAN, videotex, cable TV, personal computer communication services and so on.

The new wave of telecommunications market seemingly will increase more and more in height. Certainly these efforts will have the effect of increasing the industry's overall income. We must, therefore, consider the problem of organizational effectiveness and efficiency focusing on NTT as well as on information industry in general. As severe competition will possibly occur in telecommunications service market because of price cutting, it would be very important for each company to reduce cost so that it may gain profit. In other words, the company that can minimize cost via proper cost control techniques and simultaneous use of organizational strategy could survive.

Next table is a list of the new common carriers and their service menu.

no.	company name	services (schedule)
1.	Daini Denden	private line and telephone
2.	Japan Telecom	Same as above
3.	Teleway Japan	Same as above
4.	Japan Satellites Com.	private line
5.	Cosmos Satellites	Same as above
6.	Tokyo Telecom. Net.	local telephone
7.	Lakecity CATV	local private line and telephone

Table 1
NCC and their services

Faced with the introduction of competitive mechanism into the industry,NTT has never given up the expansion of its business. It has had to cope with severe competition and begun carrying out its adaptive behaivior. NTT, for the first time, begun aggressive sales promotion of newly developed services. The public-telephone-card, high-speed digital transmmision service, video-conferencing are some of the new services being offered currently and gaining in popularity.

Furthermore, as price strategy, NTT has curtailed the price of telephone cards by 5%, discounted the contract charge of car telephone by 33%, and reduced the charge for long-distance telephone call.

More important thing worth mentioning is the NTT's several organizational strategies shown below:

1) To adopt and enforce the divisions system----reorganizing the headquarters into 9 divisions (principally by services offered) and making the regional communicationcenters into 11 divisional branch offices. Twenty Divisional Managers are granted more authority over personnel and budget affairs than before. Divisional performances will be appraised by such criteria as profit (or loss) and income/outlay data in the near future, but for the time being operating profit is used as an appraisal measure.

2) To establish subsidiaries and affiliated companies----for example, NTT has established Nippon Information and Communication Corp (NI+C), a 50-50 partnership with IBM Japan, Ltd. Besides this, NTT has formed 31 companies for the sake of commercial using of internal technologies. With the diversification strategy, NTT has tried to reduce its exessive number of workers by transferring them to the subsidiaries.

3) To establish the Labor-Management Committee----the aim of this committee contains the attainment of better communications between Union and top-management, for instance.

4) Other new plan----new payroll system, promotion system,and training system were proposed and reached agreement between labor and management.

HOW TO CUT THE COST IN COMPETITIVE MARKET

Cost which is inevitable for communication has not been affected so much by the organization reform. Cost of network equipment (depreciation expense and maintenance cost, etc.) is an example. The costs which would be increased or decreased on account of the reformation are considered to be policy-dependent operating and administration expenses. These costs include those items which could be reduced suprisingly by enhancing productivity and cutting material purchase rates. Now that main purposes of the administration reform and privatization are to make an economical and flexible organization, excessive and needless expenses must by all means be eliminated. As a private company,NTT is expected to endeavor for cost reduction and it is just on this point that the success of NTT reform policy depends.

Meanwhile, NTT had to pay the following cost newly incurred:

(a) income tax (nation, local), enterprise tax, stamp duty.
(b) provision for pensions and bonuses, allowance for doubtful account.
(c) non-tax-deductable expenses (entertainment and contribution)
(d) change of calculation on depreciation cost.

Privatized NTT will have to do better only after controlling these cost increases and amplifying sales.

For all that, NTT has got several benefits on financial side owing to the deregulation policy. For example, new NTT has been able to invest their excess funds freely and to borrow money from anyone at a moderate rate. NTT will also be able to list its shares on the Tokyo Stock Exchange and to raise new funds by issuing new common stock. Of course, It has begun investing money in many newly-formed subsidiaries and joint ventures hoping for future profits.

Table 2 is the final Profit and Loss Statement of NTT as a public corporation.

	billion yens	million dollars
SALES and REVENUE	¥4547	$22735
OPERATING EXPENSES	3783	18915
OPERATING INCOME	764	3820
OTHER INCOME	127	635
OTHER EXPENSES	394	1970
ORDINARY INCOME	497	2485
EXTRAORDINARY INCOME	82	410
EXTRAORDINARY CHARGE	251	1255
NET INCOME	¥328	$1640

Table 2

P / L (1984.4.1-1985.3.31) NTT Public Corp.

One year has passed since new NTT was born, but its annual report has not yet been published. We can use, however, the interim data and prospectus data as Table 3.

	interim report	prospectus data
OPERATING REVENUE	¥2555	¥5530
OPERATING EXPENSES	2143	---
OPERATING INCOME	357	---
ORDINARY INCOME	171	340
NET INCOME AFTER TAX	77	183
	(1985.9.30)	(1986.3.31)

Table 3
Selected Interim & Prospectus Data
(yens in billions)

If prospectus data are reliable, we can compare these data with those of Table 2. Operating revenue would increase by ¥983 billion but ordinary income would decrease by ¥157 billion, and in consideration of the newly accrued costs, new NTT would attain the same performance as they did last year.

However, we cannot say for sure that NTT's reform has been effective only by these data. The new commer has not yet been operating, and a couple of years later when the competitors begin the telephone business, NTT's revenue would be likely to fall down sharply as ATT experienced. NTT President and other men said,"The interim results are better than the prospectus and this is because of strengthening our selling power and drastic cost cutting." Even if this is true today, this will not be the same tomorrow.
Under the competitive environment, NTT could not expect more income than they are enjoying today. Thus, it is necessary for the new NTT to adopt the organization strategy, which should, we think, encompass a drastic increase of productivity and organizational effectiveness and efficiency.

CONCLUDING REMARKS

Finally, we should give several suggestions to enhance productivity and organizational effectiveness and efficiency of new NTT:

1. Enriching R&D activity so as to cope with severer external competitions.
2. Removing bureaucratic predisposition and thinking and enhancing the image of the organization.
3. Teaching and training employees with strategic thinking in their mind.
4. Constructing the strategic information system and using it for decisions such as pricing, R&D, corporate planning, marketing, resources purchasing, and sales.
5. Promoting QC/circle activity, developing Total Quality Control System, and introducing a Creativity Development Program, and finally enforcing productivity enhancement activity.

REFERENCES:

(1) K.M.Howe & E.F.Rasmussen, Public Utility Economics and Finance (Prentice-Hall, 1982).

(2) Barry R. Armandi, Organizational Structure /and Efficiency, (University Press of America,1981).

(3) Akira Ishikawa, Principles of QC Circle Activities and Effects on Productivity in Japan: A Corporate Analysis, Management International Review, Vol.25, (1985).

(4) Muneya Sato, The Cost Structure of Telecommunications Companies, Journal of Public Utility Economics, Vol. 37, No.2. (1985) 1-30.

Human Factors in Organizational Design and Management - II
O. Brown, Jr. and H.W. Hendrick (Editors)

GROUP DECISION MAKING IN AN ERGONOMICS PROGRAM IN A U.S. AUTOMOTIVE PLANT: CORRELATES OF SUCCESSFUL MEETINGS

Bradley S. Joseph, Jeffrey K. Liker and Thomas J. Armstrong
The University of Michigan
Industrial and Operations Engineering
Center for Ergonomics
Ann Arbor, MI. 48109.

The use of cross-functional task forces for ergonomic design and implementation of workstations was tested in an automotive parts plant. One key issue in the program was the effectiveness of meetings from the perspective of participants. This paper compares more to less successful meetings, according to participant ratings, and examines correlates of success.

INTRODUCTION

Despite years of ergonomic research and an abundance of technical information, industries, by and large, have not adequately incorporated ergonomic principles into the workplace. Consequently, many industries suffer from, among other things, high incidence of musclo-skeletal disorders, decreased productivity and worker dissatisfaction.

Adequate use of ergonomic principles, a multi-disciplinary science, requires the input of many diverse specialities into job design or redesign. However, the traditional organization in many industries creates barriers that prevent successful mixing of those disciplines necessary in designing and implementing ergonomically sound workstations. These barriers include a lack of specific job knowledge, communication breakdowns between disciplines, and sub-unit interest in decision making (Liker et. al., 1984). One approach to overcoming these barriers is through ergonomic task forces that bring together people representing a variety of functions and management levels.

There are many ways to measure the success of group decision making in task forces. One measure of success, participants' perceptions, is an important measure for three reasons:

First, participants' perceptions and satisfaction may affect the longevity of the group (Hackman and Oldham, 1980). Common sense reasoning would lead one to expect that participants who feel they are wasting their time in ergonomic meetings will be neither committed to, nor enthusiastic about contributing to the effort. Since group decision making in ergonomic design is often a voluntary commitment, participants who lack commitment and consistently fail to show up can seriously hamper the success of the group.

Second, participants can tell us things about their meetings that outside observers can only infer. Theories of group decision making emphasize that broad participation is a prerequisite to the most effective decisions and dissenting viewpoints should be carefully considered (Maier, 1967; Janis, 1971). Therefore, if a particular group of people censor their ideas, causing the effectiveness of the decision making to be lessened, then careful analysis of participants' perceptions may uncover the reason for the problem.

Third, ergonomics programs using group decision making, like quality of worklife programs, have dual objectives--organizational success of the group and organizational success of the members (Hackman and Oldham, 1980). Therefore, if the participative approach to ergonomics accomplishes its objectives and implements ergonomic designs to workstations (organizational success satisfied) by aggravating the members of the group (members success not satisfied), then the group decision making program is not completely successful.

A participative ergonomics program, that used group decision making to implement ergonomic changes to workstations, was developed in a large automotive parts plant. This paper identifies correlates of the success of group meetings from the viewpoints of participants'. While the program was not restricted to activities in meetings, this was a major arena in which ergonomic problem solving and group decision making took place.

THE STUDY

A new organizational subsystem, based on participation in group decision making and established to work within the constraints of the existing organization, was implemented at an automotive parts plant. The organization was designed to overcome barriers to success of the program by using group decision making to bring together diverse specialities and implement sound ergonomic design. In preparation, all participants were given a training course designed to increase ergonomic awareness and provide a basic understanding of ergonomic considerations in job design. An evaluation of the training is presented by Liker et al. in this volume.

The plant was organized as two autonomous areas (A and B) with central management and staff acting as resources and setting major plant policy. There were four groups in the organization that regularly met to discuss ergonomic considerations in job design:

a) *The advisory committee*-made up of top plant management and union officials, their function was to set and continually monitor all goals and policies of the ergonomics program.
b) *The Area A task force*-made up of middle management and area union representatives, their major responsibility was to monitor and steer the Area A worker group into areas of excessive ergonomic risk, serve as a clearinghouse for job redesign proposals and provide resources for implementation.
c) *The Area A worker group*-composed of machine operators, supervisors and skilled trades, their role was to identify stressful jobs and determine ways to redesign the workstation to reduce or eliminate stress.
d) *The Area B task force/worker group*- who, because they combined the resources and personnel from the task force and worker group, was responsible for the functions of both the task force and worker group as outlined for area A.

Three data sources contributed to the analysis of correlates of participant perceptions. First, the outcomes variables of participants' perceptions of meetings is based on assessments taken immediately after the meetings using a self administered form. The attributes assessed using this form were:

Goals of meeting	Quality of decisions made at meeting
Goals of next meeting	Diagnosis of ergonomic problems
Participation of individual	Individual's influence on the meeting
Participation of group	Leadership at meeting
Participant knowledge of discussion topics	Satisfaction with the meeting
	Satisfaction with program

In addition to these individual attributes, a composite score (mean score) provides a summary assessment of the meeting.

The second source of data characterizes attributes of the meeting independent of the participants' assessments. Meeting observations, made by the author who attended all meetings, resulted in the creation of the variables:*meeting attendance and meeting activity*. As discussed in the introduction, satisfaction may affect and be affected by absenteeism. Activities were measured to see if some activities were more satisfying than others.

The final source of information, gathered early in the program, was the basic background data on each participant. In this paper, we focus on job status as a correlate of participant reactions. Studies of participatory organizations often report some resistance to participation by "middle managers" (Kantor, 1983; Hackman and Oldham, 1980) and this hypothesis was tested.

RESULTS

The relationship of attendance to participant perceptions.
Table 1 presents correlations between the average assessment (across all attendees) of meetings and attendance at those meetings. While there are few significant correlations, there is a consistent pattern of positive correlations between attendance and positive participant perceptions. In particular, the Area B task force/worker group, Area A worker group, and the advisory committee had consistent positive correlations between attendance and the clarity of meeting goals (.312, .371, .352 respectively). Significant correlations (.05 level) were found for the Area A task force, the advisory committee and the Area B task force/worker group. For the Area A task force, significant correlations were found between attendance and "goal setting for the next meeting" (.663) and "importance of the decision" (.595). The advisory committee had a significant correlation for "satisfaction with the program" and attendance (.667). Finally, a significant correlation was found between "knowledge of the topics discussed" and attendance for the Area B task force/worker group (.531).

Table 1 Correlation between Percent Meeting Attendance and Meeting Assessment Scores by Group (Meeting scores based on average of all participants in meeting).

		Area A		Area B
	Advisory Committee (n=20)+	Task Force (n=13)+	Worker Group (n=14)+	Task Force/ Worker Group (n=14)+
Clarity of Meeting Goals (Goals1)	.352	.122	.371	.312
Clarity of Goals for Next Meet. (Goals2)	.040	.663**	.080	.230
Individual Participation (Part1)	.012	-.396	.156	.325
Percept. of Groups Partic. (Part2)	-.024	-.229	-.112	-.049
Know. Relev. to Discus. (Know)	-.249	-.396	.284	.531**
Importance of Decision (Decis)	.026	.595**	.248	.097
Diagnosis of Ergo. Prob. (Ergo.)	.038	.000	.214	.240
Influence over Decisions (Influ.)	-.170	-.123	-.199	.204
Leadership Quality (Lead.)	.289	-.202	.107	.109
Satisfact. with Meeting (Sat.1)	-.074	.236	.166	.156
Satisfact. with Program (Sat.2)	.667***	.026	-.155	.120
Mean Score (Mean)	.094	.065	.174	.220

** significant correlation at .05 level
*** significant correlation at .01 level

+Number of Meetings for each Group

It is interesting to note that attendance and goals (goals of this and the next meeting) were consistently related for each study group. Since the program stressed the participative approach, the groups may have been less able to set clear and meaningful goals when a large percentage of the group were absent from the meeting.

The perceived "quality of decisions" was related to attendance for only the Area A task force. This might be a function of this group's role within the ergonomic program. Their main functions included guiding the Area A worker group to problems areas, deciding which alternative solutions were acceptable and giving support in implementing changes to the floor. Completing these tasks required constant decision making and coordination of several diverse specialities within the task force. Since the task force membership included only those specialities necessary in making informed decisions on the functions outlined above, absence of any member would seriously hamper their ability to make good decisions.

Finally, participant's "knowledge relevent to the discussion" was significantly correlated to attendance for only the Area B task force/worker group. It is speculated that this may be a function of the manufacturing process the group worked on. In contrast to area A (who worked on an existing manufacturing process), this group worked exclusively on a new process during the implementation (launch) phase. Successful launch of the line depends, among other things, on the coordination of many resources in different engineering specialities and in the personnel responsible for developing and installing the equipment (e.g. job setters, skilled trades, vendors). The launch superintendent, who is responsible for coordinating these resources, was a member of the group and was familiar with all the information pertaining to the launch. In addition, because successful implementation of ergonomic efforts was closely related to the launch, membership on the spindle group included representatives from each resource in the launch. Thus, when members were absent from a meeting, the subsequent loss of information was crucial and adversely affected the performance of the group.

Relationship of meeting activities to participant perceptions.
Five generic activities were identified: *program maintenance* (action plan development and group maintenance activities), *problem identification* (activities involved in determining ergonomic stress on jobs), *workplace redesign* (redesign activities to reduce or eliminate ergonomic stress), *implementation* (activities involved in implementing or installing the workplace redesign solutions) and *non-ergonomic* (activities that do not have to do with the program's content or maintenance).

Table 2 presents correlates between the percent of the meeting devoted to each meeting activity and the mean meeting assessment score (averaged across attendees and assessment dimensions). There is a significant and strong positive relationship between implementation activities and mean scores for the Area A worker group and task force groups and a strong, but not significant relationship, for the Area B task force/worker group. That is, participants scored meetings more positively when significant portions of the meeting focused on implementing the workplace designs. On the other hand, the advisory committee was more negative in their evaluation when they worked on implementation activities, indicating implementation or related activities were not beneficial to the group. Correlations between program maintenance activities and assessments were small but negative for the task force and postive for the advisory committee. Finally, correlations with problem identification were small and negative for the advisory committee and task force and significant and negative for the Area A worker group.

Table 2 Correlation between Individual Activities and Mean Meeting Score (Meeting scores average of all participants in meeting) by Group.

		Area A		Area B
	Advisory Committee (n=20)+	Task Force (n=13)+	Worker Group (n=14)+	Task Force/ Worker Group (n=14)+
Program Maintenance	.142	-.285	-.054	.060
Problem Identification	-.066	-.186	-.600**	.229
Workplace Redesign	-.152	.185	-.120	.147
Implementation	-.421	.477*	.676***	.308
Non-Ergonomic Activities	.000	-.126	.250	.596**

* significant correlation at .1 level
** significant correlation at .05 level
*** significant correlation at .01 level

+Number of Meetings for each group

The significant negative correlation between problem identification and assessments for the Area A worker group was partly due to their using a formal job analysis system. A central activity for the program, and especially for the carrier and spindle groups is problem identification. To facilitate the process, the carrier group responded by developing and using an ergonomic job analysis checklist. This formal system may have made meetings less dynamic and even frustrating for participants.

Clearly, implementation was something that the operational groups enjoyed. Implementation discussions tended to be more informal and dynamic and provided a feeling that things were getting done. The only exception to this trend was the advisory committee. They, in fact, attempted to design and implement workstation changes early in the program, before they clearly defined their role. The result was disastrous--of the 14 ideas for redesign, none were implemented. This experience prompted the advisory group to carefully define the mission of the program and their role in helping achieve the mission. From that point on, they did not get involved in the nitty-gritty of workplace design and implementation and their meetings were much more effective.

The relationship of job status to participant perceptions.
Figure 1 presents participants' perceptions by job status. Regardless of ergonomic group in the program, middle managers had significantly lower scores than other levels in the organization. This supports Kantor's (1983) depiction of middle management as trapped between management expectations and pressures of production.

Figure 1 Meeting Assessments by Job Status+ (Averaged across all groups; N=40; Blue Collar=12, Middle Managers= 23, Managers=5)

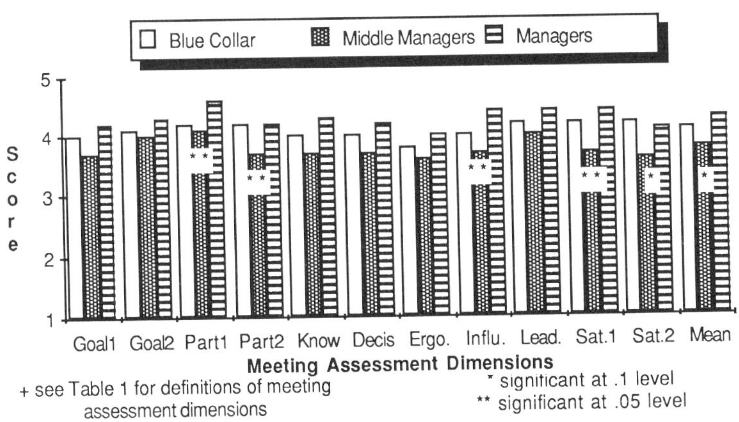

CONCLUSIONS

The use of cross-functional task forces for ergonomic design and implementation can reduce many of the communication barriers that prevent persons with relevant knowledge and interests from working together. However, such programs, if they are to be voluntary, depend on the goodwill and commitment of participants. The data presented here provide insights into what it takes to maintain high levels of energy and involvement.

First, the data supports the view that maintaining participation is in fact an important issue. When some members fail to show up to meetings, this may adversely affect the progress of the meeting. The Area A worker group devoted a considerable portion of meetings with low attendance to discussing this problem and working on ways to increase future attendance. As a result of these proactive group maintenance efforts, the overall effectiveness of the group appears to have surpassed their counterparts (Joseph, 1986).

Second, some activities are inherently more satisfying to group members. The operational groups responsible for design and implementation found that planning and discussing implementation was the most satisfying activity. Other activities (e.g. group maintenance and problem identification) did not provide a sense of accomplishment and satisfaction. The implication is that special attention must be devoted to making inherently less enjoyable activities more satisfying. Perhaps taking time to review progress at the end of each meeting will provide feedback to participants on those accomplishments that don't have tangible outcomes. Also, it is important to come to closure on projects to ensure feelings of accomplishment and success. Wherther (1982) and others, writing on quality circles, suggest that it is particularly important in the start-up phase to choose easy projects to complete and implement them as quickly as possible. This reinforces participant commitment and motivation. Note that implementation was not satisfying for the advisory committee. This group was not set up to implement and did not have the personnel or intimate shop floor knowledge necessary for this task. The steering committee has the most ambiguous role in the program and it is important that they clarify their role and duties early in the program.

Finally, middle management were least satisfied with the progress of the program. As Kantor (1983) observed, participation is what the top tells the middle to do to the bottom. Thus, middle management is caught between management's hopes and desires for program success and the short-term responsibilities for getting product out the door. We can simply reiterate what others have said--special care must be taken to bring middle management into the process and make their participation a positive experience.

Literature Cited:

Hackman, R.J. and G.R. Oldham, *Work Redesign.* (Addison-Wesley, 1980).

Janis, I.L., Group Think, *Psychology today.* No. 6 (1971) 43-44,46,74-76.

Joseph. B.S., *A Participative Ergonomic Control Program in a U.S. Automotive Plant: Evaluation and Implications.* Ph.D. dissertation, Department of Industrial Engineering- Center for Ergonomics, Ann Arbor: The University of Michigan (1986).

Kantor, R. M., *The Change Masters-Dilemmas of Managing Participation.* (Simon and Schuster,New York, 1983).

Liker, J.K., Joseph B.S. and Armstrong T.J., "From Ergonomic Theory to Practice: Organizational Factors Affecting the Utilization of Ergonomic Knowledge." in H.W. Hendrick and O. Brown, Jr. (eds.) *Human Factors in Organizational Design and Management.* (North-Holland,Amsterdam, 1984).

Maier, N.R.F., "Assets and Liabilities in Group Problem Solving: The Need for an Integrative Function," *The Psychological Review.* No. 4 (1967) 239-249.

Werther, W.B., Quality Circles: Key Executive Issues, *Journal of Contempory Buisness.* Vol.11 No.2 (1982).

Human Factors in Organizational Design and Management - II
O. Brown, Jr. and H.W. Hendrick (Editors)

EVALUATING AN ORGANIZATION'S POTENTIAL FOR ADOPTING COMPUTER INTEGRATED MANUFACTURING

Daniel Levi
Charles Slem
California Poltechnic State University
San Luis Obispo

Andrew Young
Northern Telecom

This case study develops a set of criteria for evaluating whether Computer Integrated Manufacturing (CIM) is an appropriate innovation for an organization. These criteria are used to analyze a telecommunications manufacturing facility's potential for adopting CIM.

The introduction of Computer Integrated Manufacturing (CIM) has had a mixed success. Some companies who have adopted CIM view it as an innovation which has markedly improved their performance; while other companies have invested large sums of money with little useful return. How can one predict whether CIM will be a successful innovation for an organization?

There are two ways to try to answer this question. First, one could examine the characteristics of innovative organizations and analyze whether a particular company had these characteristics. Second, one could examine the unique characteristics of CIM and analyze whether they matched a company's needs and capabilities. This paper outlines an attempt to do this for a large telecommunications manufacturing facility.

CHARACTERISTICS OF INNOVATIVE ORGANIZATIONS

There are a variety of characteristics which describe the differences between innovative and non-innovative organizations. They can be used to help analyze whether an organization has a culture which promotes innovation and to identify potential problem areas within an organization.

Culture

Innovative organizations tend to have articulated superordinate goals which are widely understood by the members of the organization [6]. They have a culture of pride, which includes both pride in the quality of the product and a value commitment to the organization [3]. The cultural themes emphasize that change is a major part of doing business.

Innovative organizations also emphasize their commitment to their employees.

Management

Innovations occur more often in organizations with a participative or collaborative style of management which emphasize coordination. Management actively promotes innovation by using broad job assignments which include change as part of the job's goals. Managers support and empower others as a way to promote innovation. This empowerment process includes backing champions of innovations, allocating resources for change, and making public commitments in support of innovation and change [3].

Organizational Structure

Innovative organizations have loose, simple structures with tight controls. They tend to have lean staffs, decentralized decision-making, and multiple power centers [4,5]. Although matrix organizations tend to be more innovative, the key issue is that innovative organizations know how to form and use temporary systems (such as task forces or project teams) to deal with problems and innovations.

Human Resources

Innovative organizations tend to be people oriented [3]. They place a lot of emphasis on training and personnel development for both technical and interpersonal skills. The threat of change is offset by a commitment to training and job security. Job rotation across functional areas is also encouraged to promote internal collaboration.

Success

Innovative organizations are usually successful before they innovate. Innovation is more likely to be successful in organizations which have slack resources and who are not undergoing crisis [5]. Successful organizations are more stable, have a longer time perspective for strategic planning, and a bias toward action [4]. These characteristics are also associated with innovativeness.

CIM - THE INNOVATING ORGANIZATION

Although CIM is a type of organizational innovation, it has characteristics which make it different from other types of innovations. While innovations are usually classified as either product, production process, organizational structure, or people, CIM is an innovation in which all of these areas change at the same time. Because of this, CIM is often referred to as a theory or philosophy rather than an innovation. It transforms an organization into an organization which is structured to innovate [2].

The magnitude of the amount of organizational change required by CIM makes it an especially difficult process. It does not fit into many of the predictors of easy innovation diffusion such as communicability or minimal complexity [7]. CIM requires a long term (and possibly permanent) change process which must include participants from all areas of the organization. The implementation stage will be a continuous redesigning task as both products and production processes evolve.

When viewing CIM, it is relatively easy to understand how large the changes will be in product development and production processes. The process of creating CIM will affect the organizational structure by forcing the organization to rely more heavily on task forces and other integrative mechanisms. The effect of increased reliance on these temporary systems will precipitate changes in the organization's structure.

What is least apparent about CIM may be the most important feature of successful adoption - the development and change in the people. Most of the companies which have successfully adopted CIM have a strong reputation for support of their employees [1]. This commitment to people reduces the threat of CIM on the employees, and helps the organization to see the necessity of investing sufficient time and resources into developing their human resources.

METHOD

Various methods were used to analyze this manufacturing facility in order to examine their potential for innovation and compatibility with CIM. These methods included a reanalysis of a coporate communications survey, a stress survey given to a sample of managers, and in-depth interviews with thirty directors, managers, and professionals in the company. After the initial analysis was completed, we re-interviewed a sample of the managers and directors to check, clarify, and expand our analysis.

DOES THIS ORGANIZATION HAVE THE CHARACTERISTICS OF AN INNOVATIVE ORGANIZATION ?

Culture

This organization has a culture which promotes innovation. It has a history of constantly changing its products and production processes. The employees recognize that change is necessary to stay competitive and they are proud to be on the "leading edge" of technological innovations. They view the rapid pace of change in the telecommunications industry primarily as a challenge rather than a threat.

Their culture also shows a great deal of commitment to the organization and its products. There is pride in the quality of the product. People are challenged by tight time schedules and the complexity of designing and producing new products. Working there is viewed as a challenge and people have a healthy problem-solving attitude toward those challenges.

Management

They have a results oriented management which allows people to solve their own problems while retaining tight controls. People recognize the need to justify their actions by demonstrating a favorable benefit/cost analysis. However, there is also a sense that management will be responsive to new ideas and solutions if it meets this criteria.

Management seems good at delegating authority and allowing employees to solve their own problems. This problem solving approach to management has forced people to communicate and coordinate with others. Although there is conflict in coordinating problems across functional areas, this is a healthy sign of active problem solving.

Organizational Structure

The overall organizational structure has many of the characteristics of successful, innovative companies. It is a lean staff with few levels of management and multiple power centers. There does not appear to be an overly bureaucratic structure or an excessive emphasis on rules or procedures.

Although this is not a matrix organization, the organization has a history of adapting its structure to meet situational demands. The establishment of Cabinet level task forces for dealing with CIM and other major innovations are good examples of their ability to structurally adapt to promote innovation.

Human Resources

The organization is making a concerted effort to improve the internal perception of its commitment to human resources. Job security is not guaranteed to employees. Although there is an internal program of job training and personnel development, many employees recognize the need for more training. Some employees are skeptical about the organization's willingness to internally develop its human resources and to promote from within the organization.

However, employee relations are in fairly good shape. There is widespread trust and respect for the company's Human Resources Department. Many employees view the organization as increasing its commitment to people which is reflected in the published goals of the organization. In addition, employees do change jobs across functional areas which helps to promote internal collaboration.

Success

This is a successful organization which has slack resources available to it and is not undergoing crisis. The company provides stability which allows it to plan and invest in the future. There is an action orientation which recognizes the necessity of change.

Conclusion

Overall, this organization provides a good environment for innovation. The culture evidences a lot of pride in the organization and values innovativeness. Management promotes independent problem solving by employees while retaining controls on the results. The organizational structure provides stability and security, while at the same time being open to the use of temporary structures to promote change. The organization is increasing its commitment to the development of its human resources. This is a successful company which has the resources and capability to provide a supportive environment for innovation.

IS CIM A GOOD INNOVATION FOR THIS ORGANIZATION ?

Positive Aspects

The promise of CIM - to integrate all of the aspects of the manufacturing process and speed the development of new product innovations - is well suited to the central perceived performance gap of the organization. The managers view producing new innovations under tight time schedules as one of their central problems. They recognize that coordination problems are one of the major obstacles to rapid product innovativeness. "How to develop a system which allows us to stay first in new product development?" is a question which managers view as crucial to their organization's survival. CIM is the appropriate type of innovation for them because it does target this central perceived performance gap.

Their culture is also well matched for CIM. The central cultural themes focus on the need for change and innovation, and pride in their products. CIM is an innovation which is well suited to these cultural values and, therefore, should be viewed as an appropriate innovation by the members of the organization.

Negative Aspects

Unfortunately, CIM has many characteristics which make innovation diffusion difficult. It is a complex innovation which is difficult to communicate. The costs of the innovation are high, and the economic benefits are difficult to calculate and occur in the future. Because the benefits of CIM are hard to quantify and observe, justifying the continuation of it will be difficult. The adoption process is being managed by a change agent, partially supported by corporate funding, which may foster the belief that it is being imposed on the organization from outside. Shifting to CIM will require an increased commitment to training and personnel development. Finally, CIM is a new innovation so there are not many successful examples to observe.

Conclusion

The concept of CIM is well suited to the needs and culture of the organization. It is the right kind of idea to satisfy their perceived performance gaps. However, many of the details of CIM show that it will be difficult to adopt and implement. This is an organziation oriented toward challenges. CIM will be a difficult challenge, but one that is appropriate.

CONCLUSION

Our evaluation of this organization showed that they were a good site for innovations and the adoption of CIM. At present, they have completed the first phase of a multi-year transition to CIM. This first phase has included the development and implementation of a number of computerized systems and automated processes. Their initial success in developing and adopting some of the components of CIM helps to confirm our analysis of their potential for successfully implementing Computer Integrated Manufacturing.

REFERENCES

[1] Filley,R.D., Roundtable participants talk about CIM myths and realities, people aspects and the future, Industrial Engineering. 17 (1985) 35-51.

[2] Harrington,J., Computer Integrated Manufacturing (Industrial Press, New York, 1973).

[3] Kanter,R.M., The Change Masters (Simon and Schuster, New York, 1983).

[4] Peters,T. and Waterman,R., In Search of Excellence (Harper and Row, New York, 1982).

[5] Rogers,E.M. and Rogers,R.A., Communication in Organizations (Free Press, New York, 1976).

[6] Snyder,R., To improve innovation, manage corporate culture, in Bennis,W.,Benne,K., and Chin,R. (eds.), The Planning of Change (Holt, Rinehart, and Winston, New York, 1985).

[7] Zaltman,G.,Duncan,R., and Holbeck,J., Innovation and Organizations (J. Wiley and Sons, New York, 1973).

Human Factors in Organizational Design and Management - II
O. Brown, Jr. and H.W. Hendrick (Editors)
Elsevier Science Publishers B.V. (North-Holland), 1986

DIFFUSION OF SOCIOTECHNICAL DESIGN FOLLOWING INTERVENTION: A CASE STUDY OF ORGANIZATIONAL LEARNING

Mark F. Levine

Department of Management
California State University
Chico, CA 95929-0005

Bela Feher

Organizational Systems Department
Navy Personnel R & D Center
San Diego, CA 92152-6800

This action research project used sociotechnical systems analysis and design to identify and implement changes in organization structure and functioning to complement a prior shop-level change to semiautonomous work teams in a shore-based Navy repair system. During the researchers' temporary disengagement from the client system, members implemented further changes applying sociotechnical concepts. These changes are examined in terms of organizational learning concepts.

INTRODUCTION

Action research as a framework for planned organizational change sets forth two objectives: to improve organizational performance and to contribute to scientific knowledge. The sociotechnical systems (STS) approach to organization design was developed in the action research framework, yielding a worthwhile interplay of practice and theory in such domains as work system design and the process of change (Clark, 1972). As one aspect of the change process, STS strongly advocates participation on the part of organization members who would be affected by subsequent redesign. The research evidence supports this position in that participation by organization members in planned change has been found to reduce resistance, enhance commitment, and decrease time required to achieve the higher performance potential of the new system design (Davis & Cherns, 1975a,b). In addition to performance benefits of participation, organizations can theoretically acquire the conceptual and experiential bases necessary to take ownership of the change efforts and carry on the process independent of external expertise. Member participation in organization change thereby provides excellent opportunities to extend scientific knowledge in the area of organizational learning. Unfortunately, these opportunities have not frequently been pursued.

In their recent review of the literature on organizational learning, Fiol and Lyles (1985) identified two difficult areas that must be addressed to overcome this deficiency: definition and measurement. With respect to definition they cite a variety of inconsistent perspectives in the literature. They define organization learning as "the development of insights, knowledge, and associations between past actions, the effectiveness of those actions, and future actions" (p. 811). They present the construct of organizational learning as having two components: (1) content, either cognitive or behavioral (or both); and (2) level of learning in the cognitive domain, from fine-tuning within an existing organizational paradigm (low-level learning) to restructuring of the paradigm itself (high-level learning).

Thus, in conjunction with organizational change, a range of potential learning can occur on the part of the organization. Changes in organizational structure and functioning (behavior)

The views expressed in this paper are those of the authors and not necessarily the Navy Department.

may or may not be accompanied by an understanding of the underlying concepts (cognitive) shared among members. The level of cognitive learning may also vary from refinements within an established interpretation system (Daft & Weick, 1984) directed toward immediate outcomes (low-level learning) to creation of new organizational paradigms or interpretive schemas that result in long-term effects on organizational structure and functioning (high-level learning). Thus, low-level cognitive development involves the integration of new events into existing organizational systems; whereas, high-level cognitive development is demonstrated by organization members using prior change episodes as initial building blocks of a new organizational paradigm. Associated with high levels of learning is the necessity of unlearning (Hedberg, 1981). In order to realize new norms, values, behaviors, and system functioning associated with high-level learning, old interpretive frameworks and practices must be displaced.

Each component of organizational learning has associated measurement problems. Ideally, cognitive development over time could be measured either by interviews or by using paper and pencil assessments. However, shared interpretation systems do not necessarily find expression in consistent organizational actions. Behavioral development can be demonstrated by actions taken within the context of planned organizational change efforts. Yet changes in organizational action do not assure that the actions were a result of shared, well-understood cognitive schemas, rather than sheer intuition.

Ideally, a measure of both cognitive and behavioral development and analysis of the relationship between the two would provide insight into the complex phenomena of organizational learning. The present case study is an initial attempt to examine organizational learning associated with an STS design program.

THE CASE

Organizational Setting

The case is set in the Navy Shore Intermediate Maintenance Activity (SIMA) in San Diego, California. This SIMA is part of a network of afloat and shore installations that provide maintenance support to the fleet. All are manned and managed by military personnel. This SIMA is the largest, with 2000 personnel in 57 repair shops organized into 5 branches.

Initial Shop-level Change at SIMA San Diego

The traditionally organized Pump Shop at SIMA, designed on the principle of job fractionation, had experienced low productivity, high rework, and difficulty in meeting production schedules. The shop was organized as two large work sections of 15-20 persons. One section was responsible for removal of pumps from the ships, disassembly, ordering parts, and farming out components to assist work centers. The other section reassembled the pumps, tested them, and reinstalled them aboard the customer ships.

SIMA command charged a new Machinery Branch Officer with the task of improving shop performance. He accomplished this by reorganizing the shop around work teams that were given "cradle to grave" responsibility for the repair of assigned pumps. Productivity, adjusted for changes in manning, increased 40% (Levine & Feher, 1985). In light of this apparent success, SIMA command requested that the Navy Personnel Research and Development Center (NAVPERSRANDCEN) evaluate the team concept used in the Pump Shop and recommend ways to improve and institutionalize it.

Collaborative Sociotechnical System Design

The project was a collaborative effort between the researchers from NAVPERSRANDCEN and the management and staff of SIMA to diagnose present organizational functioning, identify problem areas, generate alternative solutions, critically evaluate and recommend preferred solutions, and facilitate implementation of approved changes. These activities were carried out by a task force whose membership included the authors and a diagonal slice of organization members who had relevant knowledge of and responsibility for pump repair.

Structured and unstructured interviews, observation, and analysis of historical records were used by the task force during the first phase of the study to gain an understanding of the pump repair process. This initial analysis ascertained that the work team design was well suited to the environmental conditions, task demands, and personnel training requirements, as demonstrated by improved output and reduced rework. However, continuing problems in the pump repair process were attributed to a limited perspective in previous changes, i.e., only the core operations in pump repair were reorganized, leaving ancillary support and assist functions largely untouched, requiring diversion of productive resources from the pump teams (Feher & Levine, 1984). Further, analysis of the pump repair system included a scan of the pump shop environment, an examination of the internal operations of pump teams within the shop, and assessment of the work relations between the shop and its primary assist work centers and staff support functions. Redesign based on these analyses produced a series of recommendations for changes to overcome these deficiencies (Feher & Levine, 1985a,b). The commanding officer approved the actions recommended and sanctioned implementation of the changes.

Changes in the Pump Shop's Relationship with the Electrical Shop

The integration of electricians from the Electrical Shop into the Pump Shop was the first change implemented in the pump repair system. This gave Pump Shop management control over the resources necessary to repair close coupled pumps (60% of shop workload). Initially, the task force called for the formation of teams in the Electrical Shop to reduce the fractionation of work on all electrical motors. They further recommended that specific teams of electricians be aligned with pump teams to integrate the work on close coupled pumps. After the commanding officer's acceptance of the recommendation, the task force requested that Electrical Branch management review the recommended change and either implement it or propose an alternative to accomplish the same end.

The response of Electrical Branch management was to develop an alternative solution that was approved and implemented over the next month. Fourteen electricians and one supervisor were reassigned to the Machinery Branch to form a motor repair team within the Pump Shop to handle all close coupled pump motors.

Changes in the Pump Shop's Relationship with the Machine Shop

Three problems were identified in analysis of the Machine Shop's role in the pump repair system: (1) lack of accountability for timely completion of all parts of a job because each part was individually scheduled through the shop; (2) time delays in moving parts between successive sections within the Machine Shop (e.g., lathe, grind); and (3) lack of any mechanism for tracking all components of a job. To address these problems the task force decided to explore the possibility of reorganizing the Machine Shop into two product-oriented clusters (pumps and motors; other components), each capable of performing all machining operations for the products it processed.

A collaborative analysis was made of shop inputs (present loading of each Machine Shop section by product source) to determine the feasibility of such a reorganization. It was concluded that the reorganization was possible, but the task force was divided over the new demands it would place upon the Machine Shop section foremen. The shopmaster, representing the section foremen, expressed strong reservations because the reorganization would require supervision across the entire range of machining operations as opposed to traditional specialization by machine type.

The pros and cons of the proposed reorganization were debated during a formal task force meeting. After two hours a consensus was reached to implement on a trial basis an alternative proposal presented by the shopmaster. The work acceptance function in the Machine Shop was changed so that the two shop planners specialized in products originating from certain shops and became responsible for scheduling and tracking the work on all parts related to each job within their domain (e.g., pumps and motors). This proposal addresses the three problems identified above, but shifts the responsibility and accountability for improvement from section foremen to the planners. So far the change appears to be successful; however, continuing increase in demands on the pump repair system might require reexamination of the concept of reorganizing the Machine Shop into product clusters.

Changes in the Pump Shop's Relationship with SIMA Support Functions

Implementation of changes in the planning of pump repair was more complex due to the involvement of a greater number of people in the task force. Evaluation of the relationship between the Planning Department and pump repair revealed two important concerns. First, considerable shop production resources were being spent in duplicating Planning Department efforts pertaining to ship checks and the procurement of technical documentation. Second, in cases involving revisions and emergent work, delays of three or four days were common due to centralization of the planning function.

Recognition of these concerns led the task force to recommend that planning after a ship's arrival should be moved to the shop. The civilian head of the Planning Department, a task force member, expressed strong reservations about decentralizing any aspect of the planning function. It is possible that such reservations arose partially due to his civil service status which relates scope of responsibility and number of subordinates to pay grade and promotion. After debate during a formal task force meeting, a consensus was reached to implement a three-month trial period for shop planners following one month of training. Shop planners became a permanent feature of the pump repair system following the trial period.

Reallocation of Space

Once the task force reached a consensus regarding the changes within the shop, between the shop and its assist work centers, and between the shop and SIMA support functions, it became apparent that additional space within the building was needed. To accomplish this, two shops unrelated to pumps had to be moved out of the building. Fortunately, a new building was just being completed in which potentially desirable space was available. Unfortunately, that space was previously designated to other SIMA functions. The task force ultimately recommended that some of the new space be redesignated to those shops that needed to be moved. The task force argued that pump repair is one of the most critical to SIMA customers and, therefore, should take priority over other intended use of the space. The commanding officer concurred and space in the building was reallocated.

Continuation of the Change Process by Organization Members

When NAVPERSRANDCEN had fulfilled its charge, it disengaged from the client system. Over the subsequent seven months, normal personnel turnover in SIMA included the Machinery Branch Officer over the Pump Shop and the Repair Officer responsible for scheduling and coordinating SIMA-wide production and customer liaison. During the interim, NAVPERSRANDCEN was charged with applying the STS analysis and design method to the entire regional system of intermediate maintenance. This led to reestablishment of relationships with SIMA, where it was discovered that system redesign activities had continued.

Noting the similarity of operational requirements between the Valve and Pump Shops, the new Machinery Branch Officer had transferred the concept of team-based organization to valve repair, including shop-level planning and scheduling and tracking of work in the Machine Shop. The Repair Officer went further than simply transferring ideas. He introduced fundamental changes, affecting organization-wide functioning, that pertained to material expediting and customer liaison. In each instance, teams were formed and assigned responsibility for a group of ships. Teams provided one point of contact for all related system functioning and improved depth of expertise in performance of these functions.

DISCUSSION AND CONCLUSIONS

The case demonstrates that participation of organization members in a sociotechnical redesign project can result in both behavioral and cognitive organizational learning. The initial change to teams in the Pump Shop was a fundamental change in organization and operation based on individual intuition. Top management's reaction to the success of the initial change to teams in the Pump Shop, implemented without an organization-wide paradigm shift, was to question the teams' long-term viability given planned high levels of personnel turnover (60% yearly). Ultimately, the success of the changes in the Pump Shop challenged the old organizational paradigm and began the process of cognitive development, involving both unlearning and replacement.

The follow-on sociotechnical redesign intervention brought about complementary changes throughout the pump repair system to capitalize on the advantages of teams (cross-training, motivation, accountability) and give them long-term viability. It also produced an important insight, helping form a new organizational paradigm: that changes at the organizational level were necessary in both assist work centers (e.g., Electrical, Machine) and in a critical staff support function (i.e., Planning) to complement the change to teams at the shop level.

When the NAVPERSRANDCEN research team disengaged from the client, there was opportunity for continued cognitive development. This occurred and involved players who were not active participants in earlier changes. Apparently STS design concepts and their potential for application diffused in the organization. New expectations regarding productivity of pump repairs easily transferred to the adjacent valve repair group through shared management. Management desire for greater depth of personnel expertise readied it to give up previous preferences for worker specialization. As a result, the new Machinery Branch Officer was able to transfer the team concept to valve repair, demonstrating low-level cognitive development.

Efforts by the new SIMA Repair Officer to create teams in related subsystems such as material expediting and customer liaison reflect a shift from the traditional bureaucratic paradigm to the STS model. These changes in perceptions and action went beyond duplication. The changes in customer liaison arose from customer demands that highlighted inadequacies of the traditional approach to customer liaison that relied on individual expertise and motivation. Teams were recognized to offer the advantages of broader expertise and geographic flexibility. Similar dynamics produced the change in material expediting. These changes are evidence of a new cognitive framework for interpreting complex system interrelationships (i.e., high-level cognitive learning).

This analysis of organizational learning highlights a number of theoretical and methodological issues. Although it is theoretically elegant, the distinction of behavioral and cognitive organizational learning is difficult to maintain in practice, requiring judgment of the extent to which action was the result of previous knowledge or of an insightful or serendipitous convergence of events. With regard to the level of learning, it is difficult to discern when a paradigm shift, defining high-level cognitive learning, occurs. Conceptually, a new paradigm must be shared throughout the organization, but total consensus must be recognized as unrealistic in terms of either necessity or feasibility. Both newness and pervasiveness of the shift are difficult to assess. Furthermore, consensus must be developed regarding delineation between organizational and individual learning. This may be particularly important in nonmilitary settings where turnover of key personnel is much lower. From another perspective, lack of mobility of key personnel may hinder the organizational unlearning required to realize high-level cognitive development. Clarification of these issues would enable more precise and thorough measurement of the construct.

With respect to methodological issues, more rigorous research into organizational learning will first require clarification of the operational definition of the construct. A methodology for monitoring cognitive learning over time by multiple participants in the client system also must be developed. However, need for such dynamic assessment raises the spector of sensitization and obtrusiveness. An alternative approach might avoid the implied aggregation issues by using true organizational indicators of learning, such as SOPs, organization myths, or responses to ambiguous informational stimuli. In the spirit of action research, it is hoped that this study provides both insight into the difficulties and encouragement toward addressing the conceptual and measurement issues.

REFERENCES

[1] Clark, P.A. (1972). *Organizational design: Theory and practice.* London: Tavistock Publications.

[2] Daft, R.L., and Weick, K.E. (1984). Toward a model of organizations as interpretation systems. *Academy of Management Review*, 9(2), 284-295.

[3] Davis, L.E., and Cherns, A. (Eds.) (1975a). *Quality of working life: Problems, prospects, and state of the art.* New York: Free Press.

[4] Davis, L.E., and Cherns, A. (Eds.) (1975b). *Quality of working life: Cases and commentary.* New York: Free Press.

[5] Feher, B., and Levine, M.F. (1984). *Enhancing structure and management of Navy intermediate maintenance activities: Phase I. Analysis of the pump repair process* (HFOSL Technical Note 72-84-4). San Diego: Navy Personnel R & D Center.

[6] Feher, B., and Levine, M.F. (1985a). *Enhancing structure and management of Navy intermediate maintenance activities: Phase II. Implementation of changes in the pump repair system* (Technical Report 85-33). San Diego: Navy Personnel R & D Center.

[7] Feher, B., and Levine, M.F. (1985b). *Organization redesign for productivity improvement: Method and case study* (Technical Report, 86-2). San Diego: Navy Personnel R & D Center.

[8] Fiol, C.M., and Lyles, M.A. (1985). Organizational learning. *Academy of Management Review*, 10(4), 803-813.

[9] Hedberg, B. (1981) How organizations learn and unlearn. In P. Nystrom & W. Starbuck (Eds.), *Handbook of organizational design*, pp. 1-27. New York: Oxford University Press.

[10] Levine, M.F., and Feher, B. (1985). Evaluation of technostructural change in a Navy repair organization. *Proceedings, American Institute of Decision Sciences, Western Region*, 14, 161-163.

Human Factors in Organizational Design and Management - II
O. Brown, Jr. and H.W. Hendrick (Editors)
Elsevier Science Publishers B.V. (North-Holland), 1986

CONTRIBUTION OF A MULTI-METHOD APPROACH TO UNDERSTANDING IMPLEMENTATION

John P. Sheposh, Ph.D.
Joyce Shettel-Neuber, Ph.D.

Navy Personnel Research and Development Center
San Diego, California

This paper describes an approach which blends grounded theory with an existing management theory to better understand the status of a newly implemented quality improvement program. Information used for the development of the grounded theory was obtained through a case study technique and artisans' perceptions of work impediments. The theoretical framework employed was Likert's Profile of Organizational Characteristics (1967). While each technique provided useful information, in combination these three sources of information made a more powerful and more reliable diagnostic tool.

Many private and public sector organizations in the U.S. are currently initiating organization-wide quality improvement efforts. Due to the large-scale investment of time, effort, and money required to implement such a program, it would be useful to identify, measure, and track those organizational characteristics which are likely to promote or inhibit the implementation of an organization-wide quality control effort.

With respect to the measurement and identification of organizational features important to implementation efforts in general, there has been a growing emphasis on the use of grounded theory (Dunn & Swierczek, 1977). Rather than applying on an a priori basis an existing theory to a specific implementation or change effort, a grounded theory approach proposes that a model or theory be developed directly from the observations of events and processes embedded in that organization. On the other side of the coin, there recently has been expressed a concern over the atheoretical, open-ended nature of evaluating implementation of organizational change and the need for the use of a theory-driven approach (the use of existing theoretical knowledge) in the service of change evaluation (Chen and Rossi, 1983). This paper reports on the application of both of these seemingly contradictory approaches in assessing the status of a newly implemented quality program.

APPROACH

The implementation effort described in this paper is presently being carried out at a Naval Air Rework Facility (NARF). One of the production divisions (Components and Metal) and its related support divisions at this facility have been the major focus of the implementation effort to this point in time. The basic components of the implementation effort, which are characteristic of a Total Quality Control (TQC) orientation, include: (a) the application of statistical tools and techniques to improve everyday work processes, (b) extensive training and retraining in these methods, (c) a greater degree of participative management, quality activities carried out in all functional disciplines, and (d) an organizational structure that consists of quality boards at all levels of the organization (Ackoff, 1981; see Figure 1). Concerning this last point, the boards established here were designed to provide a mechanism to facilitate both vertical and horizontal communication concerning work processes and quality activities necessary to the implementation of TQC. This system of interlinking boards was adopted to insure that the quality improvement effort would be focused and yet extend across all departments within the Components and Metal building.

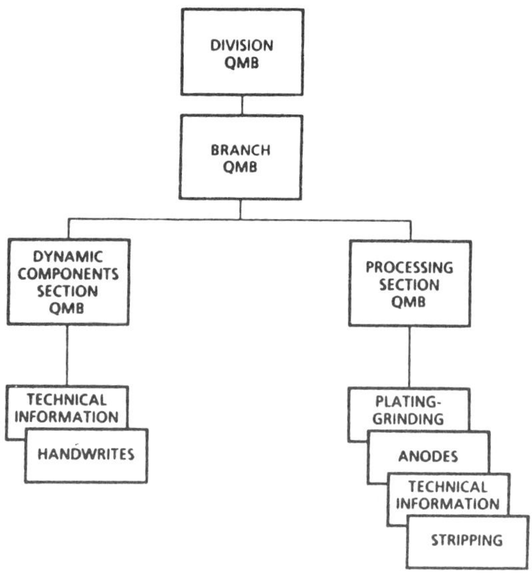

Figure 1
Components and Metal Building
Implementation Structure

A TQC implementation effort, such as that being undertaken by the NARF, has as its aim the cultivation of certain organizational characteristics (Ishikawa, 1985). These include (a) quality as the primary aim, (b) a consumer orientation, (c) the removal of sectionalism, (d) the use of statistical methods, (e) the use of participatory management, and (f) cross-functional management. It is evident that attaining these TQC characteristics may require a fundamental organizational change in the direction of more participatory management (i.e. individuals are asked their opinions and allowed to contribute at all stages of the process). The organizational change anticipated with the implementation of TQC is compatible with some existing organizational theory. One theory that is particularly relevant is Likert's (1967) concept of organizations. The organizational characteristics recommended for an environment conducive to TQC are very similar to the ones characterized in Likert's most highly evolved management system, as measured by his Profile of Organizational Characteristics (1967). The Profile of Organizational Characteristics (POC), therefore, could be a useful instrument to use in the assessment of a TQC implementation. We reasoned that an organization that has difficulties in implementing this type of program would be characterized by a lower level management system. Furthermore, the greater the progress in implementing this program, the more likely there would be a shift to a higher level management system. As part of a series of assessments of the management system, the first assessment was given in May 1985 to 33 supervisors representing seven departments which perform the work processes in the Components and Metal Division.

In order to collect information that might be useful in the development of a grounded theory of TQC in this particular setting, two sources of information were tapped: (1) perceptions of 59 artisans concerning the aspects of the organization that impede work performance, and (2) information bearing on the day-to-day implementation effort which was obtained by means of a case study approach. The case studies also enabled us to document the outcomes resulting from implementation activities. The case studies described three of the project teams (Plating-Grinding, Anodes, and Technical Information), that are at the lowest level of the board structure (see Figure 1). The authors attended meetings and documented the progress of these groups from May 1985 through January 1986.

FINDINGS

The case study analysis of the project teams provided information regarding the results achieved by the teams and the progress of the implementation effort. Table 1 presents the outcomes obtained by the three project teams. All three teams realized some improvements as a result of their efforts. Some of the benefits derived from the projects were: the reduction of additional reworking of aircraft components, greater control over work procedures, and an enhanced atmosphere of interdepartmental and cross functional cooperation. This final result is not surprising considering the cooperation that is indicated by the variety and number of work units required to address each issue, as shown in Table 1.

Table 1
Project Team Outcomes

Project Team	Objective	Work Units	Results
Plating-Grinding	Reduce reprocessing in grind-plate-grind cycle	Engineering Quality Assurance Production Control Production Engineering Production	o Better communication between Plating and Grinding o Significant reduction in process variability o More accurate identification of rejects
Anode Cleaning	Develop improved method for cleaning utility anodes	Engineering Production Engineering Production	o New method found to be effective (Savings: $300 per replacement anode) o Anticipated improvements in reliability of chrome application
Technical Information	Improve control over technical information	Quality Assurance Production Engineering Tech Data Center Production	o Better communication channels and procedures established o Tech data in E&R shop updated o Procedures established for continual updating of tech data

A second set of results relate to indicators of the progress of the implementation. We felt that one set of indicators would be the nature of the actions taken by these teams. The extent to which actions such as these were carried out in the project team areas would be indicative of the vitality of the TQC implementation. As indicated in Table 2, the teams were moderately to highly successful on the following: they established communication links where necessary, developed procedures designed to reduce variation in their work processes, and attempted to employ measures of the process. All of these activities represent a departure from

Table 2
Indicators of Implementation Progress

Project Team	Actions Taken	Difficulties Encountered
Plating-Grinding	o Measurement system for process monitoring established o Communication link established between shops	o Insufficient collection and application of objective data o Reluctance to use obtained data to improve work processes
Anode Cleaning	o Tested new work method to improve process of utility anode cleaning o Established procedure to pull and clean anodes every 6 weeks	o Reliance on on-the-job experience rather than on data o Lack of process measurement and monitoring
Technical Information	o Work material organized and updated o Procedures established for better control over process o Communication link established	o Difficulty in linking concept of quality to non-production area o Difficulty in determining scope of the project o Less than optimal use of measurement o Insufficient communication between project team and upper boards

the way they typically conducted their work and are representative of a TQC approach. The case studies also revealed areas that are problematic for this approach. As indicated in Table 2, there have been difficulties in establishing the routine use of measurement, both for monitoring, control, and improvement of purposes. In addition, the projects are not clearly tied to a larger, overall effort. This is exemplified by the fact that the findings and recommendations from all of the project teams have not been presented to boards above the Section QMB level. The status of the boards was partially a result of resistance and partially a lack of planning regarding the implementation. The problems then appear to be at the management level. Management, either through the QMBs or the regular communication channels, did not provide adequate support and information regarding the implementation effort to personnel. The purpose, breadth, and elements of the TQC effort were not fully understood by personnel. Management did not display the visible active, and consistent involvement that is required to indicate a high degree of commitment. The data concerning artisans' views of work impediments are pertinent here (see Table 3). It is interesting to note that when several months earlier we assessed workers' perceptions of their jobs, the impediments most frequently cited as keeping them from doing their best work were systems problems (e.g., poor planning, waiting for parts and supplies, supervisors not listening, and

Table 3
Percentage of Artisans Indicating Factors Which Were Impediments to the Performance of Their Job[a]

1	Poor planning	64.5[b]
2	Stopping one thing to start another	62.6
3	Disciplinary standards inconsistent	59.4
4	Waiting for parts/supplies	50.9
5	Not having right equipment/tools	47.5
6	Others not doing their job	47.5
7	Mismatched parts/paperwork	47.5
8	Poor on-the-job training	45.8
9	Supervisors not listening	44.1
10	Too much expected during 8-hour shift	42.4
11	Unclear/conflicting orders/instructions	40.7
12	Poor working conditions	39.1
13	Paperwork unclear/different	37.3

[a] Total number of impediments were 26
[b] Percentage who marked "a great deal" or above on the scale

unclear or conflicting orders and instructions), which are in the province of management.

Our characterization of management's role regarding the status of the implementation--their commitment, degree of understanding, and acceptance of TQC--was enhanced by the results of the POC, which were obtained several months earlier. The means for the 49 items on the POC can be identified as following into one of the four types of management systems identified by the POC (see Figure 2). In this administration, 37 of the means fell in the lower half of System 3 (Consultative Style) and 12 were in System 2 (Benevolent Authoritative Style). In general, the pattern can be characterized as falling short of higher levels that theoretically should be compatible with TQC. The profile illustrated in Figure 2 describes a system in which: responsibility for achieving organizational goals rests with management, there is a moderate amount of cooperative teamwork, and

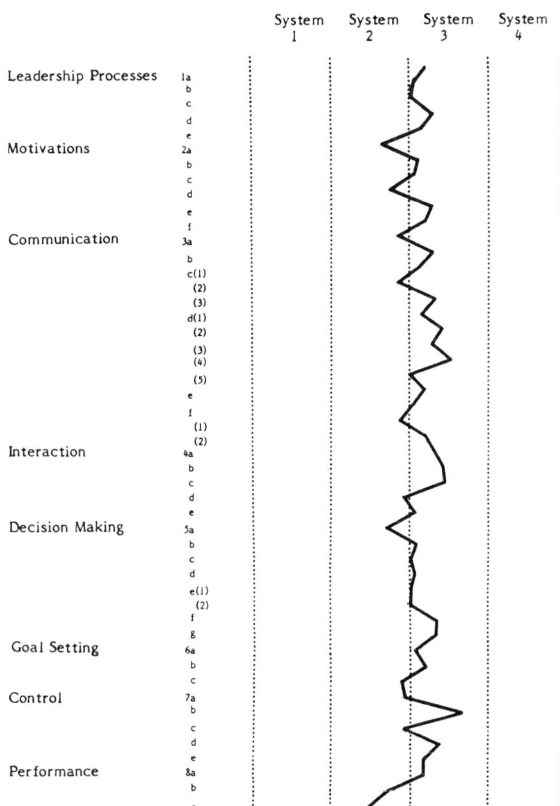

Figure 2
Mean Responses of Managers to POC

subordinates have some influence on goals, methods, and activities in their units. The initial assessment indicates that the management system operating at that time set practices that were not best suited for adopting and carrying out the implementation of TQC.

IMPLICATIONS

The information obtained through case studies, the responses of artisans to the impediments questionnaire, and the results from the POC were each useful. In combination, however, the three sources of information make a more powerful diagnostic tool. They have suggested areas for further investigation as well as indicated some of the areas that need a greater degree of attention in the implementation. From an evaluation perspective, these findings can serve a formative function by providing feedback that would be useful to those involved in the implementation effort. Further, we are in the process of collecting a second wave of data from these three sources. If we find, for example, that the perceived impediments have been reduced in a significant fashion, this is evidence not only for the success of the implementation to this point in time but also bodes well for continued progress. Overall, the results of this continuing assessment may help to determine the interrelationships between such a quality effort and aspects of the organization, leading to the development of a theoretical model and the identification of important components of such an effort. Information obtained in such an assessment effort (a) may aid in the identification of conditions necessary for a broad-based, sustained change, (b) provides a framework with which to test alternate hypotheses regarding the organizational effects of implementing TQC , and (c) may identify aspects of change accompanying TQC which have implications for other areas of organizational change.

REFERENCES

(1) Dunn, W.N., & Swierczek, F.W., Planned organizational change: Toward grounded theory, Journal of Applied Behavioral Science, 13(2) (1977) 135-157.

(2) Chen, H-T., & Rossi, P.H., Evaluating with sense: The theory-driven approach, Evaluation Review 7(3) (1983) 283-302.

(3) Ackoff, R.A., Creating the corporate future (John Wiley & Sons, New York, 1981).

(4) Ishikawa, K., What is total quality control? (Prentice-Hall, Englewood Cliffs, N.J., 1985).

(5) Likert, R., The human organization: Its management and value (McGraw-Hill Book Company, New York, 1967).

Human Factors in Organizational Design and Management - II
O. Brown, Jr. and H.W. Hendrick (Editors)

ORGANIZATION PATTERNS IN A SMALL CITY POLICE DEPARTMENT

Stuart Shaffer, Ph.D., President, Shaffer Psychological Institute
James W. Carrington, Chief of Police (Ret.), Compton Police Department
Robert O. Price, Chief of Police, Bakersfield Police Department
Alastair A. Stunden, Ph.D., Principal, Shaffer Psychological Institute

This paper describes the analysis and correction of organization and communication patterns in a small city police department. Psychosocial, cultural, and technological factors influencing the interaction between and within the city management system, police organization, and the police officers are reviewed. Recommendations for the relief of identified problems are discussed as is the role of the human factors specialist as a consultant to police management.

PURPOSE:

The problems associated with faulty organization and communication patterns in small governmental organizations have been poorly described, if at all, in the human factors literature. Case studies which analyze these defective interactions, and provide solutions which are then put to practical use are even less available and rarely reported. This study is a report of such an activity in a police department having 25 sworn personnel and whose organizational network included approximately 25 additional persons each of whom had significant impact on both the problems to be identified and their solutions.

The police agency, responsible for a city with a population of approximately 20,000 and a geographic area of about 20 square miles, is located in a rural section of a large county in California. The local newspaper which had taken an active advocacy position to improve conditions within the police department, summarized the problems by stating:

> "As most readers know, the sergeants and patrol officers have for years been complaining that something was seriously wrong with the department management. And for years council members and a former city manager have answered the men's complaints with a simple, 'We back the management'."

These issues finally came to a head when the Police Association threatened to erect a billboard at the edge of the city declaring it to be 'unsafe'. This act resulted in a solicitation by the city for consultants who could assist in resolving what had now become a patently critical issue both in terms of safety for the city and its inhabitants and for the organizational integrity of the political unit responsible for providing such services. Because of familiarity with the work Shaffer Psychological Institute (S P I) was doing with other cities and police and fire agencies within the same county, S P I was asked by the concerned City Manager to consider what steps might be taken to assist in resolving the problem. An exploration of the issues was made during which the role of S P I as a consultant to the city was defined. Meetings were subsequently held with both the City Manager and the City Council. Initially, some attempt was made by the city government to seek relief for its own role confusions which were negatively influencing resolving the problems with the police department. However, it was clear that if the city were to find relief for the problem it had identified within its own structure, i.e. that of a problem police department, it could not also try to resolve its own internal issues simultaneously. Trying to do everything at once would only result in a dilution of resolve, energy and ultimately an avoidance of doing anything. If change was going to take place it was necessary to define a pivotal point around which the work of the consultant could revolve with some possibility of success.

At the conclusion of these important preliminary discussions a decision was made by the city government to direct S P I examine the police department **for the purpose of identifying those factors which had led to the apparent dissension and demoralization within the agency.**

SUMMARY OF METHODS AND PROCEDURES:

The primary method employed to gather data in this study was the semi-structured interview. This data collection instrument was reduced to one form that would be applicable regardless of the personnel interviewed or the nature of the content to be developed. The decision as to the content of the interview was made after discussions, both individual and group, were held with the City Manager and City Council. The following content areas for the interview were agreed upon.

- o - Work Satisfaction/Morale
- o - Team Work/Group Cooperation
- o - Supervisor/Employee Interpersonal Relations
- o - Communications
- o - Department Motivation
- o - Training
- o - Working Conditions and Equipment
- o - Officer Recognition/Praise and Discipline
- o - Attitudes about Management/City Government
- o - Supervisor's Competence, Fairness, Support
- o - Department/Professional Pride and Identification
- o - Growth and Advancement/Individual's Goals/Promotions
- o - Work Load/Job Demands/Job Pressure
- o - Job Security
- o - Suggestions for Improving Working Conditions
- o - Reactions to the Survey/Interview

In addition to the content, the methods by which the data were to be collected were also agreed upon. For example, it was clear from these early discussions that the magnitude of the problem was such and the number of people involved were so limited that a 100 percent sample of respondents was both feasible and desirable. In order to enhance the validity of the interview procedures, as well as to profit from the additional data offered by redundancy, it was also decided to interview each of the police department personnel twice. The first interview was to be conducted by one of the consulting police chiefs and the second interview by one of the two psychologists. This tactic offered the further advantage of utilizing someone with specific knowledge of the unique psychosocial, cultural and technological factors typically found in a police agency. It is axiomatic that the effective consultant is a credible consultant. And, that the ability of the consultant to influence the system by bringing forces to bear that will move the organization in the right direction is crucial, particularly in a politically sensitive organization such as a police department. The use of police chiefs in this role made accessible to the human factors specialist a level of special knowledge and expertise that would otherwise have been unavailable or too time consuming to develop. Thus, the many management interactions significant to the appropriate functioning of a police agency were easily accessible, and quickly integrated into the final report.

All interviewees were guaranteed confidentiality and the data collection instruments were coded so that names or other identifying information did not appear on the interview sheet. Regular discussions were held between S P I staff members to review progress, and to exercise quality control over the interview procedure itself. It was the impression of the interviewers that the data collected from the interviews were given in an open, forthright and cooperative manner, providing both accurate and relevant information for the subsequent analysis. As indicated, the format for the interview was semi-structured with the respondent being encouraged to discuss the issue being presented in as open and candid a manner as possible. All the interviewers were trained to separate attitudes, i.e. reports of feelings, from behavior. Respondents were encouraged to speak of things they had done or seen, not situations they had heard of but not directly experienced. This format encouraged the collection of **critical incidents** which historically has generated the most reliable data base for the analysis of organizational problems of this type. Thus the interviewers were actively encouraged to collect critical incident data to the extent possible.

The interviews, taking from one to two hours each, were conducted over a three day period during the regular working hours of the employees. As police agencies provide services on a 24 hour basis, it was necessary to make S P I personnel available around the clock in order to complete the interview process within a reasonable period of time. In all, 30 man days were employed in conducting the interviews. In this manner, then, all of the police management and supervisory personnel; sergeants, lieutenants, and Chief of Police were interviewed, as were the members of each patrol unit. The dispatchers, the Animal Control Officer and the Chief's secretary were also interviewed by two members of the S P I team.

The data from the interviews were subsequently coded, collated and integrated into a final report which included the following sections:

o - Dispatchers
o - Patrol Division
o - Police Management
o - General Recommendations

RESULTS:

The results of the survey were incorporated into a final report which was discussed with the Chief of Police prior to its being presented to the City Manager and City Council. These results are summarized below:

Dispatchers: The primary recommendation focussed on the need to establish and maintain a coherent chain of command which would make possible consistency in supervision and discipline, assignment to deployment period, and the establishment of operational policies which could be adhered to. Specific suggestions were also offered regarding the setting up of communication between the dispatcher's supervisor and the patrol officers. Over the years an ad hoc policy had developed which permitted the 'acting' Lieutenant to supervise the police dispatchers despite the presence of a civilian supervisor. The supervisor's former role was restored and training provided that would prevent the previous problem from reoccurring.

Patrol Division: In this area it was recommended that a formal training policy be initiated to upgrade the skill levels of the patrol officers; that formal performance evaluations be instituted, that the departmental policy manual be completely rewritten so that 'ad hoc' policy decisions would no longer be made, that communication channels be opened by having police management attend roll call and other briefings and that procedures be instituted to deal with personnel complaints, grievances and suggestions.

Police Management: Three levels of management were identified as presenting significant problems to be solved. The first, the Sergeants, were noted to have little supervisory training both professionally/technically and in interpersonal areas which would enable them to direct successfully the activities of their subordinates. Specific recommendations in each of these areas were made as well as in improving communication skills, e.g. report writing, and oral briefings. Many of the Sergeant's deficiencies in training resulted from their role confusion. They were expected to supervise the patrol officers but the Sergeants also held membership in the Police Association which was the bargaining unit between the police officers and the city. It was not possible nor practical to remove them from the Police Association because they had no other formal organizational home which offered them some protection from the vagaries often associated with small town politics. Also, they had a long history with the Association which made change more difficult. Thus the training provided to them required a level of consistency not typically required in other settings. Concepts such as motivating subordinates, conducting performance evaluations were taught but with great difficulty and emotional stress. However, as the training structure imposed began to take hold, the value of the concepts, their effectiveness in reducing tension and improving productivity began to yield positive results.

The second level, the Lieutenants, presented an unique problem. No Lieutenants actually existed within the department. Instead, 'acting' Lieutenants had been appointed from the Sergeant rank. Thus, there were no full time managers within the department other than the Chief of Police. And, as has previously been noted, a police agency is a 24 hour operation requiring the capacity for immediate decision making, sometimes of the most critical kind. It was concluded that the absence of an 'on the spot' decision maker with only 'acting' authority and responsibility, could easily be understood to account for much of the management and policy confusion which was negatively influencing the behavior of the patrol officers. It was recommended that the Lieutenant positions be filled immediately and with personnel from the 'outside' so that new perspectives and professional insights could be immediately inserted into the departmental mix.

The third level, that of Chief of Police, was also identified as lacking in management skills both at interpersonal and professional levels. Specific recommendations for training were offered in areas of administration/management and in communication skills. It was also recommended that the Police Chief begin to receive regular performance evaluations from the City Manager in keeping with standard city policy.

General Recommendations: Another set of recommendations dealt with the relationship of outside influences on the Police Department, e.g. the City Manager, the members of the City Council, the Chamber of Commerce, local citizens and related law enforcement agencies (Sheriff's Department, and District Attorney's Office). Thus, it became important to clarify the police department's interactions with other police agencies within the County, and to clarify the communication links within city government, including the development and adherence to a formal chain of command so as to reduce to a minimum interference by other city departments with internal police operations.

OCCUPATIONAL IMPLICATIONS;

Most of the research that has been done in the field of police management strongly supports the conclusions that the stress related factors that appear to be intrinsic in police activities are most often caused by police management itself rather than the factors intrinsic in the responsibilities inherent in the job. Thus, one finds that police officers tend to develop more job related stress, incur more citizen complaints and internal affairs investigations as a function of how they are being treated by management rather than in the nature of the job. Where it is seen that officers are being treated unfairly, seemingly punished in an unjust manner, where policies and regulations appear to have minimal relationship to the requirements of the job, there is not only higher turnover (a relatively healthy sign when people cannot work under adverse conditions) but those officers that do remain have considerably more of the problems and complaints noted above. Its interesting that, like most clinically depressed patients, officers working in a sick organization see no opportunities for growth, see themselves as being treated unfairly or more like victims, see themselves as the targets of investigations, even when there are no investigations going on, and, generally, render community services that are considered to be less effective as measured by the community. On the other hand, when the organization becomes healthy, as is the case in the police department discussed above, employee attitudes change considerably. For example, whereas in the past this police department had difficulty securing candidates for management level jobs, once the organization turned around, the communication with the outside world was such that the department has had minimal difficulty securing candidates for these positions. (In the instance of this police department, because of its special knowledge of the department and its management requirements, S P I was able to aid significantly in the search for suitable management candidates. Criteria were developed in concert with the Chief of Police and the City Manager and selection instruments were utilized which most competently revealed the police management characteristics desired.) Secondly, officers who have previously applied for transfers have rescinded their requests and are now choosing to work longer hours, with greater productivity because they feel better about working for the organization. Not only do they feel there is a future, but they tend to feel more in control of themselves, their job and other activities.

Finally, there has been a significant change in the role perceptions the police officers have about themselves, their work and the department employing them. Police work is one of those occupations where people inside the organization are generally treated as low level or "know-nothing" employees by management. However, once the police officer leaves the confines of the police department he/she is treated by the public as a person who is expected to be in charge of the most difficult of situations, and to be a competent decision maker. This role, of course, is exactly the opposite of what is expected of the officer while working within the confines of the police organization. Frequently this conflict becomes so intense that the officers involved cannot tolerate the discrepancy and dissonance which ensues. This results not only in inefficient performance but in the occurrence of stress related disorders the cost of which is not only high for the officer but for the governmental agency for which the officer works. The changes made in the police organization described herein, resulted in a decrease in the perceived discrepancy between how officers were treated by management and their role in the field. These changes have also contributed to the improvement in officer performance noted above.

LATE PAPERS*

*These papers arrived too late to be included under their appropriate session headings. Their correct positioning is given by the superscript numbering at the end of their titles in the contents list.

Human Factors in Organizational Design and Management - II
O. Brown, Jr. and H.W. Hendrick (Editors)

EXPERTS KNOWLEDGE TRANSFER CYCLE(EKTC) FOR SYSTEM SIMULATION - CONCEPT, METHODOLOGY AND TOOL -

Ichiro INOUE[1)], Masahiko FUYUKI[2)] and Yoshiyasu YAMADA[3)]

1) C&C systems Research Laboratories, NEC Corporation
2) Dept. of Industrial Engineering, Kansai University
Suita, Osaka 564, JAPAN
3) Dept. of Management and Informatics, SANNO College
Isehara 259-11, JAPAN

Under severe conditions, such as fierce competition among enterprises, frequent changes in demands, specifications and production technologies, production systems are constantly required to extensively improve their productivity and flexibility, pursuing more multi-product-small-quantity production.(1)

Simulation is one of the most effective methods for evaluating complex and complicated production system. In fact, simulation has been increasingly utilized as a means of problem-solving in various practical fields and a number of simulation systems have been developed.(2,3,4 etc.) A "tools-system"(5), which we developed, has been applied to various production fields and practiced case studies in production system evaluation. When carrying on the simulation activities, kinds of know-how are involved in various forms in various phases. When attempting to solve problems in a production field using simulation, simulation tools-system software is supplied to the field. The furnished data includes software programs, manuals, instructions, and experts consultation, covering different forms of know-how.

This process can be described as transferring know-how from a system builder to a system practitioner. In know-how transfer in simulation technology, modelling is the most delicate and consequently the most labor consuming facet. That is, in order to make simulation analysis effective, documents are not sufficient to convey all the necessary know-how information. Thus, it is necessary that a simulation expert give a hand in the process. This requires much time and labor on the part of the expert.

The demands for simulation analysis are rapidly increasing and there are very few specialists. To cope with this situation, it is necessary that the simulation technology transfer process be systematically and well designed. For enhancing simulation technology transfer quality and efficiency, it is important that support tools be developed and a substitute be found for the labor involved in the simulation.

An expert system is a kind of the system, which deals with very human and delicate subjects, such as knowledge and know-how. Therefore, it is expected that, when these expert systems are introduced in a practical activity field and started to put in daily operation, they would easily produce quite a few problems, such as rejection from a practical activity field, disregard of the system and difficulties in system improvement. It is important to understand these points and the problems have to be carefully discussed in advance. This paper first discussed the importance of making most use of expert knowledge and know-how data, focusing on production system simulation as a production management support method. EKTC (Experts Knowledge Transfer Cycle) concept/methodology is proposed. Then, discussion is centered on modelling, which is the most delicate and important activity in simulation. Further, a concept and methodology for an expert system on a basis of "Tools-System" concept, that supports for enhancing production management support effectiveness, by the experts knowledge transfer cycling between system builders and system practitioners, are proposed and a system is developed.

1. ACTIVITY SUPPORT AND EXPERT SYSTEMS

From a functional viewpoint, a production system is considered to be composed of a set of functions as such as production management, design, production engineering, testing, manufacturing, etc. Needless to say, men are activity promoters to realize the functions in a practical production system. For enhancing men's activities in quality and efficiency, various kinds of support systems have been developed.

Whatever the support form a system may have, the system possesses a system life cycle, such as new development, implementation to a practical field, operation and improvement (Figure 1) (6).

When observing that a system, which needs expert knowledge in use, reaches the implementation stage from the system development stage, system builders develop its system, and supply a system practitioner not only with the system but also with knowledge such as how to use it at the same time, as well as give necessary instructions, education and training.

The more expert the system becomes, the more difficult the problems are, which are often seen in system implementation and operation. Some of the problems are as follow.

- System is not accepted by system practitioners in a production field.
- System can not be made perform its functions.
- Even a system, introduced to the field and functioning at full efficiency, will have difficulties in use as time passes.

When the above-listed problems occur, system builders tend to attribute these problems to the system practitioners side (7, 8, 9). That is,

- The system itself is perfectly designed, and has all necessary functions. However, system practitioners are too technically immature to understand and handle the system in a pertinent way.

On the contrary, system practitioners tend to attribute the problems to the system builders side.

- The system is not well designed in regard to handling ease, because system builders are flooded with minute technical details.
- None of the functions, which the system provides, are practically suitable to a real system, although they are prepared in variety.
- It is extremely difficult or practically impossible to improve the system in order to adjust it for practical use, since it is already too large and complex.

It should be admitted that it is quite natural that system builders and system practitioners have different opinions, because they have different standpoints and different viewpoints in the system life cycle of system development, system implementation, system improvement, etc.

System builders and system practitioners, play different roles in the organization of a production system and have different purpose settings in actual production activities. Consequently, the knowledge, know-how to be accumulated, and techniques and technologies to be polished up in daily activities are different for each of them. In pursuing the activities in system building, system operation, etc., their specialities are naturally different. Therefore, cognitive patterns start to differ among them and limits start to be created in their activities.

A system builder thinks of the usage pattern for the system developed and requests a system practitioner to follow the usage pattern, from the standpoint of a system builder, by showing instructions and examples. However, once the system is introduced and started to be put into practical use, the system often becomes independent from the original intention of the system builder and starts to establish new meanings in the practical field. Instead, better usage than the original could be created, since the system is started to be used in a trial and error manner by system practitioners as specialists. Even if the system is put in use against the system builder's intention and brings some kinds of convenience to the production

Figure 1. System life cycle

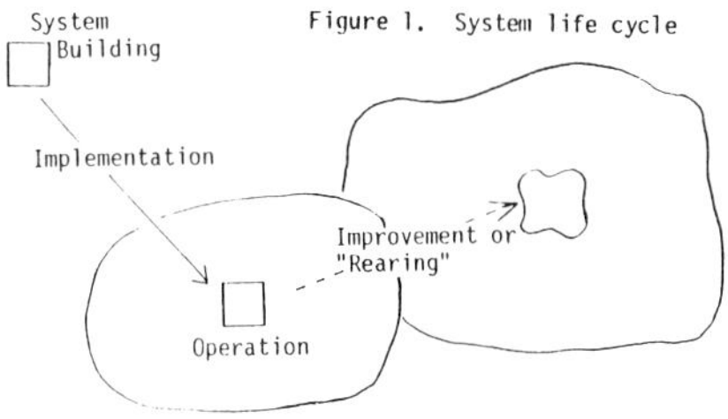

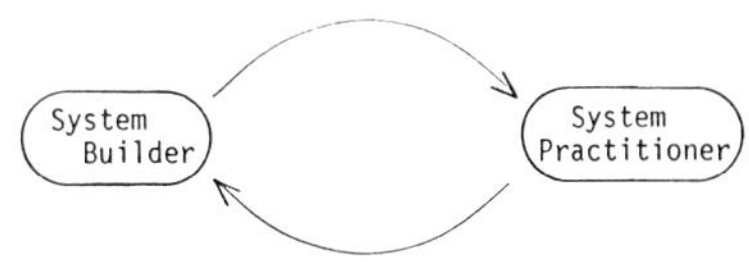

Figure 2. EKTC (Experts Knowledge Transfer Cycle)

activities, the system is significant enough to carry on. Thus, system practitioners are specialists in system practice in practical fields and in the position to create expert knowledge regarding system practice and to be capable to supply them.

As shown up to now, in order to build a "good" system, so that it will function at full efficiency and to bring about effectiveness and efficiency in the practical activity fields, it is very important that system builders obtain expert knowledge on system practice. Then, the "good" system is built knowledge about how to use it. On the other hand, system practitioners come to operate them, utilizing the provided system and knowledge. It is also important that they accumulate and/or polish the expert knowledge and know-how on system practice in daily activities and feed back the pertinent information to system builders.

To sum up, it should be pointed out that system builders and system practitioners have to create and develop each of their knowledge as experts in their activity fields, have to establish a knowledge transfer cycle and then to rotate the knowledge smoothly, as shown in Figure 2.

2. EKTC: EXPERTS KNOWLEDGE TRANSFER CYCLE AND ITS SUPPORT

It is usually difficult to form an experts knowledge transfer cycle. Some of the causes for that can be listed, from technical and socio-cultural viewpoints, are as follow.

- System builders have a firm belief that they understand everything about production, because of their higher position in the management hierarchy. In other words, they do not listen to system practitioner's opinion, which is a kind of expert knowledge about system practice, because they think that they have enough expert knowledge, even about system practice.
- There is no official route to be used to submit an improvement proposal or the route is not clear.
- Even when the route is established, an extra workload is required in order to make a report to system builders.
- The information is often distorted in the report making process.
- System practitioners are so busy with their on-going daily activities that the report making business is often put aside.

To cope with the previously mentioned problems, smooth cycling in EKTC is essential. The following factors are important for that purpose.

- First of all, system builders have to realize the importance of the knowledge that system practitioners possess and produce in daily activities.
- For putting the information obtained from a field in an easily utilized format, the information has to be easily collected from the field. Further, it must be edited in an easy form to be analyzed.
- The necessary data has to be collected without system practitioner's extra work,or minimum work if any, because they are usually too busy with their on-going daily activities to make their report. It is one possibility for the data to be collected automatically in the on-going system operation.

- When system practitioners find that the instructions and know-how, offered from system builders, are inconvenient or insuitable, the knowledge has to be easily modified by their own hands, in order to fit them to the field.
- System builders make system buildings/improvements, on the basis of the knowledge obtained from system practitioners. Then, they provide system practitioners, in return, with the knowledge on how to operate the system, how to compose a model, etc.
- It is more effective to prepare "explanations with dynamic expression" such as a computer assisted guidance, rather than static ones such as manuals, documents. Further, a two way information flow method, such as a Q&A method is recommended for enhancing know-how exchange effectiveness.

3. EXPERTS KNOWLEDGE TRANSFER IN SIMULATION ANALYSIS ACTIVITY

Modelling is a most delicate and important issue in simulation analysis. This section focuses on the modelling in simulation. Methods to establish the experts knowledge transfer cycle and to make the transferring smooth and further, its support is described.

It is natural that system builders be experts in simulation system building and in general theory of simulation modelling. Also, they can play an expert role in system usage in the beginning of system introduction to a practical field. However, it is impossible for them to retain their position as the expert. On the other hand, system practitioners are naturally amateur in regard to system building, but are expert in system practice. Usually, they become expert and invent more effective usage than the one that the system builder proposes in the process of system operation, even if they are not so good in handling the system at the system introduction stage.

In the simulation analysis activity cycle, system builders provide system practitioners with knowledge on how to concretize simulation purposes, on how to examine the simulation analysis effectiveness and on how to compose a model. This knowledge is supplied to them in guidance form. On the contrary, system practitioners are requested to provide system builders with knowledge, such as on how the system is being utilized, what kind of new tools or systems are to be introduced, etc.

When focusing on modelling in simulation analysis activities, it is noted that system builders are experts to discuss modelling in general, but system practitioners become rather experts as they accumulate their modelling experiences in their practical field.

Expert knowledge that system practitioners create in various practical fields are essential for system builders to establish better general principle and better instructions based on it. At the same time, the better general principle and the better instructions give useful suggestions to enable system practitioners to compose models in practical fields. Thus, system builders have to collect as many cases of practical system practice as possible, from system practitioners, in order to construct a better general principle and to give a better guidance for system operation. At the same time, system practitioners have to obtain general principle and methods, form the system builders side, in order to enhance the modelling quality.

On the basis of this understanding, methodology and tools have to be constructed in a systematic manner. This brings about multiple good effects in production activities.

4. EKTC SUPPORT TOOLS-SYSTEM

On the basis of EKTC concept/methodology, a tools-system has been formulated for supporting the expert knowledge transfer cycling. Figure 3 shows a conceptual scheme of the tools-system. Some of the tools are listed in the figure.

A guidance tool, modelling support tools, including Q&A format sheets, and input data preparation support tool, have been developed, as components in the modelling EKTC support tools-system. (Figure 4)

The modelling support tools-system has been applied to simulation analysis for plant design, production management/control activities in printed circuit board production system, assembly lines, and LSI production system. Knowledge transfer from system builders to system practitioners and vice versa have been promoted to a great extent. This EKTC establishment has resulted in enhancing simulation analysis effectiveness and efficiency and further productivity in production systems.

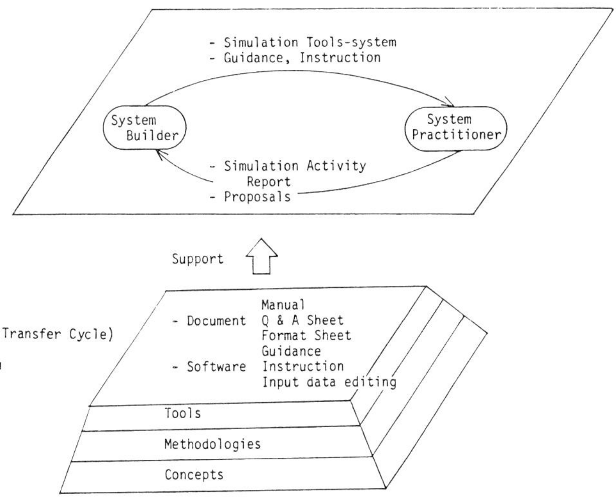

Figure 3.

EKTC (Experts Knowledge Transfer Cycle) and support Tools-System

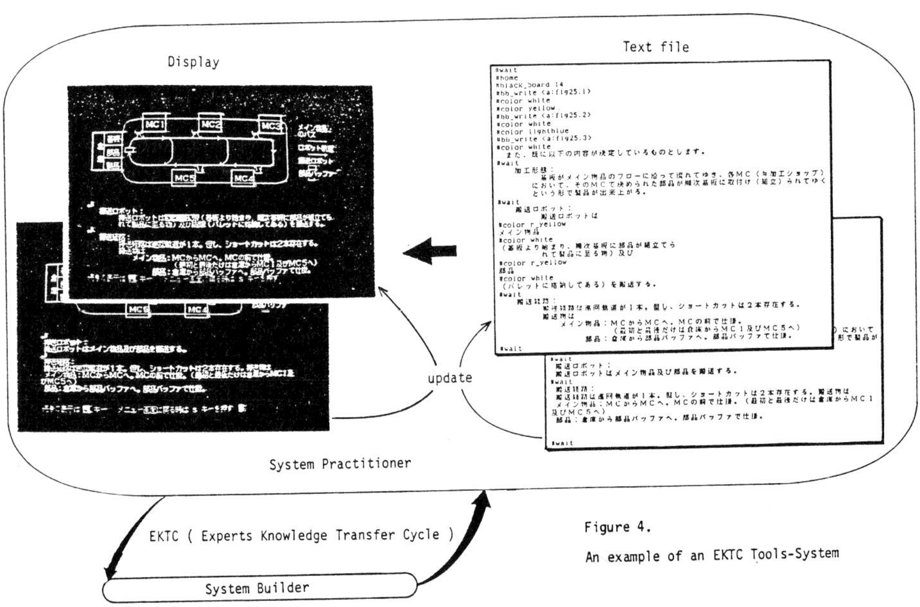

Figure 4.

An example of an EKTC Tools-System

CONCLUSION

Under circumstances where multi-products-small-quantity production with higher productivity and higher flexibility is persistently pursued, evaluation activity in production system becomes routine work (10) and the importance of simulation analysis has been continuously growing. The system science view is absolutely necessary for enhancing simulation analysis effectiveness and efficiency. With the wider view, there are roughly two different standpoints, related to those who are engaged in simulation analysis, system builders and system practitioners. The expert knowledge that each of them produces in their daily activities has to be smoothly transferred between them. On the basis of these idea, EKTC (Experts Knowledge Transfer Cycle) concept has been proposed and some practical applications have been discussed.

Production systems are becoming more complex and more sophisticated. More functional diversification and more specialization are being further developed in production systems. In this situation, it is especially important to further advance system-scientific consideration and research efforts on production system activities themselves and their support ways and means.

ACKNOWLEDGEMENT

The authors would like to thank Mr. T. Kida and Mr. T. Nakahira of Kansai University for their cooperation in developing the tools Further, the authors would like to thank Dr. M. Naniwada of the Production Engineering Laboratory, NEC for suggestions, and Prof. I. Ito of Kansai University for encouragement.

REFERENCES

1. Falster, P. and Mazumder, R.B. (edited), Modelling Production Management Systems, North Holland, 1985
2. Bollinger, I.G., Crookall, J.R., The Role of Simulation in Design Teaching of manufacturing Systems, Annals of the CIRP, vol. 20/2/1981, pp 525-532
3. Inoue, I., PLUS-Applications to Job Shop Production Systems, Proc. Conf. JIMA (Japan Industrial Management Association) 1979 (in Japanese)
4. Pegden, C.D., Pritsker, A.A.B., SLAM: Simulation Language for Alternative Modelling, Simulation, Nov. 1979, pp. 145-157
5. Inoue, I., Yamada, Y. and Adachi, T., A Tools-System in Decentralized Production Management Systems, Computers in Industry, 6(6), 1985, pp. 465-476
6. Inoue, I., Yamada, K. and Kobayashi, H., A Framework of Production System Life-Cycle Technology, Proc. Conf. ORSJ (The Operations Research Society of Japan), 1984 (in Japanese)
7. Inoue, I. and Yamada, Y. A Total Evaluation Model/Methodology of Production Systems with the Consideration of Socio-cultural Aspects, Modelling Production Managements Systems, North Holland 1985, pp. 123-134
8. Argyris, C., Management Information Systems, The Challenge of Rationality and Emotionality, Management Science, 17 (6), 1971
9. Kling, R., Social Analyses of Computing: Theoretical Perspectives in Recent Empirical Research, Computing Surveys, 12 (1), 1980
10. Inoue, I. and Yamada, Y., An Evaluation Process Cycle Model in Production Systems, Proc. Conf. ORSJ (The Operations Research Society of Japan) 1983 (in Japanese)

Human Factors in Organizational Design and Management - II
O. Brown, Jr. and H.W. Hendrick (Editors)

COMPUTER ASSISTED DESIGN OF MANUAL MATERIALS HANDLING TASKS

Ross A. Jackson
Ontario Hydro
700 University Avenue
Toronto, Ontario
Canada

Norah A. Jackson
Countryman and Krolick
1794 Kingston Road
Pickering, Ontario
Canada

In spite of modern labour-saving devices, many tasks still involve strenuous lifting, pushing and pulling. These tasks often lead to fatigue, low productivity and back injuries. Computerized biomechanical modelling can predict workload and stress on the lower back when a plant, tool or work method is still at the design stage. Proper use of these techniques can reduce injuries and increase productivity by reducing workload and stress. This is illustrated by two case studies. In the first, the risk of back injury from tree trimming operations was determined, and two new tools were evaluated. In the second, a lifting task was redesigned to reduce stress on the lower back.

Introduction

Ontario Hydro is a large electrical utility responsible for the generation, transmission and some of the distribution of electricity to homes and businesses in the Province of Ontario, Canada. The distribution system and the environment it exists in introduce severe challenges to the design of tools and work methods. Frequently the type of work encountered leads to back injuries. In 1984 alone, Ontario Hydro lost over four employee-years to back injuries[1]. Injuries (including back injuries) can be reduced through the use of better tools, work methods and work places; however, it is difficult to tell what is an improvement and what is not. Until recently, there was no fast and economical method of evaluating the design of tools, work stations or work methods. Static link-segment coplanar models (eg, Chaffin and Anderson[2]) are adequate for slow, smooth sagital plane lifts, but become inaccurate when the lifts are fast or jerky[3,4], as is often the case in the work environment. Dynamic modelling is necessary for this type of lift. Unfortunately, the calculations required for dynamic modelling are too time-consuming to be done by hand. To overcome this problem, Ontario Hydro purchased a number of biomechanical modelling computer programs from the University of Waterloo, and then modified them to enable us to do the following:

1. Video-tape a manual materials handling task, then develop a link segment model from the video tape.
2. Dynamically model the reaction forces on the major joints of the body
3. Calculate both instantaneous and cumulative workload and power requirements for most manual materials handling tasks.
4. Calculate reaction forces, workload, power, and energy requirements for new jobs at the design stage through computer simulation. This lets us estimate the risks at the design stage rather than after a facility has been built.
5. Graphically compare the relative magnitudes of the major factors that influence the risk of back injury.

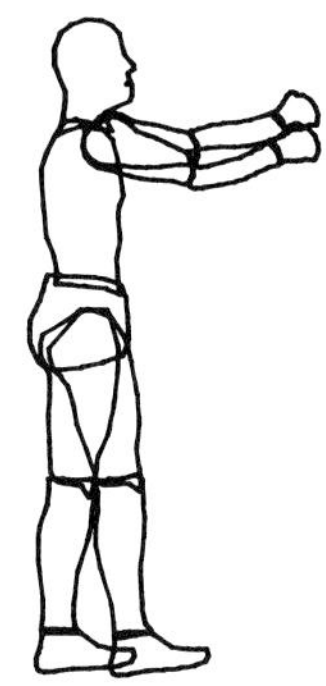

Figure 1: The linked Segment Model used for the Analysis.

The advantages of this type of system will be illustrated with two case studies. In the first, the risk from present methods of tree trimming was determined, and compared with the risks from using two new tools. In the second example, a lifting task was analyzed, and the potential advantages of a new piece of equipment were determined.

Methodology

The risk of back injury is determined by calculating the reaction forces at the lower back (L4/L5). This is done by using a rigid linked segment model to simulate the worker (Figure 1). Typically the work is video-taped. The coordinates of the link segment markers are then measured using a FOR-A Video Position Analyzer. This reads the coordinates directly from the screen and transmits them to a Hewlet Packard 9836 computer. It is necessary to digitize a number of frames for each trial to do a dynamic calculation. A static calculation only requires one frame. The calculations are started at the load, which may be a mass, a force (a push or a pull), or a combination of the two. The reaction forces on the next segment are then calculated, including the forces resulting from the mass of the segment. The process is continued until the lumbrosacral joint is reached. Since there are no guidelines for dynamic reaction forces, the forces at L4/L5 are compared with the guidelines for compressive force established by NIOSH[5]. A guideline of 500 N for maximum sagital plane shear has been adopted until standards are developed. This guideline was arrived at through informal discussions with several researchers, and is based primarily on personal opinion. Standard statistical methods are used to calculate the mean compressive and shear forces, upper confidence limits on these forces, and the percentage of operations that fall above the limits. We have no guidelines for shear in other planes, or for torque; however, we can compare these forces for different jobs and say which would result in the lowest forces. Until researchers develop sound guidelines for these forces, the effectiveness of practitioners will be severely limited.

<u>Example 1</u>

Trees beside power lines are frequently trimmed by workers in aerial bucket trucks using chain saws and hydraulic pruners. The saws and pruners range in mass from 5.9 kg to 10.0 kg, and the workers frequently lean as far forward as possible to reach branches. A dynamic biomechanical analysis was performed to determine if there is a significant risk of back injury for these workers.

The results of the analysis are listed in Tables 1 and 2. Since the compressive force at L4/L5 fell between the Action Limit (AL) and the Maximum Permissible Limit (MPL) in 23% to 48% of the operations, either administrative or engineering controls are required. This conclusion is supported by the fact that shear forces at L4/L5 exceeded 500 N in 5% to 10% of the operations.

Table 1: Compressive Force at L4/L5 - Example 1

		Compressive Force (N)		% of Operations Above AL		% of Operations Above MPL	
Tool	Mass (kg)	Mean	Upper 95% Confidence Limit	Best Estimate	Upper 95% Confidence Limit	Best Estimate	Upper 95% Confidence Limit
Saw 1	5.9	2841	3312	23%	44%	0%	0%
Saw 2	6.8	2990	3474	30%	52%	0%	0%
Saw 3	8.2	3197	3697	39%	62%	0%	0%
Saw 4	10.0	3484	4007	48%	75%	0%	0%
Pruner	6.8	3111	3639	36%	59%	0%	0%
Circular Saw*	6.1	2799	3269	22%	42%	0%	0%
Hydraulic Chain Saw*	5.4	2452	2871	9%	22%	0%	0%

*Simulated

Table 2: Shear Force at L4/L5 - Example 1

		Shear Force (N)		% of Operations Above 500N	
	Mass (kg)	Mean	Upper 95% Confidence Limit	Best Estimate	Upper 95% Confidence Limit
Saw 1	5.9	254	351	5%	16%
Saw 2	6.8	272	362	7%	18%
Saw 3	8.2	279	371	8%	21%
Saw 4	10.0	290	385	10%	24%
Pruner	6.8	281	360	5%	15%
Circular Saw*	6.1	257	330	3%	10%
Hydraulic Chain Saw*	5.4	220	290	1%	4%

*Simulated

The nature of the work prevented us from employing the more common remedies to this type of problem. Mechanical lifting aids are not practical near power lines. The job cannot be redesigned to reduce bending because the angle of the worker is determined by the location of the branch, the length of the boom on the bucket truck, and the location of the truck (which is determined by the terrain). Furthermore, we cannot limit the work to people who are in above average physical condition. The only remaining alternative was to find lighter tools. Two potentially useful saws were located. One had a hydraulically operated circular saw on the end of a fiberglass pole. The other had a hydraulically-operated chain saw on the end of a fiberglass pole. The units weighed 6.1 and 5.4 kg respectively. Since the actual tools were not available, the stresses on the lower back were calculated by computer simulation using the accelerations that were measured for the hydraulic pruner. The results are shown in Tables 1 and 2. Compression and shear forces appear to be lower with the cutter with the chain saw than with any of the other tools and were below or close to the guidelines for most operations. This saw is presently being used on a trial basis by one of our crews to determine if it is suitable from an operations point of view. In the interim, we have recommended that only the light saws (ie, saws 1 and 2) be used.

Example 2

Protection and Control Technicians at Ontario Hydro frequently transport test equipment to a variety of locations. The equipment ranges from a 2.3 kg meter to a 24.5 kg oscillograph. It is loaded by hand into the back of a station wagon or the trunk of a car. Compression and shear forces were calculated using a static model because only slow, smooth lifts were involved. The results for the oscillograph are summarized in Tables 3 and 4.

Table 3:
Compressive Forces Generated at L4/L5 While Lifting the Oscillograph

Vehicle	Compressive Force (N)		% of Operations Above AL		% of Operations Above MPL	
	Mean	Upper 95% Confidence Limit	Best Estimate	Upper 95% Confidence Limit	Best Estimate	Upper 95% Confidence Limit
Car	6384	6860	>99%	>99%	50%	72%
Station Wagon	4653	5871	71%	88%	21%	41%
Station Wagon with Rack*	1953	2135	0%	0%	0%	0%

*Simulated

Table 4: Shear Forces Generated at L4/L5 While Lifting the Oscillograph

Vehicle	Shear Force (N)		% of Operations Above 500N	
	Mean	Upper 95% Confidence Limit	Best Estimate	Upper 95% Confidence Limit
Car	554	582	87%	96%
Station Wagon	549	583	79%	92%
Station Wagon with Rack*	210	235	0%	0%

*Simulated

The compressive forces exceeded NIOSH's action limit most of the time, and NIOSH's maximum permissible limit was frequently exceeded. Further analysis of the data showed that the most important factor contributing to the compressive and shear forces was the horizontal distance between the load and the worker's centre of mass. We examined the effect of using sliding racks to reduce the horizontal distance by manipulating an anatomically correct two dimensional manikin on a scale drawing of a station wagon with sliding racks. This allowed the range of postures that a worker was likely to adapt to be determined. The coordinates of the body segments were read directly from the grid paper and input into the computer manually. The results (Tables 3 and 4), demonstrate that compressive and shear forces at L4/L5 would fall within acceptable levels if racks were used.

These examples clearly illustrate how computer modelling of human factors issues can be incorporated into the development and evaluation of tools and work methods, allowing the designer to determine the risk of back injury before the actual tools or work stations are built and before work procedures are in place. In addition, it can reduce the number of accidents and increase productivity once the work methods are in place by helping the supervisor through design ensure that the risk of back injury is kept to a minimum.

References

1. Cogen, C.F., McCarron, L.L., Pimpinella, A.A. and Stainback, D.G., Ontario Hydro Health and Safety Performance 1984. Internal Ontario Hydro publication, 1985.

2. Chaffin, D.B. and Anderson, G.B.J., Occupational Biomechanics. New York: John Wiley and Sons, 1984.

3. McGill, S.M. and Norman, R.W., "Dynamically Determined Low Back Moments During Lifting", J. Biomechanics, December 1985, 18 (12), pages 877-885.

4. Leskinen, T.P.J., Stalhammer, H.R., Kuorinka, I.A.A. and Troup, J.D.G., "The Effect of Inertial Factors on Spinal Stress When Lifting", Engineering in Medicine, 1983, 12 (2), pages 87-89.

5. National Institute for Occupational Safety and Health, Work Practices Guide for Manual Lifting, U.S. Department of Health, 1981. DHHS (NIOSH) Publication No. 81-122. NTIS Publication No. PB82-178948.

Human Factors in Organizational Design and Management - II
O. Brown, Jr. and H.W. Hendrick (Editors)

CASE STUDY OF A VOICE MAIL SYSTEM: ENVIRONMENTAL CONSIDERATIONS IN THE DESIGN PROCESS

Joan M. Roemer, Wayne L. Pendley,[1]
Mark O. Stempski, Mark C. Borgstrom
IBM Corporation, General Products Division
Tucson, AZ 85744

In a field study of a computer-based voice mail system within an organization, users reported relatively low usage and satisfaction with the system. Based on their ratings and comments, we make several design recommendations. We also discuss how environmental factors may have had an impact on system usability and acceptance. Designers must consider environmental factors early in the development process, and must validate a system's usability in the customer's environment.

INTRODUCTION

Various researchers in the area of human-computer interaction have outlined behavioral principles for designers to follow in developing user interfaces. These principles include gaining an understanding of who the users are, prototyping and early testing of the interface with users, and iteratively redesigning and testing the interface throughout the development process [1,2]. Presumably, such an effort will yield a highly usable product.

We studied the use of a computer-based voice mail system in an industrial organization, and users' attitudes toward it. The system allows users to send and receive voice messages using pushbutton telephones. It offers editing, filing, retrieval, distribution, and control functions. The product was based on an experimental tool that had been developed using human factors principles. The design process included behavioral analyses of users' communications requirements, prototyping, iterative evaluations, and empirical research. The experimental system was judged successful from the standpoints of usability, usefulness, and user satisfaction [3].

The purpose of this study was to evaluate the released product in a working environment. (Note: the product is no longer marketed).

METHOD

Site Background

We studied users at a development and manufacturing plant that employs about 5,000 people, with over 3,000 of them being subscribers to the voice mail system. There were actually four different systems, each having the capacity to accommodate 1,000 users. Subscribers were professional (technical and nontechnical), managerial, and administrative personnel.

---

[1] Located at Communication Products Division, Raleigh, NC.

Questionnaire

About a year and a half after the introduction of the voice mail system to the plant, we developed a questionnaire to assess usage of the product, overall level of satisfaction, usability, usefulness of various functions and features, reaction to possible nuisance situations, and preferences for alternative methods of communication. The questionnaire consisted of rating scales, paired comparisons, fill-in-the-blank, and open-ended questions. We distributed it to 200 randomly selected subscribers and to 50 randomly selected ex-subscribers (users who had requested that they be removed from the system). Responses were received from 121 subscribers and 26 ex-subscribers.

RESULTS

Usage of Voice Mail System

On average, subscribers signed on to the system 6.9 times per week. A fairly large percentage (44.8%) signed on only after being notified that they had a message waiting. If they failed to reach someone on the phone directly, they called the system only 21% of the time. Subscribers claimed that if they had been automatically connected to the system after failing to reach someone, they would have used it 59% of the time. Subscribers and ex-subscribers gave open-ended reasons for not using the system, the most frequent of which are listed below under "Nuisance Situations."

Satisfaction, Usability, and Usefulness Ratings

Respondents rated their satisfaction with the voice mail system on a "percent satisfied" scale (see Figure 1). Not surprisingly, ex-subscribers gave very low ratings. A rather large percentage of subscribers also responded using the low end of the scale.

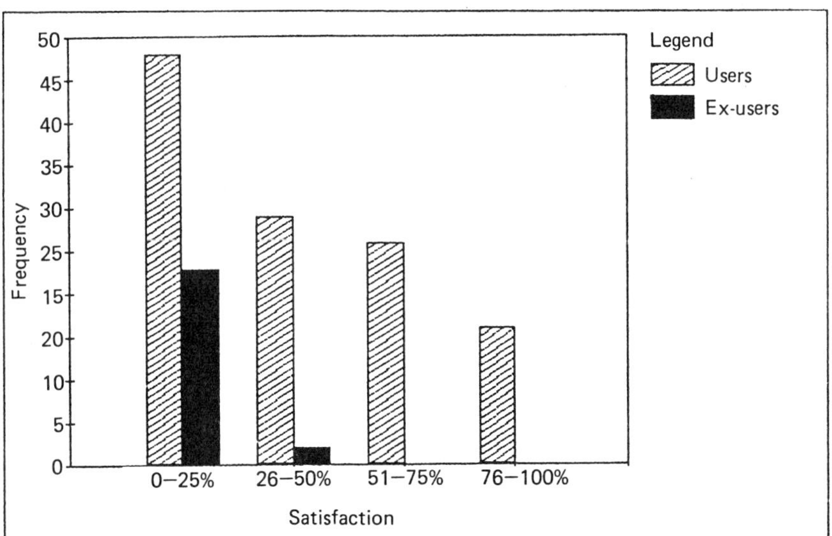

Figure 1
Ratings of Satisfaction with the Voice Mail System

"Ease of learning" was negatively rated by 31.1% of subscribers and 52% of ex-subscribers; "ease of use after learning" was rated negatively by 60% of ex-subscribers and positively by 61% of subscribers. Regression analyses showed that the "ease of use" rating was the best predictor of the satisfaction rating.

Users indicated very low frequency of use of advanced functions, with the exception of the review/edit function. The functions or features rated most useful were: automatic notification of waiting messages, review/edit, privacy, and security. These features also turned out to be what was most liked about the voice mail system, along with remote use capability and ability to use the system at any time of day. Users gave very high usefulness ratings to the proposed features of automatic forwarding to the system, and accommodation of rotary and pulse dialing.

Potential Nuisance Situations

The worst-rated situations were: not being automatically connected to the system, external callers having difficulty using the system, and having four different systems. "Need for exact spelling" was rated negatively by 72.7% of the ex-subscribers. The most frequent categories of response to the question "Why don't you use the system?" were the first three nuisance situations above, as well as the following: "difficult, cumbersome, too many keystrokes, time-consuming; "person to be called not a subscriber;" "other subscribers don't use system -- fewer calls returned;" "prefer to use other available message systems."

Preferred Message System

Respondents comparatively rated the voice mail system and alternative means of leaving and receiving messages. Table 1 shows their rankings. Surprisingly, ex-subscribers gave the highest ratings to recorders; we had speculated that users who were dissatisfied with the voice mail system probably would not like recorders either. Ex-subscribers received most of their messages through another person or by recorder (47.8% and 43.6%) and left most of their messages by the same means (49.9% and 36%). Subscribers received most of their messages through another person or by the voice mail system (35.2% and 34.4%), but left only 19.8% of their messages through voice mail.

Table 1. Preferred Message System

Subscribers	Ex-Subscribers
1 - person	1 - recorder
2 - computer	2 - person
3 - recorder	3 - computer
4 - voice mail	4 - voice mail

DISCUSSION

Design Implications

Although the voice mail system was intended to be a voice-store-and forward system and not a telephone-answering system, users felt that the product should provide the function of automatically forwarding them to the system when they failed to reach someone by phone. Apparently, they had more use for this function than they did for deliberately setting out to send a noninteractive communication. The amount of effort required to leave a message probably colored users' view of the product's overall ease of use (having to hang up, redial the system, key in name and password, listen to system messages, key in the other subscriber's name and the appropriate commands).

According to respondents, external callers usually either refused to use the system or were confused by it. Callers could not leave messages if they were using rotary or pulse telephones. Subscribers who wanted to use advanced functions could not if the other person was on a different system.

Some obvious design recommendations follow. First, provide automatic forwarding to the voice mail system--leaving a message in this circumstance should be as easy as using a recorder; the system should be transparent to callers. In other words, eliminate or minimize the need for keystrokes. Prompt callers to leave a message, let them talk and hang up. Second, for a given organization, provide one system, at least from the users' point of view. (Note that the plant studied is in the process of converting to a ROLM phone system and will be using PHONEMAIL, which has these characteristics.)

Other recommendations based on the questionnaire responses are: for all functions, minimize the number of required keystrokes; provide the capability to define macros or function keys for frequently performed tasks; eliminate or minimize the need for spelling. Careful consideration should be given to what aspects of the user inteface system administrators can customize. For example, it is possible that they might undo the usability of the system by overriding default prompts with lengthy, difficult prompts spoken by unprofessional speakers.

Design Process Implications

Understanding the users, prototyping, and iterative design and testing are necessary, but not sufficient. This case study illustrates that validation in a customer's environment is also essential. It is unlikely that every design flaw and problem situation will be identified in a test setting--and it is very likely that users of an experimental system will have more positive and accepting attitudes toward the system. Besides, the developers might understand who the intended users are and how the product should be used, but ultimately, they do not determine who the users will be and may have little control over what users expect of the product, and how they use or "misuse" it.

If it is necessary to validate the system in a customer environment, the corollary is that it is necessary to understand, early in the design process, what environmental factors might have an impact on the system's usability and acceptance. Some of these environmental factors are organizational or social. Not only do designers need to understand individual user requirements for a system like this, they also need to understand the requirements of an entire community of users. If a segment of that community chooses not to use a voice mail system, the system can be rendered ineffective. Many of the questionnaire respondents stated that they did not use the system because the people they wanted to communicate with didn't use it.

Users objected to there being four different systems. The assignment of people to each system reflected the functional organization of the plant; while the communication requirements within a function may have been met, the communication requirements between functions were not.

Another environmental factor is the availability of related resources, facilities, and tools. There were several competing modes of communication available at this plant; the survey results showed that all had significant usage, and that users often chose alternatives to the voice mail system. For example, all managers had secretaries, and all secretaries used an electronic mail system; therefore, managers had no real need for either the basic or advanced functions of the voice mail system.

Lack of resources can also be a problem. Before the voice mail system was installed, operators in centrally located message centers took phone messages for professional-level employees. To reduce costs, most of the message centers were eliminated. Although the questionnaire did not tap this directly, it was apparent that many employees were unhappy that message-center support had been discontinued, and thought that the voice mail system had been brought in to replace it. Therefore, some hostility toward the voice mail system for this reason is understandable, especially when users found that the system was more difficult to use than the message centers.

Implementation or tailoring of a system is another environmental factor. We already discussed the potentially negative effects of customized system prompts and messages. Another local option was the definition of user identifications (userids). The plant used surnames, and some users had difficulty with spelling. Numeric codes (for instance, the four-digit phone extension) might have worked better, given that userids were required at all.

The content and administration of training is often a customer option, whether or not training packages are available with the product. Initial training at the plant we studied was probably not very effective; it consisted of a presentation of the system's features and functions to several hundred people at a time. Later on, the system administrator developed and distributed documentation to supplement the user's guide that came with the product. One researcher found that a major factor in the degree of acceptance of another voice mail system among several customers was the quality of the training provided [4]. A possible way to ensure the availability of a good training package to all users would be to integrate it into the design of the product as an online tutorial. At the very least, it should be part of the user's guide.

A final environmental consideration is the intended use or application of the system. If the requirement at the plant had been for a means of noninteractive communication, as opposed to telephone answering, the voice mail system would have received better acceptance. Some applications are highly specialized and might or not be successful depending on how well the system meets special requirements.

A successful application was the Olympic Message System (OMS), a voice mail system developed for the 1984 Olympic athletes [5,6]. We think that several factors contributed to its success, some of which support the points made here. For example, the developers knew exactly who the intended users were and were able to establish their requirements fairly precisely. Not only did they iteratively design and test prototypes, they also conducted a realistic field evaluation at a pre-Olympic event. As a result, they made many improvements to the user interface, eliminating most of the advanced functions in favor of simplicity. They added a reply function to eliminate spelling requirements.

Other aspects of the OMS design that led to good acceptance included sufficient computing power to accommodate all users on one system, operator assistance so that guest callers were not required to press any keys, and a training videodisk integrated into the message stations. Another key factor for its success was that OMS met the communication requirements of a community of users in an environment where there were limited communication facilities available (the athletes did not have personal phones), where the athletes were difficult to locate, and where the guest callers (who were in 100 different countries) did not necessarily have access to pushbutton phones.

CONCLUSION

We would restate the behavioral principles for designing user interfaces as follows: understand the users, understand the users' environment, prototype the system and test it with representative users at an early stage, redesign and retest as necessary, and validate the usability of the released product in a customer environment.

REFERENCES

[1] Al-Awar, J., Chapanis, A., and Ford, W. R., Tutorials for the first-time computer user, IEEE Trans. Professional Communication PC-24 (1), (1981) 30-37.

[2] Gould, J. D. and Lewis, C., Designing for usability--key principles and what designers think, Communications of the ACM 28 (1985) 300-311.

[3] Gould, J. D. and Boies, S. J., Human factors challenges in creating a principal support office system--the Speech Filing System approach, ACM Trans. Office Information Systems 1 (4), (1983) 273-298.

[4] Erlich, C., Personal communication, 1986.

[5] Gould, J. D., The Olympic Message System, Human Factors Society Bulletin 27 (9), September 1984, pp.1-2.

[6] Gould, J. D. and Boies, S. J., Human factors of the 1984 Olympic Message System, Proc. Human Factors Society 28th Annual Meeting, (1984) 547-551.

Human Factors in Organizational Design and Management - II
O. Brown, Jr. and H.W. Hendrick (Editors)

A METHOD TO DEVELOP CASE-STUDY-BUILDING SKILLS TO USE IN MANAGING AND REDESIGNING HIGH TECHNOLOGY WORK

Margaret B. Miller Vaughan, Professor
Augustana College, Business and Economics
Sioux Falls, South Dakota
U.S.A.

Our "utilitarian" demand for technologies of information control is leading to an ignorance of human values in business/management higher education. Management functions, and the social-science concepts on which they are founded, may be better learned if the interpersonal skills used in successful management of complex problems are carefully built into study assignments for technical people. This report shows one successful method used; the historic rationale, the steps to follow, and the skills ensured by its practice are described. Those skills are: boundary spanning, integrative lateral communications, learning/study skills, flexibility, support for a climate of inquiry, monitoring accuracy, and reporting/spokesman skills.

Most industries in the US now rely on knowledge from enterprises that are in the leading edge of technical development, either directly or indirectly. The predictions that these years would be the age of information control used to protect and also disseminate technical knowledge were given as a bellwether by Orwell in 1894 and refined by J. Naisbet and P. Aburdene (Reinventing the Corporation, New York, Warner, 1985). Ritterbush has assembled new predictive arguments that technology is an institution that prescribes social and cultural "causal texture" (below). Schools and training programs are often trying to catch up by employing outdated educational models that do not contain the means for people to acquire some of the most critical concepts and skills indicated by these predictions.

This paper is a brief descriptive report of one successful approach, used with graduate and undergraduate management students, to provide direct experience in those skills specifically indicated as necessary by authors of such predictions. The skills of greatest importance here are the ones that are usually ignored in standard analytic and case-study approaches to instruction.

They are:

- problem-naming
- problem-construction
- adjusting the design of inquiry to the situation's commodities
- critically assuming the rôle of direct monitor.

A quick review of literature shows that since Orwell, only the last 15 years have produced clear indications of some of the high-rated skills that remain least addressed by higher education in business/management. Formative works describe recent changes in the components of organizational goals and environments, and linkages of these to realities that are recognized as outcomes of the changes, indicating the skills needed to answer the demands of those outcomes.
Some key linkages are summarized here:

COMPONENTS OF THE INTERNAL AND EXTERNAL ENVIRONMENTS	COMPONENT-RELATED OUTCOMES	SKILLS DEMANDED FOR ADAPTATION
Complexity Real-time managing Uncertainty	Dual structures: mechanistic and organic co-existing Ambiguity and conflict	Boundary spanning Integrative lateral skills Study skills

For <u>Complexity, Uncertainty and Real-Time</u>, see:

Adams, J.S.	"The Structure and Dynamics of Behavior in Boundary Roles," in Marvin Dunnette (Ed.), <u>Handbook of Industrial and Organizational Psychology</u>, New York, Wiley, 1983.
Mintzberg, H.	<u>The Nature of Managerial Work</u>, New York, Harper & Row, 1973.
Perrow, C.	<u>Complex Organizations</u>, Scott Foresman, Glenview, Ill., 1972.
Price, J. and Mueller, C.	<u>Handbook of Organizational Measurement</u>, New York, Longman, 1986.
Sayles, L.	<u>Leadership: what effective managers really do and how they do it</u>, New York, McGraw Hill, 1979.
------, & Chandler, M.	<u>Managing Large Systems</u>, New York, Harper & Row, 1971.
Thamhain, H. and Wileman, D.	"Diagnosing conflict determinants in project management," <u>IEEE, Transactions on Engineering Management</u>, February 1975, <u>EM-22</u>, pp. 35-44.

Technical Competitiveness	Quick turn-around time High value for knowledge of trial/error and innovation feedback -- often diffuse, elaborate, and ambiguous Much sunk capital via R & D Expensive errors	Programs for inquiry Climate of inquiry Flexibility Support for adaptive subsystems

For Technical Competitiveness, see:

Daft, R. and MacIntosh,	"A New Approach to Design and Use of Management Information," California Management Review, Fall 1978, pp. 82-92.
Pelz, D. and Andrews, F.	Scientists in Organizations: Productive Climates for Research and Development, (Revised), ISR, Univ. of Michigan, Ann Arbor, 1976.
Peters, T. and Waterman, R.,	In Search of Excellence, New York, Harper & Row, 1982
Ritterbush, P.	Technology as Institutionally Related to Human Values, Acropolis, Wash. D.C., 1974.
Zaltman, G. et al.,	Innovations and Organizations, New York, Wiley, 1973.

Global Horizons	Ambiguity Conflict	Learning/study skills
High Mobility	Plethora of constituents Personnel turnover	Monitoring and spokesman skills

For Global Horizons and Mobility, see:

Brandt, W. and J. Hulbert	"Patterns of communications in the multinational corporation, an empirical study," Journal of Internat. Business Studies, Sp. 1976, pp. 57-64.
C. Cooper and Payne (Eds.)	Current Concerns in Occupational Stress, Wiley and Sons, London, 1980.
Kahn, R. et al.	Organizational Stress, New York, Wiley, 1964.
Livingstone, J.	The International Enterprise, New York, Wiley, 1975.
Mintzberg,	Op. cit.

The process of case-study analysis consists of:

(1) study of technical and theoretic concepts;
(2) reading of a case developed earlier, in a representative firm;
(3) analysis (group or individual) of the case using its problem contents, auxiliary information, and the concepts studied, all compared to the case information;
(4) proposing solutions or next methods to apply to get solutions.

The new approach reported here includes (1), (3), and (4) of these steps; it substitutes several more in place of step (2) above:

(2) a. selecting a situation that is actual and uncovering the interesting variables in it (people, events, trends, strengths, flaws); stating the problems;
(2) b. devising a means to observe objective case-history information about the situation and people, and to keep accurate records;
(2) c. making the observations;
(2) d. recording the observations.

Specific instructions given to students to accomplish these seven steps, are reproduced in the Appendix:

Some drawbacks have been noted, giving heightened awareness about the difficulties so common in social science research, and especially where the inquirer is probing for information that is often proprietary and also laden with technical terminology which is very specialized;

(1) the inquirer is seen as incompetent to receive the desired information, or to judge it correctly;
(2) the inquirer is seen as a "plant," having a covert strategy to do harm;
(3) the subjects are insecure and unable to be candid in responses;
(4) the problems are multiply caused and multiply expressed -- so that an unbiased inquiry seems too difficult to try;
(5) the information finally received is so contradictory, that the inquiry itself seems to have caused new problems.

All professional consultants are familiar with these stumbling blocks. They are to be accepted as instructive, rather than as obstructive. As the inquirer moves into the situations giving such unintended outcomes, there is greater likelihood of developing strategies of assurance: about competence, trustworthiness, problem segmentation into integral parts, and intent of follow-through. Against such blocks, are to be compared the many other more easily obtained instructive outcomes, ensured by the process of case-study composition:

(1) the inquirer must use language carefully, precisely, for users;
(2) the inquirer must know facets of theory well enough to choose just those sections which are relevant to the concrete case chosen;
(3) the inquirer must conceive and then sell the inquiry as useful and/or interesting to the participant organizations, before, during, and after the inquiry.
(4) the inquirer must confront contingencies directly and choose them as bases for problems.

To experience the combined use of the humanities skills with technical systems is to learn about human values in science within the "vita activa," as contrasted with the contemplative life. Anticipated socio/technical responsibilities are very interesting to students. Ways must be designed and delivered to practice those responsibilities before assignment outside the place of learning. This process here described is such a design, an individualized way to combine the "classical" with the "utilitarian" curriculum.

We find, over and over again, and especially among technically-trained professionals in the sciences and various branches of engineering, that communicative and interactive skills are developed more surely in the expanded creative process: boundary spanning, integrative lateral skills, learning/study skills, flexibility, support for adaptive subsystems--a climate of inquiry, monitoring accuracy, and reporting/ spokesman skills.

APPENDIX

A. Steps to Follow

I. Survey the topics of a human-systems theory of goal-directed activity, with thorough coverage of Organizational Structures, Roles, Relations, Leadership, Communications, and Group/Intergroup Effectiveness.

II. During the survey period of lectures and discussion, the student will develop illustrations of topics, as they are covered. This is done through note-taking, assigned exercises, and the student's own outside readings on topics that are of concrete interest to the student. These are reported, either by outlines handed in for commentary or by oral reports to the instructor, shared perhaps with the whole student group.

III. The student chooses several that are the most interesting and considered by the instructor to be the most do-able. The criteria of do-ability are:
- does the student really have an interest in it?
- is it a bona fide human-systems topic?
- can the student get direct access to some specific data? Actual situations and people in them are best: particular specialized literature or records might substitute for raw data from fieldwork.

IV. The student gradually finalizes a problem proposal (or several of these):
Tentative naming of the problem;
Alternative ways of studying this;
Parts of texts that particularly pertain to this;
Key contextual features of the firm and the work group: a "portrait" of the structure of the group.

V. An interview of questions (structured or unstructured) is prepared by the student, to administer to the focal person(s) in the selected situation. A careful construction of this interview questionnaire has as its purpose the avoidance of attributions that are unfair, to get an objective scanning of actual events (current and historic) that constitute the case.

VI. An important step may not be possible in large classes or with short time frames: that is to report in seminar fashion their preliminary plans among eachother as constructive critics. For good research, this is an essential step: multiple perspectives are included as a vital part of case development, as of case analysis.

VII. The student proceeds to create a refined case outline to serve as a Table of Contents and as a guide for the writeup. There must also be developed a literature survey to provide theoretic ideas. The student (1) gives the interviews, (2) writes up their content (which constitutes the Case History), (3) discusses the literature surveyed and compares it to the History as explanation of possible cause/effect relationships in the situation interviewed, and (4) describes any conclusions reached.

Note: Conclusions may include: (a) X caused Y; or (b) either X, L, M could have caused Y; or (c) X caused Y and here is how to stop that from happening; or (d) no conclusions can be made from this method--what is needed is knowledge about (or the resources of) Z; or (e) the situation of X can be assuredly classified as a Type 0 situation from the theory.

Also, technologic information (details about the technical functions other than human systems) must be included only if they serve this emphasis on human factors. The technologist-as-student is often reluctant to leave aside old habits of concentrating on quantitative information that represents the results of human endeavor, rather than the human endeavor itself.

B. <u>The Case Study Format</u>

Here is how the case should look, in general, in its organization:

I. History
   A. Introduction: a brief abstract of the entire study.
   B. A brief description of the situation described.
   C. A brief description of the person (persons) who is (are) focal to the case.
   D. A history of the persons; their background (observed data).
   E. A history of the situation: its background (observed data).

II. Analysis
   A. A review of the literature's theoretic foundations that pertain to these histories and to the current situation; a comparison of earlier concepts, studies, findings, ideas, facts, to the observed data, point by point.
   B. Conclusions, or conclusive statements.
   C. Appendices: printed material from the firm; notes from interviews; the interview questions used to extract primary data.
   D. End Notes: references specifically identified.

AUTHOR INDEX

AUTHOR INDEX